ENGINEERING MATHEMATICS EXPOSED

ENGINEERING MATHEMATICS EXPOSED

Mary Attenborough

McGRAW-HILL BOOK COMPANY

London · New York · St Louis · San Francisco · Auckland
Bogotá · Caracas · Lisbon · Madrid · Mexico · Milan
Montreal · New Delhi · Panama · Paris · San Juan
São Paulo · Singapore · Sydney · Tokyo · Toronto

Published by
McGRAW-HILL Book Company Europe
Shoppenhangers Road, Maidenhead, Berkshire, SL6 2QL England
Telephone 0628 23432
Fax 0628 770224

British Library Cataloguing in Publication Data

Attenborough, Mary
  Engineering Mathematics Exposed
  I. Title
  510

  ISBN 0-07-707975-2

Library of Congress Cataloging-in-Publication Data

Attenborough, Mary (Mary Patricia)
    Engineering mathematics exposed / Mary Attenborough.
       p.   cm.
    Includes bibliographical references and index.
    ISBN 0-07-707975-2
    1. Engineering mathematics.   I. Title.
  TA330.A83 1994   94-802
  620′.0051--dc20   CIP

12345 CL 97654

Typeset by P & R Typesetters Ltd, Salisbury, Wilts
and printed and bound in Great Britain by Clays Ltd, St Ives plc

To my partner
Michael Gallagher

CONTENTS

PREFACE

This book is based on my notes from lectures to students of electrical, electronic and computer engineering at South Bank University. It presents the background mathematics and a basic first year degree/diploma course. It encompasses the requirements not only of students with a good maths grounding, but also of those who with enthusiasm and motivation can make up the necessary knowledge. Engineering applications are integrated at each opportunity and ideas are also given for projects and investigations. Situations where a computer should be used to perform calculations are indicated and 'hand' calculations are encouraged only in order to illustrate methods and important special cases. The software package provided with this book gives opportunities to use more complex methods and to interpret problems and their solutions graphically easily. In this way there is an emphasis on understanding and applying mathematics rather than on technical abilities.

Developments in the field of engineering, particularly the extensive use of computers and microprocessors, have changed the necessary subject emphasis within mathematics. This has meant incorporating areas such as Boolean algebra, graph theory and logic into the content of engineering mathematics. A particular growth area of interest is digital signal processing, with applications as diverse as medical, control and structural engineering, non-destructive testing and geophysics. An important consideration when writing this book was to give more prominence to the treatment of discrete functions (sequences), solutions of difference equations and z transforms, and also to contextualize the mathematics within a systems approach to engineering problems.

Mary Attenborough

ACKNOWLEDGEMENTS

I should like to thank my former colleagues in the Electrical and Electronic Engineering Department at South Bank University who supported and encouraged me with my attempts to re-think approaches to the teaching of engineering mathematics. I also must thank all of my students for teaching me everything they didn't know about mathematics and many of the things that they did know about engineering. Particular thanks are due to the full-time HNC students for their support and the part-time degree students for their brutal honesty.

I should like to thank all the reviewers for their invaluable comments, particularly Professor Michael Yates and Dr Stephen Ryrie, and the editorial staff at McGraw-Hill, particularly Camilla Myers. Thanks to my employers, Simon Petroleum Technology, for their flexibility during some of the more demanding hours in the writing of this material. Thanks to my ex-colleague at SPT, Tony James, who provided interesting comments on parts of the manuscript and to both him and Andrew Marlow for the enlightening discussions on aspects of this book, despite our hazardous daily commuting between London and Marlow. Above all, I must thank Gabrielle Dolan for checking the manuscript, providing answers to the exercises and writing most of the tutor's manual.

BACKGROUND MATHEMATICS

NUMBERS AND OPERATIONS

1.1 INTRODUCTION

Numbers and their operations are used to represent real-life situations. For example, a typical addition could be: there were three cows in a field and they were joined by five cows, making eight in all. The use of numbers allows us to think of different situations as represented by the same mathematical problem, $3+5=8$. The arithmetic remains the same whether the animals are cows, sheep or pigs. Representing a real-life situation by a mathematical one is called mathematical modelling.

Thinking of a simple real-life problem can often help to work out what the result of an operation should be. Supposing we have forgotten how to divide fractions and come across the problem

$$2 \div \tfrac{1}{3}$$

We could invent a simple problem to help out in that situation. 'There are two bars of chocolate and each one is divided into thirds. How many pieces are there?' We could then clearly see that the answer should be six. However, this approach will not always help. For instance, it is difficult to think of a simple everyday problem that would lead to the multiplication

$$(-3)(-6)=?$$

In these cases, where the result is not obvious, we need to remember established rules for the outcome. In this chapter we shall examine the rules of the operations of addition, subtraction, multiplication and division on real numbers. The real numbers are all those numbers, including positive and negative whole numbers and fractional and decimal numbers, that represent points on the real number line. When we draw a number line we usually mark only the whole numbers. This does not mean that the others do not concern us. The line is assumed to continue indefinitely in both directions, therefore representing all possible real numbers. A real number line is shown in Fig. 1.1.

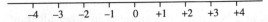

Figure 1.1 A real number line.

1.2 ALGEBRA: USING LETTERS FOR NUMBERS

There are a number of reasons that we might use a letter to represent a number. One situation is for a scientific formula, for example Ohm's law, $V = IR$, which gives the voltage across a pure resistor given the current and resistance. This formula is a statement which is supposed to be true in all relevant situations. Hence the letters are used to represent any physically possible values for the quantities, voltage, current and resistance.

Letters can be used in the same way to state mathematical laws. For instance, you may wish to state that it does not matter whether you swap the order of an addition: the result will be the same. This can be expressed by the statement:

For any two numbers a and b, $a + b = b + a$

This statement sums up all possibilities for adding any two numbers and a and b could be replaced by any numbers, for example:

$$3 + 5 = 5 + 3$$

$$8 + 7 = 7 + 8$$

$$1 + 2 = 2 + 1$$

$$2.1 + 8.62 = 8.62 + 2.1$$

$a + b = b + a$ is a generalization of all these examples.

Algebra is the study of the rules of operations. It involves using letters because we need to generalize the rules to all possible values. In this chapter we shall only concern ourselves with the algebra of numbers. The algebra of sets or matrices, for instance, could result in different rules. We need to decide on the rules and be able to use them in practice. For an engineer the important thing is to recognize that the rules of manipulating numbers and their operations exist, and that the rules must not be broken.

An operation on numbers is a way of combining two numbers to give a single number. However, we often need to write down expressions that involve more than a single operation. For instance, the expression $3 + 4 + 5$ involves two additions, while the expression $3(x + 4)$ involves one addition and one multiplication. When more than two numbers are to be combined in this way the order in which the operations are performed may be important. Brackets can be used to indicate which operation should be performed first. However, in order to simplify expressions involving more than one operation we need to know all the rules concerning which operation should be done first and how we can change the order of performing operations.

Before we go on to discuss the rules of numbers and their operations we should be aware that we shall be using letters to represent all possible numbers. This is different from using a letter to represent some unknown quantity which will take only a certain number of fixed values. For instance, $x + 3 = 5$ is called an equation. It is only true for $x = 2$ and is not true for all possible values of x. A short study of equations is given in Chapter 3.

Some shorthand is used in expressions. The '.' or $\times$ indicating multiplication is often left out, particularly in an expression involving letters; i.e. ab is taken to mean $a.b$ or $a \times b$.

1.3 RULES FOR ADDITION AND MULTIPLICATION

Some of the rules of numbers have special names, and because they can recur in other algebras we will list these here. These basic rules can be used to establish the everyday rules of manipulating expressions. Here a, b and c can be any numbers.

$a+b=b+a$	$a.b=b.a$	commutative laws
$(a+b)+c=a+(b+c)$	$(a.b).c=a.(b.c)$	associative laws
$a+0=a$	$a.1=a$	identity laws
$a.(b+c)=a.b+a.c$		distributive law
$a.0=0$		multiplication by zero

The use of these laws becomes automatic with practice. The practical rules that we use are as follows:

1. In an expression involving only multiplication or addition, terms can be swapped around to any order you like. This can aid calculation:

$$2.3.5=2.5.3=10.3=30$$

$$2+5+8+5=2+8+5+5=10+10=20$$

This rule comes from the commutative and associative laws.
2. Adding 0 to any number has no effect. Also, multiplying by 1 leaves numbers unchanged. The number 0 is called the additive identity and 1 the multiplicative identity.
3. The distributive law is the rule used when expanding expressions.

$$a(b+c)=ab+ac$$

If two bracketed terms are multiplied together then the law is used repeatedly to expand the expression.

Example 1.1 Expand $(x+y)(2x+3y)$.

SOLUTION First, expand the second bracket

$$(x+y)(2x+3y)=(x+y)2x+(x+y)3y$$

and now expand again

$$(x+y)2x+(x+y)3y=x2x+y2x+x3y+y3y$$

Use the commutative law to swap the order of the multiplications and the fact that $x.x=x^2$ to give

$$x2x+y2x+x3y+y3y=2x^2+2xy+3xy+3y^2$$

Collecting together common terms gives

$$2x^2+5xy+3y^2$$

This process can be shortened by directly multiplying out the brackets, remembering that each term in the first bracket must be multiplied by each term in the second bracket:

$$(x+y)(2x+3y) = x2x + y2x + x3y + y3y$$

Continuing as before this simplifies to

$$2x^2 + 5xy + 3y^2$$

If we swap the sides of the law so that it reads

$$a.b + a.c = a.(b+c)$$

then it becomes the rule which allows us to factorize, and we have taken out a common factor of a.

Example 1.2 Factorize $3(2x+1) + x(2x+1)$.

SOLUTION There is a common factor of $(2x+1)$:

$$3(2x+1) + x(2x+1) = (3+x)(2x+1)$$

4. If more than one operation is involved in an expression, brackets, ( ), should be used to indicate the order in which they are performed. There are two exceptions to this requirement:
   (i) The associative laws mean that any number of additions, or any number of multiplications can be performed without order being important.
   (ii) There is a convention used to specify order. The convention is summed up in the acronym BoDMAS, which stands for **B**rackets, **D**ivision, **M**ultiplication, **A**ddition and **S**ubtraction.

Example 1.3 Find $6 \times 2 \times 3 \times 4 + 5 \times 6 \times 3 - 3 \times (1+2)/6$.

SOLUTION

$$6 \times 2 \times 3 \times 4 + 5 \times 6 \times 3 - 3x(1+2)/6 = 6 \times 2 \times 3 \times 4 + 5 \times 6 \times 3 - 3 \times 3/6$$

(performing the operation in the brackets first)

$$6 \times 2 \times 3 \times 4 + 5 \times 6 \times 3 - 3 \times 3/6 = 6 \times 2 \times 3 \times 4 + 5 \times 6 \times 3 - 3 \times 0.5$$

(performing the division next)

$$6 \times 2 \times 3 \times 4 + 5 \times 6 \times 3 - 3 \times 0.5 = 144 + 90 - 1.5$$

(performing the multiplications)

$$144 + 90 - 1.5 = 234 - 1.5$$

(performing the addition)

$$234 - 1.5 = 232.5$$

(performing the subtraction)

Hence $6 \times 2 \times 3 \times 4 + 5 \times 6 \times 3 - 3 \times (1 + 2)/6 = 232.5$

Example 1.4 Remove brackets where possible:

$$(a/(b/c)) - ((pq)(rt)) + (a(b+c))$$

SOLUTION Using the BoDMAS convention

$$(a/(b/c)) - ((pq)(rt)) + (a(b+c)) = a/(b/c) - pqrt + a(b+c)$$

The brackets are still needed in the first term to indicate that the division b/c should be performed first and a should then be divided by the result of b/c. The brackets are needed in the term $a(b+c)$ to indicate that $b+c$ should be performed before multiplying the result by a. All other brackets may be left out, as the order is either unimportant or is specified by the BoDMAS convention.

If in doubt use brackets to indicate order as they never do any harm.

Before we look at the rules for subtraction we will look more closely at addition and multiplication in particular, in order to understand the behaviour of negative numbers.

1.4 POSITIVE AND NEGATIVE NUMBERS

Every number, except 0, is either positive or negative. This is indicated by placing a sign in front of the number. Positive numbers may have their sign omitted. On the number line, as shown in Fig. 1.1, numbers to the left of 0 are negative numbers and numbers to the right are positive numbers. Here the symbols $+$ and $-$ are being used in a different way from that in Section 1.3. In Section 1.3 the symbol $+$ is an operation, an order to perform the task of addition. Similarly, $-$ can be used as an order to perform the task of subtraction. In this section we are using the same symbols to indicate whether the number is positive or negative. To stress this distinction, for the moment we shall indicate the sign of a number by placing it slightly raised in front of the number, like this: $^-3$, $^+2$.

We could explain how to do addition by using the number line. Start at the position on the number line representing the first number and move to the right or left depending on whether you are adding a positive or negative number. This is illustrated in Fig. 1.2. How do we perform $^+3 + {}^-2$?

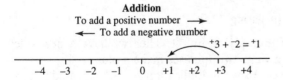

Figure 1.2 Performing addition on a number line.

To add $^-2$ we move 2 units to the left. Hence

$$^+3 + {}^-2 = {}^+1$$

$$^-3 + {}^-2 = {}^-5$$

$$^+1 + {}^-4 = {}^-3$$

$$^+1 + {}^+2 = {}^+3$$

This has given us a method of adding all numbers, positive and negative. What about multiplication of positive and negative numbers? First, we decide on the result of multiplying any number by $^-1$ and then use that to establish how to multiply any two numbers. One way of thinking of multiplication is as repeated addition:

$$^+3.^+2 = {}^+2 + {}^+2 + {}^+2 = {}^+6$$

What is $^+3.^-1$? Using repeated addition we get

$$^+3.^-1 = {}^-1 + {}^-1 + {}^-1 = {}^-3$$

As $^+3.^-1 = {}^-1.^+3$ (the commutative law), then $^-1.^+3 = {}^-3$.

Multiplication by $^-1$ changes the sign of the number and is called negating the number. Thus multiplication by $^-1$ has the effect of reflecting the position of the number on the number line about 0. Fold the line at 0 and any number will meet its negated value. This is illustrated in Fig. 1.3.

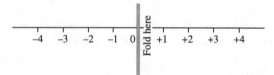

Figure 1.3 A number is negated (multiplied by $^-1$) by reflecting it about 0.

Example 1.5

$$^-1.^+3 = {}^-3 \qquad {}^-1.^-3 = {}^+3$$

$$^-1.^-2 = {}^+2 \qquad {}^-1.^-1 = {}^+1$$

A shorthand for $^-1.x$ is $-x$, where x can take any value. Thus x negated is $-x$.

A negative number can be treated as -1 times its positive value and this is used when performing multiplication. The effect of the signs is considered last after the multiplication of the positive numbers has been performed:

$$^-5.^-3 = {}^-1.5.^-1.3$$

$$= {}^-1.^-1.5.3 \qquad \text{(rearranging the terms)}$$

$$= {}^-1.^-1.15$$

$$= {}^-1.^-15 \qquad \text{(negating 15)}$$

$$= 15 \qquad \text{(negating } {}^-15)$$

Multiplying by $^-1$ twice has the effect of multiplying by $^+1$. Thus we have a new rule governing multiplication of negative numbers, giving: 'two minuses make a plus'.

We can extend this idea by noticing that successive multiplication by -1 repeatedly flips the sign from positive to negative.

$$^-1.^+3 = {}^-3$$

$$^-1.^-1.^+3 = {}^+3$$

$$^-1.^-1.^-1.^+3 = {}^-3$$

$$^-1.^-1.^-1.^-1.^+3 = {}^+3$$

To multiply a string of numbers together, count the number of negative numbers. If it is even, the result is positive; if it is odd the result is negative.

Example 1.6

$$^+3.^-2.^-1.^-5 = ^-30 \qquad \text{(3 minuses)}$$

$$^-2.^+6.^-8.^+1 = ^+96 \qquad \text{(2 minuses)}$$

1.5 RULES FOR SUBTRACTION

Subtraction is the opposite (more mathematically called the inverse) of addition. Subtraction can be performed on a number line by starting with the first number and then moving to the left or right, depending on whether you are subtracting a positive or a negative number, that is by moving in the opposite direction to addition. This is shown in Fig. 1.4.

Compare the following two problems: first, the addition $^-2 + ^-3 = ?$, which is pictured in Fig. 1.5(a), and second the subtraction $^-2 - ^+3 = ?$, which is pictured in Fig. 1.5(b). We can see that they both result in $^-5$. This example illustrates that subtraction of any number is the same as adding its negated value. This rule is true because subtraction is the inverse of addition, and it can be expressed by

$$a - b = a + (-b) \qquad (1.1)$$

where $-b$ is the negated value of b.

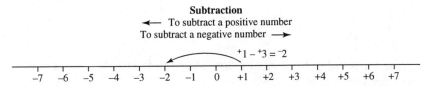

Figure 1.4 Subtraction.

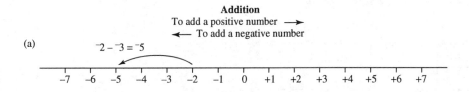

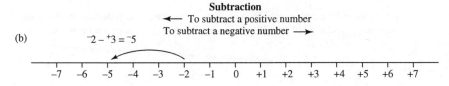

Figure 1.5 (a) The addition $^-2 + ^-3 = ?$ and (b) the subtraction $^-2 - ^+3 = ?$

The negated value of b can also be called its additive inverse because if you add b to anything and then add $-b$ you get back to where you started. In other words, adding b followed by adding $-b$ results in adding 0 because $b + (-b) = 0$.

If we substitute a negative value for b into Eq. (1.1), for instance $b = ^-3$, then as the negated value of $^-3$ is $^+3$ we find

$$a - ^-3 = a + ^+3$$

This result can be reproduced for any negative value of b and leads to a second version of the 'two minuses make a plus' rule.

The relationship between addition and subtraction is very convenient because it allows us to replace any subtraction by an addition. We deal with any minus signs, as in the expression $a-b$, by rewriting them as $a+(-b)$. Then we are able to rearrange expressions using the rules of addition as before. This rule can be summed up as: 'attach the sign to the number it is in front of before shuffling around'.

For example:

$$a+b-c=a-c+b=-c+a+b$$

Here we have treated $-c$ as $+(-c)$.

The only remaining problem with subtraction is how to cope with a minus sign in front of a bracket. This is treated as $^-1$ multiplying the whole bracket; for example:

$$2-(5-2)=2+{}^-1.(5+{}^-2)$$

$$={}^+2+{}^-5+{}^+2$$

$$={}^-1$$

As we have now established the link between negated numbers and subtraction we can drop the need to raise the sign to indicate positive and negative numbers.

1.6 DIVISION

Just as there is a very close relationship between addition and subtraction, there is, of course, a similar relationship between multiplication and division.

Multiplication of whole numbers can be interpreted as repeated addition. The rectangle in Fig. 1.6 has an area which can be found by adding each of the areas of the horizontal strips. This gives an area of $4 \times 3 = 12$ square units.

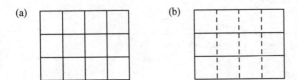

(a) (b)

Figure 1.6 (a) A 4×3 rectangle has an area of 12 square units given by $4 \times 3 = 12$. (b) The problem 'How many horizontal strips of area 3 square units are there in the rectangle of area 12 square units?' is solved by using the inverse operation of division: $12/3 = 4$.

Division is then the inverse process, and can be thought of by the problem, 'how many horizontal strips of area 3 square units are there in a rectangle of area 12 square units?'. Here we are thinking of division arising from the problem of finding the value x such that

$$x \times 3 = 12$$

giving the solution $12/3 = 4$.

This reasoning tells us that division is almost the inverse operation to multiplication. It is only 'almost' because of the problem with zero. Division by zero is not defined. If we think of division as reversing the process of multiplication then it is clear that there is no way of reversing multiplication by zero. Multiplication by 0 always gives 0, for instance

$$0 \times 6 = 0$$

$$0 \times 12 = 0$$

$$0 \times 9 = 0$$

All of these give the same result and therefore given the problem $0.x = 0$ we cannot determine the value of x. Hence division by zero is not defined.

For all non-zero numbers, instead of performing division we can perform an equivalent multiplication. This is a familiar idea in the language we use to describe division. Division of a cake into two pieces is to find a half of the cake. Division by 3 is the same as finding one third of something. Notice that 2 and $\frac{1}{2}$ are **reciprocals** of one another, as are 3 and $\frac{1}{3}$. To find the reciprocal of a number write it as a fraction and turn the fraction upside down. The number 2 can be written as $2/1$; therefore its reciprocal is $\frac{1}{2}$. The reciprocal of x is the number $1/x$.

The reciprocal of a number is also called its multiplicative inverse because if you multiply anything by x and then by $1/x$ you get back to where you first started. In other words, if you multiply by x and then by $1/x$ you have multiplied by 1 overall because

$$x . \frac{1}{x} = 1$$

In many situations, division can be replaced by multiplication using the multiplicative inverse. However, all the rules of division cannot be so conveniently discarded, as we did for subtraction, since we want to be able to use fractional expressions. We will concern ourselves more with coping with fractional expressions in the next chapter.

Note the following properties *not* obeyed by division:

$$a/b \neq b/a \qquad \text{division is not commutative}$$

$$(a/b)/c \neq a/(b/c) \qquad \text{division is not associative}$$

$$c/(a+b) \neq c/a + c/b \qquad \text{division is not left distributive}$$

However, division does obey the right distributive law:

$$(a+b)/c = a/c + b/c$$

1.7 EQUALITY OF EXPRESSIONS

If we state that two expressions are equal, then in general we mean they are equal for all possible values; that is, the letters should be able to be replaced by any number.

As there are an unlimited number of numbers it is not a good idea to start substituting in order to show that the expressions are equal, as this would require an infinite amount of work. To show that two expressions are equal we must use the established rules for the behaviour of numbers and their operations.

There are three possible ways to go about showing that some equality is correct:

1. Start with the left-hand side (LHS) of the equals sign and try to manipulate it into the same form as the right-hand side (RHS).
2. Start with the right-hand side and try to get it into the same form as the left-hand side.
3. Rearrange both sides of the statement to show that they are both equal to a third expression.

If you suspect, however, that some equality is false, there is a quick method to show this: find any numbers which, when substituted for the letters, lead to a false statement.

Example 1.7 Show that $2(a+b)-2a=2b$.

SOLUTION Remove the brackets on the left-hand side of the equality:

$$2(a+b)-2a=2a+2b-2a=2b$$

This is now equal to the RHS of the original statement.

Example 1.8 Show that $a.b+c=a.(b+c)$ is a false statement.

SOLUTION Substitute any values for a, b and c, for instance, $a=2$, $b=3$, $c=1$. This gives:

LHS: $a.b+c=2.3+1=7$

RHS: $a.(b+c)=2.(3+1)=2.4=8$

Since the RHS is different from the LHS then the statement is false.

1.8 SUMMARY

1. The order of operations follows BoDMAS (Brackets, Division, Multiplication, Addition, Subtraction).
2. In expressions involving only addition or only multiplication the order is not important.
3. The distributive law

$$a(b+c)=ab+bc$$

is used to expand expressions or to take out a factor.
4. To negate a number, flip it over to the opposite side of 0 on the number line, i.e. multiply by -1. x and $-x$ are additive inverses, and they add up to 0.
5. To avoid any problems with subtraction convert to addition: 'attach the sign to the number it is in front of before shuffling around'.
6. Turn a number upside down to find its reciprocal (also called its multiplicative inverse). That is, x and $1/x$ are multiplicative inverses and they multiply together to give 1.
7. Two minuses make a plus:

$$a-(-b)=a+b$$
$$(-a)(-b)=ab$$

8. Division by zero is undefined.
9. To prove some statement involving an equality use the established rules of how numbers behave to rearrange one or both sides.
10. To disprove some statement involving an equality substitute particular numbers to give a false statement.

1.9 EXERCISES

1.1 For each part use letters to represent a rule that generalizes all of the given inequalities. Do you think your rule is true?

(a) $3\times5=5\times3$, $4\times6=6\times4$, $3\times7=7\times3$

(b) $2 \times (4+2) = 2 \times 4 + 2 \times 2$
  $3 \times (4+6) = 3 \times 4 + 3 \times 6$
(c) $3 + (4+2) = (3+4) + 2$
(d) $1/(1/3) = 3,$ $1/(1/4) = 4,$ $1/(1/5) = 5$

1.2 Use the commutative, associative and distributive laws only to justify the following statement:

$$a((c+d)b) = abc + adb$$

1.3 Give the negated values of the following:

(a) $^+5$ (b) $^-10$ (c) $-x$ (d) $\dfrac{1}{-x}$

1.4 Simplify the following:

(a) $a - 5b - (3b - 5a) - 3(a+b)$
(b) $-c + a - 2b + (5 - (3b - 2a)).(-2)$

1.5 Does subtraction obey the following laws?

(a) $(a-b) - c = a - (b-c)$ associative law
(b) $a - 0 = 0 - a = a$ identity law
(c) $a(b-c) = ab - ac$ distributive law

1.6 Does division have a right identity, i.e. is there a number x such that for all possible values of a then $a/x = a$?
1.7 Does division have a left identity, i.e. is there a number x such that for all possible values of a then $x/a = a$?
1.8 What is the reciprocal of $1/x$?
1.9 Calculate:

(a) $3 - 4 \times 2 + 5(4+6)/2$
(b) $7 - 3 - 2 + 4(3+7) - (5-2)$
(c) $(4-2)(3+2)(3 \times 2)$
(d) $(5-3)/(-2+4) - 7$

1.10 Remove the brackets and combine like terms:

(a) $(p + 2q) + (p - 3q)$
(b) $3(2x - y) - 2(3x + y)$
(c) $(y+3)(y+2)$
(d) $2(a+b)^2 + 3(a+b)^2$

1.11 State whether the following are true or false and justify your answer, where p, q, x and y can take any values.

(a) $x(5-y) - 3(5-y) = (5-y)(x-3)$
(b) $2p - 6 + pq - 3q = (p-3)(2+q)$
(c) $p^2 - 2p + 3 = -p^2 + 3$
(d) $(p^2 - 1)/(p+1) = p - 1$ where $p \neq -1$
(e) $5pq^2 - 2p^2q + pq = -3p(p+q)$
(f) $-4p - 6 \times 2q + 8(p - 3 \times 3q) = 4p - 9 \times 9q$

1.12 Give the additive inverse of each of the following:

(a) -3 (b) $\frac{1}{7}$

(c) $-a$ (d) $x - y$

(e) $-(x-y)$

1.13 Give the multiplicative inverse of each of the following:

(a) 0.25 (b) -0.25

(c) $\frac{9}{14}$ (d) a

(e) $\dfrac{1}{a}$ (f) $a-4$

(g) a^2

1.14 Simplify

(a) $3(2x-y)-2(3x+y)$
(b) $a(a+2)-a(a-3)-4$

1.15 The reliability of a system with three units in parallel is given by $1-(1-R_1)(1-R_2)(1-R_3)$, where R_1, R_2, R_3 are the individual reliabilities of the three parallel units. Show that this expression is equal to

$$R_1+R_2+R_3-R_1R_2-R_1R_3-R_2R_3+R_1R_2R_3$$

1.16 A body moving under constant acceleration a with initial velocity u has velocity v after time t given by $v=u+at$. The distance travelled by the body is given by

$$s=\frac{(u+v)t}{2}$$

Substitute for v in the expression for s to show that s can be found in terms of u, a and t by the formula

$$s=ut+\tfrac{1}{2}at^2$$

FRACTIONS AND FRACTIONAL EXPRESSIONS

2.1 INTRODUCTION

As mentioned in Section 1.6, we need to understand the rules governing division because of the need to have fractional expressions. In this text we shall not attempt to go through all the rules of fractions by looking at cakes divided into parts. This is an excellent way of checking that rules involving fractions do make sense, but we assume that you have developed sufficient graphical skills to make up such examples for yourself, and we are aiming here to give a summary of the results and how they are used.

A fraction, or fractional expression, has names associated with the top and bottom lines.

numerator

↓

$$\frac{a}{b}$$

↑

denominator

2.2 FRACTIONAL NUMBERS

If a number is expressed as a fraction, in general the best approach is to convert it to a decimal by performing the division on a calculator. This will usually give an approximate value for the number, as only fractions which, in their lowest form, have denominators with factors 2 and 5 can be expressed exactly in decimal notation. Decimal numbers are easy to multiply and add etc. Depending on the accuracy of your calculator, however, you can end up with some interesting results by always using decimal form to perform a calculation. Try the following:

$$\frac{3}{9} \times 9$$

You could well get the result 2.999999 and then decide to round that to 3. Of course in this case a quick look at the problem should tell you that the answer is 3. This is an example of a simple problem that is better done without a calculator by using the rules of fractions.

In more complicated calculations it is important to watch out for the order of performing the operations:

$$\frac{3\frac{4}{9}+4}{-6-2.5\times\frac{3}{7}}$$

Notice that $3\frac{4}{9}$ means $3+\frac{4}{9}$. The easiest way to find the value of the entire expression is to calculate the bottom line first, store this in memory, and then calculate the top line. Finally divide the top by the bottom line. Do this yourself – you should get -1.0527 to four decimal places.

For the rest of this chapter we shall concern ourselves with the rules of manipulating fractions, mainly in order to be able to simplify fractional expressions.

2.3 FRACTIONS WITH DENOMINATOR 1

Divide anything by 1 and it will not be changed. To go back to our modelling ideas introduced in Chapter 1, we could convince ourselves of this by thinking of a problem such as 'There were three cream cakes on the trolley and only one person left in the dining room'. Such a situation would result in the poor unfortunate person eating all three cakes, hence $\frac{3}{1}=3$. In general, this leads to $\frac{a}{1}=a$.

This rule can be very useful to get rid of denominators of 1, or it can be used to write any simple expression as a fractional one, should we so wish.

2.4 MULTIPLYING FRACTIONS

To multiply fractions, multiply the numerators together and the denominators together.

$$\frac{a}{b}.\frac{c}{d}=\frac{ac}{bd}$$

By using the result of Section 2.3 we can now multiply any two expressions together, whether fractional or not.

$$\frac{x+1}{x}.(x-5)=\frac{x+1}{x}.\frac{x-5}{1}=\frac{(x+1)(x-5)}{x}$$

Here we have written $(x-5)$ as a fraction with denominator 1, and then performed the multiplication.

We can also rewrite expressions such as $\frac{3}{4}a$ as $\frac{3a}{4}$ by using this same argument.

2.5 EQUIVALENT FRACTIONS

From the identity law any number multiplied by 1 remains unchanged:

$$a.1=a$$

Also, as any number divided by itself gives 1, there are many ways of writing the number 1:

$$1=\tfrac{1}{1}=\tfrac{2}{2}=\tfrac{3}{3}=\tfrac{4}{4}=\tfrac{5}{5}=\cdots=\tfrac{1000}{1000}=\tfrac{8.2}{8.2}$$

Any fraction will be unchanged in value if both the top and bottom lines are multiplied by the same number, because we are only multiplying by 1, e.g.

$$\tfrac{3}{4} = \tfrac{3 \times 2}{4 \times 2} = \tfrac{3 \times 3}{4 \times 3} = \tfrac{3 \times 10}{4 \times 10}$$

hence

$$\tfrac{3}{4} = \tfrac{6}{8} = \tfrac{9}{12} = \tfrac{30}{40}$$

This can be extended to say that a fractional expression is not changed by multiplying the top and bottom lines by the same thing (in general, however, be careful to avoid something that could equal 0).

Considering a/b where $b \neq 0$; then

$$\frac{a}{b} = \frac{a.a}{b.a} \qquad \text{if } a \neq 0$$

and

$$\frac{a}{b} = \frac{a(b-1)}{b(b-1)} \qquad \text{if } b \neq 1$$

In the same way, a fractional expression is not changed by cancelling the same expression from the top and bottom lines. Be careful though: whatever is cancelled must be a factor of all of the numerator and of the denominator. For example:

$$\tfrac{15}{24} = \tfrac{5}{8}$$

Here the factor 3 was cancelled from the top and bottom lines. Also

$$\frac{3a+3b}{3c} = \frac{3(a+b)}{3c} = \frac{a+b}{c}$$

where we have cancelled the common factor of 3 and

$$\frac{(x-1)x}{x(x-2)} = \frac{x-1}{x-2} \qquad \text{where } x \neq 0$$

where we have cancelled the common factor of x. This last equality is only true if $x \neq 0$, because the left-hand side has no defined value for $x = 0$.

Be careful when cancelling to ensure that you cancel correctly. For instance, note the following:

$$\frac{a+3b}{3a} \neq \frac{a+b}{a}$$

3 is not a factor of the whole of the top line of the fraction so no cancellation can be performed in this case.

2.6 THE DISTRIBUTIVE LAW, OR SPLITTING THE LINE

As established in Section 1.6,

$$\frac{a+b}{c} = \frac{a}{c} + \frac{b}{c}$$

This is the rule which allows fractions with a sum on the top line to be split into the sum of two fractions:

$$\tfrac{1+2}{3} = \tfrac{1}{3} + \tfrac{2}{3}$$

Similarly, it gives the rule for summing two fractions with the same denominator:

$$\tfrac{1}{3} + \tfrac{2}{3} = \tfrac{1+2}{3}$$

The following example, where $x \neq 0$, uses the rule in both ways:

$$\frac{x+1}{x} + \frac{x-2}{x} = \frac{x+1+x-2}{x}$$

$$= \frac{2x-1}{x}$$

$$= \frac{2x}{x} - \frac{1}{x}$$

$$= 2 - \frac{1}{x}$$

Be careful, because the horizontal line acts in the same way as a bracket, which can lead to complications with minus signs. An example of this is

$$-\frac{a+b}{c} = \frac{-(a+b)}{c} = \frac{-a-b}{c} = -\frac{a}{c} - \frac{b}{c}$$

Always remember that the bottom line cannot be split in the same way as the top line, i.e.

$$\frac{a}{b+c} \neq \frac{a}{b} + \frac{a}{c}.$$

2.7 ADDING FRACTIONS

The method of the previous section allows us to add fractions with a common denominator, but we need a rule to be able to sum any two fractions or fractional expressions. This is done by 'finding the common denominator', e.g.

$$\tfrac{3}{4} + \tfrac{5}{6}$$

The smallest number that both 4 and 6 will divide into is 12 (the lowest common denominator). In practice it may be easier simply to multiply the denominators together (giving 24) and use that for the common denominator.

Now write the two fractions in terms of the common denominator. We solve two puzzles. First,

$$\frac{3}{4} = \frac{?}{12}$$

Here the question mark must be 9, as we have multiplied the top and bottom lines by 3.

$$\frac{5}{6} = \frac{?}{12}$$

Here the question mark must be 10, as we have multiplied the top and bottom lines by 2.

We then get

$$\tfrac{3}{4}+\tfrac{5}{9}=\tfrac{9}{12}+\tfrac{10}{12}=\tfrac{9+10}{12}=\tfrac{19}{12}$$

As the fractions now have a common denominator they have been combined to give a sum of

$$\tfrac{19}{12}=\tfrac{12}{12}+\tfrac{7}{12}=1\tfrac{7}{12}$$

Combining two fractional expressions is very similar. For example:

$$\frac{x-1}{x+2}+\frac{x}{x-3}\qquad\text{where }x\neq-2\text{ and }x\neq3$$

The common denominator is $(x+2)(x-3)$ and we rewrite both fractions:

$$\frac{x-1}{x+2}=\frac{(x-1)(x-3)}{(x+2)(x-3)}\qquad\text{(multiplying the top and bottom by }(x-3))$$

and

$$\frac{x}{x-3}=\frac{x(x+2)}{(x-3)(x+2)}\qquad\text{(multiplying the top and bottom by }(x+2))$$

Hence

$$\frac{x-1}{x+2}+\frac{x}{x-3}=\frac{(x-1)(x-3)+x(x+2)}{(x+2)(x-3)}$$

2.8 DIVIDING FRACTIONS

Again we use the idea of equivalent fractions to perform division of fractions. For example, we wish to write

$$\frac{\tfrac{2}{3}}{\tfrac{3}{5}}$$

as a single fraction. Before changing this expression, it is important to understand the order of operations in this expression. The longer line means that there are implied brackets on the top and bottom of that line, i.e.

$$\frac{\tfrac{2}{3}}{\tfrac{3}{5}}$$

means

$$\frac{\left(\tfrac{2}{3}\right)}{\left(\tfrac{3}{5}\right)}$$

To make this into one simple fraction we get rid of the denominator by multiplying by the multiplicative inverse of 3/5 i.e. the reciprocal, 5/3. To do this without changing the value of the expression we must multiply both the numerator and the denominator, hence using the idea of equivalent fractions.

$$\frac{\tfrac{2}{3}}{\tfrac{3}{5}}=\frac{\tfrac{2}{3}\tfrac{5}{3}}{\tfrac{3}{5}\tfrac{5}{3}}=\frac{\tfrac{10}{9}}{1}=\frac{10}{9}$$

The same approach will work for a fractional expression. To simplify

$$\frac{\dfrac{3}{y-3}}{\dfrac{y}{y+1}}$$

multiply the numerator and denominator by $(y+1)/y$, which gives

$$\frac{\dfrac{3}{y-3}}{\dfrac{y}{y+1}} = \frac{\dfrac{3}{y-3}\dfrac{y+1}{y}}{\dfrac{y}{y+1}\dfrac{y+1}{y}} = \frac{\dfrac{3(y+1)}{y(y-3)}}{1}$$

$$= \frac{3(y+1)}{y(y-3)}$$

2.9 SUMMARY

1. To deal with a complicated numerical problem involving fractions, convert to decimal form by using a calculator. This will, in general, give only an approximate result.
2. It is necessary to know the rules of fractions in order to simplify fractional expressions.
3. To multiply two fractions multiply the top lines (numerators) and the bottom lines (denominators):

$$\frac{a}{b} \times \frac{c}{d} = \frac{ac}{bd}$$

4. Any simple expression can be written as a fraction with denominator 1:

$$a = \frac{a}{1}$$

5. A fraction will not change its value if multiplied by the same expression on the top and bottom lines because this is the same as multiplying by 1. However, zero must be avoided.

$$\frac{a}{b} = \frac{ax}{bx} \qquad x \neq 0$$

6. The distributive law for division means that a fraction with a sum in the numerator can be split into the sum of two fractions.

$$\frac{a+b}{c} = \frac{a}{c} + \frac{b}{c}$$

This cannot be done with the denominator:

$$\frac{a}{b+c} \neq \frac{a}{b} + \frac{a}{c}$$

7. The sum of two fractions can be expressed as a single fraction by writing them over a common denominator:

$$\frac{a}{b} + \frac{c}{d} = \frac{ad + bc}{bd}$$

8. To divide fractions, multiply the top and bottom lines by the reciprocal of the bottom line. By point 5 above, this does not change the value of the expression.

$$\frac{\dfrac{a}{b}}{\dfrac{c}{d}} = \frac{\dfrac{a}{b}\dfrac{d}{c}}{\dfrac{c}{d}\dfrac{d}{c}} = \frac{\dfrac{ad}{bc}}{1} = \frac{ad}{bc}$$

2.10 EXERCISES

2.1 Calculate the following:

(a) $4\frac{27}{9} + \frac{10}{3} \times 9$

(b) $12.6 - \dfrac{4 - \frac{5}{8}}{\frac{3}{7} + 2\frac{12}{17}}$

2.2 Simplify

$$\frac{\dfrac{a}{1} + \dfrac{1}{b}}{1} + a$$

2.3 Simplify

$$5\frac{2b}{a+2}(a+3)$$

2.4 Write $\frac{ac}{ba}$ as the product of two terms, in as many different ways as you can think of.

2.5 Simplify, specifying those values for which the expression is not defined:

(a) $\dfrac{x + xy}{x(x+1)}$

(b) $\dfrac{a + ba + ac}{ad + ae + af}$

2.6 Simplify

$$-\frac{3x-2}{x(x-1)} + \frac{1}{x(x-1)} \qquad \text{where } x \neq 0 \text{ and } x \neq 1$$

2.7 Express as a single fraction:

(a) $\dfrac{x}{x-1} - \dfrac{x-1}{x}$

(b) $\dfrac{a}{(a-b)(a-c)} + \dfrac{b}{(a-b)(a-d)} + \dfrac{c}{(a-b)}$

2.8 Simplify, where $x \neq 0.5$ and $x \neq 0$:

$$\frac{\left(\dfrac{x+1}{2x-1}\right)}{\left(\dfrac{4}{2x}\right)}$$

2.9 Find the following:

(a) $4\frac{2}{5} + \frac{3}{5} - \frac{4}{6} + \frac{3}{6}$

(b) $4 - \frac{1}{4} + 2\frac{5}{8} - \frac{11}{7}$

(c) $\dfrac{4-2}{8} + \dfrac{-6}{-3} - \dfrac{18}{3}\left(\dfrac{1}{3} - \dfrac{7}{16}\right)$

2.10 Express the following as single fractions:

(a) $\dfrac{2}{y+1} - \dfrac{1}{y+3}$

(b) $\dfrac{3}{y-8} - \dfrac{2}{8-y}$

(c) $\dfrac{y-1}{y(y+1)} - \dfrac{y}{(y+1)(y+2)}$

(d) $\dfrac{2}{3y} - \dfrac{3}{2y}$

(e) $\dfrac{y+2}{y+3} - \dfrac{y+1}{y+2}$

2.11 Simplify the following:

(a) $\dfrac{\frac{3}{8}}{\frac{9}{6}}$

(b) $\dfrac{\frac{1}{3}}{y}$

(c) $\dfrac{\dfrac{y+1}{y+2}}{\dfrac{y}{y-2}}$

(d) $\dfrac{\dfrac{y-3}{y+1}}{\dfrac{1}{y} + \dfrac{2}{y+1}}$

(e) $\dfrac{\dfrac{3}{y} - \dfrac{2}{y^2}}{y+1}$

2.12 State whether the following are true or false and justify your answer:

(a) $\dfrac{1}{x-2} = \dfrac{1}{x} - \dfrac{1}{2}$ where $x \neq 2$ and $x \neq 0$

(b) $-\dfrac{x-2}{x} = \dfrac{2}{x} - 1$ where $x \neq 0$

(c) $\dfrac{x+3}{(x+1)(x+3)} = 1 + \dfrac{1}{x+1}$ where $x \neq -1$ and $x \neq -3$

(d) $\dfrac{1}{x+1} - \dfrac{1}{x-1} = \dfrac{-2}{x^2-1}$ where $x \neq -1$ and $x \neq 1$

2.13 The inductance, $L\mu H$, of a circular cross-sectional coil can be estimated by the formula

$$L = \frac{0.0787 N^2 d^2}{3.5d + 8l}$$

where $d =$ diameter of the coil (cm), $l =$ length of winding (cm) and $N =$ number of turns. If $d = 1\frac{2}{3}$, $l = 3\frac{3}{4}$ and $N = 220$, calculate the inductance L.

2.14 The circumferential strain on a compressed air cylinder is given by

$$\varepsilon_c = \frac{\sigma_c}{E} + v\frac{\sigma_r}{E} - v\frac{\sigma_l}{E}$$

where σ_c, σ_l, σ_r are the circumferential, longitudinal and radial stresses, respectively and v and E are Poisson's ratio and the modulus of elasticity for the material. If $v = 0.28$, $\sigma_c = 78$ MN m^{-2}, $\sigma_l = 38$ MN m^{-2}, $\sigma_r = 26$ MN m^{-2} and $E = 198$ GN m^{-2}, calculate ε_c.

THREE

EQUATIONS

3.1 INTRODUCTION

An equation is a statement, e.g.

$$x + 2 = 5$$

which may or may not be true, depending on the value which is substituted for the 'unknown', in this case x. In the above example, if we substituted $x = 3$ then the equation becomes

$$3 + 2 = 5$$

which is true. However, if we substitute $x = 4$ we get

$$4 + 2 = 5$$

which is false.

The value(s) which make(s) the equation into a true statement are called the solution(s). To solve an equation means to find all the solutions. Not all equations can be solved by the use of algebraic manipulation. Only in a few simple cases is this possible. Most solution methods involve using a numerical method and therefore are generally performed using a computer. It is, however, useful to know how to solve equations in those simple cases where it can be done by algebraic manipulation. Another thing to watch out for is that not all equations have solutions. A quadratic equation like $x^2 = -4$ has no real solutions.

One way to solve an equation is to take a guess at the solution and substitute in the guess to see if it is correct. However, this can be a long process and may run into difficulties if we do not know how many solutions there are. We therefore present some methods for solving simple equations.

3.2 INVERSE OPERATIONS

One way to solve a simple equation is to think of an equation as a puzzle. There are two people A and B, and A proposes the problem

$$2x + 3 = 9$$

by saying, 'I have thought of a number, multiplied it by 2 and then I added 3, the result is 9. What is the number I first thought of?'.

B immediately realizes that the original number can be found by going through the whole process backwards. Hence B thinks, 'I start with 9, subtract 3 (giving 6) and divided by 2 (giving 3). Hence the original number, x, is 3'.

This method uses the idea of inverse operations to solve the equation. The equation can be viewed as a flow diagram with input x and boxes indicating the operations. The inverse flow diagram is then found by reversing the order and replacing the original operations by their inverses. A flow diagram for $2x+3=9$ and the inverse flow diagram giving its solution is shown in Fig. 3.1. This method of using flow diagrams illustrates the use of inverse operations in the process of solving simple equations.

Equations are more usually solved in stages. At each stage a new equation is found which is equivalent to the original equation.

(a)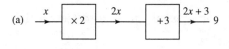

(b)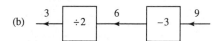

Figure 3.1 (a) A flow diagram describing the equation $2x+3=9$ and (b) the inverse diagram, giving the solution as $x=(9-3)/2$.

3.3 EQUIVALENT EQUATIONS

An equation can be viewed as a pair of scales. The scales start off in an equilibrium position with the left-hand weight exactly balancing the right-hand weight. In order to maintain the balance we can add weights, subtract weights or multiply or divide by some value, but only if we do exactly the same to both sides. If we perform the same operation on both sides of the equals sign and we maintain the balance of the equation we say that the new equation is equivalent to the original equation. Equivalence can be indicated by using the symbols $\Leftrightarrow$ or $\equiv$.

Example 3.1 $2x+3=9$ can be viewed as the situation where two unknown weights are on one side of the scales with three other one unit weights. On the other side are nine one unit weights. This is pictured in Fig. 3.2.

To find x, first remove 3 one unit weights from both sides, as in Fig. 3.3, showing that $2x+3=9\Leftrightarrow2x=6$. Now divide both sides by 2, giving Fig. 3.4, showing that $2x=6\Leftrightarrow x=3$. Here we have found the solution, $x=3$.

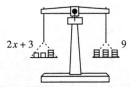

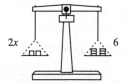

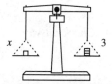

Figure 3.2 The equation $2x+3=9$ can be viewed as the situation where two unknown weights are on one side of the scales with three other one unit weights. On the other side are nine one unit weights and the scales are balanced.

Figure 3.3 The scales pictured in Fig. 3.2 have had three one unit weights removed from both sides, so they still balance and represent the equation $2x=6$. This shows that $2x+3=9\Leftrightarrow2x=6$.

Figure 3.4 The scales pictured in Fig. 3.3 have had the weights in both the scale pans halved so that they still balance and represent the equation $x=3$. This shows that $2x=6\Leftrightarrow x=3$.

We solved the equation in a number of steps, ensuring that at each stage we have an equation equivalent to the original equation. Equivalent equations were found by 'doing the same thing to both sides'. To decide *what* to do to both sides we use the idea of inverse operations. The aim is to reach the position when only the unknown quantity remains on one side of the equation. For convenience we shall call the unknown quantity x. The coefficient of x is the number multiplying x.

How to solve a simple equation

1. Remove any brackets where necessary.
2. 'Collect' everything involving x on one side of the equation and anything not involving x on the other side. (This is done by a number of additions and subtractions to both sides of the equation.)
3. Combine all the terms involving x so that there is a single coefficient of x.
4. Divide both sides of the equation by the coefficient of x.
5. Check that the answer is correct by substituting into the original equation to see whether the statement is true with this value of x.

Example 3.2 Solve the equation $2(3-3x)+5x=-3x$.

SOLUTION
1. *Remove brackets*

$$6-6x+5x=-3x$$

$$\Leftrightarrow 6-x=-3x$$

2. *Collect terms*
Collect the x terms on the left-hand side of the equation (by adding $3x$ to both sides of the equation) and all other terms on the right-hand side of the equation (by subtracting 6 from both sides of the equation). Here we write out the intermediate stages:

$$6-x+3x=-3x+3x \qquad \text{(adding } 3x \text{ to both sides)}$$

$$\Leftrightarrow 6+2x=0 \qquad \text{(combining terms)}$$

$$\Leftrightarrow 6+2x-6=0-6 \qquad \text{(subtracting 6 from both sides)}$$

$$\Leftrightarrow 2x=-6$$

3. *Divide by the coefficient of x*

$$\frac{2x}{2}=\frac{-6}{2} \qquad \text{(Divide both sides by 2)}$$

$$\Leftrightarrow x=-3$$

4. *Check answer*
Substitute $x=-3$ into the original equation:

$$2(3-3x)+5x=-3x$$

giving

$$2(3-3(-3))+5(-3)=-3(-3)$$

$$\Leftrightarrow 2(3+9)-15=9$$

$$\Leftrightarrow 24-15=9$$

which is true. Therefore $x = 3$ is the solution to the equation

$$2(3 - 3x) + 5x = -3x$$

The equations we have met so far are called linear equations because they only involve terms in x and constant values. Linear equations have exactly one solution. Solving quadratic equations is not so straightforward.

3.4 SOLVING QUADRATIC EQUATIONS

Quadratic equations are those that involve x^2 and possibly x but no higher powers of x. Quadratic equations cannot always be solved by using inverse operations as described so far. To understand which this is so, try thinking of the puzzle approach which we used in Section 3.3. Person A proposes the puzzle

$$x^2 - 2x = 3$$

by saying, 'I thought of a number, multiplied it by itself then subtracted twice the original number, this gave me the answer 3. What is the number I first thought of?'.

Person B now tries to go through the puzzle backwards by saying, 'Start with 3, add on twice the number that A first thought of...', here B is stuck because it is not possible to add on twice a value which has not yet been discovered. This is pictured as a flow diagram in Fig. 3.5.

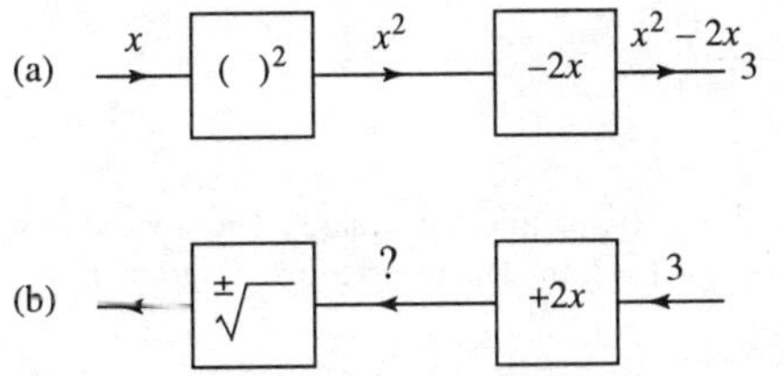

Figure 3.5 An attempt to solve the equation $x^2 - 2x = 3$ by using a flow diagram. (a) The flow diagram for $x^2 - 2x = 3$. (b) The inverse flow diagram cannot be completed because the output from the '$+2x$' operation requires a knowledge of the still unknown quantity x.

Supposing A, taking pity on B, proposed a simpler problem:

$$x^2 - 4 = 5$$

Now A says, 'I have thought of a number, multiplied it by itself, subtracted 4 and the result is 5'.

B, wishing to appear disheartened, has another go at going through the problem backwards: 'Start with 5, add on 4 (giving 9), now take the square root of 9, giving $+3$ or -3'. Notice here that B knows that the 'inverse' of squaring a number is taking the square root. It is not an exact inverse because both 3^2 and $(-3)^2$ give 9, so there are two possibilities when taking the square root of 9.

Jubilantly B announces to A that the original number was $+3$ or -3. Of course A could be awkward and insist that B states exactly what the original number was and that a choice of two numbers is not good enough. In such circumstances B would lose because it is impossible to say exactly the number that A started with. However, let us assume that A accepts the idea of two solutions to the puzzle. Why then was it possible to solve $x^2 - 4 = 5$ by this method, when it was not possible to solve $x^2 - 2x = 3$ by the same method? The problem occurs because of the term in x.

Completing the square

There is a method called 'completing the square' which gets round the problem of solving a quadratic equation by writing all the x^2 terms and the x terms into an exact square. Just for interest, we shall look at $x^2 - 2x = 3$ using this method. The left-hand side of the equation can be rewritten using a complete square and a constant value:

$$x^2 - 2x = (x-1)^2 - 1$$

This last expression is true, whatever the value of x. Check this by multiplying out the brackets. Now we can rewrite the original problem as

$$(x-1)^2 - 1 = 3$$

Now A's original puzzle has been reworded to 'I thought of a number and subtracted 1, then squared the result and subtracted 1. Finally I had the number 3. What number did I first think of?'.

B can now go through the problem backwards. 'Start with 3, add on 1 (giving 4), take the square root (giving $+2$ or -2) then add on 1, giving $+3$ or -1'. Hence B has solved the puzzle. This process can be shown in a flow diagram, as in Fig. 3.6.

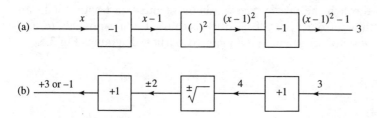

Figure 3.6 The equation $x^2 - 2x = 3$ has been rewritten as $(x-1)^2 - 1 = 3$ by 'completing the square'. The inverse flow diagram of this equation can be completed. (a) The flow diagram for $(x-1)^2 - 1 = 3$. (b) The inverse flow diagram, giving the solution as $\pm\sqrt{3+1} + 1$.

Rather than go through this process for every single quadratic equation, the square has been completed for the general quadratic equation $ax^2 + bx + c = 0$, giving the formula:

$$x = \frac{-b \pm \sqrt{b^2 - 4ac}}{2a}$$

Note that no real number squared is negative, and hence the square root in this expression cannot be found if $b^2 - 4ac$ is negative. There are no real solutions to a quadratic equation in this case. Also, if $b^2 - 4ac$ is zero then there is only one unique solution to the quadratic equation instead of two.

This then gives our first method for solving quadratic equations:

1. Take out all unnecessary brackets and collect all the terms on one side of the equation leaving zero on the other side.
2. Use the formula if $ax^2 + bx + c = 0$ then

$$x = \frac{-b \pm \sqrt{b^2 - 4ac}}{2a}$$

3. Substitute any solutions found into the original equation to check they are correct.

Example 3.3 Solve $x(x-2)=3$.

SOLUTION
1. *Remove brackets and collect terms*
   $$x(x-2)=3 \Leftrightarrow x^2-2x=3 \Leftrightarrow x^2-2x-3=0$$
2. *Use the formula*
   $a=1$, $b=-2$, $c=-3$, giving

$$x=\frac{-(-2)\pm\sqrt{(-2)^2-4(1)(-3)}}{2}$$

$$\Leftrightarrow x=\frac{2\pm\sqrt{16}}{2}$$

$$\Leftrightarrow x=\frac{2+4}{2} \quad \text{or} \quad x=\frac{2-4}{2}$$

$$x=3 \quad \text{or} \quad x=-1$$

3. *Check the solutions*
   Substitute $x=3$ into the original equation $x(x-2)=3$, giving $(3)(3-2)=3$, which is true.
   Substitute $x=-1$ into $x(x-2)=3$, giving $(-1)(-1-2)=3$, which is true.
   Both solutions are correct, so $x=3$ or $x=-1$ are solutions to the equation $x(x-2)=3$.

The second method of solving quadratic equations uses a completely different approach and is based on the fact that if two numbers or expressions multiply together to give 0 then one or the other must be 0. This method is only used for simple quadratic equations where it is easy to spot a factorization.

1. Take out all unnecessary brackets and collect all the terms on one side of the equation, leaving 0 on the other side.
2. Factorize into two brackets, each one containing an x term.
3. Use the fact that if two things multiplied together give 0 then one or the other must be 0 to reduce the equation to two linear equations.
4. Solve the two linear equations.
5. Substitute both solutions into the original equation to check that they are correct.

Example 3.4 Solve $x(x-2)=3$.

1. *Remove brackets and collect terms*

   $$x(x-2)=3 \Leftrightarrow x^2-2x=3 \Leftrightarrow x^2-2x-3=0$$

2. *Factorize*

   $$x^2-2x-3=0 \Leftrightarrow (x-3)(x+1)=0$$

3. *Reduce to two linear equations*

   $$(x-3)(x+1)=0 \Leftrightarrow x-3=0 \text{ or } x+1=0$$

4. *Solve the linear equations*

   $$x-3=0 \text{ or } x+1=0 \Leftrightarrow x=3 \text{ or } x=-1$$

5. *Check the solutions*

Substitute $x = 3$ into the original equation, $x(x-2) = 3$, giving $(3)(3-2) = 3$ which is true.

Substitute $x = -1$ into $x(x-2) = 3$, giving $(-1)(-1-2) = 3$ which is true.

Both solutions are correct, so $x = 3$ or $x = -1$ are solutions to the equation $x(x-2) = 3$.

3.5 SOLVING EQUATIONS INVOLVING FRACTIONAL EXPRESSIONS

Some equations with fractional expressions will transform to linear or quadratic equations. To solve such equations multiply the whole equation by a common denominator. The equation can then be solved in the ways we have already met.

Example 3.5 Solve the following for x:

$$\frac{1}{3} = \frac{5}{x-1} + \frac{6}{x-2}$$

SOLUTION

$$\frac{1}{3} = \frac{5}{x-1} + \frac{6}{x-2}$$

Multiply by $3(x-1)(x-2)$. Note that because the equation involves fractional expressions involving x, values which make the denominators 0 would give a division by 0, which is undefined. Such values must be excluded from the possible solutions.

$$\frac{1}{3} = \frac{5}{x-1} + \frac{6}{x-2}$$

$$\Leftrightarrow 3(x-1)(x-2)\frac{1}{3} = 3(x-1)(x-2)\left(\frac{5}{x-1} + \frac{6}{x-2}\right) \quad \text{where } x \neq 1 \text{ and } x \neq 2$$

$$\Leftrightarrow (x-1)(x-2) = 15(x-2) + 18(x-1)$$

$$\Leftrightarrow x^2 - x - 2x + 2 = 15x - 30 + 18x - 18$$

$$\Leftrightarrow x^2 - 3x + 2 = 33x - 48$$

$$\Leftrightarrow x^2 - 3x - 33x + 2 + 48 = 0$$

$$\Leftrightarrow x^2 - 36x + 50 = 0$$

$$\Leftrightarrow x = \frac{36 \pm \sqrt{1296 - 200}}{2}$$

$$\Leftrightarrow x = \frac{36 \pm \sqrt{1096}}{2}$$

$$\Rightarrow x \approx \frac{36 + 33.106}{2} \quad \text{or} \quad x \approx \frac{36 - 33.106}{2} \quad \text{(to 3 d.p.)}$$

$$\Leftrightarrow x \approx 34.553 \quad \text{or} \quad x \approx 1.447 \quad \text{(to 3 d.p.)}$$

Check

Substitute $x = 34.553$ into the original equation:

$$\frac{1}{3} = \frac{5}{x-1} + \frac{6}{x-2}$$

$$\frac{1}{3} = \frac{5}{34.553-1} + \frac{6}{34.553-2}$$

which gives $0.333 = 0.149 + 0.184$, which is true.

Substitute $x = 1.447$ into the original equation

$$\frac{1}{3} = \frac{5}{1.447-1} + \frac{6}{1.447-2}$$

giving $0.333 = 11.1857 - 10.8499$, which is correct to one decimal place.

3.6 REARRANGING FORMULAE

The same ideas we have used to solve equations can be used to rearrange formulae.

Example 3.6 Two resistors, of resistances R_1 and R_2, arranged in an electrical circuit in parallel, have equivalent resistance R where

$$\frac{1}{R} = \frac{1}{R_1} + \frac{1}{R_2}$$

Express R_1 as the subject of the formula.

SOLUTION Multiply by the common denominator RR_1R_2:

$$\frac{1}{R} = \frac{1}{R_1} + \frac{1}{R_2}$$

$$\Leftrightarrow RR_1R_2\left(\frac{1}{R}\right) = RR_1R_2\left(\frac{1}{R_1} + \frac{1}{R_2}\right)$$

where R, R_1, R_2 are non-zero

$$\Leftrightarrow R_1R_2 = R_1R_2 + R_1R_2$$

Collect all the terms involving R_1 on one side of the equation and other terms on the other side of the equation:

$$R_1R_2 = RR_2 + RR_1 \Leftrightarrow R_1R_2 - RR_1 = RR_2$$

Factorize, taking out of the common bracket the common factor of R_1:

$$\Leftrightarrow R_1(R_2 - R) = RR_2$$

Now divide by the term multiplying R_1, i.e. $(R_2 - R)$:

$$\frac{R_1(R_2 - R)}{(R_2 - R)} = \frac{RR_2}{(R_2 - R)} \qquad \text{where } (R_2 - R) \neq 0 \text{ (i.e. } R_2 \neq R)$$

$$R_1 = \frac{RR_2}{R_2 - R}$$

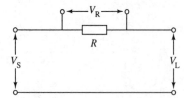

Figure 3.7 The circuit for Example 3.7.

Example 3.7 In the high frequency a.c. circuit described in Fig. 3.7, the power into a load, P, can be expressed in terms of the voltage of the supply, V_S, the voltage across the load, V_L, the voltage across the resistor V_R and the resistance R, by the expression

$$P = \frac{V_S^2 - (V_L^2 + V_R^2)}{2R}$$

Express V_L as the subject of the formula.

SOLUTION

$$P = \frac{V_S^2 - (V_L^2 + V_R^2)}{2R}$$

$$\Leftrightarrow 2RP = V_S^2 - (V_L^2 + V_R^2) \qquad \text{(multiplying both sides by } 2R)$$

$$\Leftrightarrow 2RP = V_S^2 - V_L^2 - V_R^2 \qquad \text{(removing the bracket)}$$

$$\Leftrightarrow V_L^2 = V_S^2 - V_R^2 - 2RP$$

(putting the term involving V_L on one side of the equation and all other terms on the other side by adding V_L^2 to both sides and subtracting $2RP$ from both sides)

$$\Leftrightarrow V_L = \pm\sqrt{V_S^2 - V_R^2 - 2RP}$$

(taking the square root of both sides).

As V_L must be positive the negative result can be discarded, giving $V_L = \sqrt{V_S^2 - V_R^2 - 2RP}$.

3.7 SUMMARY

1. An equation is a statement, involving some unknown, x, which may be true or false.
2. A solution is a value for x which makes the equation into a true statement.
3. Equations are solved if we know all the solutions.
4. Equations remain equivalent if the same operation is performed to both sides of the equation. The symbols for equivalence are $\Leftrightarrow$ and $\equiv$.
5. Linear equations can be solved using inverse operations and by doing the same thing to both sides.
6. Quadratic equations, $ax^2 + bx + c = 0$, can be solved by using the formula

$$x = \frac{-b \pm \sqrt{b^2 - 4ac}}{2a}$$

or by using factorization.
7. Many equations involving fractional expressions can be transformed to linear or quadratic equations by multiplying by a common denominator.
8. Formulae may be rearranged to change the subject of the formula using the same techniques used for solving equations.

3.8 EXERCISES

3.1 Solve the following equations:

(a) $12 = 6 - 3x$

(b) $y = 19 - 3y$

(c) $4(z - 2) = 12$

(d) $5(a + 3) = 25$

(e) $\dfrac{x}{9} = 18$

(f) $\dfrac{28}{a} = 7$

(g) $4p - 2 = 3p + 3$

(h) $\dfrac{x}{6} + \dfrac{x - 2}{5} = 4$

(i) $\dfrac{p}{2} + \dfrac{p + 4}{3} = 18$

(j) $\dfrac{x - 6}{9} = \dfrac{3 + x}{6}$

(k) $\dfrac{3x + 3}{5} - \dfrac{2x - 3}{15} = -16$

(l) $\dfrac{p}{p + 1} = \dfrac{1}{p} + 1$

(m) $\dfrac{y - 1}{y} = \dfrac{y}{y - 3}$

(n) $\dfrac{x}{x + 2} - \dfrac{1}{x - 2} = 1$

3.2 Solve the following equations:

(a) $y^2 + 17y = 60$

(b) $z^2 - 3z = 10$

(c) $p^2 = 8p - 15$

(d) $(x - 2)(x + 1) = 18$

(e) $\dfrac{y}{y - 3} = y - 4$

(f) $\dfrac{6}{y} = y + 5$

(g) $z + \dfrac{12}{z} = 7$

(h) $4x^2 + 10x + 3 = 0$

(i) $y^2 - 5y + 2 = 0$

(j) $z^2 + 3z - 5 = 0$

(k) $q^2 - q - 3 = 0$

(l) $x + \dfrac{1}{x} = 3$

(m) $(y + 1)(y + 2) = 28$

3.3 The lens maker formula states that $\frac{1}{v} + \frac{1}{u} = \frac{1}{f}$ where v is the distance of the object from the lens, u is the distance of the image from the lens and f is the focal length. Write u as the subject of the formula.

3.4 An object moving at constant acceleration, a, moves a distance s in time t when its initial velocity is u. s is given by $s = ut + \frac{1}{2}at^2$.

Make t the subject of the formula and, assuming that t is positive, find t when

(a) $s = 100$ m, $u = 1$ m s^{-1}, $a = 3$ m s^{-2}
(b) $s = 205$ m, $u = 10$ m s^{-1}, $a = 1$ m s^{-2}

3.5 A laterally insulated metal bar, of constant cross-sectional area A and of length x, is maintained at a temperature T_1 at one end and at a temperature T_2 at the other end (see Fig. 3.8). The heat Q crossing any cross-section of the bar in time t is given by

$$\frac{Q}{t} = \lambda A \left(\frac{T_2 - T_1}{x} \right)$$

where λ is the thermal conductivity of the bar. Express T_2 as the subject of the formula.

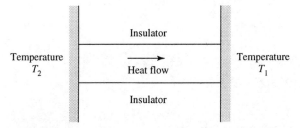

Figure 3.8 The laterally insulated metal bar for Exercise 3.5.

POWERS AND LOGARITHMS

4.1 INTRODUCTION

Just as multiplication by a whole number can be thought of as repeated addition, for example

$$5 \times 4 = 4 + 4 + 4 + 4 + 4$$

'raising to the power of' can be thought of as repeated multiplication. For instance, 4 raised to the power 5 gives

$$4^5 = 4 \times 4 \times 4 \times 4 \times 4$$

Here 4 is the base and 5 is the exponent, power or index (plural indices).

4^5 is read as '4 to the power 5'. If the index is 2, e.g. 4^2, this can be read as '4 to the power 2' or '4 squared'. If the index is 3, e.g. 4^3, this can be read as '4 to the power 3' or '4 cubed'.

In this chapter we look at extending the idea of powers so that the index does not necessarily need to be a whole positive number. We look at the properties of powers and the 'inverse' operations of roots and logarithms.

4.2 THE RULES OF INDICES

The rules of indices can be justified quite easily for simple cases. The main rules are listed below with an example. Here a can be any positive number.

Multiplication – add indices

$$a^m \times a^n = a^{m+n}$$

Example 4.1

$$2^3 \times 2^2 = 2^5$$

because

$$2^3 \times 2^2 = (2 \times 2 \times 2) \times (2 \times 2) = 2^5$$

Division – subtract indices

$$\frac{a^m}{a^n} = a^{m-n}$$

Example 4.2

$$\frac{3^5}{3^2} = 3^3$$

because

$$\frac{3^5}{3^2} = \frac{3 \times 3 \times 3 \times 3 \times 3}{3 \times 3}$$

$$= \frac{3 \times 3 \times 3 \times \cancel{3} \times \cancel{3}}{1 \times \cancel{3} \times \cancel{3}} = 3 \times 3 \times 3 = 3^3$$

Negative indices

$$a^{-n} = \frac{1}{a^n}$$

Example 4.3

$$2^{-3} = \frac{1}{2^3}$$

This takes a few steps to justify. Compare the following:

$$2^4 \times 2^{-3} = 2^{4+(-3)} = 2^1 = 2 \tag{4.1}$$

(by using the multiplication rule) and

$$2^4 \times \frac{1}{2^3} = \frac{2^4}{2^3} = 2^{4-3} = 2^1 = 2 \tag{4.2}$$

(by using the division rule). Hence, comparing Eq. (4.1) and Eq. (4.2)

$$2^4 \times 2^{-3} = 2^4 \times \frac{1}{2^3}$$

so

$$2^{-3} = \frac{1}{2^3}$$

Fractional indices

$$a^{1/n} = \sqrt[n]{a}$$

Example 4.4

$$2^{1/3} = \sqrt[3]{2}$$

Compare the following:

$$2^{1/3} \times 2^{1/3} \times 2^{1/3} = 2^{(1/3 + 1/3 + 1/3)} = 2^1 = 2 \qquad (4.3)$$

(using the multiplication rule) and

$$\sqrt[3]{2} \times \sqrt[3]{2} \times \sqrt[3]{2} = 2 \qquad (4.4)$$

by the definition of the cubed root. Hence

$$2^{1/3} \times 2^{1/3} \times 2^{1/3} = \sqrt[3]{2} \times \sqrt[3]{2} \times \sqrt[3]{2} = 2$$

Comparing Eq. (4.3) and Eq. (4.4) gives

$$2^{1/3} = \sqrt[3]{2}$$

The zeroth power

$$a^0 = 1$$

Example 4.5

$$2^0 = 1$$

Compare the following:

$$2^0 \times 2^2 = 2^{0+2} = 2^2 \qquad (4.5)$$

(by the multiplication law) and

$$1 \times 2^2 = 2^2 \qquad (4.6)$$

Comparing Eq. (4.5) and Eq. (4.6) gives $2^0 \times 2^2 = 1 \times 2^2$ so

$$2^0 = 1$$

Taking the power of a power – multiply indices

$$(a^m)^n = a^{mn}$$

Example 4.6

$$(4^2)^3 = 4^6$$

because

$$(4^2)^3 = (4 \times 4)^3 = (4 \times 4) \times (4 \times 4) \times (4 \times 4) = 4^6$$

A common error is to attempt to combine an addition of powers into a single power. This cannot usually be done. For example

$$2^3 + 2^2$$

cannot be directly expressed as a power of 2.

$$2^3 + 2^2 = 8 + 4 = 12$$

and 12 is not an exact power of 2.

However, if there is repetition of the same term then they can be combined by using the fact that multiplication is repeated addition:

$$2^2 + 2^2 + 2^2 + 2^2 = 4 \times 2^2 = 2^2 \times 2^2 = 2^4$$

Example 4.7 Express as a single power $(2^3)^2 \times 2^{-2}$.

SOLUTION

$$(2^3)^2 \times 2^{-2} = 2^6 \times 2^{-2} = 2^4$$

Example 4.8 Express as a single power

$$\frac{(4^3)^{\frac{1}{2}}}{4^{\frac{1}{2}}} + 4$$

SOLUTION

$$\frac{(4^3)^{\frac{1}{2}}}{4^{\frac{1}{2}}} + 4 = \frac{4^{\frac{3}{2}}}{4^{\frac{1}{2}}} + 4 = 4^{\frac{3}{2} - \frac{1}{2}} + 4$$

$$= 4^1 + 4 = 2 \times 4 = 2 \times 2^2 = 2^3$$

The rules we have looked at for indices apply to any positive base. The power of a negative base is not always defined (among real numbers). For instance, try on your calculator to calculate $(-3)^{\frac{1}{4}}$ and you will find that if answers something like '–E–' which is the response it gives to an undefined operation. $(-3)^{\frac{1}{4}} = \sqrt[4]{(-3)}$ and this means find a number which when raised to the fourth power gives -3. We have already discussed that the square of any number is always positive, and this is also true for any other even power. It is not possible to find a number which raised to the fourth power would be negative, so $(-3)^{\frac{1}{4}}$ is not defined.

There are two powers that get an individual mention on a scientific calculator: 10^x and e^x. The reason for 10^x being of particular importance clearly comes from the fact that we perform day to day arithmetic to base 10. $e = 2.7182818$ to 7 decimal places, and its special relevance comes from its use in solving problems of exponential growth and decay, which will be discussed in Chapter 14.

4.3 THE INVERSE OPERATIONS

In Chapter 1 we discussed subtraction as the inverse operation to addition and division as (almost) the inverse of multiplication (only almost, because of the problem with 0).

Before thinking about the inverse of taking a power, notice that

$$3^2 \neq 2^3$$

because $3^2 = 3 \times 3 = 9$ and $2^3 = 2 \times 2 \times 2 = 8$. In general,

$$a^b \neq b^a$$

If we swap round the numbers, it does not give the same result. It is this unfortunate fact which leads to there being two inverses. To understand this, it will be easier to think about some simple equations involving powers, and to bring back person A and person B.

Person A proposes the problem $x^4 = 625$ to B by saying, 'I have thought of a number and raised it to the power of 4. The answer I get is 625; what is the number I first thought of?'.

Luckily B, whose mental arithmetic is not all that good, happens to have a calculator. B also knows that the 'inverse' of raising to the power of 4 is taking the 4th root, and calculates $\sqrt[4]{625}$ by using the fact that this is the same as $625^{\frac{1}{4}}$. This gives the answer 5, but from experience with square roots (in Chapter 3) B knows to check for the possibility that -5 is also a solution. Sure enough, both 5^4 and $(-5)^4$ would result in 625.

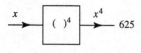

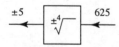

Figure 4.1 The equation $x^4 = 625$ pictured using a flow diagram. The inverse flow diagram is found by moving in the opposite direction, replacing the operation by its 'inverse'. This leads to the solution $x = \pm 5$.

This problem can be pictured using flow diagrams, as used in Chapter 3, and is shown in Fig. 4.1.

B then announces to A that the number that A first thought of was 5 or ⁻5. A accepts this answer and goes on to propose a new problem, $4^x = 1024$, by saying, 'I have thought of a number and raised 4 to the power of this unknown number. The answer I get is 1024. What is the number I first thought of?'.

Notice that the unknown is now the exponent, not the base. The inverse in this case is different. B, of course checks this by trying out the 4th root. The 4th root of 1024 is 5.6569 to 4 decimal places, but this certainly does not satisfy the equation, as $4^{5.6569} = 2545$. B needs a new operation to solve the problem. The operation needed is called the logarithm, base 4. The answer to the problem is $\log_4(1024)$. Unfortunately, B is still stuck because this operation is not on the calculator. She only has a choice of log, which is shorthand for the logarithm base 10, or ln, which is shorthand for the logarithm base e.

B explains the dilemma to A, who therefore proposes the simpler problem, $10^x = 10\,000$, by saying, 'I have thought of a number and raised 10 to the power of this known number. The answer I get is 10 000, what is the number I first thought of?'.

B is now able to answer this problem by finding $\log_{10}(10\,000)$, giving the answer 4. This problem and its solution is pictured in a flow diagram in Fig. 4.2. B checks that $10^4 = 10\,000$, which is correct, and is able to give A the answer. However, B still has some work to do to find the answer to the original problem of $4^x = 1024$. We shall return to this later.

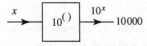

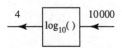

Figure 4.2 The equation $10^x = 10\,000$ pictured using a flow diagram. The inverse flow diagram is found by moving in the opposite direction, replacing the operation by its inverse. This leads to the solution $x = 4$.

These problems, which result in us needing logarithms, lead us to a definition of the logarithm. If some number y is the result of raising a to the power x, i.e.

$$y = a^x \tag{4.7}$$

then the relationship between x and y can also be expressed as

$$x = \log_a(y) \tag{4.8}$$

That is, $\log_a$ is defined as an inverse operation.

Another idea about inverses was that if one operation is performed on some number, then the inverse operation, we should get back to the number we first started with. For instance, in Chapter 1 we started with 5, added 3 and then subtracted 3 (the inverse of adding), which took us back to 5. Similarly, if we start with 4, find 10^4, and then take $\log_{10}$, we should get back to 4. Check this on the calculator. This means that $\log_{10}(10^4)=4$, and, in general, replacing 4 by any number, x, and using a to represent any base, we get

$$\log_a(a^x)=x.$$

Here a should be a positive number, else things can get tricky. x can take any value.

This will also work the other way round: if you take $\log_a$ first and raise a to the power of the result you should get back to your original number. Hence

$$a^{\log_a(x)}=x$$

However, this way round, not only should a be positive but also x must be a positive number. The logarithm of 0 and logarithms of negative numbers are not defined. Your calculator responds with the error condition '–E–' if you try to calculate the logarithm of a negative number.

As logarithms and powers are so closely linked, not surprisingly there are rules for logarithms that match each of the rules given in Section 4.2.

4.4 RULES OF LOGARITHMS

Check the example in each case.

Log of a product – add logs

$$\log_a(xy)=\log_a(x)+\log_a(y)$$

Example 4.9

$$\log_{10}(10\times100)=\log_{10}(10)+\log_{10}(100)$$

Log of a division – subtract logs

$$\log_a\left(\frac{m}{n}\right)=\log_a(m)-\log_a(n)$$

Example 4.10

$$\log_{10}\left(\frac{1000}{100}\right)=\log_{10}(1000)-\log_{10}(100)$$

Log of a reciprocal

$$\log_a\left(\frac{1}{x}\right)=-\log_a(x)$$

Example 4.11

$$\log_{10}\left(\frac{1}{10\,000}\right)=-\log_{10}(10\,000)$$

The inverse rule

$$\log_a(a^x)=x$$

and it follows that

$$\log_a(a) = 1$$

Example 4.12

$$\log_{10}(10^4) = 4$$

The log of 1

$$\log_a(1) = 0$$

Example 4.13

$$\log_{10}(1) = 0$$

The log of a power

$$\log_a(x^n) = n \log_a(x)$$

Example 4.14

$$\log_{10}(100^4) = 4 \log_{10}(100)$$

Example 4.15 Simplify

$$\log_{10}\left(\frac{100^2}{10^3}\right)$$

SOLUTION

$$\log_{10}\left(\frac{100^2}{10^3}\right) = \log_{10}(100^2) - \log_{10}(10^3)$$

$$= \log_{10}((10^2)^2) - 3 \log_{10}(10)$$

$$= \log_{10}(10^4) - 3 = 4 - 3 = 1$$

The change of base rule

$$\log_a(x) = \frac{\log_b(x)}{\log_b(a)}$$

Example 4.16

$$\log_{100}(10\,000) = \frac{\log_{10}(10\,000)}{\log_{10}(100)}$$

Example 4.17 We are now able to solve the problem that A posed to B earlier: find x if $4^x = 1024$.

SOLUTION Take the logarithm, base 4, of both sides, giving

$$\log_4(4^x) = \log_4(1024)$$

As $\log_4(\ )$ is the inverse operation to $4^{(\ )}$, the left-hand side gives x, so we have

$$x = \log_4(1024)$$

Suppose that our mental arithmetic is not up to calculating the value of the right-hand side. We can use the change of base rule to find

$$\log_4(1024) = \frac{\log_{10}(1024)}{\log_{10}(4)}$$

And using a calculator this gives the answer 5.

4.5 SUMMARY

1. a^b is read as 'a raised to the power b'. It can be thought of as repeated multiplication when b is a whole positive number.
2. There are many rules for indices, which are summarized in Section 4.2.
3. The 'inverse' of raising to the power of n is to take the nth root. If $y = x^n$ then $x = \sqrt[n]{y}$, providing x is positive.
4. The inverse of exponentiation is taking the logarithm. If $y = a^x$ then $x = \log_a(y)$, where a is positive.
5. There are many rules for logarithms, which are summarized in Section 4.4.

4.6 EXERCISES

4.1 Explain, using the idea of the inverse, why $\sqrt[n]{x^n} = x$, where x is a positive number and $n \neq 0$. Test this out on a calculator using various values of x and n. Investigate what happens if x is allowed to take negative values.

4.2 (a) Find $(-27a^{-2})^{\frac{1}{3}}$ when $a = 8$.

    (b) Find $(3b^{-3})^{-1}$ when $b = 2$.

    (c) Evaluate $2^{-3}ab^2c^{-1}$ when $a = 2$, $b = 3$, $c = 4$.

    (d) Find $(a^3b^{-1})^{\frac{1}{2}}$ when $a = 3$ and $b = 2$.

4.3 Evaluate:

(a) $16^{3/2}3^{-1}$

(b) $9^{3/4} \times 3^{-1/2}$

(c) $\dfrac{2^{1/3}2^{4/3}}{4}$

4.4 Express the following as single powers, where possible:

(a) $\sqrt{a} \times a^3$ (b) $\dfrac{3^7 \times 3^2}{3^4}$ (c) $(2^4)^5$

(d) $\dfrac{2^{18}}{2^{16}}$ (e) $\sqrt[3]{a^6b^9}$ (f) $(9a^2)^2$

(g) $\sqrt{2^{16}}$ (h) $10\,000$ (i) $\sqrt{y} \times \sqrt[3]{y}$

(j) $y^2 + y^3$

4.5 Express as a single logarithm:

(a) $\log(x^3) + 3\log(x)$ (b) $\log(2xy) - \log(x) - \log(y)$

(c) $2\log(x^3) - \log(x^2) - 3\log(x)$

4.6 Use the change of base rule to express the following in terms of $\log_{10}$, and hence calculate their values:

(a) $\log_8(6)$ (b) $\log_3(12)$ (c) $\log_4(1000)$

4.7 Solve the following equations:

(a) $x^4 = 16$ (b) $x^3 = -27$ (c) $x^6 = -64$

(d) $x^{1/2} = 4$ (e) $x^{3/5} = 125$ (f) $\log_{10}(x) = 0$

(g) $\log_2(x) = 3$ (h) $10^x - 1 = 0$ (i) $2\log_{10}(x) - 4 = 0$

4.8 A voltage-sensitive resistor has a current/voltage relationship of the form $I = kV^n$ where I is expressed in mA, V is in volts and k and n are constants. Find k and n if $I = 10$ mA when $V = 1.7$ V and $I = 2$ mA when $V = 1$ V.

MEASUREMENT AND CALCULATION

5.1 INTRODUCTION

Engineering measurements cannot be guaranteed to be 100 per cent accurate. The amount of possible inaccuracy is referred to as the maximum error in the measurement. Performing calculations can also introduce (hopefully small) errors into a result. All but the simplest of calculations are performed either by using a calculator or a computer and these can accommodate only a limited number of digits in each number. After each stage of the calculation the result may need to be shortened so that it can again fit into the number of digits available. For instance 1/3 gives the result 0.333 333 333 3... (or 0.$\dot{3}$, which reads as '0.3 recurring'), but a calculator typically gives the result 0.333 333 3, shortening the result to seven decimal places. This shortening process introduces a rounding error.

The purpose of this chapter is to present ideas to be used in expressing errors in measurement and procedures that should be used to ensure a sensible numerical calculation. It is not our aim to discuss the details of which button to press on a calculator or in which order (for this you should refer to the operation manual provided). Many of the ideas concerning calculation are equally relevant to using a computer.

5.2 ROUNDING A SINGLE NUMBER

There are two common ways of rounding a single number: by expressing it to a specified number of decimal places or to a specified number of significant figures.

Decimal places

Example 5.1 Write 2.457 82 to two decimal places.

SOLUTION Count two places after the decimal point: this would give 2.45. The result of 2.45 is what we get if we chop the number after two decimal places. A more accurate way to express it is to round the number by looking at the next digit, in this case a 7. If the digit is equal to or greater than 5 then we round upwards, i.e. increase the previous digit by 1, while if it is less than 5 we round downwards, giving the same answer as if we chopped the number after the second decimal place. In this case, we round upwards because 2.457 82 is nearer to 2.46 than it is to 2.45, as can be seen in Fig. 5.1. Thus, 2.457 82 = 2.46 to two decimal places.

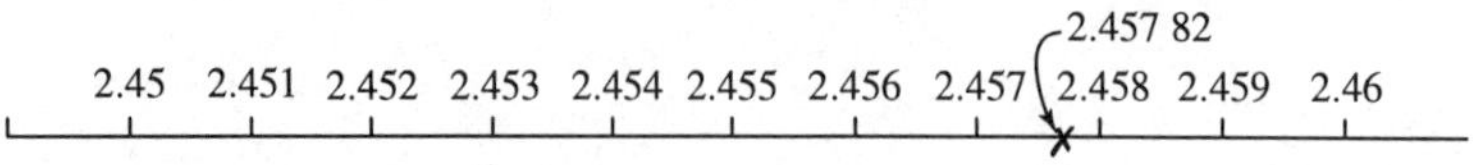

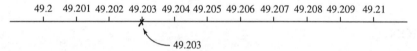

Figure 5.1 2.457 82 rounded to 2 d.p. gives 2.46, as the number is nearer to 2.46 than to 2.45.

Example 5.2 Write 49.203 to two decimal places.

SOLUTION Count two places after the decimal point: this would give 49.20 = 49.2. Looking at the next digit, in this case a 3, we see that we do not need to round upwards. Thus 49.203 = 49.2 to two decimal places (see Fig. 5.2).

49.2   49.201 49.202 49.203 49.204 49.205 49.206 49.207 49.208 49.209 49.21

— 49.203

Figure 5.2 49.203 rounded to 2 d.p. gives 49.2, as the number is nearer to 49.2 than to 49.21.

Significant figures

Example 5.3 Write 2.457 82 to four significant figures.

SOLUTION Count four figures from the left of the number, ignoring any leading zeros; this would give 2.457. The result of 2.457 is that we get if we chop the number after four significant figures. Again we round the number by looking at the next digit, in this case a 8. If the digit is equal to or greater than 5 then we round upwards, but if it is less than 5 then we round downwards. In this case we round upwards because 2.457 82 is nearer to 2.458 than it is to 2.457. Thus 2.457 82 = 2.458 to four significant figures.

Example 5.4 Write 0.000 492 03 to two significant figures.

SOLUTION Count two figures from the left of the number, ignoring any leading zeros: this would give 0.000 49. Again we round the number by looking at the next digit, in this case a 2. As this is less than 5 we round downwards to give 0.000 49. Thus 0.000 492 03 = 0.000 49 to two significant figures.

5.3 ERROR FROM ROUNDING

By rounding numbers as in Section 5.2 we introduce an error into the result. This error can be studied by looking at the greatest and least possible values of a quantity expressed to a specified number of significant figures.

Example 5.5 A value is given as 1.7 to two significant figures. What is the greatest and least values that it could take? What is the maximum error in the given value?

SOLUTION Consider a number line near the value of 1.7, as in Fig. 5.3, marking numbers with up to two significant figures and halfway points between them. To be nearer to 1.7 than 1.6 the number must be just greater than 1.65. To be nearer to 1.7 than 1.8 then it must be less than 1.75. The greatest value it can take is (just less than) 1.75 and the least value is 1.65. The maximum error can be found by halving the range of the possible values, that is:

$$\text{range of possible values:} \quad 1.75 - 1.65 = 0.1$$

$$\text{maximum error} = \text{range}/2 = 0.1/2 = 0.05$$

The value could be given as 1.7 ± 0.05.

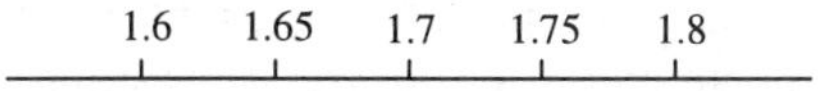

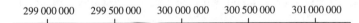

Figure 5.3 A value is given as 1.7 to 2 s.f. It must therefore lie in the range 1.65 to 1.75.

Example 5.6 The velocity of light is given as 300 000 000 m s^{-2} to three significant figures. What are the greatest and least values it could be?

SOLUTION Consider a number line around the value of 300 000 000, as in Fig. 5.4, marking numbers with up to three significant figures and halfway points between them. The least and greatest values that the velocity could be are 299 500 000 and (just less than) 300 500 000.

$$\text{Range of possible values} = 300\,500\,000 - 299\,500\,000 = 1\,000\,000$$

$$\text{Maximum error} = \text{range}/2 = 1\,000\,000/2 = 500\,000$$

The value could be given as $300\,000\,000 \pm 500\,000$.

| 299 000 000 | 299 500 000 | 300 000 000 | 300 500 000 | 301 000 000 |

Figure 5.4 The velocity of light is given as 300 000 000 to 3 s.f. It must therefore lie in the range 299 500 000 to 300 500 000.

5.4 ERROR FROM MEASUREMENT

Consider using a ruler to measure the size of a table. The smallest measure on the ruler is a millimetre; hence the result is only likely to be correct to the nearest millimetre and probably (as you may have moved slightly while taking the measurement) even less accurate than that. If we used the same ruler to measure the length of a football pitch then errors would build up at each stage and we would not expect the answer to be correct to the last millimetre and not even to the last centimetre.

The error in a calculation or measurement cannot be known exactly. In the absence of any clear idea of how to estimate the error then one way to proceed is to repeat the measurement several times. Then take the average of all the measurements as the measured value and the range of values gives an idea of the possible maximum error. More detailed methods of analysing the error are looked at in Chapter 24 on probability and statistics.

Suppose that measuring the table leads to the values 1.48 m, 1.46 m, 1.47 m, 1.49 m, 1.495 m. The average of the values gives 1.479 m. The range of values is $1.495 - 1.46 = 0.035$, giving an estimate of the error as $0.035/2 = 0.0175$ m. Therefore the length of the table is 1.479 ± 0.0175 m. The error expressed in this form is called the *absolute* error. If it is expressed as a fraction or percentage of the true value then it is called the relative error. Using the average value as an estimate of the true value gives the percentage error as

$$(0.0175/1.479) \times 100 = 1.2\%$$

Then the length of the table is given as $1.479 \pm 1.2\%$ and the relative error is $\pm 1.2\%$.

Sometimes the possible errors can be found by using known information about the measuring apparatus.

If we measure the current flowing in a circuit using an ammeter then the presence of the ammeter increases the resistance in the circuit. This in turn will introduce an error into the reading. This error cannot be avoided, but it can be kept to a minimum by choosing an ammeter whose resistance is very much less than that of the circuit, say less than 1% of the circuit resistance. The error introduced by the ammeter is proportional to the size of the current being measured and also to the ratio of the internal resistance of the ammeter to the circuit resistance (Fig. 5.5).

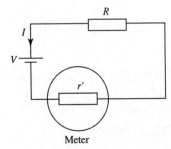

Figure 5.5 Using a meter to measure the current in a circuit. R is the resistance of the circuit and r' is the internal resistance of the ammeter.

If the meter is chosen such that the internal resistance is very much less than the resistance to be measured then the effect of this error cannot be too serious. For instance, if the circuit resistance is known to be around 10 KΩ then a meter with an internal resistance of about 100 Ω will introduce an error of about $100/10\,000 = 1/100 = 1\%$.

It is usually true that small errors are roughly proportional to the size of the quantity being measured. For this reason it is more useful to indicate the relative error. In the example just considered, suppose we measure the current to be 12.4 mA. As we think the meter has introduced an error of up to 1% (in excess), and considering other possible errors in the measurement (caused, for instance, by the presence of magnetic fields, friction of moving parts, error in reading from the meter) as introducing smaller errors then we could adjust the reading down by 1% to take account of the effect of the internal resistance and estimate the error as being of the order of 1%. This gives an estimate of the current as 12.4 mA $-$ 1% of 12.4 mA $= 12.4 - 0.124 = 12.276$ mA. This gives the current as $12.276 \pm 1\%$ (or 12.276 mA ± 0.123 mA or between 12.153 mA and 12.399 mA).

If this measurement is now used to calculate other circuit values then the final result should only be expressed to a consistant level of accuracy. As the original measurement is correct to no more than three significant figures then the final calculated result should be expressed to no more than three significant figures, with an indication that the maximum error will be at least as big as the original error.

Example 5.7 An object is weighted on four different scales, giving the values 1.01 kg, 0.99 kg, 0.97 kg, 1.02 kg. Estimate the weight of the object, indicating an error range.

SOLUTION Take the average of the given values for the weight. This gives weight $=$ $(1.01 + 0.99 + 0.97 + 1.02)/4 = 0.9975$.

The range of values is $1.02 - 0.97 = 0.5$, giving an estimate of the maximum error as $0.5/2 = 0.25$. This gives the weight of the object as 0.9975 ± 0.25 kg.

Example 5.8 A voltmeter with an internal resistance of $1\,\text{M}\Omega$ is used to measure the voltage across a 5 kΩ resistor (see Fig. 5.6). The voltage recorded is 20.1 V, estimate the voltage across the resistor.

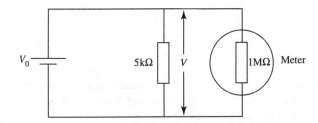

Figure 5.6 Using a voltmeter to measure the voltage across a 5 kΩ resistor where the internal resistance of the voltmeter is 1 MΩ.

SOLUTION A voltmeter will underestimate the voltage by an amount given by the ratio of the resistance of the resistor divided by the internal resistance of the voltmeter. Therefore a better estimate of the resistance is the measured value $20.1 - (5000/1\,000\,000)$ of 20.1. This gives $20.1 - 0.5\% = 20.1 - 0.1005 = 19.9995$. Therefore we could give the voltage as $19.9995 \pm 0.5\% = 19.9995 \pm 0.1005 = 19.899$ to 20.1 V. This also takes account of other possible errors in the measurement, as referred to when considering the ammeter.

5.5 SCIENTIFIC NOTATION

The display of most calculators is limited to about eight digits. This therefore presents a problem when trying to enter a number of smaller magnitude than 0.000 000 1 or bigger than 999 999 99. To overcome this problem scientific notation is used. The number 105 000 000 can be represented as 1.05×10^8 and in this form the number can be entered into the calculator.

A number is in scientific notation when it is expressed as a number between 1 and 10 (i.e. with only one digit before the decimal point) multiplied by some power of 10. To calculate the correct power of 10 count the number of places that the decimal point has moved.

Example 5.9 Express 10 873.5 in scientific notation.

SOLUTION

$$10\,873.5 = 1.087\,35 \times 10^4$$

To move the decimal point from its original position to after the first digit requires a movement of four places. The power must be positive because 1.087 35 is smaller than 10 873.5.

Example 5.10 Express 0.000 056 328 9 in scientific notation.

SOLUTION

$$0.000\,056\,328\,9 = 5.632\,89 \times 10^{-5}$$

To move the decimal point from its original position to after the first non-zero digit requires a movement of five places. The power must be negative because 5.632 89 is bigger than 0.000 056 328 9.

Machines store floating point numbers in a manner similar to scientific notation. Machine numbers are expressed as binary (base 2) numbers rather than decimal numbers (base 10) which we use for hand calculations. However they still have a limited amount of space to store the number. The result of a calculation on most calculators can only have an accuracy of seven or eight significant figures at the most. Most computer calculations have a similar accuracy.

Another reason for using scientific notation is to aid checking a calculation. This we look at in Section 5.7.

5.6 DIMENSIONS AND UNITS

The dimensions of a physical quantity give the way it is related to the fundamental quantities of mass, length, time, electric current, temperature, amount of substance and luminous intensity. The metric units for these are given in the Système International d'Unités (SI) as:

Quantity	Unit	Symbol
Length	metre	m
Mass	kilogram	kg
Time	second	s
Electric current	ampere	A
Temperature	kelvin	K
Amount of substance	mole	mol
Luminous intensity	candela	cd

From these other units can be derived, some of the most common being:

Quantity	Unit	Symbol
Force	newton	$N = kg\ m\ s^{-2}$
Work, energy, heat	joule	$J = N\ m$
Power	watt	$W = J\ s^{-1}$
Electric charge	coulomb	$C = A\ s$
Electric potential	volt	$V = W\ A^{-1}$
Electric resistance	ohm	$\Omega = V\ A^{-1}$
Inductance	henry	$H = V\ s\ A^{-1}$
Frequency	hertz	$Hz = s^{-1}$
Magnetic flux	weber	$Wb = V\ s$
Magnetic flux density	tesla	$T = Wb\ m^{-2}$

There are other ways of expressing these 'compound units'; for instance, one newton $= kg\ m\ s^{-2}$ can be written as $kg\ m/s^2$ and is read as 'kilogram metres per second per second' or 'kilogram metres seconds to the minus 2'.

When performing a calculation the units of the result can be found from the units of the values in the formula. For instance if velocity $=$ distance/time, then given that the distance travelled is 20 m in 5 s then velocity $= 20$ m/5 s $= 4$ m s^{-1} or 4 m/s (4 metres per second).

Notice that we performed the calculation on the units and this must give the correct units for the quantity being calculated. If a formula contains an addition or a subtraction then each part of the formula must have the same units. For example, if an object is moving under constant acceleration then the distance travelled, s, can be related to the time t, initial velocity u and the acceleration a by the formula:

$$s = ut + \tfrac{1}{2}at^2$$

If $u = 12$ m s^{-1}, $t = 5$ s, $a = 6$ m s^{-2} then

$$s = 12 \times 5(\text{m s}^{-1} \times \text{s}) + 6 \times 5 \times 5(\text{m s}^{-2}) \times \text{s} \times \text{s}$$

$$= 60 \text{ m} + 75 \text{ m} = 135 \text{ m}$$

Notice that both parts of the formula, ut and $\tfrac{1}{2}at^2$ give units of metres, which is correct as it is a distance that we are calculating.

Prefixes indicating multiples and sub-multiples of units

The most common of these are:

Multiplication factor	Prefix	Symbol
10^9	giga	G
10^6	mega	M
10^3	kilo	k
10^{-2}	centi	c
10^{-3}	milli	m
10^{-6}	micro	μ
10^{-9}	nano	n
10^{-12}	pico	p

5.7 ROUGH CALCULATIONS

Rough calculations should be used as a way of checking longer calculations either by hand or on a calculator. First, write all the numbers in scientific notation. Separate out all the powers of 10. Approximate the numbers, possibly to only one significant figure or some other convenient amount and perform the calculation by hand or mentally. If the rough calculation is more or less the same as the value found from using the calculator when it can be assumed that the calculated value is correct.

> **Example 5.11** The conductivity of a length of conducting material of uniform cross-section can be obtained from the expression
>
> $$\sigma = \frac{IL}{VA}$$
>
> where I is the current, L is the length of the section, A is the cross-sectional area and V is the voltage drop across the material.
> A certain section of conducting material in an integrated circuit device has a current passed across it of 0.004 A, producing a voltage drop of 84 mV. The length is 3.5 mm and it has a cross-section which is a rectangle of dimensions 1 micrometre by 3.7 micrometres. Determine its conductivity.

SOLUTION First, write all the quantities in consistent units and in scientific notation.

$$I = 4 \times 10^{-3} \text{ A}$$

$$L = 3.5 \times 10^{-3} \text{ m}$$

$$V = 84 \times 10^{-3} \text{ V} = 8.4 \times 10^{-2} \text{ V}$$

$$A = 1 \times 10^{-6} \times 3.7 \times 10^{-6} \text{ m}^2$$

This gives

$$\sigma = \frac{4 \times 10^{-3} \times 3.5 \times 10^{-3}}{1 \times 10^{-6} \times 3.7 \times 10^{-6} \times 8.4 \times 10^{-2}} \frac{\text{A m}}{\text{V m}^2}$$

This can be approximated to

$$\frac{4 \times 4}{1 \times 4 \times 8} \times \frac{10^{-3-3}}{10^{-6-6-2}} = \frac{1}{2} \times \frac{10^{-6}}{10^{-14}} = 0.5 \times 10^8 = 5 \times 10^7$$

The result, using a calculator, is $4.504\,504 \times 10^7$. The approximate result confirms that the result given using a calculator is a sensible one. The units are

$$\frac{\text{A m}}{\text{V m}^2} = \frac{\text{A}}{\text{V m}} = \frac{1}{\Omega \text{ m}} = (\Omega \text{ m})^{-1}$$

However, we should finally notice that the calculator has produced a result with seven significant figures in the result. This is unreasonable considering that we should assume that the original numbers had no greater accuracy than two significant figures. We therefore represent the result to two significant figures, giving the conductivity as $4.5 \times 10^7 \ (\Omega \text{ m})^{-1}$.

5.8 PERFORMING CALCULATIONS ON A CALCULATOR

Each stage in a calculation may introduce a rounding error into the result. The larger the number of significant figures used to perform the calculation the smaller the rounding error will be. The best accuracy is obtained by performing all calculations to the maximum number of significant figures available on the calculator. Even in the case where our original values were only accurate to, say, three significant figures, we should perform calculations to the maximum number of significant figures and then round at the end to a number of figures consistent with our original measurements. To see an example of the problem introduced by large rounding errors, perform a calculation like $(9.1111/9) \times 9$.

We can see immediately that as we are dividing a number by 9 and then multiplying again by 9 we should get back to the original number of 9.1111.

Now perform this on a calculator, writing down any interim results to the maximum number of significant figures.

$$9.1111/9 = 1.0123444 \text{ to eight significant figures}$$

Multiply this by 9 to get 9.1110996, which, rounded to five significant figures, gives 9.1111, the correct result.

Now do the same thing again, but only writing the interim result to five significant figures.

$$9.1111/9 = 1.0123 \text{ to five significant figures}$$

Multiply this by 9 to get 9.1107, which is no longer correct to five significant figures.

A relatively large error has been introduced by not using the largest possible number of significant figures in the course of doing the calculation.

Use of the memory on the calculator can prevent the need to write down a lot of long numbers. However, it is a good idea to write down the result after each four or five operations in order to be able to go back and check the result, should it appear not to be correct.

Example 5.12 Three capacitors are in series in a circuit (Fig. 5.7). The equivalent capacitance is given by

$$C = \frac{C_1 C_2 C_3}{C_1 C_2 + C_2 C_3 + C_1 C_3}$$

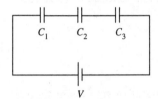

V   **Figure 5.7** Three capacitors in series.

If the three capacitors have capacitances of $C_1 = 4.2$ µF, $C_2 = 5.6$ µF and $C_3 = 2.7$ µF then calculate the equivalent capacitance *C*.

SOLUTION

$$C = \frac{4.2 \times 5.6 \times 2.7}{4.2 \times 5.6 + 5.6 \times 2.7 + 4.2 \times 2.7} \times \frac{\mu F \times \mu F \times \mu F}{\mu F \times \mu F}$$

$$= 1.270\,588\,2 \text{ µF}$$

An approximate calculation gives

$$\frac{4 \times 6 \times 3}{4 \times 6 + 6 \times 3 + 4 \times 3} \text{ µF} = \frac{72}{54} \text{ µF} \approx 1.3 \text{ µF}$$

This approximate result confirms that the result given by the calculator is a sensible one. However, a result with eight significant figures does not seem reasonable in the circumstances, as we could assume the original measurements are only accurate to two significant figures. Finally, then, we round the result to two significant figures, giving 1.3 µF for the equivalent capacitance.

Example 5.13 The quantity of heat that will raise a substance from temperature T_1 °C to T_2 °C is given by $Q = mc(T_2 - T_1)$ joules, where *m* is the mass of the substance (given in kg), *c* is the specific heat capacity in J (kg^{-1} °C^{-1}). A mass of 3.2 kg of aluminium absorbs 30 405 joules of heat and has a final temperature of 30 °C. Find the original temperature of the block. (The specific heat capacity of aluminium is 950 J (kg^{-1} °C^{-1}) for temperatures around room temperature.)

SOLUTION Rearranging the expression for the heat absorption so that the original temperature is the subject of the formula gives

$$T_1 = T_2 - \frac{Q}{mc}$$

$$= 30 - \frac{30\,450}{3.2 \times 950}\ °C = 30 - 10.016\,447\ °C = 19.983\,553\ °C$$

A rough calculation gives

$$T_1 = 30 - \frac{300\,00}{3 \times 1000}\ °C = 30 - 10 = 20\ °C$$

The rough calculation gives a figure near the calculated value, therefore indicating that the value is reasonable. Finally, we look at the accuracy of the original values. They are apparently correct to two significant figures, so we round the value of the initial temperature to 19.983 553 to two significant figures, giving 20 °C.

5.9 SUMMARY

1. A number can be approximated by expressing it to a specified number of significant figures or decimal places.
2. Errors are introduced into a result from errors in measurement and rounding errors.
3. Any measurement should be accompanied by an idea of the possible error in the measurement.
4. Scientific notation allows the representation of large and small numbers using only a limited number of digits.
5. Rough calculations should be used to check any calculated results.
6. To keep the effect of rounding errors to a minimum, always record interim calculations to the maximum number of significant figures or use the memory on the calculator.
7. The final value given after a calculation should be rounded to a number of significant figures consistent with the initial measurement(s).

5.10 EXERCISES

5.1 Express the following to the number of decimal places (d.p.) indicated:

(a) 2.567 (2 d.p.)   (b) 105 879.225 (1 d.p.)

(c) 0.000 007 (1 d.p.)   (d) 38.999 99 (2 d.p.)

5.2 Express the following to the number of significant figure (s.f.) indicated:

(a) 23.846 (3 s.f.)   (b) 10 456 879.22 (5 s.f.)

(c) 1.000 56 (2 s.f.)   (d) 0.000 000 056 789 (2 s.f.)

5.3 The following numbers are approximate. Estimate the maximum and least possible values they could be and express the maximum error as a percentage.

(a) 1.6   (b) 0.355

(c) 0.000 046   (d) 2 100 000 000

5.4 The distance between two cities is recorded as 270 kilometres. What are the greatest and least possible values for the distance?

5.5 The following measurements are given with an indicated percentage error. Give the greatest and least possible values of the measured value.

(a) $5.1\,\Omega \pm 3\%$ (b) $0.0061\,s \pm 0.5\%$

(c) $12.9\,kg \pm 5\%$ (d) $583\,000\,m \pm 4\%$

5.6 The following are sets of recorded measurements of the same quantity. Give an estimate of the value and an error range, also expressing the maximum error as a percentage.

(a) The length of a metal bar: 12.2 cm, 11.95 cm, 12.05 cm, 12.15 cm, 11.9 cm.
(b) A force: 5.12 N, 5.3 N, 5.6 N, 5 N.
(c) The settling time of an electric circuit after the power supply has been switched on: 0.12 s, 0.1 s, 0.15 s, 0.143 s, 0.11 s.

5.7 Write the following numbers in scientific notation:

(a) $12\,000\,000$ (b) $0.000\,000\,786\,9$ (c) 46.789   (d) 9.0005

5.8 Calculate the following, checking each calculation by an approximation and writing the answer to an appropriate number of significant figures:

(a) $49.2 + 5.21 + 8.64 - 12.2$

(b) $\dfrac{8.7 \times 9.6}{12} - \dfrac{6.2 - 1.4}{9.6}$

(c) $\dfrac{(16.27 - 0.782)13.21}{19.46 + 17.21} + 12.68$

(d) $\dfrac{0.0078}{360\,000} + \dfrac{0.0063}{720\,000} + \dfrac{6.8}{25}$

5.9 A resistor, R_1, is quoted as $10\,\Omega$ with 10% tolerance and another resistor, R_2, is quoted as $25\,\Omega$ with 5% tolerance. When in parallel the combined resistance is given by R, where

$$\frac{1}{R} = \frac{1}{R_1} + \frac{1}{R_2}$$

Calculate R and the tolerance for R when R_1 and R_2 are arranged in parallel.

5.10 The resistance of a length of copper wire is given by

$$R = \frac{\rho l}{a}$$

where ρ is the resistivity of copper $= 1.7 \times 10^{-8}\,\Omega m$ at room temperature, $l =$ length of wire and $a =$ cross-sectional area of the wire $= \pi r^2$; r is the radius of the wire.

   Find the resistance of a piece of copper wire of length 0.2 m and diameter 1 mm at room temperature.

5.11 The manner in which the resistance of any material varies with temperature is approximately given by the formula $R = R_0(1 + \alpha T)$, where T is the temperature in °C, α is the temperature coefficient of resistance and R_0 is the resistance at 0 °C. α for carbon is $-0.000\,48\,°C^{-1}$. If a length of carbon wire has a resistance of $0.01\,\Omega$ at 20 °C, find its resistance at 0 °C.

RIGHT-ANGLED TRIANGLES

6.1 INTRODUCTION

Right-angled triangles are triangles in which one angle is 90°, that is one quarter of a complete rotation. If the length of one side and the size of one angle (as well as the right angle) are known, or if two of the sides are known, then any other angle or side can be calculated. The ratios of any two sides of a right-angled triangle are called trigonometric ratios. The most important of these are the sine, cosine and tangent. The value of the ratios is found to depend only on the angles in the triangle; hence they can be tabulated against the angles and found using a calculator. Pythagoras's theorem relates the lengths of the sides to each other and is used to find a third side if two of the lengths are known.

Right-angled triangles are important for understanding vectors (see Chapter 15) and complex numbers (Chapter 16). Often other shapes can be split into a number of right-angled triangles and the properties of these used to find lengths and angles in the more complex shapes.

6.2 THE ANGLE SUM OF A TRIANGLE

The sum of the internal angles of a triangle is 180°. This means that if we have a right-angled triangle and we know the value of one of the other angles then the third angle can be found (see Fig. 6.1). In Fig. 6.1(a), $20° + 90° + x = 180° \Leftrightarrow x = 180° - 90° - 20° \Leftrightarrow x = 70°$. In (b), $40° + 90° + x = 180° \Leftrightarrow x = 180° - 90° - 40° \Leftrightarrow x = 50°$. In (c), $65° + 90° + x = 180° \Leftrightarrow x = 180° - 90° - 65° \Leftrightarrow x = 35°$.

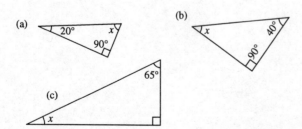

Figure 6.1 The angle sum of a triangle is 180°.

Note that for right-angled triangles the same result can be found by subtracting the other known angle from 90°. Thus: (a) $x = 90° - 20° = 70°$; (b) $x = 90° - 40° = 50°$; (c) $x = 90° - 65° = 35°$.

6.3 SIMILAR SHAPES

Objects are similar if they are the same shape but not necessarily the same size (Fig. 6.2). Similar shapes will also have the same angles.

A similar shape is found by performing a change of scale: by magnifying or miniaturizing. If, for instance, one side is doubled in length, then so are all the others. The new shape so formed will be similar to the original. As the sides have been magnified or miniaturized by the same amount then there can be no change in the relative lengths of any two sides. The ratios of the sides are not changed by magnification or miniaturization.

Similar triangles can be nested to show that the corresponding angles are equal (see Fig. 6.3). As any two right-angled triangles containing the same angle must be similar then the ratios of any two sides in a right-angled triangle can be found if we know the angles in the triangle.

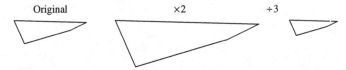

Original   ×2   ÷3

Figure 6.2 These shapes are similar. Notice that corresponding angles are equal. Also notice that the ratio between any two sides is constant.

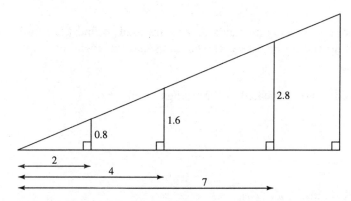

Figure 6.3 Nested similar right-angled triangles. The ratio of any two sides in each of the triangles is the same. The ⌐ symbol indicates a right angle. Notice that the ratio of the height to the base in each triangle is the same. $\frac{0.8}{2} = \frac{1.6}{4} = \frac{2.8}{7} = 0.4$.

Example 6.1 A right-angled triangle has an internal angle of 30°. Find the ratio of the side opposite this angle to the hypotenuse.

SOLUTION First, draw any triangle with a 30° angle and a right angle (90°), as in Fig. 6.4. The hypotenuse is the side opposite the right angle. The triangle in Fig. 6.4 has a side opposite the 30° angle of length 1.75 cm and a hypotenuse of length 3.5 cm. The ratio is therefore $1.75/3.5 = 0.5$.

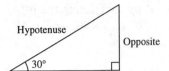

Figure 6.4 A right-angled triangle with a 30° angle.

6.4 THE RATIOS OF THE SIDES

Each of the ratios of a right-angled triangle has a name. The most important to remember are the sine, cosine and tangent; the others, cosecant, secant and cotangent, are the reciprocals of these. The names are abbreviated to sin, cos, tan, cosec, sec and cotan. To see how these are all defined refer to Fig. 6.5.

$$\sin(\alpha) = \frac{\text{opposite}}{\text{hypotenuse}} \qquad \cos(\alpha) = \frac{\text{adjacent}}{\text{hypotenuse}} \qquad \tan(\alpha) = \frac{\text{opposite}}{\text{adjacent}}$$

$$\text{cosec}(\alpha) = \frac{\text{hypotenuse}}{\text{opposite}} \qquad \sec(\alpha) = \frac{\text{hypotenuse}}{\text{adjacent}} \qquad \text{cotan}(\alpha) = \frac{\text{adjacent}}{\text{opposite}}$$

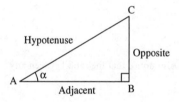

Figure 6.5 A right-angled triangle and the names given to the ratios of the sides (trigonometric ratios).

The values of these ratios can be found by using a calculator and are used to find the length of a side of the triangle when one side and an angle are known or to find the angle when two of the sides are known.

Example 6.2 Find the length of the side indicated in the triangle in Fig. 6.6.

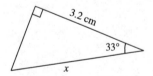

Figure 6.6 The triangle for Example 6.2.

SOLUTION The unknown side, x, is the hypotenuse. The marked side of 3.2 cm is adjacent to the given angle of 33°. One trigonometric ratio containing the adjacent and hypotenuse is the cosine.

$$\cos 33° = \frac{\text{adjacent}}{\text{hypotenuse}} = \frac{3.2}{x}$$

$$\Rightarrow 0.8387 = \frac{3.2}{x}$$

$$\Leftrightarrow x = \frac{3.2}{0.8387} \approx 3.82 \text{ to two decimal places}$$

So the length of the side is 3.82 cm to two decimal places.

Example 6.3 Find the length of the side indicated in the triangle in Fig. 6.7.

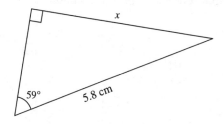

Figure 6.7 The triangle for Example 6.3.

SOLUTION The unknown side, x, is opposite the given angle of 59°. The hypotenuse is given as 5.8 cm. One trigonometric ratio containing the opposite and hypotenuse is the sine, giving

$$\sin 59° = \frac{\text{opposite}}{\text{hypotenuse}} = \frac{x}{5.8}$$

$$0.8572 = \frac{x}{5.8} \Rightarrow x = 4.97 \text{ to two decimal places.}$$

Therefore the length of the side is 4.97 cm.

Example 6.4 Find the angle A in the triangle in Fig. 6.8.

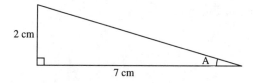

Figure 6.8 The triangle for Example 6.4.

SOLUTION The side marked as 2 cm is opposite the unknown angle A. The side marked as 7 cm is adjacent to the unknown angle A. One trigonometric ratio containing the opposite and adjacent is the tangent:

$$\tan(A) = \frac{\text{opposite}}{\text{adjacent}} = \frac{2}{7} = 0.2857$$

Use the inverse tangent function button to get A = 16°; hence the unknown angle A is 16°.

We can use the ratios of the sides to solve problems where an angle and the length of one side are known, or where two sides are known and we want to find the value of an angle. However, we need some other way to find the third side when two of the sides are known.

6.5 PYTHAGORAS'S THEOREM

Pythagoras's theorem states that the square on the hypotenuse is equal to the sum of the squares on the other two sides. This is expressed as $c^2 = a^2 + b^2$, and the sides are labelled in Fig. 6.9.

Pythagoras's theorem is easy to justify and is done so by consideration of the area of a square of side $a + b$, as shown in Fig. 6.10. The area of the total square is given by the product of the sides $(a + b)^2$. Inside the square there are four triangles of total area $4 \times \frac{1}{2}(a \times b) = 2ab$ (the area of a right-angled triangle is half base × height). The area of the square in the middle is c^2.

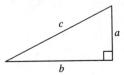

Figure 6.9 Pythagoras's theorem gives that $c^2 = a^2 + b^2$.

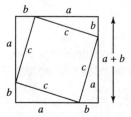

Figure 6.10 To justify Pythagoras's theorem, draw a square of side $a + b$ and place triangles of sides a, b and c in each corner.

As the total area of the four triangles and the square of side c must equal the area of the big square of side $a + b$ we have

$$(a + b)^2 = 2ab + c^2$$

$$\Leftrightarrow a^2 + 2ab + b^2 = 2ab + c^2$$

Subtracting $2ab$ from both sides gives $a^2 + b^2 = c^2$, which is Pythagoras's theorem.

Pythagoras's theorem can now be used to find the third side given the length of the other two.

Example 6.5 Find the length of the side indicated in Fig. 6.11.

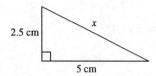

Figure 6.11 The triangle for Example 6.5.

SOLUTION x is the hypotenuse. It therefore takes the place of c in Pythagoras's formula:

$$a^2 + b^2 = c^2 \qquad \text{and } a = 2.5, \ b = 5$$

$$\Leftrightarrow (2.5)^2 + 5^2 = x^2$$

$$\Leftrightarrow x^2 = 6.25 + 25$$

$$= 31.25$$

$$\Rightarrow x = \sqrt{31.25} \approx 5.6 \qquad \text{(where, as } x \text{ represents a length, } x \geqslant 0)$$

The length of the hypotenuse is 5.6 cm to two significant figures.

Example 6.6 Find the length of the side indicated in Fig. 6.12.

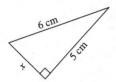

Figure 6.12 The triangle for Example 6.6.

SOLUTION x is not the hypotenuse, and therefore takes the place of a or b in Pythagoras's formula, while 6 takes the place of c, giving

$$x^2 + 5^2 = 6^2$$
$$\Leftrightarrow x^2 + 25 = 36$$
$$\Leftrightarrow x^2 = 11$$
$$\Rightarrow x = \sqrt{11} \approx 3.32 \qquad \text{(where } x \geqslant 0)$$

The length of the side is 3.32 cm to two decimal places.

Example 6.7 A ladder of length 6 m is placed against a wall. The greatest angle it can make to the ground and still be stable is 70°.

(a) Find the maximum reach of the ladder vertically up the wall.
(b) What length of ladder would be needed to reach a window at a height of 20 m when the foot of the ladder cannot be placed any nearer than 2 m away from the wall?

These problems can be pictured as in Fig. 6.13.

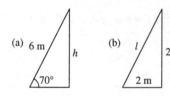

Figure 6.13 (a) The maximum angle of the ladder is 70°. (b) A ladder is needed to reach a window at a height of 20 m.

SOLUTION

(a) The unknown height is marked as h in Fig. 6.13(a). h is opposite the 70° angle and the ladder of length 6 m forms the hypotenuse. Using the sine we get

$$\sin(70°) = \frac{h}{6} \Leftrightarrow h = 6 \sin(70°) \Rightarrow h \approx 5.64 \text{ to two d.p.}$$

The height reached by the ladder is 5.64 m to two decimal places.

(b) The length of the ladder, marked as l, forms the hypotenuse in Fig. 6.13(b). As we know the lengths of the other two sides we use Pythagoras's theorem, giving

$$l^2 = 2 + 20$$
$$\Leftrightarrow l^2 = 4 + 400 \Leftrightarrow l^2 = 404$$
$$l = 20.1 \text{ to two d.p.}$$

The length of the ladder is 20.1 m to two decimal places.

Example 6.8 In the triangle shown in Fig. 6.14, find the length of the side marked c.

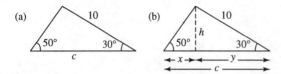

Figure 6.14 (a) The triangle for Example 6.8. We wish to find the length marked c. (b) The same triangle with a perpendicular drawn and unknown lengths, h, x and y marked. Notice that $c = x + y$.

SOLUTION Split the triangle into two right-angled triangles by dropping a perpendicular as in Fig. 6.14(b). The length, c, is given by $c = x + y$; however, h needs also to be calculated in order to find x.

From the right-angled triangle with the 30° angle:

$$\sin(30°) = \frac{h}{10} \Leftrightarrow h = 10 \sin(30°) \Leftrightarrow h = 5$$

Also from Pythagoras:

$$h^2 + y^2 = 10^2$$

Substituting for h gives

$$5^2 + y^2 = 10^2 \Leftrightarrow y^2 = 100 - 25$$

$$\Leftrightarrow y^2 = 75 \Rightarrow y \approx 8.66 \text{ to two d.p.}$$

From the triangle with the 50° angle:

$$\tan 50° = \frac{h}{x} \Leftrightarrow x = \frac{5}{\tan 50°} \Rightarrow x \approx 4.2$$

We are now able to find the length of the side c, using $c = x + y$, giving $c = 8.66 + 4.2 = 12.86$. The length of the side marked c in the triangle is 12.86.

6.6 SUMMARY

1. Right-angled triangles can be solved if one angle (other than the right angle) is known together with one side, or if two sides are known.
2. The trigonometric ratios are used to solve problems where one angle and one side are known and we wish to find the length of one of the other sides. They are also used to find an angle when the lengths of two sides are known.
3. The most important trigonometric ratios are:

$$\sin(\alpha) = \frac{\text{opposite}}{\text{hypotenuse}} \qquad \cos(\alpha) = \frac{\text{adjacent}}{\text{hypotenuse}} \qquad \tan(\alpha) = \frac{\text{opposite}}{\text{adjacent}}$$

and the others are given in the caption to Fig. 6.5.
4. Pythagoras's theorem is used to find the third side in a right-angled triangle when the other two are known. It is expressed by

$$a^2 + b^2 = c^2$$

where c is the hypotenuse.
5. Many other geometrical problems can be solved by splitting shapes into right-angled triangles.

6.7 EXERCISES

6.1 Find the sides of the triangles shown in Fig. 6.15.

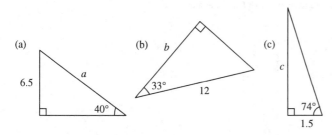

(a)

(b)

(c)

Figure 6.15 Triangles for Exercise 6.1.

6.2 Find the angles in the triangles shown in Fig. 6.16.

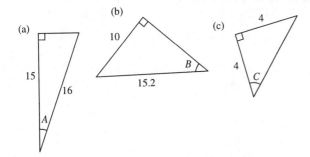

(a)

(b)

(c)

Figure 6.16 Triangles for Exercise 6.2.

6.3 Find the unknown side of the triangles shown in Fig. 6.17.

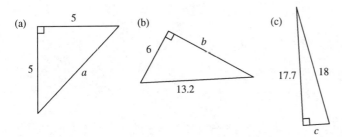

(a)

(b)

(c)

Figure 6.17 Triangles for Exercise 6.3.

6.4 A road has a gradient of 1 in 10 (that is it rises 1 m for every 10 m horizontally). If a car travels 30 m along the road, how far has it travelled in a horizontal direction and how far has it risen?

6.5 A statue is such that at a certain time of day, when the sun makes an angle of approximately 63° with the ground, the length of its shadow is 10.2 m. Estimate the height of the statue.

6.6 Two forces, F_1 and F_2, are at right angles and the magnitude of the resultant force is given by the length of the diagonal, as shown in Fig. 6.18. If $F_1 = 3.6$ N and $F_2 = 5.4$ N find the magnitude of F_R.

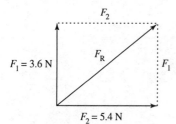

Figure 6.18 Forces for Exercise 6.6. F_R is the resultant of forces F_1 and F_2 at right angles.

6.7 A generator of frequency 20 Hz is applied to a series circuit as shown in Fig. 6.19(a). The resistor has a resistance of 5.3 Ω and at this frequency the inductor is found to have a reactance of 294.1 Ω. The resultant voltage V can be found by solving the triangle as shown in Fig. 6.19(b) where V_L is the voltage across the inductor ($V_L = 294.1I$) and V_R is the voltage across the resistor ($V_R = 5.3I$), where I is the current in the circuit. Find the size of V in terms of I and also find the phase ϕ, which gives the angle by which the resultant voltage leads the current.

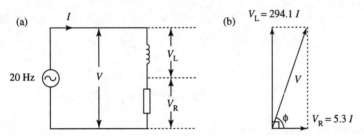

Figure 6.19 (a) Circuit for Exercise 6.7 and (b) triangle of voltages with resultant V.

TWO

SETS, FUNCTIONS
AND CALCULUS

SEVEN

SETS AND FUNCTIONS

7.1 INTRODUCTION

Finding relationships between quantities is of central importance in engineering. For instance, we know that given a simple circuit with a $1000\,\Omega$ resistance then the relationship between current and voltage is given by Ohm's law, $I = V/1000$. For any value of the voltage V we can give an associated value of I. This relationship means that I is a function of V. From this simple idea there are many other questions that need clarifying, some of which are:

1. Are all values of V permitted? For instance, a very high value of the voltage could change the nature of the material in the resistor and the expression would no longer hold.
2. Supposing the voltage V is the equivalent voltage found from considering a larger network. Then V is itself a function of other voltage values in the network (see Fig. 7.1). How can we combine the functions to get the relationship between this current we are interested in and the actual voltages in the network?

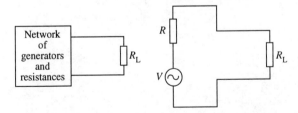

Figure 7.1 The volatage V is an equivalent voltage found by considering the combined effect of circuit elements in the rest of the network.

3. Supposing we know the voltage in the circuit and would like to know the associated current. Given the function that defines how current depends on the voltage, can we find a function that defines how the voltage depends on the current? In the case where $I = V/1000$ it is clear that $V = 1000I$. This is called the inverse function.

Another reason exists for better understanding of the nature of functions. In Chapters 12 and 13 we shall study differentiation and integration, which look at the way that functions change. A good understanding of functions and how to combine them will help considerably in those chapters.

The values that are permitted as inputs to a function are grouped together. A collection of objects is called a set. The idea of a set is very simple, but studying sets can help not only in understanding functions, but also in understanding the properties of logic circuits, as discussed in Chapter 10.

7.2 SETS

A **set** is a collection of objects, called **elements**, in which the order is not important and an object cannot appear twice in the same set.

Example 7.1 Explicit definitions of sets, i.e. where each element is listed, are:

$$A = \{a,b,c\}$$

$$B = \{3,4,6,7,8,9\}$$

$$C = \{\text{Linda, Raka, Sue, Joe, Nigel, Mary}\}$$

a∈A means 'a is an element of A' or 'a belongs to A'. Therefore in the above examples:

$$3 \in B$$

$$\text{Linda} \in C$$

The **universal set** is the set of all objects we are interested in and will depend on the problem under consideration. It is represented by $\mathscr{E}$.

The **empty set** (or **null set**) is the set with no elements. It is represented by $\varnothing$ or $\{\ \}$.

Sets can be represented diagrammatically, generally as circular shapes. The universal set is represented as a rectangle. Such diagrams are called **Venn diagrams**.

Example 7.2

$$\mathscr{E} = \{a,b,c,d,e,f,g\} \qquad A = \{a,b,c\} \qquad B = \{d,e\}$$

This can be shown as in Fig. 7.2.

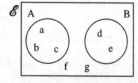

Figure 7.2 A Venn diagram of the sets $\mathscr{E} = \{a,b,c,d,e,f,g\}$, $A = \{a,b,c\}$ and $B = \{d,e\}$.

We shall mainly be concerned with sets of numbers as these are more often used as inputs to functions.

Some important sets of numbers are (where '...' means 'continue in the same manner'):

1. The set of **natural numbers** $\mathbb{N} = \{0,1,2,3,4,5,\ldots\}$
2. The set of **integers** $\mathbb{Z} = \{\ldots -3,-2,-1,0,1,2,3,\ldots\}$

3. The set of **rationals** (which includes fractional numbers), $\mathbb{Q}$
4. The set of **reals** (all the numbers necessary to represent points on a line), $\mathbb{R}$.

Sets can also be defined using some rule, instead of explicitly.

Example 7.3 Define the set A explicitly where $\mathscr{E} = \mathbb{N}$ and $A = \{x | x < 3\}$.

SOLUTION The $A = \{x | x < 3\}$ is read as 'A is the set of elements x, such that x is less than 3'. Therefore, as the universal set is the set of natural numbers, $A = \{0,1,2\}$.

Example 7.4 $\mathscr{E} =$ days of the week and $A = \{x | x$ is after Thursday and before Sunday$\}$. Then $A = \{$Friday, Saturday$\}$.

Subsets

We may wish to refer to only a part of some set. This is said to be a subset of the original set.
$A \subseteq B$ is read as 'A is a subset of B' and it means that every element of A is an element of B.

Example 7.5

$$\mathscr{E} = \mathbb{N}$$

$$A = \{1,2,3\} \qquad B = \{1,2,3,4,5\}$$

Then $A \subseteq B$.

Note the following points:

1. All sets must be subsets of the universal set, i.e. $A \subseteq \mathscr{E}$ and $B \subseteq \mathscr{E}$.
2. A set is a subset of itself, i.e. $A \subseteq A$.
3. If $A \subseteq B$ and $B \subseteq A$ then $A = B$.

Proper subsets

$A \subset B$ is read as 'A is a proper subset of B' and means that A is a subset of B but A is not equal to B. Hence $A \subset B$ and simultaneously $B \subset A$ are impossible.
A proper subset can be shown on a Venn diagram as in Fig. 7.3.

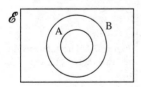

Figure 7.3 A Venn diagram of a proper subset of B: $A \subset B$.

7.3 OPERATIONS ON SETS

In Chapter 1 we studied the rules obeyed by numbers when using operations like negation, multiplication and addition. Sets can be combined in various ways using set operations. Sets and their operations form a Boolean algebra, which we shall look at in greater detail in Chapter 10, particularly its application to digital design. The most important set operations are given in this section.

Complement

$\bar{A}$ or A' represent the complement of the set A. The complement of A is the set of everything in the universal set which is not in A, as pictured in Fig. 7.4.

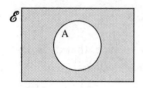

Figure 7.4 The shaded area is the complement, A', of the set A.

Example 7.6

$$\mathscr{E} = \mathbb{N}$$

$$A = \{x | x > 5\}$$

Then $A' = \{0,1,2,3,4,5\}$.

Example 7.7 The universal set is the set of real numbers, represented by a real number line.

If A is the set of numbers less than 5, $A = \{x | x < 5\}$, then A' is the set of numbers greater than or equal to 5: $A' = \{x | x \geqslant 5\}$. These sets are shown in Fig. 7.5.

Figure 7.5 $A = \{x | x < 5\}$ and $A' = \{x | x \geqslant 5\}$.

Intersection

$A \cap B$ represents the intersection of the sets A and B. The intersection contains those elements that are in A and also in B. This can be represented as in Fig. 7.6, and examples are given in Figs 7.7–7.9.

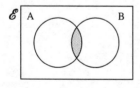

Figure 7.6 The shaded area represents the intersection of A and B.

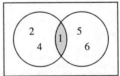

Figure 7.7 The intersection of two sets: $\{1,2,4\} \cap \{1,5, 6\} = \{1\}$.

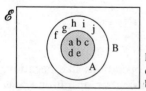

Figure 7.8 The intersection of two sets: $\{a,b,c,d,e\} \cap \{a, b,c,d,e,f,g,h,i\} = \{a,b,c,d,e\}$.

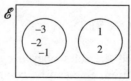

Figure 7.9 The intersection of two sets: $\{-3,-2,-1\} \cap \{1,2\} = \varnothing$, the empty set, as they have no elements in common.

Note the following important points:

1. If $A \subseteq B$ then $A \cap B = A$. This is the situation in the example given in Fig. 7.8.

2. If A and B have no elements in common then $A \cap B = \varnothing$ and they are called **disjoint**. This is the situation given in the example in Fig. 7.9. Two sets which are known to be disjoint can be shown on the Venn diagram as in Fig. 7.10.

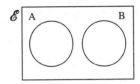

Figure 7.10 Disjoint sets A and B.

Union

$A \cup B$ represents the union of A and B, i.e. the set containing elements which are in A or B or in both A and B. On a Venn diagram the union can be shown as in Fig. 7.11, and examples are given in Figs 7.12–7.14.

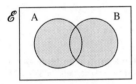

Figure 7.11 The shaded area represents the union of sets A and B.

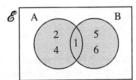

Figure 7.12 The union of two sets: $\{1,2,4\} \cup \{1,5,6\} = \{1,2,4,5,6\}$

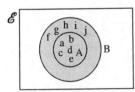

Figure 7.13 The union of two sets: $\{a,b,c,d,e\} \cup \{a,b,d, e,f,g,h,i,j\} = \{a,b,c,d,e,f,g,h,i,j\}$.

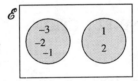

Figure 7.14 The union of two sets: $\{-3,-2,-1\} \cup \{1, 2\} = \{-3,-2,-1,1,2\}$.

Note the following important points:

1. If $A \subseteq B$ then $A \cup B = B$. This is the situation in the example given in Fig. 7.13.
2. The union of any set with its complement gives the universal set, i.e. $A \cup A' = \mathscr{E}$, the universal set. This is pictured in Fig. 7.15.

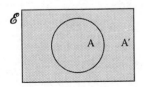

Figure 7.15 The shaded area represents the union of a set with its complement, giving the universal set.

Cardinality of a finite set

The number of elements in a set is called the cardinality of the set and is written as $n(A)$ or $|A|$.

Example 7.8

$$n(\varnothing) = 0 \qquad n(\{2\}) = 1 \qquad n(\{a,b\}) = 2$$

For finite sets the cardinality must be a natural number.

Example 7.9 In a survey, 100 people were students and 720 owned a video recorder. 794 people owned a video recorder or were students. How many students owned a video recorder?

$$\mathcal{E} = \{x | x \text{ is a person included in the survey}\}$$

Setting

$$S = \{x | x \text{ is a student}\}$$

$$V = \{x | x \text{ owns a video recorder}\}$$

we can solve this problem using a Venn diagram, as in Fig. 7.16.

x is the number of students who own a video recorder. From the diagram we get

$$100 - x + x + 720 - x = 794$$

$$\Leftrightarrow 820 - x = 794$$

$$\Leftrightarrow x = 26$$

Therefore 26 students own a video recorder.

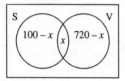

Figure 7.16 S is the set of students in a survey and V is the set of people who own a video recorder. The numbers in the sets give the cardinality of the sets: $n(S) = 100$, $n(S \cup V) = 794$, $n(V) = 720$, $n(S \cap V) = x$.

7.4 RELATIONS AND FUNCTIONS

Relations

A relation is a way of pairing up members of two sets. This is just like the idea of family relations. For instance, a child can be paired with its mother, brothers can be paired with sisters etc. A relation is such that it may not always be possible to find a suitable partner for each element in the first set, whereas sometimes there will be more than one. For instance, if we try to pair every boy with his sister there will be some boys that have no sisters and some boys that have several. This is pictured in Fig. 7.17.

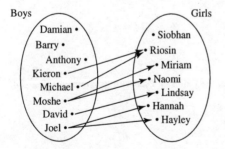

Figure 7.17 The relation boy→sister. Some boys have more than one sister and some have none at all.

Functions

Functions are relations where pairing is always possible. Functions are like mathematical machines. For each input value there is always exactly one output value.

Calculators output function values. For instance, input 2 into a calculator, press $1/x$, and the calculator will display the number 0.5. The output value is called the **image** of the input value. The set of input values is called the domain and the set containing all the images is called the codomain.

The function $y = 1/x$ is displayed in Fig. 7.18 using arrows to link input values with output values.

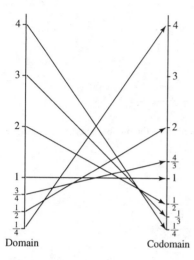

Domain   Codomain

Figure 7.18 An arrow diagram of the function $y = 1/x$.

Functions can be represented by letters. If the function of the above example is given the letter f to represent it, then we can write

$$f : x \mapsto \frac{1}{x}$$

This can be read as 'f is the function which when input a value for x gives the output value $1/x$'. Another way of giving the same information is

$$f(x) = \frac{1}{x}$$

$f(x)$ represents the image of x under the function f and is read as 'f of x'. It does not mean the same as f times x. $f(x) = 1/x$ means 'the image of x under the function f is given by $1/x$', but is usually read as 'f of x equals $1/x$'.

Even more simply, we usually use the letter y to represent the output value, the image, and x to represent the input value. The function is therefore summed up by $y = 1/x$.

x is a variable because it can take any value from the set of values in the domain. y is also a variable, but its value is fixed once x is known. So x is called the **independent variable** and y is called the **dependent variable**.

The letters used to define a function are not important. $y = 1/x$ is the same as $z = 1/t$ is the same as $p = 1/q$ provided that the same input values (for x, t or q) are allowed in each case.

More examples of functions are given in arrow diagrams in Figs 7.19(a) and 7.20(a). Functions are more usually drawn using a graph, rather than by using an arrow diagram. To get the graph the codomain is moved to be at right angles to the domain and input and output values are marked by a point at the position (x, y). Graphs are given in Figs 7.19(b) and 7.20(b).

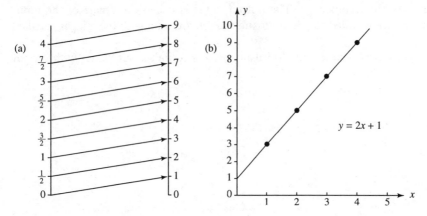

Figure 7.19 The function $y = 2x + 1$, where x can take any real value (any number on the number line). (a) is the arrow diagram and (b) is the graph.

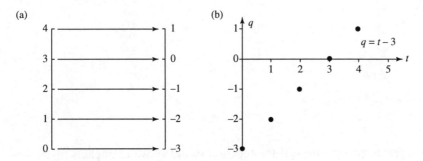

Figure 7.20 The function $q = t - 3$, where t can take any integer value. (a) is the arrow diagram and (b) is the graph.

Continuous functions and discrete functions applied to signals

Functions of particular interest to engineers are either functions of a real number or functions of an integer. The function given in Fig. 7.19 is an example of a real function and the function given in Fig. 7.20 is an example of a function of an integer, also called a discrete function.

Often we are concerned with functions of time. A variable voltage source can be described by giving the voltage as it depends on time, as also can the current. Other examples are the position of a moving robot arm, the extension or compression of car shock absorbers and the heat emission of a thermostatically controlled heating system. A voltage or current varying with time can be used to control instrumentation or to convey information. For this reason it is called a signal. Telecommunication signals may be radio waves or voltages along a transmission line or light signals along an optical fibre.

Time, t, can be represented by a real number, usually non-negative. Time is usually taken to be positive because it is measured from some reference instant, e.g. when a circuit switch is closed. If time is used to describe relative events then it can make sense to refer to negative time. If lightning is seen 1 s before a thunderclap is heard, then this can be described by saying the lightning happened at -1 s, or alternatively that the thunderclap was heard at 1 s. In the two cases the time origin has been chosen differently. If time is taken to be continuous and represented by a real variable then functions of time will be continuous or piecewise continuous. Examples of graphs of such functions are given in Fig. 7.21.

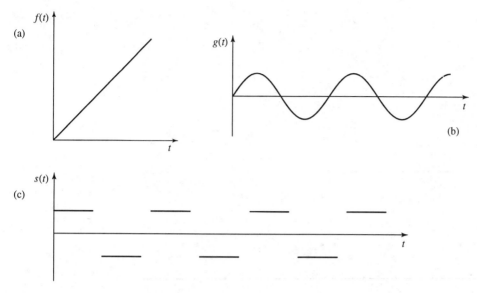

Figure 7.21 Continuous and piecewise functions where time is represented by a real number >0: (a) a ramp function; (b) a wave; (c) a square wave. (a) and (b) are continuous, while (c) is piecewise continuous.

A continuous function is one whose graph can be drawn without taking your pen off the paper. A piecewise continuous function has continuous bits with a limited number of jumps. Figs 7.21(a) and (b) are continuous functions and Fig. 7.21(c) is a piecewise continuous function. If we have a digital signal, then its values are only known at discrete moments of time. Digital signals can be obtained by using an analogue to digital convertor on an originally continuous signal. Digital signals are represented by discrete functions, as in Fig. 7.22.

A digital signal has a sampling interval, T, which is the length of time between successive values. A digital signal is represented by a discrete function. For example, in Fig. 7.22(a) the digital ramp can be represented by the numbers

$$0, 1, 2, 3, 4, 5, \ldots$$

If the sampling interval T is different from 1 then the values would be

$$0, T, 2T, 3T, 4T, 5T \ldots$$

This is a discrete function, also called a sequence. It can be represented by the expression $f(t) = t$, where $t = 0, 1, 2, 3, 4, 5, 6 \ldots$, or, using the sampling interval, T, by $g(n) = nT$, where $n = 0, 1, 2, 3, 4, 5, 6 \ldots$.

Yet another common way of representing a sequence is by using a subscript on the letter representing the image, giving

$$f_n = n \qquad \text{where } n = 0, 1, 2, 3, 4, 5 \ldots$$

or, using the letter a for the image values,

$$a_n = n \qquad \text{where } n = 0, 1, 2, 3, 4, 5 \ldots$$

Substituting some values for n into the above gives

$$a_0 = 0, a_1 = 1, a_2 = 2, a_3 = 3 \ldots$$

As a sequence is a function of the natural numbers (or if negative input values are allowed, the integers) there is no need to specify the input values and it is possible merely to list the output values in order. Hence the ramp function can be expressed by $0, 1, 2, 3, 4, 5, 6 \ldots$.

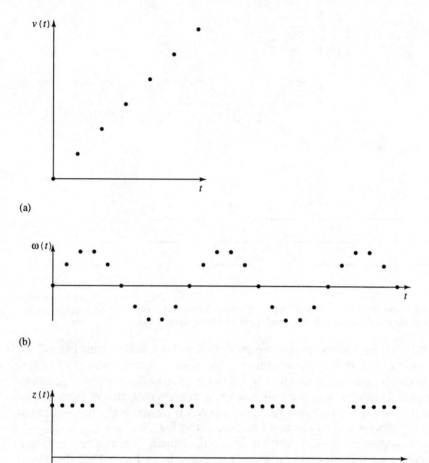

(a)

(b)

(c)

Figure 7.22 Examples of discrete functions: (a) a digital ramp; (b) digital wave; (c) a digital square wave.

Time sequences are often referred to as 'series'. This terminology is not usual in mathematics books, however, as the description 'series' is reserved for describing the sum of a sequence. Sequences and series are dealt with in more detail in Chapter 18.

Example 7.10 Plot the following analogue signals over the values of t given (t real):

(a) $x = t^3 \qquad t \geqslant 0$

(b) $y = \begin{cases} 0 & t \leqslant 3 \\ t - 3 & 3 < t \leqslant 5 \\ 2 & t > 5 \end{cases}$

(c) $z = \dfrac{1}{t^2} \qquad t > 0$

SOLUTION In each case choose some values of t and calculate the function values at those points. Plot the points and join them.

(a)

t	0	0.5	1	1.5	2	2.5	3	3.5
$x=t^3$	0	0.125	1	3.375	8	15.625	27	42.875

These values are plotted in Fig. 7.23(a).

(b)

t	1	1.5	2	2.5	3	3.5	4	4.5	5	5.5	6	6.5	7
y	0	0	0	0	0	0.5	1	1.5	2	2	2	2	2

$$\underbrace{\qquad\qquad}_{y=0}\qquad\underbrace{\qquad\qquad}_{y=t-3}\qquad\underbrace{\qquad\qquad}_{y=2}$$

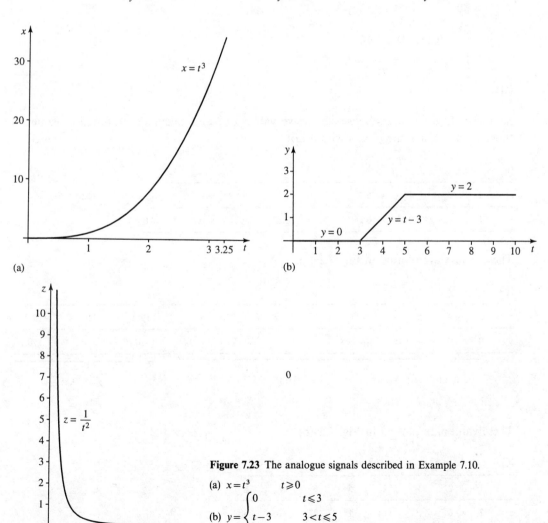

Figure 7.23 The analogue signals described in Example 7.10.

(a) $x=t^3 \qquad t\geqslant 0$

(b) $y=\begin{cases} 0 & t\leqslant 3 \\ t-3 & 3<t\leqslant 5 \\ 2 & t>5 \end{cases}$

(c) $z=1/t^2 \qquad t>0$

These values are plotted in Fig. 7.23(b).

(c)

t	0.0001	0.001	0.01	0.1	1	10	100	1000	10 000
z	10^8	10^6	10^4	100	1	0.01	10^{-4}	10^{-6}	10^{-8}

These values are plotted in Fig. 7.23(c).

Example 7.11 Plot the following discrete signals over the values of t given (t an integer):

(a) $x = \dfrac{1}{t-1} \qquad t > 2$

(b) $y = \begin{cases} 0 & t \leqslant 4 \\ \dfrac{1}{t} - 0.25 & 4 < t < 10 \\ -0.15 & t \geqslant 10 \end{cases}$

(c) $z = 4t - 2 \qquad t > 0$

SOLUTION In each case, choose successive values of t and calculate the function values at those points. Mark the points with a dot.

(a)

t	2	3	4	5	6	7	8	9	10
x	1	0.5	0.33	0.25	0.2	0.17	0.14	0.13	0.11

These values are plotted in Fig. 7.24(a).

(b)

t	3	4	5	6	7	8	9	10	11	12
y	0	0	−0.05	−0.08	−0.11	−0.12	−0.14	−0.15	−0.15	−0.15

$$y = 0 \qquad\qquad y = \frac{1}{t} - 0.15 \qquad\qquad y = -0.15$$

These values are plotted in Fig. 7.24(b).

(c)

t	1	2	3	4	5	6	7	8
z	2	6	10	14	18	22	26	30

These values are plotted in Fig. 7.24(c).

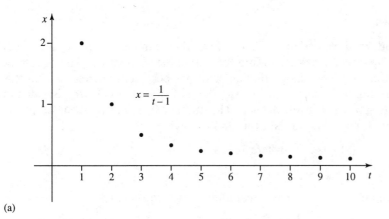

(a)

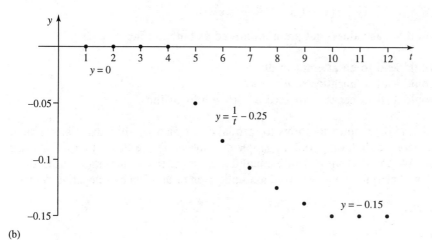

(b)

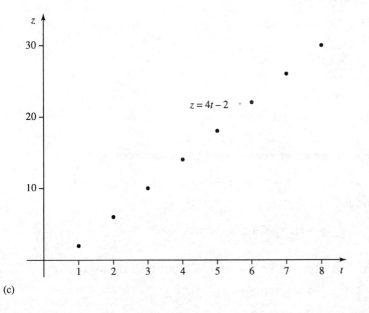

(c)

Figure 7.24 The digital signals described in Example 7.11.

(a) $x = \dfrac{1}{t-1}$ $t \geqslant 2$

(b) $y = \begin{cases} 0 & t < 4 \\ \dfrac{1}{t} - 0.25 & 4 < t < 10 \\ -0.15 & t \geqslant 10 \end{cases}$

(c) $z = 4t - 2$ $t > 0$

Undefined function values

Some functions have 'undefined values', that is numbers that cannot be input into them successfully. For instance input 0 on a calculator and try getting the value of $1/x$. The calculator complains (usually displaying '–E–') indicating that an error has occurred. The reason that this is an error is that we are trying to find the value of $1/0$, that is 1 divided by 0. We discussed in Chapter 1 the fact that division by 0 is not defined. The number 0 cannot be included in the domain of the function $f(x) = 1/x$. This can be expressed by saying

$$f(x) = 1/x \qquad \text{where } x \in \mathbb{R} \text{ and } x \neq 0$$

which is read as

'f of x equals $1/x$ where x is a real number not equal to 0'.

Often we assume that we are considering functions of a real variable and only need to indicate the values that are not allowed as inputs for the function. So we may write

$$f(x) = 1/x \qquad \text{where } x \neq 0$$

Things to look out for as values that are not allowed as function inputs are:

1. Numbers that would lead to an attempt to divide by zero
2. Numbers that would lead to negative square roots
3. Numbers that would lead to negative or zero inputs to a logarithm

Examples 7.12(a) and (b) require solutions to inequalities, which we shall discuss in greater detail in Chapter 8. Here we shall only look at simple examples and use the same rules as used for solving equations. We can find equivalent inequalities by doing the same thing to both sides, with the extra rule that, for the moment, we avoid multiplication or division by a negative number.

Example 7.12 Find the values that cannot be input to the following functions, where the independent variable (x or r) is real:

(a) $y = 3\sqrt{x-2} + 5$

(b) $y = 3 \log_{10}(2 - 4x)$

(c) $R = \dfrac{(r + 1000)}{1000(r - 2)}$

SOLUTIONS

(a) $y = 3\sqrt{x-2} + 5$

Here $x - 2$ cannot be negative as we need to take the square root of it.

$$x - 2 \geq 0 \Rightarrow x \geq 2$$

Therefore the function is

$$y = 3\sqrt{x-2} + 5 \qquad \text{where } x \geq 2$$

(b) $y = 3 \log_{10}(2 - 4x)$

Here $2 - 4x$ cannot be 0 or negative, or else we could not take the logarithm.

$$2 - 4x > 0 \Rightarrow 2 > 4x \Rightarrow 2/4 > x$$

or equivalently $x < \frac{1}{2}$. So the function is

$$y = 3 \log_{10}(2 - 4x) \qquad \text{where } x < 0.5$$

(c) $R = \dfrac{(r + 1000)}{1000(r - 2)}$

Here $1000(r-2)$ cannot be 0 else we would be trying to divide by 0.

Solve the equation for the values that r cannot take

$$1000(r - 2) = 0$$
$$r - 2 = 0$$
$$r = 2$$

The function is

$$R = \frac{(r + 1000)}{1000(r - 2)} \qquad \text{where } r \neq 2$$

Example 7.13 Find the values that can be input to the following discrete functions, where the independent variable is an integer:

(a) $y = \dfrac{1}{(k - 4)} \qquad \text{where } k \in \mathbb{Z}$

(b) $f(k) = \dfrac{1}{(k - 3)(k - 2.2)} \qquad \text{where } k \in \mathbb{Z}$

(c) $a_n = n^2 \qquad \text{where } n \in \mathbb{Z}$

SOLUTIONS

(a) $y = \dfrac{1}{(k - 4)}$

Here $k - 4$ cannot be 0 else there would be an attempt to divide by 0. We get $k - 4 = 0$ when $k = 4$, so the function is:

$$y = \frac{1}{(k - 4)} \qquad \text{where } k \neq 4 \text{ and } k \in \mathbb{Z}$$

(b) $f(k) = \dfrac{1}{(k - 3)(k - 2.2)} \qquad \text{where } k \in \mathbb{Z}$

Solve for $(k - 3)(k - 2.2) = 0$ giving $k = 3$ or $k = 2.2$. As 2.2 is not an integer there is no need to exclude it specifically from the function's input values, so the function is

$$f(k) = \frac{1}{(k - 3)(k - 2.2)} \qquad \text{where } k \neq 3 \text{ and } k \in \mathbb{Z}$$

(c) $a_n = n^2 \qquad n \in \mathbb{Z}$

Here there are no problems with the function, as any integer can be squared. There are no excluded values from the input of the function.

Using a recurrence relation to define a discrete function

Values in a discrete function can also be described in terms of the values for preceding integers.

Example 7.14 Find a table of values for the function defined by the recurrence relation

$$f(n) = f(n-1) + 2 \tag{7.1}$$

where $f(0) = 0$.

SOLUTION Assuming that the function is defined for $n = 0, 1, 2, \ldots$, then we can take successive values of n and find the values taken by the function.

$n = 0$ gives $f(0) = 0$, as given.

Substituting $n = 1$ into Eq. (7.1) gives

$$f(1) = f(1-1) + 2$$
$$\Leftrightarrow f(1) = f(0) + 2 = 0 + 2 = 2 \quad \text{(using } f(0) = 0)$$

Hence $f(1) = 2$.

Substituting $n = 2$ into Eq. (7.1) gives

$$f(2) = f(2-1) + 2$$
$$\Leftrightarrow f(2) = f(1) + 2$$
$$\Leftrightarrow f(2) = f(1) + 2 = 2 + 2 = 4 \quad \text{(using } f(1) = 2)$$

Hence $f(2) = 4$.

Substituting $n = 3$ into Eq. (7.1) gives

$$f(3) = f(3-1) + 3$$
$$\Leftrightarrow f(3) = f(2) + 2 = 4 + 2 \quad \text{(using } f(2) = 4)$$

Hence $f(3) = 6$.

Continuing in the same manner gives the following table:

n	0	1	2	3	4	5	6	7	8	9	10	...	n	...
f	0	2	4	6	8	10	12	14	16	18	20	...	$2n$	...

Notice we have filled in the general term $f(n) = 2n$. This was found in this case by simple guesswork.

7.5 COMBINING FUNCTIONS

The sum, difference, product and quotient of two functions, f and g

Two functions with $\mathbb{R}$ as their domain and codomain can be combined using arithmetic operations. We can define the sum of f and g by

$$(f + g):x \mapsto f(x) + g(x)$$

The other operations are defined as follows:

$$(f - g):x \mapsto f(x) - g(x) \quad \text{difference}$$
$$(f \times g):x \mapsto f(x) \times g(x) \quad \text{product}$$
$$(f/g):x \mapsto \frac{f(x)}{g(x)} \quad \text{quotient}$$

Example 7.15 Find the sum, difference, product and quotient of the functions

$$f:x \mapsto x^2 \text{ and } g:x \mapsto x^6.$$

SOLUTIONS

$$(f+g):x \mapsto x^2 + x^6$$
$$(f-g):x \mapsto x^2 - x^6$$
$$(f \times g):x \mapsto x^2 \times x^6 = x^8$$
$$(f/g):x \mapsto \frac{x^2}{x^6} = x^{-4}$$

The specification of the domain of the quotient is not straightforward. This is because of the difficulty which occurs when $g(x)=0$. When $g(x)=0$, the quotient function is undefined and we must remove such elements from its domain. The domain of f/g is $\mathbb{R}$ with the values where $g(x)=0$ omitted.

Composition of functions

This method of combining functions is fundamentally different from the arithmetical combinations of the previous section. The composition of two functions is the action of performing one function followed by the other, i.e. a function of a function.

Example 7.16 A post office worker has a scale expressed in kilograms which gives the cost of a parcel depending on its weight. There is also an approximate formula for conversion from pounds (lb) to kilograms. The worker wishes to find out the cost of a parcel which weighs 3 lb.

The two functions involved are:

$$a:\text{kilograms} \rightarrow \text{money}$$

and

$$c:\text{lbs} \rightarrow \text{kilograms}$$

a is defined by Fig. 7.25 and the function c is given by

$$c:x \mapsto x/2.2$$

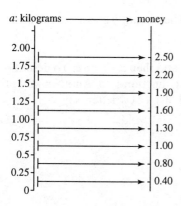

Figure 7.25 The function a:kilograms→money used in Example 7.16.

SOLUTION The composition '$a \circ c$' will be a function from lb to money. Hence 3 lb after the function c gives 1.364 and 1.364 after the function a gives £1.90, and therefore $(a \circ c)(3) = £1.90$.

Example 7.17 Suppose $f(x) = 2x + 1$ and $g(x) = x^2$. Then we can combine the functions in two ways.

1. A composite function can be formed by performing f first and then g, i.e. $g \circ f$. To describe this function we want to find what happens to x under the function $g \circ f$. Another way of saying this is that we need to find $g(f(x))$. To do this, call $f(x)$ a new letter, say y.

$$y = f(x) = 2x + 1$$

Rewrite g as a function of y:

$$g(y) = y^2$$

Now substitute $y = 2x + 1$, giving

$$g(2y + 1) = (2x + 1)^2$$

Hence

$$g(f(x)) = (2x + 1)^2$$
$$(g \circ f)(x) = (2x + 1)^2$$

2. A composite function can be formed by performing g first and then f, i.e. $f \circ g$. To describe this function we want to find what happens to x under the function $f \circ g$. Another way of saying this is that we need to find $f(g(x))$. To do this, call $g(x)$ a new letter, say y.

$$y = g(x) = x^2$$

Rewrite f as a function of y:

$$f(y) = 2y + 1$$

Now substitute $y = x^2$, giving

$$g(y^2) = 2x^2 + 1$$

Hence

$$f(g(x)) = 2x^2 + 1$$
$$(f \circ g)(x) = 2x^2 + 1$$

Example 7.18 Suppose that $u(t) = 1/(t - 2)$ and $v(t) = 3 - t$. Then again we can combine the functions in two ways.

1. A composite function can be formed by performing u first and then v, i.e. $v \circ u$. To describe this function we want to find what happens to t under the function $v \circ u$. Another way of saying this is that we need to find $v(u(t))$. To do this, call $u(t)$ a new letter, say y.

$$y = u(t) = \frac{1}{t - 2}$$

Rewrite v as a function of y:

$$v(y) = 3 - y$$

Now substitute $y = 1/(t-2)$, giving

$$v\left(\frac{1}{t-2}\right) = 3 - \frac{1}{t-2}$$

$$= \frac{3(t-2)-1}{t-2} \quad \text{(rewriting the expression over a common denominator)}$$

$$= \frac{3t-6-1}{t-2} = \frac{3t-7}{t-2} = \frac{3t-7}{t-2}$$

Hence

$$v(u(t)) = \frac{3t-7}{t-2}$$

$$(v \circ u)(t) = \frac{3t-7}{t-2}$$

2. A composite function can be formed by performing v first and then u, i.e. $u \circ v$. To describe this function we want to find what happens to t under the function $u \circ v$. Another way of saying this is that we need to find $u(v(t))$. To find this, call $v(t)$ a new letter, say y.

$$y = v(t) = 3 - t$$

Rewrite u as a function of y:

$$u(y) = \frac{1}{y-2}$$

Now substitute $y = 3 - t$, giving

$$v(3-t) = \frac{1}{(3-t)-2} = \frac{1}{1-t}$$

Hence

$$u(v(t)) = \frac{1}{1-t}$$

$$u \circ v(t) = \frac{1}{1-t}$$

Decomposing functions

In order to calculate the value of a function either by hand or using a calculator we need to understand how it decomposes. That is, we need to understand the order of the operations in the function expression.

Example 7.19 Calculate $y = (2x+1)^3$ when $x = 2$.

SOLUTION Remember the order of operations discussed in Chapter 1. The operations are performed in the following order. Start with $x=2$:

$$2x=4$$

$$2x+1=5$$

$$(2x+1)^3=125$$

So there are three operations involved:

1. Multiply by 2
2. Add on 1
3. Take the cube

This way of breaking down functions can be pictured using boxes to represent each operation that makes up the function, as was used to represent equations in Chapter 3. The whole function can be thought of as a machine, represented by a box. For each value x from the domain of the function that enters the machine, there is a resulting image, y, which comes out of it. This is pictured in Fig. 7.26.

Inside the box we can write the name of the function or the expression which gives the function rule. A composite function box can be broken into different stages, each represented by its own box. The function $y=(2x+1)^3$ breaks down as in Fig. 7.27. $y=(3x-4)^4$ can be broken down as in Fig. 7.28.

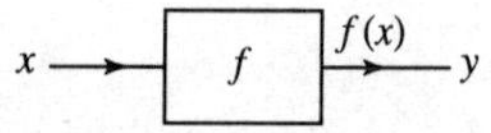

Figure 7.26 A function pictured as a machine represented by a box. x represents the input value, any value of the domain, and y represents the output, the image of x under the function.

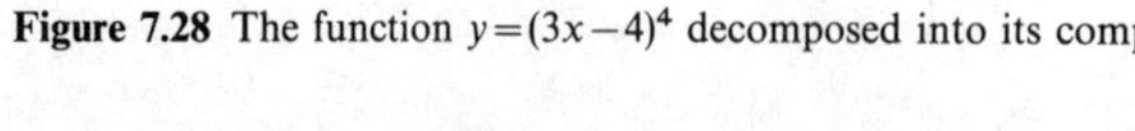

Figure 7.27 The function $y=(2x+1)^3$ decomposed into its composite operations.

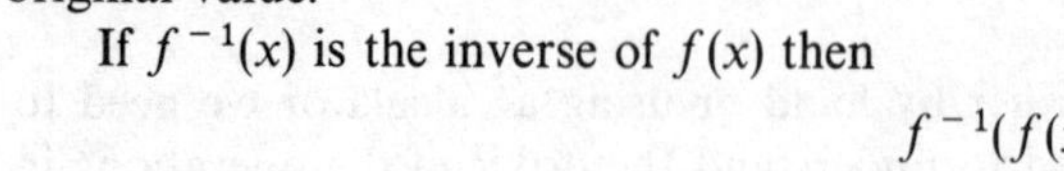

Figure 7.28 The function $y=(3x-4)^4$ decomposed into its composite operations.

The inverse of a function

The inverse of a function is a function which will take the image under the function back to its original value.

If $f^{-1}(x)$ is the inverse of $f(x)$ then

$$f^{-1}(f(x))=x$$

$$(f^{-1}\circ f):x\mapsto x$$

Example 7.20

$$f(x)=2x+1 \qquad f^{-1}(x)=\frac{(x-1)}{2}$$

To show this is true, look at the combined function

$$f^{-1}(f(x)) = \frac{2x+1-1}{2} = x$$

Finding the inverse of a linear function

One simple way of finding the inverse of a linear function is to:

1. Decompose the operations of the function
2. Combine the inverse operations (performed in the reverse order) to give the inverse function.

This is a method similar to that used to solve linear equations in Chapter 3.

Example 7.21 Find the inverse of the function $f(x) = 5x - 2$. The method of solution is given in Fig. 7.29.

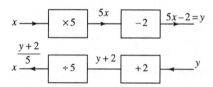

Figure 7.29 The top line represents the function $f(x) = 5x - 2$ (read from left to right) and the bottom line the inverse function.

The inverse operations give

$$x = \frac{y+2}{5}$$

Here y is the input value of the inverse function and x is the output value. To use x and y in the more usual way, where x is the input and y the output, swap the letters, giving the inverse function as $y = (x+2)/5$.

This result can be achieved more quickly by rearranging the expression so that x is the subject of the formula and then swapping x and y.

Example 7.22 Find the inverse of $f(x) = 5x - 2$.

$$y = 5x - 2$$

$$y = 5x - 2 \Leftrightarrow y + 2 = 5x \Leftrightarrow \frac{y+2}{5} = x \Leftrightarrow x = \frac{y+2}{5}$$

Now swap x and y to give $y = (x+2)/5$. Therefore $f^{-1}(x) = (x+2)/5$.

Example 7.23 Find the inverse of $g(x) = 1/(2-x)$, where $x \neq 2$.

$$y = \frac{1}{2-x} \Leftrightarrow y(2-x) = 1 \Leftrightarrow 2y - xy = 1$$

$$\Leftrightarrow 2y = 1 + xy \Leftrightarrow 2y - 1 = xy \Leftrightarrow xy = 2y - 1$$

$$\Leftrightarrow x = \frac{2y-1}{y} \qquad \text{where } y \neq 0$$

$$\Leftrightarrow x = 2 - \frac{1}{y}$$

Swap x and y to give

$$y = 2 - \frac{1}{x}$$

So

$$g^{-1}(x) = 2 - \frac{1}{x}$$

To check, try a couple of values of x. For $x = 4$:

$$g(x) = \frac{1}{2-x} = \frac{1}{2-4} = -\frac{1}{2}$$

Perform g^{-1} on the output value $-\frac{1}{2}$ by substituting $g(4) = -\frac{1}{2}$ into $g^{-1}(x)$:

$$g^{-1}(-\tfrac{1}{2}) = 2 - \frac{1}{(-1/2)} = 2 + 2 = 4$$

The function followed by its inverse has given us the original value of x.

7.6 SUMMARY

1. Functions are used to express relationships between physical quantities.
2. The allowed inputs to a function are grouped into a set, called the domain of the function. The set including all the outputs is called the codomain.
3. A set is a collection of objects called elements.
4. $\mathscr{E}$ is the universal set, the set of all objects we are interested in.
5. $\varnothing$ is the empty set, the set with no elements.
6. The three most important operations on sets are:

   (a) Intersection: $A \cap B$ is the set containing every element in both A and B.
   (b) Union: $A \cup B$ is the set of elements in A or in B or both.
   (c) Complement: A' is the set of everything in the universal set that is not in A.

7. A relation is a way of pairing members of two sets.
8. Functions are a special type of relation which can be thought of as mathematical machines. For each input value there is exactly one output value.
9. Many functions of interest are functions of time, used to represent signals. Analogue signals can be represented by functions of a real variable and digital signals by functions of an integer (discrete functions). Functions of an integer are also called sequences and can be defined using a recurrence relation.
10. To find the domain of a real or discrete function exclude values that could lead to a divide by zero, negative square roots or negative or zero logarithms or other undefined values.
11. Functions can be combined in various ways including sum, difference, product and quotient. A special operation of functions is composition. A composite function is found by performing a second function on the result of the first.

12. The inverse of a function is a function which will take the image under the function back to its original value.

7.7 EXERCISES

7.1 Given $\mathscr{E} = \{a,b,c,d,e,f,g\}$, $A = \{a,b,e\}$, $B = \{b,c,d,f\}$ and $C = \{c,d,e\}$, write down the following sets:

(a) $A \cap B$ (b) $A \cup B$ (c) $A \cap C'$
(d) $(A \cup B) \cap C$ (e) $(A \cap C) \cup (B \cap C)$ (f) $(A \cap B) \cup C$
(g) $(A \cup C) \cap (B \cup C)$ (h) $(A \cap C)'$ (i) $A' \cup C'$

7.2 Use Venn diagrams to show that:

(a) $(A \cap B) \cap C = A \cap (B \cap C)$
(b) $(A \cup B) \cup C = A \cup (B \cup C)$
(c) $(A \cap B) \cup C = (A \cup C) \cap (B \cup C)$
(d) $(A \cup B) \cap C = (A \cap C) \cup (B \cap C)$
(e) $(A \cap B)' = A' \cup B'$
(f) $(A \cup B)' = A' \cap B'$

7.3 Let $\mathscr{E} = \{0,1,2,3,4,5,6,7,8,9\}$ and given $P = \{x \mid x < 5\}$ and $Q = \{x \mid x \geqslant 3\}$ find explicitly:

(a) P (b) Q (c) $P \cup Q$ (d) P'
(e) $P' \cap Q$

7.4 Below are various assertions for any sets A, B. Write true or false for each statement and give a counter-example if you think the statement is false.

(a) $(A \cap B)' = A' \cap B'$ (b) $(A \cap B)' \subseteq A$
(c) $A \cap B = B \cap A$ (d) $A \cap B' = B \cap A'$

7.5 Using a Venn diagram simplify the following:

(a) $A \cap (A \cup B)$ (b) $A \cup (B \cap A')$ (c) $A \cap (B \cup A')$

7.6 A computer screen has 80 columns and 25 rows:

(a) Define the set of positions on the screen.
(b) Taking the origin as the top left-hand corner define

    (i) The set of positions in the lower half of the screen as shown in Fig. 7.30(a).
    (ii) The set of positions lying on or below the diagonal as shown in Fig. 7.30(b).

(a) 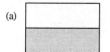   (b)

Figure 7.30 (a) Points lying in the shaded area represent the set of positions on the lower half of the computer screen in Exercise 7.6(a). (b) Points on the diagonal line and lying in the shaded area represent the set of positions for Exercise 7.6(b).

7.7 A certain computer system breaks down in two main ways: faults on the network and power supply faults. Of the last 50 breakdowns, 42 involved network faults and 20 power failures. In 13 cases both the power supply and the network were faulty. How many breakdowns were attributable to other kinds of failure?

7.8 Draw arrow diagrams and graphs of the following functions:

(a) $f(t)=(t-1)^2$ $t\in\{0,1,2,3,4\}$

(b) $g(z)=\dfrac{1}{z}$ $z\in\{-1,-0.5,0.5,1,1.5,2\}$

(c) $y=\begin{cases} x & x\in\{-2,-1\} \\ 2x & x\in\{0,1,2,3\} \end{cases}$

(d) $h:t\mapsto 3-t$ $t\in\{5,6,7,8,9,10\}$

7.9 Given that $f:x\mapsto 2x-1$, $g:x\mapsto\dfrac{1}{3}x^2$ $h:x\mapsto\dfrac{3}{x}$

(a) Find the following:

    (i) $f(2)$ (ii) $g(3)$ (iii) $h(5)$ (iv) $h(2)+g(2)$

    (v) $\dfrac{h}{g}(5)$ (vi) $h\times g(2)$ (vii) $h(g(2))$ (viii) $h(h(3))$

(b) Find the following functions:

    (i) $f\circ g$ (ii) $g\circ f$ (iii) $h\circ g$ (iv) f^{-1} (v) h^{-1}

(c) Confirm the following:

    (i) $(f^{-1}\circ f):x\mapsto x$ (ii) $(h^{-1}\circ h):x\mapsto x$ (iii) $(f\circ f^{-1}):x\mapsto x$

(d) Using the results from parts (b) and (c) find the following:

    (i) $(h^{-1}\circ h)(1)$ (ii) $h(g(5))$ (iii) $g(f(4))$

7.10 An analogue signal is sampled using an A/D convertor and represented using only integer values. The original signal is represented by $g(t)$ and the digital signal by $h(t)$ sampled at $t\in\{2,3,4,5,6,7,8,9,10\}$.
    The definitions of g and h are as below:

$$g(t)=\begin{cases} t-2.25 & t<5 \\ 6.8-t & t\geqslant 5 \end{cases}$$

$h:2\mapsto 0$ $h:3\mapsto 1$ $h:4\mapsto 2$ $h:5\mapsto 2$ $h:6\mapsto 1$ $h:7\mapsto 0$

$h:8\mapsto -1$ $h:9\mapsto -2$ $h:10\mapsto -3$

If $e(t)$ is the error function (called quantization error), defined at the sample points, find $e(t)$ and represent it on a graph.

EIGHT

FUNCTIONS AND THEIR GRAPHS

8.1 INTRODUCTION

The ability to produce a picture of a problem is an important step towards solving it. From the graph of a function, $y = f(x)$, we are able to predict such things as the number of solutions to the equation $f(x) = 0$, the regions over which it is increasing or decreasing, and the points where it is not defined.

Recognizing the shape of functions is an important and useful skill. Oscilloscopes give a graphical representation of voltage against time, from which we may be able to predict an expression for the voltage. The increasing use of signal processing means that many problems involve analysing how functions of time are effected by passing through some mechanical or electrical system.

In order to draw graphs of a large number of functions we need only remember a few key graphs and appreciate simple ideas about transformations. A sketch of a graph is one which is not necessarily drawn strictly to scale but shows its important features. We shall start by looking at special properties of the straight line (linear function) and the quadratic. Then we look at the graphs of $y = x$, $y = x^2$, $y = 1/x$ and $y = a^x$ and how to transform these graphs to get graphs of functions like $y = 4x - 2$, $y = (x-2)^2$, $y = 3/x$ and $y = a^{-x}$.

8.2 THE STRAIGHT LINE: $y = mx + c$

$y = mx + c$ is called a linear function because its graph is a straight line. Notice that there are only two terms in the function: the x term, mx, where m is called the coefficient of x, and c, which is the constant term. m and c have special significance. m is the gradient, or the slope, of the line and c is the value of y when $x = 0$, i.e. when the graph crosses the y-axis. This graph is shown in Fig. 8.1(a), with two particular examples in Figs 8.1(b) and 8.1(c).

The gradient of a straight line

The gradient gives an idea of how steep the climb is as we travel along the line of the graph. If the gradient is positive then we are travelling uphill as we move from left to right and if the

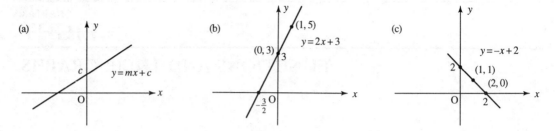

Figure 8.1 (a) The graph of the function $y = mx + c$. m is the slope of the line. If m is positive, then travelling from left to right along the line of the function is an uphill climb. If m is negative, then the journey is downhill. The constant c is where the graph crosses the y-axis. (b) $m = 2$ and $c = 3$. (c) $m = -1$ and $c = 2$.

gradient is negative then we are travelling downhill. The gradient gives the amount that y increases when x increases by 1 unit. A straight line always has the same slope at whatever point it is measured. To show that, in the expression $y = mx + c$, m is the gradient, we begin with a couple of examples, as in Figs 8.1(b) and (c).

In Fig. 8.1(b), we have the graph of $y = 2x + 3$. Take any two values of x which differ by 1 unit, e.g. $x = 0$ and $x = 1$. When $x = 0$, $y = 2 \times 0 + 3 = 3$, and when $x = 1$, $y = 2 \times 1 + 3 = 5$. The increase in y is $5 - 3 = 2$, and this is the same as the coefficient of x in the function expression.

In Fig. 8.1(c), we see the graph of $y = -x + 2$. Take any two values of x which differ by 1 unit, e.g. $x = 1$ and $x = 2$. When $x = 1$, $y = -(1) + 2 = 1$, and when $x = 2$, $y = -(2) + 2 = 0$. The increase in y is $0 - 1 = -1$ and this is the same as the coefficient of x in the function expression.

In general case, $y = mx + c$, take any two values of x which differ by 1 unit, e.g. $x = x_0$ and $x = x_0 + 1$. When $x = x_0$, $y = mx_0 + c$ and, when $x = x_0 + 1$, $y = m(x + 1) + c = mx + m + c$. The increase in y is $mx + m + c - (mx + c) = m$.

We know that every time x increases by 1 unit, y increases by m. However, we do not need always to consider an increase of exactly 1 unit in x. The gradient gives the ratio of the increase in y to the increase in x. Therefore, if we only have a graph and we need to find the gradient then we can use any two points that lie on the line.

To find the gradient of the line, take any two points on the line, (x_1, y_1) and (x_2, y_2):

$$\text{gradient} = \frac{\text{change in } y}{\text{change in } x} = \frac{y_2 - y_1}{x_2 - x_1}$$

Example 8.1 Find the gradient of the lines given in Figs 8.2(a), (b) and (c) and the equation for the line in each case.

SOLUTION

(a) We are given the co-ordinates of two points that lie on the straight line in Fig. 8.2(a) as $(0,3)$ and $(2,5)$.

$$\text{gradient} = \frac{\text{change in } y}{\text{change in } x} = \frac{5 - 3}{2 - 0} = \frac{2}{2} = 1$$

To find the constant term in the expression $y = mx + c$ we find the value of y when the line crosses the y-axis. From the graph this is 3, so the equation is $y = mx + c$, where $m = 1$ and $c = 3$, giving

$$y = x + 3$$

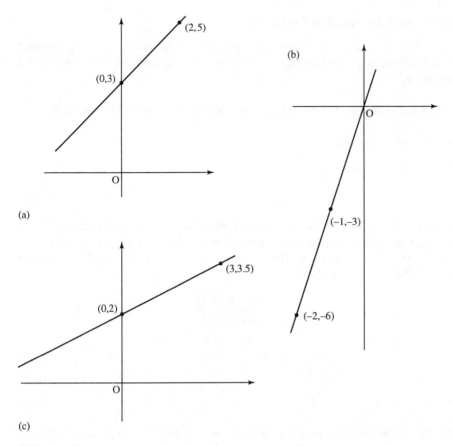

Figure 8.2 Graphs for Example 8.1.

(b) Two points that lie on the line in Fig. 8.2(b) are $(-1, -3)$ and $(-2, -6)$.

$$\text{gradient} = \frac{\text{change in } y}{\text{change in } x} = \frac{-6-(-3)}{-2-(-1)} = \frac{-3}{-1} = 3$$

To find the constant term in the expression $y = mx + c$ we find the value of y when the line crosses the y-axis. From the graph this is 0, so the equation is $y = mx + c$ where $m = 3$ and $c = 0$, giving

$$y = 3x$$

(c) Two points that lie on the line in Fig. 8.2(c) are $(0,2)$ and $(3,3.5)$.

$$\text{gradient} = \frac{\text{change in } y}{\text{change in } x} = \frac{2-3.5}{3-0} = \frac{-1.5}{3} = -0.5$$

To find the constant term in the expression $y = mx + c$ we find the value of y when the line crosses the y-axis. From the graph this is 2, so the equation is $y = mx + c$ where $m = -0.5$ and $c = 2$, giving

$$y = -0.5x + 2$$

Finding the gradient from the equation for the line

To find the gradient from the equation of the line we look for the value of m, the number multiplying x in the equation. The constant term gives the value of y when the graph crosses the y-axis, i.e. when $x=0$.

Example 8.2 Find the gradient and the value of y when $x=0$ for the following lines:

(a) $y=2x+3$ (b) $3x-4y=2$

(c) $x-2y=4$ (d) $\dfrac{x-1}{2}=1-\dfrac{y}{3}$

SOLUTIONS

(a) In the equation $y=2x+3$ the value of m, the gradient, is 2, as this is the coefficient of x. $c=3$, which is the value of y when the graph crosses the y-axis, i.e. when $x=0$.

(b) In the equation $3x-4y=2$ we rewrite the equation with y as the subject of the formula in order to find the value of m and c.

$$3x-4y=2 \Leftrightarrow 3x=2+4y$$

$$\Leftrightarrow 3x-2=4y$$

$$\Leftrightarrow \frac{3x}{4}-\frac{2}{4}=y$$

$$\Leftrightarrow y=\frac{3x}{4}-\frac{1}{2}$$

We can see, by comparing the expression with $y=mx+c$, that m, the gradient, is $\frac{3}{4}$ and $c=-\frac{1}{2}$.

(c) Write y as the subject of the formula:

$$x-2y=4 \Leftrightarrow x=4+2y$$

$$\Leftrightarrow x-4=2y$$

$$\Leftrightarrow 2y=x-4$$

$$\Leftrightarrow y=\frac{x}{2}-2$$

We can see, by comparing the expression with $y=mx+c$, that m, the gradient, is $\frac{1}{2}$ and $c=-2$.

(d) Write y as the subject of the formula:

$$\frac{x-1}{2}=1-\frac{y}{3} \Leftrightarrow \frac{x}{2}-\frac{1}{2}=1-\frac{y}{3}$$

$$\Leftrightarrow \frac{3x}{2}-\frac{3}{2}=3-y$$

$$\Leftrightarrow y=3-\left(\frac{3x}{2}-\frac{3}{2}\right)$$

$$\Leftrightarrow y=-\frac{3x}{2}+\frac{9}{2}$$

We can see, by comparing the expression with $y=mx+c$, that m, the gradient, is $-\frac{3}{2}$ and $c=\frac{9}{2}$.

Sketching a straight line graph

The quickest way to sketch a straight line graph is to find the points where it crosses the axes. However, any two points on the graph can be used, and sometimes it is more convenient to find other points.

Example 8.3
(a) Sketch the graph of $y=4x-2$.

To find where the graph crosses the y-axis substitute $x=0$ into the equation of the line: $y=4(0)+2=2$. This means that the graph passes through the point (0,2).

To find where the graph crosses the x-axis, substitute $y=0$, i.e.

$$4x-2=0$$

$$4x=2$$

$$x=\frac{2}{4}=0.5$$

Therefore the graph passes through (0.5,0).

Mark these points, (0,2) and (0.5,0), on the x- and y-axes and join the two points. This is done in Fig. 8.3(a).

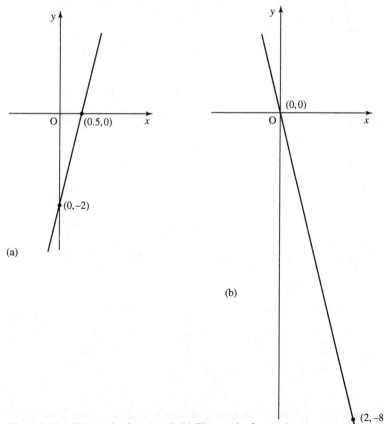

Figure 8.3 (a) The graph of $y=4x-2$. (b) The graph of $y=-4x$.

(b) Sketch the graph of $y = -4x$.

When $x = 0$ we get $y = 0$, i.e. the graph goes through the point (0,0). In this case, as the graph passes through the origin, we need to choose a different value for x for the second point. Taking $x = 2$ gives $y = -8$, so another point is $(2, -8)$. These points are marked on the graph and joined to give the graph as in Fig. 8.3(b).

8.3 THE QUADRATIC FUNCTION: $y = ax^2 + bx + c$

$y = ax^2 + bx + c$ is a general way of writing a function in which the highest power of x is a squared term. This is called the quadratic function and its graph is called a parabola, as shown in Fig. 8.4.

All the graphs in this figure cross the y-axis at $(0, c)$. To find where they cross the x-axis can be more difficult. These values, where $f(x) = 0$, are called the roots of the equation. There is a quick way to discover whether the function crosses the x-axis, only touches the x-axis, or does not cross or touch it. In the latter case there are no solutions to the equation $f(x) = 0$. The three possibilities are given in Fig. 8.4.

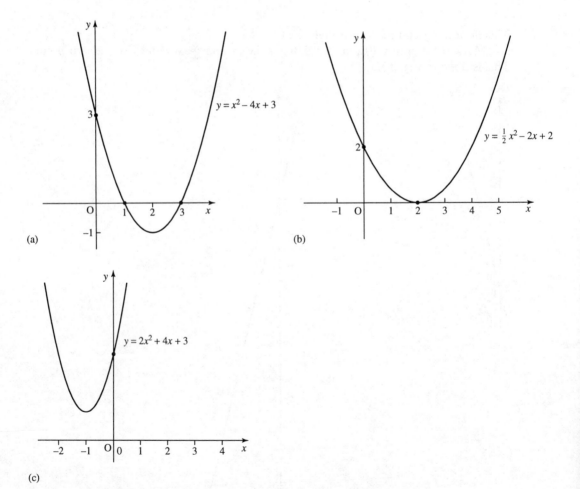

Figure 8.4 (a) The function $y = ax^2 + bx + c$. (a) Case 1, where there are two solutions to $f(x) = 0$. (b) Case 2, where there is only one solution to $f(x) = 0$. (c) Case 3, where there are no real solutions to $f(x) = 0$.

Crossing the *x*-axis

The function $y = ax^2 + bx + c$ crosses the *x*-axis when $y = 0$, i.e. when $ax^2 + bx + c = 0$.
  The solutions to $ax^2 + bx + c = 0$ were looked at in Chapter 3, and are given by the formula

$$x = \frac{-b \pm \sqrt{b^2 - 4ac}}{2a}$$

From the graph we can see that there are three possibilities.

1. In Fig. 8.4(a) where there are two solutions, i.e. the graph crosses the *x*-axis for two values of *x*. For this to happen the square root part of the formula above must be greater than zero:

$$b^2 - 4ac > 0$$

  Examples are given in Fig. 8.5.
2. Only one unique solution, as in Fig. 8.4(b). The graph touches the *x*-axis in one place only. For this to happen the square root part of the formula must be exactly 0. Examples of this are given in Fig. 8.6.
3. No real solutions, i.e. the graph does not cross the *x*-axis. Examples of these are given in Fig. 8.7.

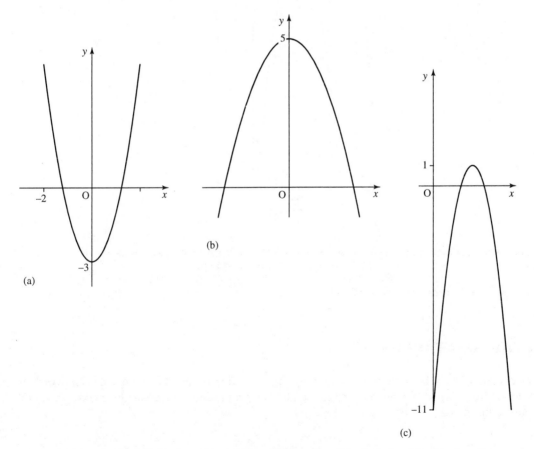

(b)

(a)

(c)

Figure 8.5 Three quadratic functions with two roots to the equation $f(x) = 0$. Each satisfies $b^2 - 4ac > 0$.
  (a) $y = 2x^2 - 3$, $a = 2$, $b = 0$, $c = -3$, $b^2 - 4ac = 0 - 4(2)(-3) = 24$.
  (b) $y = -x^2 + 5$, $a = -1$, $b = 0$, $c = 5$, $b^2 - 4ac = 0 - 4(-1)(5) = 20$.
  (c) $y = -3(x - 2)^2 + 1 \Leftrightarrow y = -3x^2 + 12x - 11$, $a = -3$, $b = 12$, $c = -11$, $b^2 - 4ac = (12)^2 - 4(-3)(-11) = 144 - 132 = 12$.

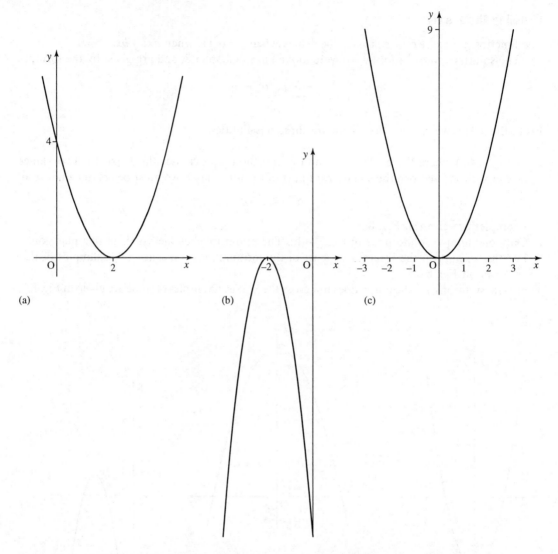

Figure 8.6 Quadratic functions with only one unique root of the equation $f(x)=0$. Each satisfies $b^2-4ac=0$.
(a) $y=x^2-4x+4$, $a=1$, $b=-4$, $c=4$, $b^2-4ac=(-4)^2-4(1)(4)=16-16=0$.
(b) $y=-3x^2-12x-12$, $a=-3$, $b=-12$, $c=-12$, $b^2-4ac=(-12)^2-4(-3)(-12)=144-144=0$.
(c) $y=x^2$, $a=1$, $b=0$, $c=0$, $b^2-4ac=(0)^2-4(1)(0)=0-0=0$.

8.4 THE FUNCTION $y=1/x$

The function $y=1/x$ has the graph shown in Fig. 8.8. This is called a hyperbola. Notice that the domain of $f(x)=1/x$ does not include $x=0$. The graph does not cross the x-axis, so there are no solutions to $1/x=0$.

8.5 THE FUNCTIONS $y=a^x$

Graphs of exponential functions, $y=a^x$, are shown in Fig. 8.9. The functions have the same

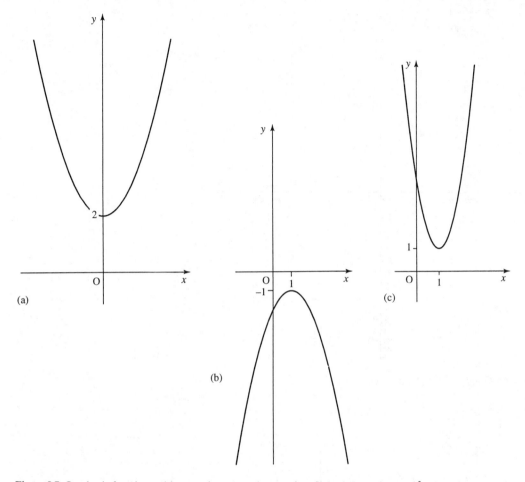

Figure 8.7 Quadratic functions with no real roots to the equation $f(x)=0$. In each case, $b^2-4ac<0$.
  (a) $y=x^2+2$, $a=1$, $b=0$, $c=2$, $b^2-4ac=(0)^2-4(1)(2)=0-8=-8$.
  (b) $y=-x^2+2x-2$, $a=-1$, $b=2$, $c=-2$, $b^2-4ac=(2)^2-4(-1)(-2)=4-8=-4$.
  (c) $y=3x^2-6x+4$, $a=3$, $b=-6$, $c=4$, $b^2-4ac=(-6)^2-4(3)(4)=36-48=-12$.

shape for all $a>1$. Notice that the function is always positive and the graph does not cross the x-axis, so there are no solutions to the equation $a^x=0$.

8.6 GRAPH SKETCHING USING SIMPLE TRANSFORMATIONS

One way of sketching graphs is to remember the graphs of simple functions and to translate, reflect or scale those graphs to get graphs of other functions. We begin with the graphs given in Fig. 8.10.

The translation $x\mapsto x+a$

If we have the graph of $y=f(x)$ then the graph of $y=f(x+a)$ is found by translating the graph of $y=f(x)$ a units to the left. Examples are given in Fig. 8.11.

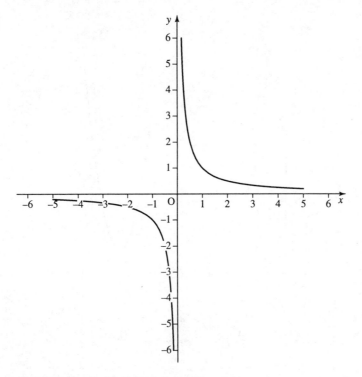

Figure 8.8 Graph of the hyperbolic function $y=1/x$.

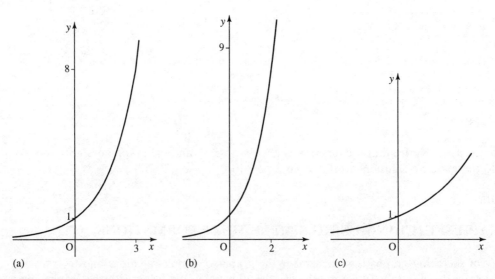

(a)

(b)

(c)

Figure 8.9 Graphs of functions $y=a^x$: (a) $y=2^x$, (b) $y=3^x$, (c) $y=(1.5)^x$.

The translation $f(x) \mapsto f(x) + A$

Adding A to the function value leads to a translation of A units upwards. Examples are given in Fig. 8.12.

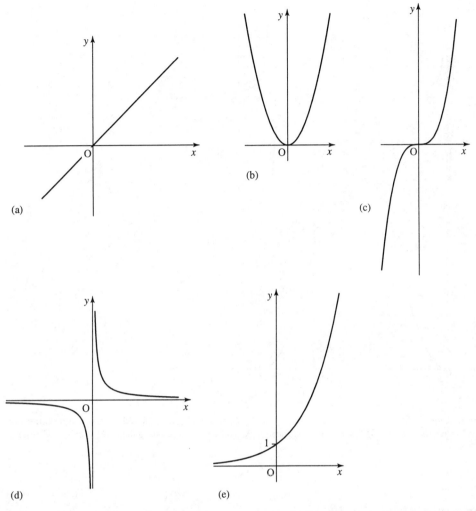

Figure 8.10 To sketch graphs using transformations we begin with known graphs. In the rest of this section we use (a) $y=x$, (b) $y=x^2$, (c) $y=x^3$, (d) $y=1/x$, (e) $y=a^x$.

Reflection about the y-axis: $x \mapsto -x$

Replacing x by $-x$ in the function has the effect of reflecting the graph in the y-axis – that is, as though a mirror has been placed along the axis and only the reflection can be seen. Examples are given in Fig. 8.13.

Reflection about the x-axis: $f(x) \mapsto -f(x)$

To find the graph of $y = -f(x)$, reflect the graph of $y = f(x)$ about the x-axis. Examples are given in Fig. 8.14.

Scaling along the x-axis: $x \mapsto ax$

Multiplying the values of x by a number, a, has the effect of squashing the graph horizontally if $a > 1$, or stretching the graph horizontally if $0 < a < 1$. Examples are given in Fig. 8.15.

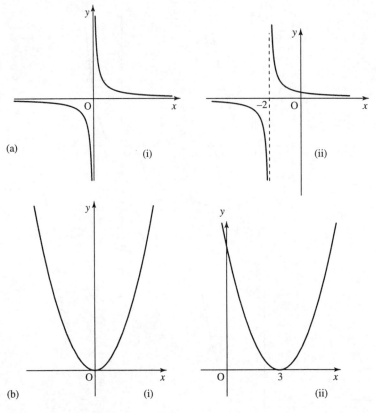

Figure 8.11 Translations $x \mapsto x + a$. (a) (i) $y = 1/x$, (ii) $y = 1/(x+2)$. Here x has been replaced by $x + 2$, translating the graph 2 units to the left. (b) (i) $y = x^2$, (ii) $y = (x-3)^2$. x has been replaced by $x - 3$, translating the graph 3 units to the right.

Scaling along the y-axis: $f(x) \mapsto Af(x)$

Multiplying the function value by a number A has the effect of stretching the graph vertically if $A > 1$, or squashing the graph vertically if $0 < A < 1$. Examples are given in Fig. 8.16.

Reflecting in the line $y = x$

If the graph of a function $y = f(x)$ is reflected in the line $y = x$ then it will give the graph of the inverse relation. Examples are given in Fig. 8.17.

In Chapter 7 we defined the inverse function as taking any image back to its original value. Check this in Fig. 8.17(a): $x = 1$ gives $y = 2$. In the inverse function, substitute 2, which gives the result of 1, which is back to the original value. However, the inverse of $y = x^2$, $y = \pm\sqrt{x}$, shown in Fig. 8.17(b), is not a function as there is more than one y value for a single value of x.

To understand this problem more fully, perform the following experiment. On a calculator enter -2 and square it (x^2), giving 4. Now take the square root. This gives the answer 2, which is not the number we first started with, and hence we can see that the square root is not a true inverse of squaring. However, we get away with calling it the inverse because it works if only positive values of x are considered. To test if the inverse of any function exists, draw a line along any value of $y = $ constant. If, wherever the line is drawn, there is ever more than one x value

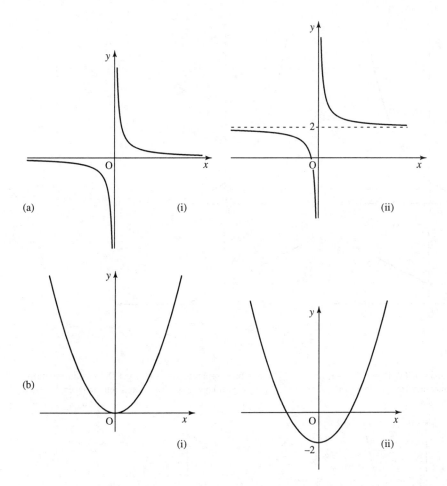

Figure 8.12 Translations $f(x) \mapsto f(x) + A$. (a) (i) $y = 1/x$, (ii) $y = 1/x + 2$. Here the function value has been increased by 2, translating the graph 2 units upwards. (b) (i) $y = x^2$, (ii) $x^2 - 2$. The function value has had 2 subtracted from it, translating the graph 2 units downwards.

which gives the same value of y then the function has no inverse function. In this situation the function is called a 'many-to-one' function. Only 'one-to-one' functions have inverses. Figure 8.18 has examples of functions with an explanation of whether they are 'one-to-one' or 'many-to-one'.

8.7 THE MODULUS FUNCTION: $y = |x|$ OR $y = \text{abs}(x)$

The modulus function, $y = |x|$, often written as $y = \text{abs}(x)$ (short for the absolute value of x) is defined by

$$y = \begin{cases} x & x \geqslant 0 \qquad (x \text{ positive or zero}) \\ -x & x < 0 \qquad (x \text{ negative}) \end{cases}$$

The output from the modulus function is always a positive number.

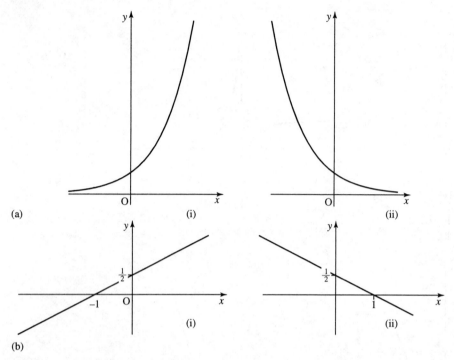

(a)

Figure 8.13 Reflections $x \mapsto -x$. (a) (i) $y=a^a$, $a>1$, (ii) $y=a^{-x}$. x has been replaced by $-x$ to get the second function. This has the effect of reflecting the graph in the y-axis. (b) (i) $y=\frac{x}{2}+\frac{1}{2}$, (ii) $y=-\frac{x}{2}+\frac{1}{2}$. x has been replaced by $-x$, reflecting the graph in the y-axis.

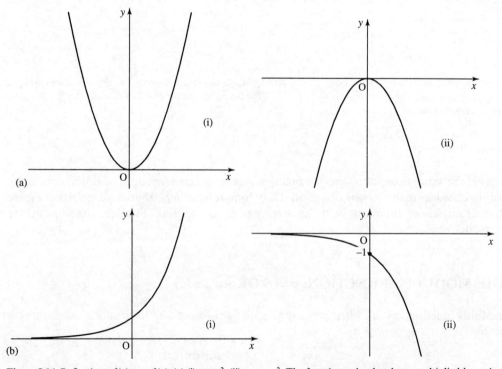

Figure 8.14 Reflections $f(x) \mapsto -f(x)$. (a) (i) $y=x^2$, (ii) $y=-x^2$. The function value has been multiplied by -1, turning the graph upside down (reflection in the x-axis). (b) (i) $y=2^x$, (ii) $y=-2^x$. The second function has been multiplied by -1, turning the graph upside down.

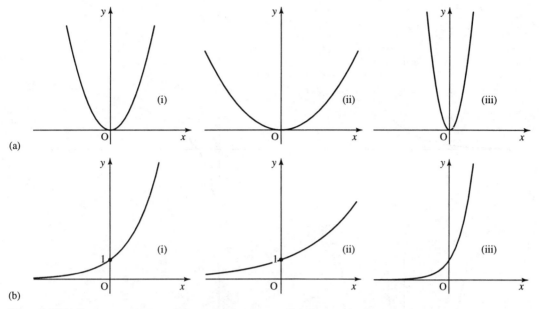

(a)

(b)

Figure 8.15 Scalings $x \mapsto ax$. () (i) $y = x^2$, (ii) $y = (\frac{1}{2}x)^2$, (iii) $y = (2x)^2$. The second function has replaced x by $\frac{1}{2}x$, which has stretched the graph horizontally (the multiplication factor is between 0 and 1). The third function has replaced x by $2x$, which has squashed the graph horizontally (the multiplication factor is greater than 1).

(b) (i) $y = 2^x$, (ii) $y = 2^{(1/2x)}$, (iii) $y = 2^x$. The second function has replaced x by $\frac{1}{2}x$, which has stretched the graph horizontally. The third function has replaced x by $2x$, which has squashed the graph horizontally.

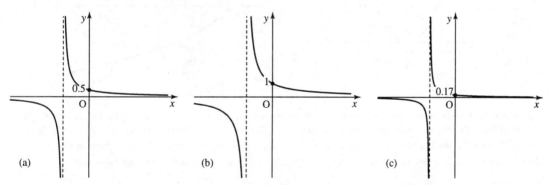

(a)

(b)

(c)

Figure 8.16 Scalings $f(x) \mapsto Af(x)$. (a) $y = 1/(x + 2)$, (b) $y = 2/(x + 2)$, (c) $y = 1/(3(x + 2))$. The second graph has the function values multiplied by 2, stretching the graph vertically. The third graph has function values multiplied by $\frac{1}{3}$, squashing the graph vertically.

Example 8.4 Find $|-3|$.

Here $x = -3$, which is negative. Therefore

$$y = -x = -(-3) = +3$$

An alternative way of thinking of it is to remember that the modulus is always positive, so simply replacing any negative sign by a positive one will give a number's modulus or absolute value.

$$|-5| = 5 \qquad |-4| = 4$$
$$|5| = 5 \qquad |4| = 4$$

The graph of the modulus function can be found from the graph of $y = x$ by reflecting the negative x part of the graph to make the function values positive. This is shown in Fig. 8.19.

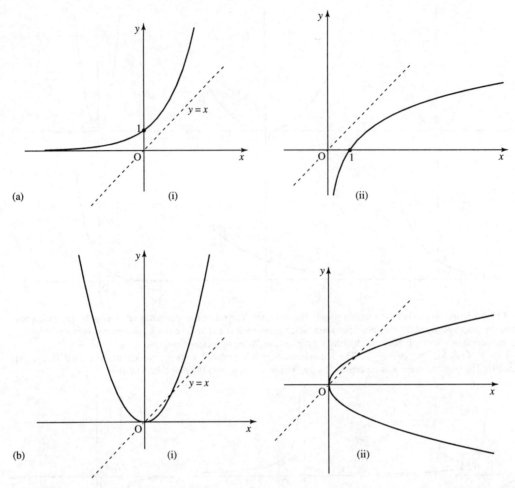

Figure 8.17 Reflections in the line $y=x$ produce the inverse relation. (a) (i) $y=2^x$, (ii) $y=\log_2(x)$. The second graph is obtained from the first by reflecting in the dotted line $y=x$. The inverse is a function, as there is only one value of y for each value of x. (b) (i) $y=x^2$, (ii) $y=\pm\sqrt{x}$. The second graph is found by reflecting the first graph in the line $y=x$. Notice that $y=\pm\sqrt{x}$ is not a function, as there is more than one possible value of y for each value of $x>0$.

8.8 SYMMETRY OF FUNCTIONS AND THEIR GRAPHS

Functions can be classified as even, odd or neither of these.

Even functions

Even functions are those that can be reflected in the y-axis and then result in the same graph. Examples of even functions are (see Fig. 8.20):

$$y=x^2, \qquad y=|x| \qquad \text{and} \qquad y=x^4$$

As previously discussed, reflecting in the y-axis results from replacing x by $-x$ in the function expression, and hence the condition for a function to be even is that substituting $-x$ for x does not change the function expression, i.e. $f(x)=f(-x)$.

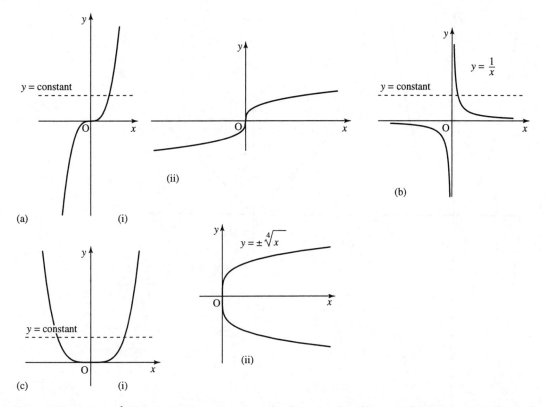

Figure 8.18 (a) (i) $y=x^3$. This function has only one x value for each value of y, as any line $y=$ constant only cuts the graph once. In this case, the function is one-to-one and it has an inverse function. (ii) $y=\sqrt[3]{x}$ is the inverse function of $y=x^3$. (b) $y=1/x$, $x\neq0$, has only one x value for each value of y, as any line $y=$ constant only cuts the graph once. It therefore is one-to-one and has an inverse function – in fact, it is its own inverse! (to see this, reflect it in the line $y=x$ and we get the same graph after the reflection). (c) (i) $y=x^4$. This function has two values of x for each value of y when y is positive (for instance, the line $y=16$ cuts the graph twice at $x=2$ and at $x=-2$). This shows that there is no inverse function, as the function is many-to-one. (ii) The inverse relation $y=\pm\sqrt[4]{x}$.

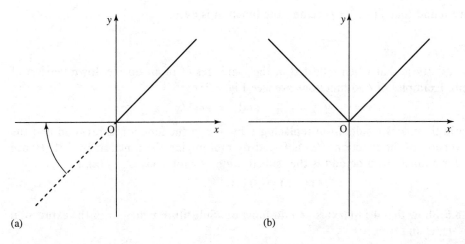

Figure 8.19 The graph of the modulus function $y=|x|$ obtained from the graph of $y=x$. (a) The graph of $y=x$ with the negative part of the graph displayed as a dotted line. This is reflected about the x-axis to give $y=-x$ for $x<0$. (b) The graph of $y=|x|$.

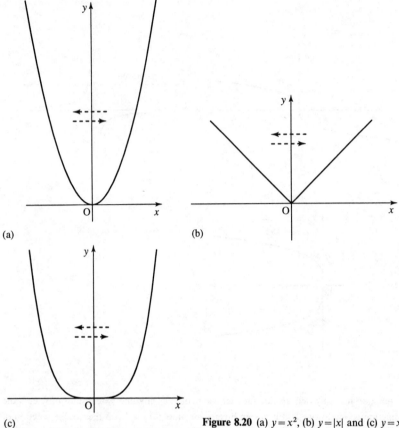

(a)

(b)

(c)

Figure 8.20 (a) $y=x^2$, (b) $y=|x|$ and (c) $y=x^4$ are even functions.

Example 8.5 Show that $3x^2-x^4$ is an even function.
Substitute $-x$ for x in the expression $f(x)=3x^2-x^4$ and we get:

$$f(-x)=3(-x)^2-(-x)^4=3(-1)^2(x)^2-(-1)^4(x)^4=3x^2-x^4$$

So we have found that $f(-x)=f(x)$, and the function is even.

Odd functions

Odd functions are those that when reflected in the y-axis result in an upside down version of the same graph. Examples of odd functions are (see Fig. 8.21):

$$y=x, \qquad y=x^3 \qquad \text{and} \qquad y=1/x$$

Reflecting in the y-axis results from replacing x by $-x$ in the function expression and the upside down version of the function $f(x)$ is found by multiplying the function by -1. Hence the condition for a function to be odd is that substituting $-x$ for x gives $-f(x)$, i.e.

$$f(-x)=-f(x)$$

Example 8.6 Show that $4x-(1/x)$ is an odd function. Substitute x for $-x$ in the expression $f(x)=4x-(1/x)$ and we get:

$$f(-x)=4(-x)-(1/-x)=-4x+(1/x)=-(4x-(1/x))$$

We have found that $f(-x)=-f(x)$, so the function is odd.

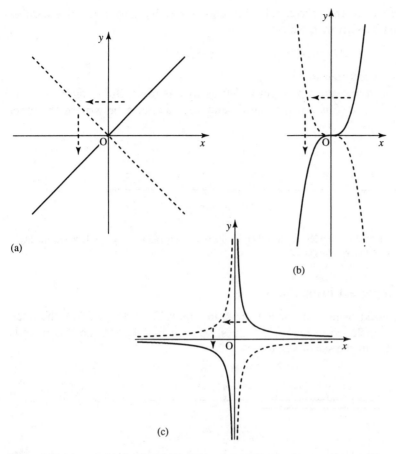

(a)

(b)

(c)

Figure 8.21 (a) $y=x$, (b) $y=x^3$ and (c) $y=1/x$ are odd functions.

8.9 SOLVING INEQUALITIES

For linear and quadratic functions, $y=f(x)$, we have discussed how to find the values where the graph of the functions crosses the x-axis, that is how to solve the equation $f(x)=0$. It is often of interest to find ranges of values of x where $f(x)$ is negative or where $f(x)$ is positive. This means solving inequalities like $f(x)<0$ or $f(x)>0$.

Like equations, inequalities can be solved by looking for equivalent inequalities. One way of finding these is by doing the same thing to both sides of the expression. There is an important exception for inequalities, which is that if both sides are multiplied or divided by a negative number then the direction of the inequality must be reversed.

To demonstrate these equivalences begin with a true proposition: $3<5$, or '3 is less than 5'. Add 2 to both sides and it is still true:

$$3+2<5+2 \quad \text{i.e. } 5<7$$

Subtract 10 from both sides and we get

$$5-10<7-10 \quad \text{i.e. } -5<-3$$

which is also true. Multiply both sides by -1. If we do not reverse the inequality we get

$$(-1)(-5)<(-1)(-3) \quad \text{i.e. } 5<3$$

which is false. However, if we use the correct rule that when multiplying by a negative number we must reverse the inequality sign then we get

$$(-1)(-5) > (-1)(-3) \qquad \text{i.e. } 5 > 3$$

which is true. This process is pictured in Fig. 8.22.

Note that inequalities can be read from right to left as well as from left to right: $3 < 5$ can be read as '3 is less than 5' or as '5 is greater than 3' and so it can also be written the other way round: $5 > 3$.

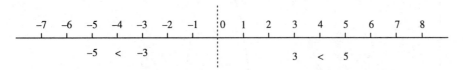

Figure 8.22 On the number line, numbers to the left are less than numbers to their right: $-5 < -3$. If the inequality is multiplied by -1 we need to reverse the sign to get $5 > 3$.

Using a number line to represent inequalities

An inequality can be expressed using a number line, as in Fig. 8.23. In Fig. 8.23(a), the open circle indicates that 3 is not included in the set of values, $t < 3$. In Fig. 8.23(b), the closed circle indicates that -2 is included in the set of values, $x \geqslant -2$.

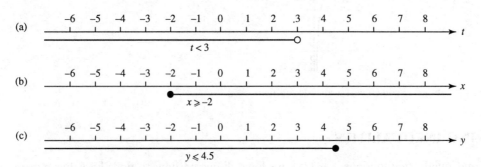

Figure 8.23 Representing inequalities on a number line.

Example 8.7 Find a range of values for t, x and y such that the following inequalities hold:

(a) $2t + 3 < t - 6$
(b) $x + 5 \geqslant 4x - 10$
(c) $16 - y > -5y$

SOLUTION
(a)
$$2t + 3 < t - 6$$

$\Leftrightarrow 2t - t + 3 < -6 \qquad$ (subtract t from both sides)

$\Leftrightarrow t < -6 - 3 \qquad$ (subtract 3 from both sides)

$\Leftrightarrow t < -9$

This solution can be represented on a number line as in Fig. 8.24.

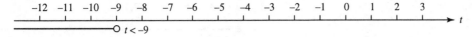

Figure 8.24 The solution to $2t+3 < t-6$ is given by $t < -9$.

(b) $$x+5 \geqslant 4x-10$$

$$\Leftrightarrow +5 \geqslant 3x-10 \qquad \text{(subtract } x \text{ from both sides)}$$

$$\Leftrightarrow 15 \geqslant 3x \qquad \text{(add 10 to both sides)}$$

$$\Leftrightarrow 5 \geqslant x \qquad \text{(divide both sides by 3)}$$

$$\Leftrightarrow x \leqslant 5$$

This solution is represented in Fig. 8.25.

Figure 8.25 The solution to $x+5 \geqslant 4x-10$ is found to be $x \leqslant 5$, here represented on a number line.

(c) $$16-y > -5y$$

$$\Leftrightarrow 16 > -4y \qquad \text{(add } y \text{ to both sides)}$$

$$\Leftrightarrow -4 < y \qquad \text{(divide by } -4 \text{ and reverse the sign)}$$

$$\Leftrightarrow y > -4$$

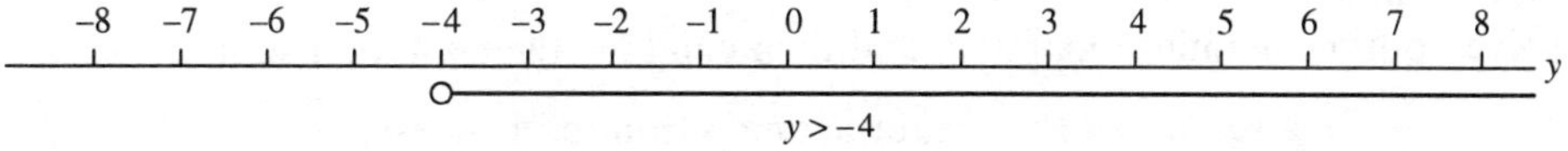

Figure 8.26 The solution to $16-y > -5y$ is found to be $t > -4$, here pictured on a number line.

Representing compound inequalities on a number line

We sometimes need a picture of the range of values given if two inequalities hold simultaneously, for instance $x \geqslant 3$ and $x < 5$. This is analysed in Fig. 8.27(a), and we can see that for both inequalities to hold simultaneously x must lie in the overlapping region where $3 \leqslant x < 5$. $3 \leqslant x < 5$ is a way of expressing that x lies between 3 and 5 or is equal to 3. In the example in Fig. 8.27(b), $x > 6$ and $x > 2$, and for them both to hold then $x > 6$.

Another possible way of combining inequalities is to say that one or another inequality holds. Examples of this are given in Fig. 8.27(c), where $x < 5$ or $x \geqslant 7$ and this gives the set of values less than 5 or greater than or equal to 7. Figure 8.27(d) gives the example where $x < 2$ or $x \geqslant 0$ and in this case it results in all numbers lying on the number line, i.e. $x \in \mathbb{R}$.

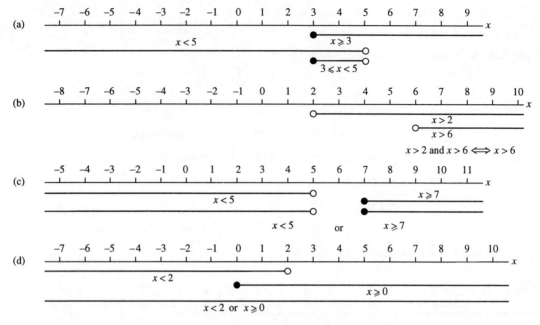

Figure 8.27 (a) $x \geqslant 3$ and $x < 5$. (b) $x > 6$ and $x > 2$ combines to give $x > 6$. (c) $x < 5$ or $x \geqslant 7$. (d) $x < 2$ or $x \geqslant 0$.

Example 8.8 Find solutions to the following combinations of inequalities and represent them on a number line:

(a) $x + 3 > 4$ and $x - 1 < 5$
(b) $1 - u < 3u + 2$ or $u + 2 \geqslant 6$
(c) $t + 5 > 12$ and $-t > 24$

SOLUTION
(a) $x + 3 > 4$ and $x - 1 < 5$
    We solve both inequalities separately and then combine their solution sets:

$$x + 3 > 4 \Leftrightarrow x > 1 \quad \text{(subtracting 3 from both sides)}$$

$$x - 1 < 5 \Leftrightarrow x < 6 \quad \text{(adding 1 to both sides)}$$

So the combined inequality giving the solution is $x > 1$ and $x < 6$, which from Fig. 8.28(a) we can see is the same as $1 < x < 6$.

(b) $1 - u < 3u + 2$ or $u + 2 \geqslant 6$
    We solve both inequalities separately and then combing their solution sets:

$$1 - u < 3u + 2 \Leftrightarrow -1 < 4u \quad \text{(subtracting 2 from both sides and adding } u\text{)}$$

$$\Leftrightarrow (-1)/4 < u \quad \text{(dividing both sides by 4)}$$

$$\Leftrightarrow u > (-1)/4$$

$$u + 2 \geqslant 6 \Leftrightarrow u \geqslant 4 \quad \text{(subtracting 2 from both sides)}$$

Combining the two solutions gives $u > (-1)/4$ or $u \geqslant 4$, and this is represented on the number line in Fig. 8.28(b), where we can see that it is the same as $u > (-1)/4$.

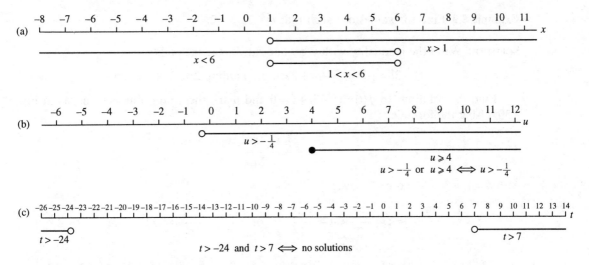

Figure 8.28 Solutions to compound inequalities as given in Example 8.8 represented on a number line.

(c) $t + 5 > 12$ and $-t > 24$

$t + 5 > 12 \Leftrightarrow t > 7$ (subtracting 5 from both sides)

$-t > 24 \Leftrightarrow t < -24$ (multiply both sides by -1 and reverse the inequality sign)

Combining the two solution sets gives $t > 7$ and $t < -24$, and we can see from Fig. 8.28(c) that this is impossible and hence there are no solutions.

Solving more difficult inequalities

To solve more difficult inequalities our ideas about equivalence are not enough on their own: we must also use our knowledge about continuous functions. In the last chapter we defined a continuous function as one which could be drawn without taking the pen off the paper. If we wish to solve the inequality $f(x) > 0$ and we know that $f(x)$ is continuous, then we can picture the problem graphically, as in Fig. 8.29. From the graph we can see that to solve the inequality we need only find the values where $f(x) = 0$ (the roots of $f(x) = 0$) and determine whether $f(x)$ is positive or negative between the values of x where $f(x) = 0$. To do this we can use any value of x between the roots. We are using the fact that as $f(x)$ is continuous it can only change from positive to negative by going through zero.

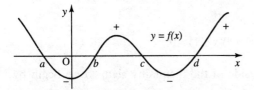

Figure 8.29 The graph of a continuous function. To solve for $f(x) > 0$ then we first find the values where $f(x) = 0$. On the graph these are marked as a, b, c and d. If the function is above the x axis then the function values are positive, if the function lies below the x axis then the function values are negative. The solution to $f(x) > 0$ is given by those values of x for which the function lies above the x axis, i.e. y positive. For the function represented in the graph the solution to $f(x) > 0$ is $x < a$ or $b < x < c$ or $x > d$.

Example 8.9 Find the values of t such that $t^2 - 3t < -2$.

SOLUTION Write the inequality with 0 on one side of the inequality sign:

$$t^2 - 3t < -2 \Leftrightarrow t^2 - 3t + 2 < 0 \qquad \text{(adding 2 to both sides)}$$

Find the solutions to $f(t) = t^2 - 3t + 2 = 0$ and mark them on a number line, as in Fig. 8.30. Using the formula

$$t = \frac{-b \pm \sqrt{b^2 - 4ac}}{2a}$$

with $a = 1$, $b = -3$ and $c = 2$, gives

$$t = \frac{3 \pm \sqrt{9 - 8}}{2} \Leftrightarrow t = \frac{3 \pm 1}{2} \Leftrightarrow t = \frac{3 + 1}{2} \quad \text{or} \quad t = \frac{3 - 1}{2} \Leftrightarrow t = 1 \quad \text{or} \quad t = 2$$

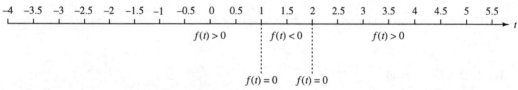

Figure 8.30 Solving $t^2 - 3t + 2 > 0$ (Example 8.9).

Substitute values for t which lie on either side of the roots of $f(t) = 0$ in order to find the sign of the function between the roots. Here we choose 0, 1.5 and 3.
When $t = 0$

$$t^2 - 3t + 2 = (0)^2 - 3(0) + 2 = 0 + 0 + 2 = 2$$

which is positive, giving $f(t) > 0$.
When $t = 1.5$

$$t^2 - 3t + 2 = (1.5)^2 - 3(1.5) + 2 = 2.25 - 4.5 + 2 = -0.25$$

which is negative, giving $f(t) < 0$.
When $t = 3$

$$t^2 - 3t + 2 = (3)^2 - 3(3) + 2 = 9 - 9 + 2 = 2$$

which is positive, giving $f(t) > 0$.
By marking the regions on the number line, given in Fig. 8.30, with $f(t) > 0$, $f(t) < 0$ or $f(t) = 0$ as appropriate we can now find the solution to our inequality $f(t) < 0$, which is given by the region where $1 < t < 2$.

Example 8.10 Find the values of x such that $(x^2 - 4)(x + 1) > 0$.

SOLUTION The inequality already has 0 on one side of the inequality sign, so we begin by finding the roots to $f(x) = 0$, i.e.

$$(x^2 - 4)(x + 1) = 0$$

Factorizing gives

$$(x^2 - 4)(x + 1) = 0 \Leftrightarrow (x - 2)(x + 2)(x + 1) = 0$$

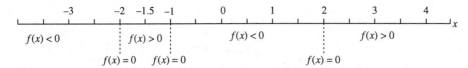

Figure 8.31 Solving $(x^2-4)(x+1)>0$ (Example 8.10).

i.e. $x=2$ or $x=-2$ or $x=-1$. So the roots are -2, -1 and 2. These roots are pictured on the number line in Fig. 8.31.

Substitute values for x which lie on either side of the roots of $f(x)$ in order to find the sign of the function between the roots. Here we choose -3, -1.5, 0 and 2.5.

When $x=-3$, $(x-2)(x+2)(x+1)=0$ gives $(-3-2)(-3+2)(-3+1)=(-5)(-1)(-2)=-10$, giving $f(x)<0$.

When $x=-1.5$, $(x-2)(x+2)(x+1)$ gives $(-1.5-2)(-1.5+2)(-1.5+1)=(-3.5)(0.5)(-0.5)=0.875$, giving $f(x)>0$.

When $x=0$, $(x-2)(x+2)(x+1)$ gives $(0-2)(0+2)(0+1)=(-2)(2)(1)=-4$, giving $f(x)<0$.

When $x=2.5$, $(x-2)(x+2)(x+1)$ gives $(2.5-2)(2.5+2)(2.5+1)=(0.5)(4.5)(3.5)=7.875$, giving $f(x)>0$.

These regions are marked on the number line as in Fig. 8.31, and the solution is given by those regions where $f(x)>0$. Looking for the regions where $f(x)>0$ gives the solution as $-2<x<-1$ or $x>2$.

8.10 USING GRAPHS TO FIND AN EXPRESSION FOR THE FUNCTION FROM EXPERIMENTAL DATA

Linear relationships

Linear relationships are the easiest ones to determine from experimental data. The points are plotted on a graph and if they appear to follow a straight line then a line can be drawn by hand and the equation can be found using the method given in Section 8.2.

Example 8.11 A spring is stretched by hanging various weights on it, and in each case the length of the spring is measured.

Mass (kg)	0.125	0.25	0.5	1	2	3
Length (m)	0.4	0.41	0.435	0.5	0.62	0.74

Approximate the length of the spring when no weight is hung from it and find the expression for the length in terms of the mass.

SOLUTION First, draw a graph of the given experimental data. This is done in Fig. 8.32. A line is fitted by eye to the data. The data does not lie exactly on a line, due to experimental error and to a slight distortion of the spring with heavier weights. From the line we have drawn we can find the gradient by choosing any two points on the line and calculating (change in y/change in x).

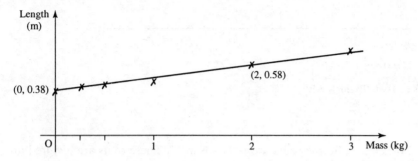

Figure 8.32 The data for length of spring against mass of the weight as given in the Example 8.11. The line is fitted by eye to the experimental data and the equation of the line can be found using the method of Section 8.2.

Taking the two points as (0,0.28) and (2,0.58) we get the gradient as

$$\frac{0.58-0.38}{2-0} = \frac{0.2}{2} = 0.1$$

The point where it crosses the y-axis, i.e. where the mass hung on the spring is 0 kg, can be found by extending the line until it crosses the y-axis. This gives 0.38 m.

Finally, the expression for the length in terms of the mass of the attached weight is given by $y=mx+c$, where y is the length, x is the mass, m is the gradient and c is the value of y when $x=0$, i.e. where there is no weight on the string. This gives

$$\text{length} = 0.1 \times \text{mass} + 0.38$$

The initial length of the spring is 0.38 m.

Exponential relationships

Many practical relationships behave exponentially, particularly those involving growth or decay. Here it is slightly less easy to find the relationship from the experimental data. However, it is simplified by using a log–linear plot. Instead of plotting the values of the dependent variable, y, we plot the values of $\log_{10}(y)$. If the relationship between y and time, t, is exponential, as we suspected, then the $\log_{10}(y)$ against t plot will be a straight line.

The reason this works can be explained as follows. Supposing $y=y_0 10^{kt}$ where y_0 is the value of y when $t=0$ and k is some constant; then, taking the log base 10 of both sides, we get

$$\log_{10}(y) = \log_{10}(y_0 10^{kt})$$
$$= \log_{10}(y_0) + \log_{10}(10^{kt})$$

As the logarithm base 10 and raising to the power of 10 are inverse operations, we get

$$\log_{10}(y) = \log_{10}(y_0) + kt$$

As y_0, the initial value of y, is a constant, and k is a constant, we can see that this expression shows that we will get a straight line if $\log_{10}(y)$ is plotted against t. The constant k is given by the gradient of the line, and $\log_{10}(y_0)$ is the value of $\log_{10}(y)$ where it crosses the vertical axis. Setting $Y=\log_{10}(y)$ and $c=\log_{10}(y_0)$, we get $Y=c+kt$, which is the equation of the straight line.

Example 8.12 A room was tested for its acoustical absorption properties by playing a single note on a trombone. Once the sound had reached its maximum intensity the player stopped and the sound intensity was measured for the next 0.2 s at regular intervals of 0.02 s. 1.0 is the initial maximum intensity at time 0. The readings were as follows:

Time	0	0.02	0.04	0.06	0.08	0.1	0.12	0.14	0.16	0.18	0.2
Intensity	1.0	0.63	0.35	0.22	0.13	0.08	0.05	0.03	0.02	0.01	0.005

Draw a graph of intensity against time and log(intensity) against time and use the latter plot to approximate the relationship between the intensity and time.

SOLUTION The graphs are plotted in Fig. 8.33, where, for the second graph (b), we take the $\log_{10}$(intensity) and use the table below:

Time	0	0.02	0.04	0.06	0.08	0.1	0.12	0.14	0.16	0.18	0.2
$\log_{10}$(intensity)	0	−0.22	−0.46	−0.66	−0.89	−1.1	−1.3	−1.5	−1.7	−2	−2.3

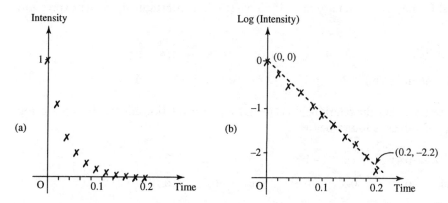

Figure 8.33 (a) Graph of sound intensity against time as given in Example 8.12. (b) Graph of $\log_{10}$(intensity) against time and a line fitted by eye to the data. The line goes through the points (0,0) and (0.2, −2.2).

We can see that the second graph is approximately a straight line and therefore we can assume that the relationship between the intensity and time is exponential and could be expressed as $I = I_0 10^{kt}$. The $\log_{10}$ of this gives

$$\log_{10}(I) = \log_{10}(I_0) + kt$$

From the graph in Fig. 8.33(b) we can measure the gradient, k. To do this we calculate (change in $\log_{10}$(intensity)/change in time), giving

$$\frac{-2.2 - 0}{0.2 - 0} = -11 = k$$

The point at which the line crosses the vertical axis gives

$$\log_{10}(I_0) = 0 \Leftrightarrow I_0 = 10^0 = 1$$

Therefore the expression $I = I_0 10^{kt}$ becomes

$$I = 10^{-11t}$$

Power relationships

Another common type of relationship between quantities is when there is a power of the independent variable involved. In this case, if $y = ax^n$, where n could be positive or negative, then the values of a and n can be found by drawing a log–log plot.

This is because taking $\log_{10}$ of both sides of $y = ax^n$ gives

$$\log_{10}(y) = \log_{10}(ax^n) = \log_{10}(a) + n \log_{10}(x)$$

Substituting $Y = \log_{10}(y)$ and $X = \log_{10}(x)$ we get $Y = \log_{10}(a) + nX$, showing that the log–log plot will give a straight line, where the slope of the line will give the power of x and the position where the line crosses the vertical axis will give $\log_{10}(a)$. Having found a and n they can be substituted back into the expression $y = ax^n$.

> **Example 8.13** The power received from a beacon antenna is thought to depend on the inverse square of the distance from the antenna to the receiver. Various measurements, given below, were taken of the power received against distance r from the antenna. Could these be used to justify the inverse square law? If so what is the constant, A, in the expression $p = A/r^2$?
>
> | Power received (W) | 0.39 | 0.1 | 0.05 | 0.025 | 0.015 | 0.01 |
> |---|---|---|---|---|---|---|
> | Distance from antenna (km) | 1 | 2 | 3 | 4 | 5 | 6 |
>
> SOLUTION To test whether the relationship is indeed a power relationship we draw a log–log plot. The table of values is found below:
>
> | $\log_{10}$(power) | −0.41 | −1 | −1.3 | −1.6 | −1.8 | −2 |
> |---|---|---|---|---|---|---|
> | $\log_{10}$(distance) | 0 | 0.3 | 0.5 | 0.6 | 0.7 | 0.78 |
>
> Graphs of power against distance and $\log_{10}$(power) against $\log_{10}$(distance) are given in Fig. 8.34.

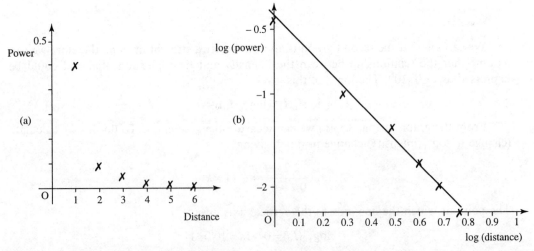

Figure 8.34 (a) Plot of power against distance (Example 8.13). (b) log(power) against log(distance). In (b), the line is fitted by eye to the data, from which the slope of the graph indicates n in the relationship $P = Ar^n$. Two points lying on the line are $(0, -0.38)$ and $(0.7, -1.8)$.

As the second graph is a straight line we can assume that the relationship is of the form $P = Ar^n$ where P is the power and r is the distance, in which case the log–log graph is

$$\log_{10}(P) = \log_{10}(A) + n \log_{10}(r)$$

We can measure the slope by calculating (change in $\log_{10}(p)$/change in $\log_{10}(r)$), and using the two points that have been found to lie on the line this gives

$$\frac{-1.8 - (-0.38)}{0.7 - 0} = -2.03$$

As this is very near to -2 then the inverse square law would appear to be justified.

The value of $\log_{10}(A)$ is given from where the graph crosses the vertical axis, and this gives

$$\log_{10}(A) = -0.38 \Leftrightarrow A = 10^{-0.38} \Leftrightarrow A = 0.42$$

So the relationship between power received and distance is approximately

$$P = 0.42 r^{-2} = \frac{0.42}{r^2}$$

8.11 SUMMARY

1. The linear function $y = mx + c$ has gradient (slope) m and crosses the y-axis at $y = c$.
2. The gradient of a straight line $y = mx + c$ is given by (change in y/change in x) and this is the same along the length of the line.
3. The graph of the quadratic function $y = ax^2 + bx + c$ is called a parabola. The graph crosses the y-axis (when $x = 0$) at $y = c$.
4. There are three possibilities for the roots of the quadratic equation $ax^2 + bx + c = 0$:

   Case I: Two real roots when $b^2 - 4ac > 0$
   Case II: Only one unique root when $b^2 - 4ac = 0$
   Case III: No real roots when $b^2 - 4ac < 0$

5. By considering the graphs of known functions $y = f(x)$, for instance those given in Fig. 8.10, and the following transformations, many other graphs can be drawn.

   (a) Replacing x by $x + a$ in the function $y = f(x)$ results in shifting the graph a units to the left.
   (b) Replacing $f(x)$ by $f(x) + A$ results in shifting the graph A units upwards.
   (c) Replacing x by $-x$ reflects the graph in the y-axis.
   (d) Replacing $f(x)$ by $-f(x)$ reflects the graph in the x-axis (turning it upside down).
   (e) Replacing x by ax squashes the graph horizontally if $a > 1$ or stretches it horizontally if $0 < a < 1$.
   (f) Replacing $f(x)$ by $Af(x)$ stretches the graph vertically if $A > 1$ or squashes it vertically if $0 < A < 1$.
   (g) Reflecting the graph of $y = f(x)$ in the line $y = x$ results in the graph of the inverse relation.

6. A function may be even or odd or neither of these.

   (a) An even function is one whose graph remains the same if reflected in the y-axis, i.e. when $x \mapsto -x$. This can also be expressed as $f(-x) = f(x)$. Examples of even functions are $y = x^2$, $y = |x|$ and $y = x^4$.

(b) An odd function is one which when reflected in the y-axis, i.e. when $x \mapsto -x$, gives an upside down version of the original graph (i.e. $-f(x)$). This can also be expressed as $f(-x) = -f(x)$. Examples of odd functions are $y = x$ and $y = x^3$.

7. Not all functions have true inverses. Only one-to-one functions have inverse functions. A function is one-to-one if any line $y = \text{constant}$ drawn on the graph $y = f(x)$ crosses the function only once. This means there is exactly one value of x which gives each value of y.

8. Simple inequalities can be solved by finding equivalent inequalities. Inequalities remain equivalent if both sides of the inequality have the same expression added or subtracted. They may also be multiplied or divided by a positive number but if they are multiplied or divided by a negative number then the direction of the inequality sign must be reversed.

9. To solve the inequalities $f(x) > 0$, $f(x) < 0$, $f(x) \leqslant 0$ or $f(x) \geqslant 0$, where $f(x)$ is a continuous function, solve $f(x) = 0$ and choose any value for x around the roots to find the sign of $f(x)$ for each region of values for x.

10. Graphs can be used to find relationships in experimental data. First, plot the data; then:

(a) If the data lies on an approximate straight line then draw a straight line through the data and find the equation of the line.

(b) If it looks exponential, then take the log of the values of the dependent variable and draw a log–linear graph. If this looks approximately like a straight line then assume there is an exponential relationship $y = y_0 10^{kt}$, where k is given by the gradient of the line and $\log_{10}(y_0)$ is the value where the graph crosses the vertical axis.

(c) If the relationship looks something like a power relationship, $y = Ax^n$, then take the log of both sets of data and draw a log–log graph. If this is approximately like a straight line then assume there is a power relationship and n is given by the gradient of the line and $\log_{10} A$ is the value where the graph crosses the vertical axis.

8.12 EXERCISES

8.1 Sketch the graphs of the following:

(a) $y = 3x - 1$ (b) $y = 2x + 1$
(c) $y = -5x$ (d) $y = \frac{1}{2}x - 3$

In each case state the gradient of the line.

8.2 A straight line passes through the pair of points given. Find the gradient of the line in each case.

(a) $(0,1)$, $(1,4)$ (b) $(1,1)$, $(2,-4)$
(c) $(-1,-1)$, $(6,3)$ (d) $(1,4)$, $(3,4)$

8.3 A straight line graph has gradient -5 and passes through $(1,6)$. Find the equation of the line.

8.4 In Fig. 8.35, various graphs are drawn to the scale 1 unit $= 1$ cm. By finding the gradients of the lines and where they cross the y-axis, find the equation of the line.

8.5 Find the values of x such that $f(x) = 0$ for the following functions:

(a) $f(x) = x^2 - 4$ (b) $f(x) = (2x - 1)(x + 1)$ (c) $f(x) = (x - 3)^2$
(d) $f(x) = (x - 4)(x + 4)$ (e) $f(x) = x^2 + x - 6$ (f) $f(x) = x^2 + 7x + 12$
(g) $f(x) = 12x^2 - 12x - 144$

Using the fact that the peak or trough in the parabola $y = f(x)$ occurs at a value of x half-way between the values where $f(x) = 0$, sketch graphs of the above quadratic functions.

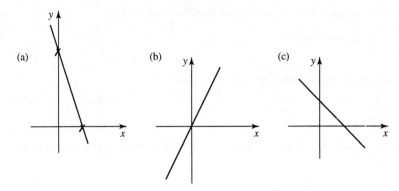

Figure 8.35 Straight line graphs for Exercise 8.4.

8.6 By considering transformations of simple functions, sketch graphs of the following:

(a) $y=\dfrac{1}{x-\frac{1}{2}}$ (b) $y=3.2^{-x}$ (c) $y=\frac{1}{2}x^3$

(d) $y=-3^{(1/2x)}$ (e) $y=(2x-1)^2$ (f) $y=(2x-1)^2-2$

(g) $y=\log_2(x+2)$ (h) $y=6-2^x$ (i) $y=4x-x^2$

8.7 Consider reflections of the graphs given in Fig. 8.36 to determine whether they are even, odd or neither of these.

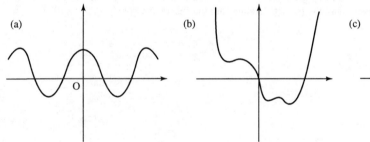

Figure 8.36 Graphs of functions for Exercise 8.7.

8.8 By substituting $x\mapsto -x$ in the following functions, determine whether they are odd or even or neither of these:

(a) $y=-x^2+\dfrac{1}{x^2}$ where $x\neq0$ (b) $y=|x^3|-x^2$

(c) $y=\dfrac{-1}{x}+\log_2(x)$ where $x>0$ (d) $y=\dfrac{-1}{x}+x+x^5$

(e) $y=6+x^2$ (f) $y=1-|x|$

8.9 Draw graphs of the following functions and draw the graph of the inverse relation in each case. Is the inverse a function?

(a) $f(t)=-t+2$ (b) $g(x)=(x-2)^2$ (c) $h(w)=\dfrac{4}{w+2}$

8.10 Find the range of values for which the following inequalities hold and represent them on a number line:

(a) $10t-2\leqslant31$ (b) $10x-3x>-2$

(c) $3-4y\geqslant11+y$ (d) $t+15<6-2t$

8.11 Find the range of values for which the following hold and represent them on a number line:

(a) $x-2>4$ or $1-x<12$ (b) $4t+2\geqslant10$ and $3-2t<1$

(c) $3u+10>16$ or $3-2u>13$

8.12 Solve the following inequalities and represent the solutions on a number line:

(a) $x^2-4<5$ (b) $(2x-3)(x+1)(x-5)>0$

(c) $t^2+4t\leqslant21$ (d) $4w^2+4w-35\geqslant0$

8.13 For the following sets of data y is thought to depend exponentially on t. Draw log–linear graphs in each case and find constants A and k such that $y=A10^{kt}$.

(a)

y	75	48	30	19	12	7
t	1	2	3	4	5	6

(b)

y	2	4.2	8.5	18	35	73
t	0.1	0.2	0.3	0.4	0.5	0.6

8.14 An experiment measuring the change in volume of a gas as the pressure is decreased gave the following measurements:

P (10^5 N m^{-2})	1.5	1.4	1.3	1.2	1.1	1
V (m^3)		0.95	1	1.05	1.1	1.16	1.24

If the gas is assumed to be ideal and the expansion is adiabatic then the relationship between pressure and volume should be $pV^\gamma=C$ where γ and C are constants, p is the pressure and V is the volume. Find reasonable values of γ and C to fit the data and from this expression find the predicted volume at atmospheric pressure, $p=1.013\times10^5$ N m^{-2}.

PROBLEM-SOLVING AND THE ART OF THE CONVINCING ARGUMENT

9.1 INTRODUCTION

Mathematics is used by engineers to solve problems. This usually involves developing a mathematical model. Just as when building a working model aeroplane we would hope to include all the important features, the same thing applies when building a mathematical model. We would also like to indicate the things we have had to leave out because they were too fiddly to deal with, and also those details that we think are irrelevant to the model. In the case of a mathematical model the things that have been left out are listed under assumptions of the model. To build a mathematical model we usually need to use scientific rules about the way things in the world behave (e.g. Newton's laws of motion, conservation of momentum and energy, Ohm's law, Kirchhoff's law for circuits etc.) and use numbers, variables, equations and inequalities to express the problem in a mathematical language.

Some problems are very easy to describe mathematically. For instance: 'Three people sitting in a room were joined by two others, how many people are there in the room in total?'. This can be described by the sum $3+2=?$, and can be solved easily as $3+2=5$.

The final stage of solving the problem is to translate it back into the original setting: the answer is, 'there are five people in the room in total'.

Assumptions were used to solve this problem. We assumed that no one else came in or left the room in the meantime and we made general assumptions about the stability of the room, e.g. the building containing it did not fall down. However, these assumptions are so obvious that they do not need to be listed. In more complex problems it is necessary to list important assumptions, as they may have relevance as to the validity of the solution.

Another example is as follows: 'There are three resistors in series in a circuit. Two of the resistors are known to have resistances of 3 and 4 ohms respectively. The voltage source is a battery of 12 volts and the current is measured as 1 amp. What is the resistance of the third resistor?'.

To help express the problem in a mathematical form, we may draw a circuit diagram, as in Fig. 9.1.

The problem can be expressed mathematically by using Ohm's law and the fact that an equivalent resistance to resistances in series is given by the sum of the individual resistances. If

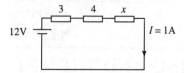

Figure 9.1 A simple circuit.

x is the unknown value of the third resistance, and $V = RI$, where $R = R_1 + R_2 + R_3$, we obtain

$$12 = (3 + 4 + x)1$$

The expression of the mathematical problem has taken the form of an equation where we now need to find x, the value of the third resistance.

The main assumptions that have been used to build this mathematical model are:

1. There are such things as pure resistors that have no capacitance or inductance.
2. Resistances remain constant and are not effected by any possible temperature changes or other environmental effects.
3. The battery gives a constant voltage which does not deteriorate with time.
4. The batter introduces no resistance to the circuit.

These assumptions are simplifications that are acceptable, because although the real world cannot behave with the simplicity of the mathematical model the amount of error introduced by making these assumptions is small.

Once we have the solution of a mathematical model, it should be tested against a real-life situation to see whether the model behaves reasonably closely to reality. Once the model has been accepted, it can be used to predict the behaviour of the system for input values other than those that it has been tested for.

The stages in solving a problem are as follows:

1. Take a real problem and express it as a mathematical one using any necessary scientific rules and assumptions about the behaviour of the system and using letters to represent any unknown quantities. Include an account of any important assumptions and simplifications made.
2. Solve the mathematical problem using your knowledge of mathematics.
3. Translate the mathematical solution back into the setting of your original problem.
4. Test the model solutions for some values to check that it behaves like the real-life problem.

Most mathematical problems are expressed by using equations or inequalities, differential or difference equations or by expressing a problem geometrically, or by a combination of all of these. We might need to incorporate a random element, which results in the need to use a probabilistic model. In many of the following chapters we shall look at the modelling process in more detail as we come across new mathematical tools and the situations in which they are used. To perform the entire modelling cycle properly we need to be able to test our results in a real-life situation in order to reconsider assumptions used in the model. This would require access to engineering situations and tools. For this reason, engineering mathematics books tend to concentrate on those models that are commonly used by engineers. Many of the applied problems presented in the following chapters, however, do present an opportunity to move from an English language description of a problem to a mathematical language description of a problem, which is an important step in the modelling process.

In this chapter we shall look at translating a problem into mathematical language, and, for the main part of the chapter, we concentrate on solving a mathematical problem and the reasoning that is involved in so doing. To solve the problem using our knowledge of mathematics we need to use the ideas of mathematical statements and how to decide whether, and express the fact that, one statement leads logically on to the next. We shall mainly use examples of solving equations and inequalities, although the same ideas apply to the solving of all problems.

9.2 DESCRIBING A PROBLEM IN MATHEMATICAL LANGUAGE

The stages in expressing a problem in mathematical language can be summarized as:

1. Assign letters to represent the unknown quantities.
2. Write down the known facts using equations and inequalities, and using drawings and diagrams where necessary.
3. Express the problem to be solved mathematically.

This is not a simple process, because it involves a great deal of interpretation of the original problem. It is useful to try to limit the number of unknowns used as much as possible, or the problem may appear more difficult than necessary.

Example 9.1 Express the following problem mathematically.

A business has a computer, computer A, which can run a certain job in 3 h and the business can also hire time on a further computer, computer B, at £20 an hour which can run the same job in 4 h. The fixed costs of the business are £1500 per day and an additional £10 an hour for casual labour if the second computer is used. In present market conditions, each job can be charged at £170. How many jobs does the company need per day to remain solvent?

SOLUTION The mathematical problem can be expressed by first, assigning letters to some of the unknown quantities and then writing down all the known facts as equations or inequalities.

First, assign letters:

Total number of jobs in a day is J.
J_A is the number of jobs run on computer A and J_B is the number of jobs run on computer B. The costs are £K per day and the profit is £P.

The known facts can be expressed as follows:

$$J = J_A + J_B$$

This expresses the fact that the total number of jobs J is made up of those run on A and those run on B.

As there are 24 hours in a day and the jobs take three hours on computer A then the maximum number of jobs that can be run on A is $24/3 = 8$. Similarly, as a job takes four hours on computer B we find the maximum number of jobs on B is 6. This gives the conditions:

$$0 \leqslant J_A \leqslant 8, \qquad 0 \leqslant J_B \leqslant 6$$

The costs are £1500 fixed costs, plus the cost of the casual labour at £10 per hour, giving £10×4 h × number of jobs on B, plus the cost of hiring computer B, £20 per hour, giving £20×4 h × number of jobs on B. This can be expressed as:

$$K = 1500 + 10 \times 4 J_B + 20 J_B$$

Next we need to relate the profit to the other variables. As the profit is £150 × the number of jobs minus the total costs, we get

$$P = 150J - K$$

Finally, we must express the mathematical problem which we would like to solve. For the business to remain solvent, the profit must not be negative; hence we get the problem expressed as: find the minimum J such that $P \geqslant 0$.

Example 9.2 Express the following in mathematical language.

A car brake pedal, as represented in Fig. 9.2(a) is pivoted at point A. What is the force in the brake cable if a constant force of 900 N is applied by the driver's foot and the pedal is stationary.

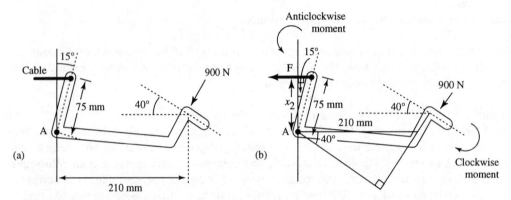

Figure 9.2 (a) A representation of a car brake pedal. (b) The same diagram as (a), with some unknown quantities marked and triangles used to formulate the problem.

SOLUTION First, we assign letters to the unknowns: let $F =$ the force in the brake cable.

In order to write down the known facts we need to consider what scientific laws can be used. As the force applied on the pedal initially provides a turning motion then we know to use the ideas of moments. The moment of a force about an axis is the product of the force F and its perpendicular distance, x, to the line of action of the force. Furthermore, as the pedal is now stationary, then the moments must be balanced, so the clockwise moment must equal the anticlockwise moment.

To use this fact we need to use two further measurements, currently unknown: the perpendicular distance from the line of action of the force provided by the driver to the axis A. This is marked as x_1 metres on the diagram in Fig. 9.2(b). The other distance is the perpendicular distance from the line of action of the force in the cable to the axis A. This is marked as x_2 metres in Fig. 9.2(b).

We can now write down the known facts, involving the unknowns x_1, x_2 and F.

From the right-angled triangle containing x_1 we have (converting 210 mm = 0.21 m):

$$\cos(40°) = \frac{x_1}{0.21}$$

From the right-angled triangle containing x_2 we have (converting 75 mm = 0.075 m):

$$\cos(15°) = \frac{x_2}{0.075}$$

The moments of the forces can now be calculated and equated. The clockwise moment is 800 x_1 and the anticlockwise moment is given by Fx_2, and hence we have:

$$800x_1 = Fx_2$$

Finally, we need to express the problem we are trying to solve. In this case it is simply 'what is F?'.

Note that in both Examples 9.1 and 9.2 certain modelling assumptions had been used in order to formulate the 'natural language' description of the problem that we were given. For instance, it is probable that the computer jobs run by the business in Example 9.1 are not all identical and therefore average figures for times and costings have been used. Similarly, in Example 9.2 no mention has been made of friction, which would provide an extra force to consider. Here we have only considered the transition from natural language and accompanying diagrams to the mathematical problem. We have implicitly assumed that the modelling process can be performed in two stages: from real-life problem to a natural language description which incorporates some simplifying assumptions, and then from there to a mathematical description. In reality, modelling a system is much more involved. We would probably repeat stages in this process if we decided that the mathematical description was too complex and would return to the real-life situation in order to make new assumptions.

We are now in a position to discuss mathematical statements and how to move from the statement of the problem to finding the desired solution.

9.3 PROPOSITIONS AND PREDICATES

When we first set up a problem to be solved we write down mathematical expressions like

$$2+3=? \tag{9.1}$$

and

$$12=(3+4+x) \tag{9.2}$$

These are mathematical statements with an unknown value. Statements containing unknowns (or variables) are called predicates. A predicate can be either true or false, depending on the value(s) substituted into it. When a value is substituted into a predicate it becomes a simple proposition. If in Eq. (9.1) we substitute 5 for the question mark we get $2+3=5 \Leftrightarrow 5=5$, which is true.

If, however, we substitute 6 we get $2+3=6 \Leftrightarrow 5=6$, which is false.

$2+3=5$ and $2+3=6$ are examples of propositions. These are simple statements that can be either true or false. They contain no unknown quantities. Notice that if we simply rewrite a proposition or predicate we use '$\equiv$' or '$\Leftrightarrow$' to mean 'is equivalent to' or 'is the same as'.

In Eq. (9.2), if we substitute 4 for x we get $12=11$ which is false, but if we substitute 5 for x we get $12=12$, which is true.

Example 9.3 Assign true or false to the following:

(a) $(3x-2)(x+5)=10$ where $x=1$
(b) $5x^2-2x+1=25$ where $x=-2$
(c) $y>5t+3$ where $y=2$ and $t=-3$

SOLUTION

(a) Substitute $x = 1$ in the expression and we get

$$(3.1 - 2)(1 + 5) = 10$$
$$\Leftrightarrow 1(6) = 10$$
$$\Leftrightarrow 6 = 10$$

which is false.

(b) Substituting $x = -2$ into $5x^2 - 2x + 1 = 25$ gives

$$5(-2)^2 - 2(-2) + 1 = 25$$
$$\Leftrightarrow 20 + 4 + 1 = 25$$
$$\Leftrightarrow 25 = 25$$

which is true.

(c) Substituting $y = 2$ and $t = -3$ into $y > 5t + 3$ gives

$$2 > 5(-3) + 3 \Leftrightarrow 2 > -15 + 3 \Leftrightarrow 2 > -12$$

which is true.

Like functions, predicates have a domain which is the set of all allowed inputs to the predicate. For instance, the predicate $1/(x - 1) = 1$ where $x \in \mathbb{R}$ has the restriction that $x \neq 1$, as letting x equal 1 would lead to an attempt to divide by 0, which is not defined. The predicate $\sqrt{x - 2} = 25$, where $x \in \mathbb{R}$ has the restriction that $x \geqslant 2$, as values of x less than 2 would lead to an attempt to take the square root of a negative number, which is not defined.

9.4 OPERATIONS ON PROPOSITIONS AND PREDICATES

Consider the problem given in Example 9.1. Notice that the conditions that we discovered when writing down the known facts must all be true in any solution that we come up with. If any one of these conditions is not true then we cannot accept the solution. The first condition must be true *and* the second *and* the third, etc.

Here we have an example of an operation on predicates. In Chapter 1, we defined an operation on numbers as a way of combining two numbers to give a single number. 'And', written as $\wedge$, is an operation on two predicates or propositions which results in a single predicate or proposition. Therefore, to express the fact that both $J \geqslant 0$ and $J = J_A + J_B$ we can write $J \geqslant 0 \wedge J = J_A + J_B$.

A compound statement is true if and only if each part is true.

As propositions can only be either true (T) or false (F), all possible outcomes of the operation can easily be listed in a small table, called a truth table. The truth table for the operation of 'and' is given in Table 9.1. p and q represent any two propositions: for instance, p and q could be defined by

$$p: \ 'J \geqslant 0'$$
$$p: \ 'J = J_A + J_B'$$

Another important operation is that of 'or'. One example of the use of this operation comes about by solving a quadratic equation. As we saw in Chapter 3, when solving quadratic equations one way of doing this is to factorize an expression which is equal to 0.

Table 9.1 Truth table for the operation 'and'. This table can also be expressed by $T \wedge T \Leftrightarrow T$, $T \wedge F \Leftrightarrow F$, $F \wedge T \Leftrightarrow F$, $F \wedge F \Leftrightarrow F$. **T stands for 'true' and F stands for 'false'.**

p	q	$p \wedge q$
T	T	T
T	F	F
F	T	F
F	F	F

Table 9.2 Truth table for 'or', $\vee$. This table can also be expressed by $T \vee T \Leftrightarrow T$, $T \vee F \Leftrightarrow T$, $F \vee T \Leftrightarrow T$, $F \vee F \Leftrightarrow F$

p	q	$p \vee q$
T	T	T
T	F	T
F	T	T
F	F	F

Table 9.3 The truth table for 'not', $\neg$. This table can also be expressed by $\neg T \Leftrightarrow F$, $\neg F \Leftrightarrow T$

p	$\neg p$
T	F
F	T

To solve $x^2 - x - 6 = 0$, the left-hand side of the equation can be factorized to give $(x - 3)(x + 2) = 0$.

Now we use the fact that for two numbers multiplied together to equal 0 then one of them, at least, must be 0 to give

$$(x - 3)(x + 2) = 0 \Leftrightarrow (x - 3) = 0 \qquad \text{or} \qquad (x + 2) = 0$$

'or' can be written using the symbol $\vee$. The compound statement is true if either $x - 3 = 0$ is true or if $x + 2 = 0$ is true. Therefore to express the statement that either $x - 3 = 0$ or $x + 2 = 0$ we can write

$$(x - 3) = 0 \vee (x + 2) = 0$$

$\vee$ is also called 'non-exclusive or' because it is also true if both parts of the compound statement are true. This usage is unlike the frequent use of 'or' in the English language, where it is often used to mean a choice: e.g. 'you may have an apple or a banana' implies either one or the other but not both. This everyday usage of the word 'or' is called 'exclusive or'.

The truth table for 'or' is given in Table 9.2.

A further operation is that of 'not', which is represented by the symbol $\neg$. For instance, we could express the sentence 'x is not bigger than 4' as $\neg(x > 4)$. The truth table for 'not' is given in Table 9.3.

Examle 9.4 Assign truth values to the following:

(a) $x - 2 = 3 \wedge x^2 = 4$ when $x = 2$
(b) $x - 2 = 3 \vee x^2 = 4$ when $x = 2$
(c) $\neg(x - 4) = 0$ when $x = 4$
(d) $\neg(a - b) = 4 \vee (a + b) = 2$ when $a = 5, b = 3$

SOLUTION
(a) $x - 2 = 3 \wedge x^2 = 4$ when $x = 2$

Substitute $x = 2$ into the predicate, $x - 2 = 3 \wedge x^2 = 4$, and we get $2 - 2 = 3 \wedge 2^2 = 4$.

The first part of the compound statement is false and the second part is true. Overall as $F \wedge T \Leftrightarrow F$ then the proposition is false.

(b) $x - 2 = 3 \lor x^2 = 4$ when $x = 2$

Substitute $x = 2$ into the predicate and we get $2 - 2 = 3 \lor 2^2 = 4$.

The first part of the compound statement is false and the second part is true. As $F \lor T \Leftrightarrow T$ then the proposition is true.

(c) $\neg(x - 4 = 0)$ when $x = 4$

When $x = 4$ the expression becomes

$$\neg(4 - 4 = 0) \Leftrightarrow \neg T \Leftrightarrow F$$

(d) $\neg((a - b) = 4) \lor (a + b) = 2$ when $a = 5$, $b = 3$

$\neg((a - b) = 4)$ when $a = 5$, $b = 3$ gives $\neg(5 - 3 = 4) \Leftrightarrow \neg F \Leftrightarrow T$.

$(a + b) = 2$ when $a = 5$, $b = 3$ gives $(5 + 3) = 2 \Leftrightarrow 8 = 2 \Leftrightarrow F$.

Overall, $T \lor F \Leftrightarrow T$, so $\neg((a - b) = 4) \lor (a + b) = 2$ when $a = 5$, $b = 3$ is true.

Example 9.5 Represent the following inequalities on a number line:

(a) $x > 2 \land x \leqslant 4$
(b) $x < 2 \lor x \geqslant 4$
(c) $\neg(x < 2)$

SOLUTION

(a) $x > 2 \land x \leqslant 4$. To represent the operation of $\land$, 'and', find where the two regions overlap (Fig. 9.3(a)). $x > 2 \land x \leqslant 4$ can also be represented by $2 < x \leqslant 4$.

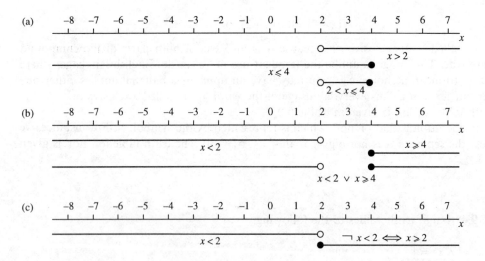

Figure 9.3 The operations (a) $\land$ (and), (b) $\lor$ (or) and (c) $\neg$ (not).

(b) $x < 2 \lor x \geqslant 4$. To represent the operation of $\lor$, 'or', take all points on the first highlighted region as well as all points in the second highlighted region and any end points (Fig. 9.3(b)).

(c) $\neg(x < 2)$. To represent the operation of 'not', take all the points on the number line not in the original region (Fig. 9.3(c)). This can also be expressed by $x \geqslant 2$.

9.5 EQUIVALENCE

We can now express an initial problem in terms of a predicate, probably an equation or a number of equations or a number of inequalities. However, to solve the problem we need to be able to move from the original expression of the problem towards the solution. In Chapter 3 we discussed how to solve various types of equations and introduced the idea of equivalent equations. In Chapter 8 we also looked at equivalent inequalities. In both cases we used the idea that in moving from one expression to an equivalent expression the set of solutions remained the same. In general, two predicates are equivalent if they are true for exactly the same set of values. We use our knowledge of mathematics to determine what operations can be performed that will maintain that equivalence. The rule that can be used to move from one equation to another was given in Chapter 3 as 'Equations remain equivalent if the same operation is performed on both sides of the equation'. In the case of quadratic equations we can also use a formula for the solution or use a factorization and the fact that $ab=0 \Leftrightarrow a=0$ or $b=0$. In passing from one equation to an equivalent equation we should use the equivalence symbol. This then makes a mathematical sentence:

$$x+5=3$$
$$\Leftrightarrow x=3-5$$

which can be read as 'The equation $x+5=3$ is equivalent to $x=3-5$'.

In all but the most obvious cases it is good practice to list a short justification for the equivalence by the side of the expression:

$$x+5=3$$
$$\Leftrightarrow x=3-5 \qquad \text{(subtracting 5 from both sides)}$$
$$\Leftrightarrow x=-2$$

Because of the possibility of making a mistake, the solution(s) should be checked by substituting the values into the original expression of the problem. To check, substitute $x=-2$ into the original equation, giving $-2+5=3$ which is true, indicating that the solution is correct.

Example 9.6 Solve the following equation:

$$x-3=5-2x$$

SOLUTION

$$x-3=5-2x$$
$$\Leftrightarrow x+2x-3=5 \qquad \text{(adding $2x$ to both sides)}$$
$$\Leftrightarrow 3x=8 \qquad \text{(adding 3 to both sides)}$$
$$\Leftrightarrow x=\tfrac{8}{3} \qquad \text{(dividing both sides by 3)}$$

Check by substituting $x=\tfrac{8}{3}$ into the original equation:

$$\tfrac{8}{3}-3=5-2\times\tfrac{8}{3}$$
$$\Leftrightarrow -\tfrac{1}{3}=-\tfrac{1}{3}$$

which is true.

We looked at methods of solving inequalities in Chapter 8. The rules for finding equivalent inequalities were 'Perform the same operation to both sides', but in the case of a negative number, when multiplying or dividing the direction of the inequality sign must be reversed. To solve more complex inequalities, such as $f(x) > 0$, $f(x) < 0$, where $f(x)$ is a continuous but non-linear function, then we solve $f(x) = 0$ and then use a number line to mark regions where $f(x)$ is positive, negative or zero. The important thing in the process is to present a short justification of the equivalence. Finally, when the set of solutions has been found, some of the solutions can be substituted into the original expression of the problem in order to check that no mistakes have been made.

Example 9.7 Solve the following inequalities:

(a) $3x - 1 < 6x + 2$
(b) $x^2 - 5x > -6$

SOLUTION

(a)
$$3x - 1 < 6x + 2$$
$$\Leftrightarrow -1 < 6x + 2 - 3x \qquad \text{(subtracting } 3x \text{ from both sides)}$$
$$\Leftrightarrow -1 - 2 < 3x \qquad \text{(subtracting 2 from both sides)}$$
$$\Leftrightarrow \frac{-3}{3} < \frac{3x}{3} \qquad \text{(dividing both sides by 3)}$$
$$\Leftrightarrow -1 < x$$
$$\Leftrightarrow x > -1$$

CHECK Test a few values from the set $x > -1$ and substitute into $3x - 1 < 6x + 2$.
 Try $x = 0$: this gives $-1 < 2 \Leftrightarrow T$.
 Try $x = 2$: this gives $3(2) - 1 < 6(2) + 2 \Leftrightarrow 5 < 14 \Leftrightarrow T$.

(b) $x^2 - 5x > -6$
 Write the inequality with 0 on one side of the inequality sign:
$$x^2 - 5x > -6 \Leftrightarrow x^2 - 5x + 6 > 0 \qquad \text{(adding 6 to both sides)}$$

 Find the solutions to $f(x) = 0$ where $f(x) = x^2 - 5x + 6$ and mark them on a number line, as in Fig. 9.4:
$$x^2 - 5x + 6 = 0 \Leftrightarrow x = \frac{5 \pm \sqrt{25 - 24}}{2}$$
(using the formula for solution of quadratic equations)
$$\Leftrightarrow x = \frac{5 + 1}{2} \vee x = \frac{5 - 1}{2}$$
$$\Leftrightarrow x = 3 \vee x = 2$$

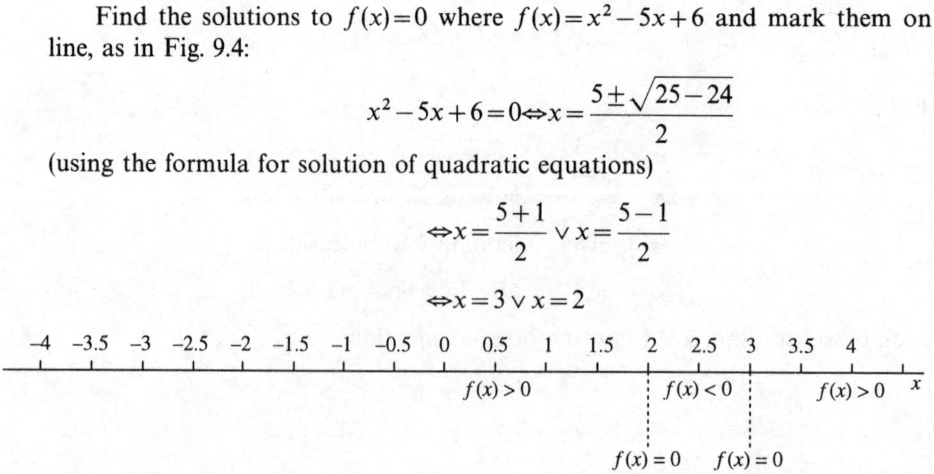

Figure 9.4 Solving $x^2 - 5x + 6 > 0$.

Using the fact that the function is continuous we can substitute values for x which lie on either side of the roots of $f(x)$ in order to find the sign of the function in that region. Here we choose 0, 2.5 and 4 and find that when $x=0$, $f(x)=x^2-5x+6=6$, so $f(x)>0$; when $x=2.5$, $f(x)=x^2-5x+6=6.25-12.5+6=-0.25$, so $f(x)<0$; and when $x=4$, $f(x)=x^2-5x+6=16-20+6=2$, so $f(x)>0$.

These regions are marked on the number line as in Fig. 9.4 and this gives the solution to $f(x)>0$ as $x<2 \vee x>3$.

CHECK   A check is to substitute some values from the solution set $x<2$, $x>3$ into the original predicate $x^2-5x>-6$. Substituting $x=1$ gives $1-5>-6 \Leftrightarrow -4>-6 \Leftrightarrow T$ and substituting $x=5$ gives $25-25>-6 \Leftrightarrow 0>-6 \Leftrightarrow T$. It therefore appears that this solution is correct.

9.6 IMPLICATION

In Chapter 3 we described one method of finding equivalent equations as that of 'doing the same thing to both sides'. This was rather simplistic, but a useful way of seeing it at the time. There are only certain things that can be 'done to both sides', such as adding, subtracting, multiplying by a non-zero expression, or dividing by a non-zero expression, that always maintain equivalence. There are also many operations that can be performed to both sides of an equation which do not give an equivalent equation but give an equation with the same solutions and yet more besides. In this situation we say that the first equation *implies* the second equation. The symbol for 'implies' is $\Rightarrow$.

An example of implication is given by squaring both sides of the equation

$$x-2=2 \Rightarrow (x-2)^2=4$$

The first predicate, $x-2=2$, has only one solution, $x=4$, but the second predicate has two solutions, $x=4$ and $x=0$. By squaring the equation we have found a new equation which includes all the solutions of the first equation and has one more besides. Implication is expressed in English by using phrases like 'If... then...'.

An expression involving an implication cannot always be turned the other way around in the same way as those involving equivalence can. An example of this is given by the following statement. It is true that: 'If I am going to work then I take the car', which can be written using the implication symbol as 'I am going to work'$\Rightarrow$'I take the car'. However, it is not true that 'If I take the car then I am going to work'. This is because there are more occasions when I take the car than simply going to work.

More examples are:

'I only clean the windows if it is sunny'

'I am cleaning the windows'$\Rightarrow$'It is sunny'

This does not mean that 'If it is sunny then I clean the windows', as there are some sunny days when I have to go to work or just laze in the garden, or I am on holiday.

An implication sign can be written, and read, from right to left:

'It is sunny'$\Leftarrow$'I am cleaning the windows'

which I can still read as 'I am cleaning the windows therefore it is sunny', or I could try rearranging the sentence as 'Only if it is sunny will I clean the windows'.

The various ways of expressing these sentiments can get quite involved. The important point to remember is that $p \Rightarrow q$ means that q must be true for all the occasions that p is true, but q

could be true on more occasions besides. Going back to equations or inequalities,

$$p \Rightarrow q$$

means that the solution set, P, of p is a subset of the solution set, Q, of q. This is pictured in Fig. 9.5.

We can now see that for two equations or inequalities to be equivalent then $p \Rightarrow q$ and $q \Rightarrow p$. This means that their solution sets are exactly the same (Fig. 9.6).

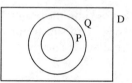

Figure 9.5 P is the solution set of p, Q is the solution set of q. $p \Rightarrow q$ means that $P \subseteq Q$. D is the domain of p and q.

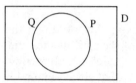

Figure 9.6 P is the solution set of p and Q is the solution set of q. Then $p \Leftrightarrow q$ means that $P = Q$.

Example 9.8 Fill in the correct symbol in each case: $\Rightarrow$, $\Leftarrow$ or $\Leftrightarrow$:

(a) $x^2 - 9 = 0 \ldots x = -3$

(b) $x = -\frac{1}{2} \ldots \dfrac{1}{(2x - 5)(x + 1)} = -\frac{1}{3}$ where $x \in \mathbb{R}$, $x \neq 5$, $x \neq -1$

(c) $(x - 3)(x - 1) > 0 \ldots x > 3 \lor x < 1$

SOLUTION
(a) $x^2 - 9 = 0 \ldots x = -3$
    Solving $x^2 - 9 = 0$ gives

$$x^2 - 9 = 0 \Leftrightarrow (x - 3)(x + 3) = 0$$

$$\Leftrightarrow \quad x = 3 \lor x = -3$$

Hence -3 is only one of the solutions of the first equation, so the correct expression is

$$x^2 - 9 = 0 \Leftarrow x = -3$$

(b) $x = -\frac{1}{2} \ldots \dfrac{1}{(2x - 5)(x + 1)} = -\frac{1}{3}$ where $x \in \mathbb{R}$, $x \neq 5$, $x \neq -1$

Solving

$$\dfrac{1}{(2x - 5)(x + 1)} = -\frac{1}{3}$$

gives

$$\dfrac{1}{(2x - 5)(x + 1)} = -\frac{1}{3} \Leftrightarrow -3 = (2x - 5)(x + 1) \quad \text{(multiplying both sides by } (2x - 5)(x + 1))$$

$$\Leftrightarrow -3 = 2x^2 - 3x - 5 \quad \text{(multiplying out the brackets)}$$

$$\Leftrightarrow 2x^2 - 3x - 2 = 0 \quad \text{(adding 3 on to both sides of the equation)}$$

$$\Leftrightarrow x = \frac{3 \pm \sqrt{9 + 16}}{4} \quad \text{(using the quadratic formula to solve the equation)}$$

$$\Leftrightarrow x = \frac{3 \pm 5}{4}$$

$$\Leftrightarrow x = 2 \vee x = -\tfrac{1}{2}$$

The second predicate,

$$\frac{1}{(2x - 5)(x + 1)} = -\tfrac{1}{3}$$

has more solutions than the first predicate, $x = -\tfrac{1}{2}$. Thus the correct expression is

$$x = -\tfrac{1}{2} \Rightarrow \frac{1}{(2x - 5)(x + 1)} = -\tfrac{1}{3} \qquad \text{where } x \in \mathbb{R}, \ x \neq 5, \ x \neq -1$$

(c) $(x - 3)(x - 1) > 0 \ldots x > 3 \vee x < 1$
  Solve the inequality on the left by first solving $f(x) = 0$:

$$(x - 3)(x - 1) = 0 \Leftrightarrow x = 3 \vee x = 1$$

Choosing values on either side of the roots, e.g. 0, 2, 4, gives

$$f(0) = (-3)(-1) = 3 \qquad \text{i.e. } f(x) > 0$$

$$f(2) = (-1)(1) = -1 \qquad \text{i.e. } f(x) < 0$$

$$f(4) = (1)(3) \qquad \text{i.e. } f(x) > 0$$

This is then marked on a number line as in Fig. 9.7. As the solution to $(x - 3)(x - 1) > 0$ is $x < 1 \vee x > 3$, we have therefore shown that $(x - 3)(x - 1) > 0 \Leftrightarrow x > 3 \vee x < 1$.

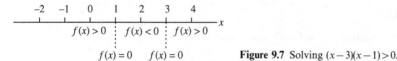

Figure 9.7 Solving $(x - 3)(x - 1) > 0$.

9.7 MAKING SWEEPING STATEMENTS

In Chapter 1, we made some statements about numbers which we stated were true for all real numbers. Some of these were the commutative laws $a + b = b + a$ and $ab = ba$, and the distributive law, $a(b + c) = ab + ac$.

There is a symbol which stands for 'for all' or 'for every' which allows these laws to be expressed in a mathematical shorthand:

$$\forall a, b \in \mathbb{R} \qquad a + b = b + a$$

$$\forall a, b \in \mathbb{R} \qquad ab = ba$$

$$\forall a, b, c \in \mathbb{R} \qquad a(b + c) = ab + ac$$

Rules, such as the commutative law, are axioms for numbers and need not be proved true. However, more involved expressions, such as

$$\forall a,b \in \mathbb{R} \qquad (a-b)(a+b) = a^2 - b^2$$

need to be justified.

If the symbol 'for all' is used with a predicate about its free variable then it becomes a simple proposition which is either true or false. We looked at how to decide on the truth or otherwise of such generalized statements in Section 1.7.

To show that an expression is true we use our knowledge of mathematics to write equivalent expressions until we come across an expression which is obviously true (like $a = a$). To prove it is false is much easier. As we have made a sweeping statement, that the expression is true for all a, b, then we only need to come across one example of numbers which make the expression false.

Example 9.9 Are the following true or false? Justify your answers.

(a) $\forall a,b \in \mathbb{R} \qquad a^3 - b^3 = (a-b)(a^2 + ab + b^2)$

(b) $\forall t \in \mathbb{R}$, where $t \neq 1$, $t \neq -1 \qquad \dfrac{1}{t+1} = \dfrac{t-1}{t^2-1}$

(c) $\forall x \in \mathbb{R}$, where $x \neq 0 \qquad \dfrac{x^2-1}{x} = x-1$

SOLUTION

(a) $\forall a,b \in \mathbb{R} \qquad a^3 - b^3 = (a-b)(a^2 + ab + b^2)$

Looking at the right-hand side of the equality we have

$$(a-b)(a^2+ab+b^2) = a(a^2+ab+b^2) - b(a^2+ab+b^2) \qquad \text{(taking out the brackets)}$$

$$= a^3 + a^2b + ab^2 - ba^2 - ab^2 - b^3 \qquad \text{(taking out the remaining brackets)}$$

$$= a^3 - b^3 \qquad \text{(simplifying)}$$

We have shown that the right-hand side is equal to the left hand-side

$$a^3 - b^3 = (a-b)(a^2+ab+b^2) \Leftrightarrow a^3 - b^3 = a^3 - b^3$$

which is true. Therefore

$$\forall a,b \in \mathbb{R} \qquad a^3 - b^3 = (a-b)(a^2+ab+b^2)$$

is true.

(b) $\forall t \in \mathbb{R}$, where $t \neq 1$, $t \neq -1 \qquad \dfrac{1}{t+1} = \dfrac{t-1}{t^2-1}$

Take the right-hand side of the equality:

$$\frac{t-1}{t^2-1} = \frac{t-1}{(t-1)(t+1)} \qquad \text{(factorizing the bottom line)}$$

$$= \frac{1}{t+1} \qquad \text{(dividing the top and bottom line by } t-1 \text{ which is allowed as } t \neq -1\text{)}$$

Hence

$$\frac{1}{t+1} = \frac{t-1}{t^2-1} \Leftrightarrow \frac{1}{t+1} = \frac{1}{t+1}$$

which is true. Thus,

$$\forall t \in \mathbb{R}, \text{ where } t \neq 1, t \neq -1 \qquad \frac{1}{t+1} = \frac{t-1}{t^2-1}$$

is true.

(c) $\forall x \in \mathbb{R}$, where $x \neq 0 \qquad \dfrac{x^2-1}{x} = x-1$

To show this is false, substitute a value for x, e.g. $x = 2$. When $x = 2$:

$$\frac{x^2-1}{x} = x-1$$

becomes

$$\frac{4-1}{2} = 2-1 \Leftrightarrow \frac{3}{2} = 1 \Leftrightarrow F$$

As the predicate fails for one value of x then

$$\forall x \in \mathbb{R}, \text{ where } x \neq 0 \qquad \frac{x^2-1}{x} = x-1$$

is false.

Another useful symbol is $\exists$, which means, 'there exists'. This can be used to express the fact that every real number has an inverse under addition. Hence we get

$$\forall a \in \mathbb{R}, \; \exists b, \; a+b = 0$$

If the symbol $\exists$ is used with a predicate about its free variable it becomes a simple proposition which is either true or false. In the case of the example given concerning the inverse, this is an axiom of the real numbers and we can just state that it is true. Other statements involving existence will need some justification. Proving existence is simpler than disproving it. If I were to state 'There exists a blue moon in the universe', to prove this to be true I only need to find one blue moon, but to disprove it I must find all the moons in the universe and show that not one of them is blue.

In other words to show that some value exists which makes a certain predicate into a true proposition then we only need to find that value and demonstrate that the resulting proposition is true. To show that no value exists is more difficult. If the domain of interest is a set of numbers we need to present an argument about any member of the set.

Example 9.10 Are the following true or false? Justify your answer.

(a) $\exists x \in \mathbb{R}, \; (x+2)(x-1) = 0$
(b) $\exists x \in \mathbb{R}, \; x^2 + 4 < 0$

SOLUTION
(a) $\exists x \in \mathbb{R}, \; (x+2)(x-1) = 0$

To show this is true we only need find one value of x which makes the equality correct. For instance take $x = -2$: when $x = -2$, $(x+2)(x-1) = 0$ becomes $(-2+2)(-2-1) = 0 \Leftrightarrow 0 = 0$ which is true. Therefore $\exists x \in \mathbb{R}, \; (x+2)(x-1) = 0$ is true.

(b) $\exists x \in \mathbb{R}, \ x^2 + 4 < 0$

Trying a few values for x (e.g. 1, 0, 20, -2) we might suspect that this statement is false. We need to present a general argument in order to convince ourselves of this.

x^2 is always positive or zero, i.e. $x^2 \geqslant 0$ for all x. If we then add 4, then for all x, $x^2 + 4 \geqslant 4$, and, as 4 is bigger than 0, $x^2 + 4 > 0$ for all x; hence $\exists x \in \mathbb{R}, \ x^2 + 4 < 0$ is false.

9.8 OTHER APPLICATIONS OF PREDICATES

Predicates are often used in software engineering. Some simple applications are:

(a) to express the condition under which a program block will be carried out (or a loop will continue execution);
(b) to express a program specification in terms of its pre-conditions and its post-conditions.

Example 9.11 Express the following in pseudo-code: print x and y if y is a multiple of x and x is an integer between 1 and 100 inclusive.

SOLUTION Pseudo-code is a system of writing algorithms which is similar to some computer languages, but is not any particular computer language. We can use any symbols we like as long as the meaning is clear.

y is a multiple of x means that if y is divided by x then the result is an integer. This can be expressed as

$$\frac{y}{x} \in \mathbb{Z}$$

The condition that x must lie between 1 and 100 can be expressed as $x \geqslant 1$ and $x \leqslant 100$. Combining these conditions gives the following interpretation for the algorithm:

if $(y/x \in \mathbb{Z} \wedge x \geqslant 1 \wedge x \leqslant 100)$ then

print x, y

endif

Example 9.12 A program is designed to take a given whole positive number, x, greater than 1, and find two factors of x, a and b, which multiplied together give x. a and b should be whole positive numbers different from 1, unless x is prime. Express the pre- and post-conditions for the program.

SOLUTION Pre-condition $x \in \mathbb{N} \wedge x > 1$.

The post-condition is slightly more difficult to express. Clearly $ab = x$ is a statement of the fact that a and b must multiply together to give x. Also a and b must be elements of $\mathbb{N}$. a and b cannot be 1 unless x is prime, this can be expressed by $(a \neq 1 \wedge b \neq 1) \vee (x$ is prime$)$. Finally we have the post-condition as

$$ab = x \wedge (a \in \mathbb{N}) \wedge (b \in \mathbb{N}) \wedge ((a \neq 1 \wedge b \neq 1) \vee x \text{ is prime})$$

9.9 SUMMARY

1. The stages in solving a real-life problem using mathematics are:

(a) Express the problem as a mathematical one, using any necessary scientific rules and assumptions about the behaviour of the system and using letters to represent any unknown quantities. This is called a mathematical model.

(b) Solve the mathematical problem by moving from one statement to an equivalent statement, justifying each stage by using relevant mathematical knowledge.

(c) Check the mathematical solution(s) by substituting them into the original formulation of the mathematical problem.

(d) Translate the mathematical solution back into the setting of the original problem.

(e) Test the model solutions for some realistic values to see how well the model predicts the behaviour of the system. If it is acceptable then the model can be used to predict more results.

2. A predicate is a mathematical statement containing a variable. Examples of predicates are equations and inequalities.

3. If values are substituted into a predicate it becomes a simple proposition which is either true or false.

4. The three main operations on predicates and propositions are $\wedge$, $\vee$, $\neg$, and these can be defined using truth tables as in Tables 9.1, 9.2 and 9.3.

5. Two predicates are equivalent if they are true for exactly the same set of values.

6. $p \Rightarrow q$ means 'p implies q', i.e. q is true whenever p is true. If p, q are equations or inequalities and $p \Rightarrow q$ then all solutions of p are also solutions of q, and q may have more solutions besides.

7. The symbol $\forall$ stands for 'for all' or 'for every' and can be used with a predicate to make it into a simple proposition, e.g. $\forall a, b \in \mathbb{R}, a^2 - b^2 = (a+b)(a-b)$, which is true.

8. The symbol $\exists$ stands for 'there exists' and can also be used with a predicate to make it into a simple proposition, e.g. $\exists x \in \mathbb{R}, 3x = 45$, which is true.

9.10 EXERCISES

9.1 Assign T or F to the following:

(a) $2x + 2 = 10$ when $x = 1$
(b) $2x + 2 = 10$ when $x = 2$
(c) $3x^2 + 3x - 6 = 0$ when $x = 1$
(d) $1 - t^2 = -3$ when $t = -2$
(e) $t - 5 = 6.5 \wedge t + 4 = 2.5$ when $t = 1.5$
(f) $u + 3 = 6 \wedge 2u - 1 = 4$ when $u = 3$
(g) $3y + 2 = -2.5 \vee 1 - y = 1$ when $y = -1.5$
(h) $\neg(x^2 - x + 2 = 0)$ when $x = -1$
(i) $\neg(t - 2 = 4 \wedge t = 3)$
(j) $\neg(t - 2 = 4) \wedge (t = 3)$
(k) $\neg(3t - 4 = 6 \vee 1 - t = -2\frac{1}{3})$ when $t = 3\frac{1}{3}$

9.2 Solve the following, justifying each stage of the solution and checking the result:

(a) $3 - 2x = -1$ (b) $1 - 2t^2 = 1 - 10t$
(c) $50t - 11 = -25t^2$ (d) $30y - 13 = 8y^2$
(e) $10t - 4 \leqslant -3$ (f) $10 - 4x > 12$

9.3 Find the range of values for which the following hold and represent them on a number line:

(a) $x + 3 \geqslant 5 \vee 1 - 2x > 3$ (b) $2 - 4t \leqslant 3 \wedge 2 - t < 1$
(c) $\neg(2x + 3 \leqslant 9)$

9.4 Fill the correct sign $\Rightarrow$, $\Leftarrow$ or $\Leftrightarrow$ or indicate none of these. Assume the domain is $\mathbb{R}$ unless indicated otherwise.

(a) $3x^2 - 1 = 0 \ldots x = \dfrac{1}{\sqrt{3}}$

(b) $\sqrt{x - 1} = 5 \ldots x = 26$ (where $x \geqslant 1$)
(c) $t^2 - 5t = 36 \ldots (t - 4)(t - 9) = 0$
(d) $\dfrac{2x - 2}{x - 3} = 1 \ldots 3x + 4 = -x$ (where $x \neq 3$)

(e) $3x=4\ldots(3x)^2=4^2$

(f) $t+1=5\ldots(t+1)^3=5^3$

(g) $(x+1)(x-3)=(x-3)(x+2)\ldots(x+1)=(x+2)$

(h) $x-1=25\ldots\sqrt{x-1}=5$ (where $x\geqslant 1$)

(i) $\dfrac{w}{w^2-1}=1\ldots\dfrac{1}{w-1}=1$ where $w\neq 1$ and $w\neq -1$

(j) $(x-1)(x-3)<0\ldots(x-3)<0\vee(x-1)<0$

(k) $x>2\vee x<-2\ldots x^2>4$

9.5 Determine whether the following statements are true or false and justify your answer:

(a) $\forall a,b\in\mathbb{R},\ a^4-b^4=(a-b)(a^3-a^2b+ab^2-b^3)$

(b) $\forall a,b\in\mathbb{R},\ a^3+b^3=(a+b)^3$

(c) $\forall x\in\mathbb{R},\ x\neq 0\quad\dfrac{1}{1/x}=x$

(d) $\exists t,t\in\mathbb{R},\ t^2-3=4$

(e) $\exists t,t\in\mathbb{R},\ t^2+3=0$

9.6 Write the following conditions using mathematical symbols:

(a) x is not divisible by 3

(b) y is a number between 3 and 60 inclusive

(c) w is an even number greater than 20

(d) t differs from t_{n-1} by less than 0.001

9.7 Express the following problems mathematically and solve them:

(a) A set of screwdrivers costs £10 and a set of hammers costs £6.50. Find the possible combinations of maximum numbers of screwdriver and hammer sets that can be bought for £40.

(b) An object is thrown vertically upwards from the ground with an initial velocity of 10 m s^{-1}. The mass of the object is 1 kg. Find the maximum height that the object can reach using:

(i) Kinetic energy (K.E.) is given by $\frac{1}{2}mv^2$, where m is its mass and v its velocity.

(ii) The potential energy (P.E.) is mgh, where m is the mass, g is the acceleration due to gravity (which can be taken as 10 m s^{-2}) and h is height.

(iii) Assuming that no energy is lost as heat due to friction, then the conservation of energy law gives K.E. + P.E. = constant.

9.8 A road has a bend with radius of curvature 100m. The road is banked at an angle of 10°. At what speed should a car take the bend in order not to experience any side thrust on the tyres? Use the following assumptions:

(a) The sideways force needed on the vehicle in order to maintain it in circular motion (called the centripetal force) $=mv^2/r$, where r is the radius of curvature of the bend, v is the velocity and m the mass of the vehicle.

(b) The only force with a component acting sideways on the vehicle is the reactive force of the ground. This acts in a direction normal to the ground. (That is, we assume no frictional force in a sideways direction.)

(c) The force due to gravity of the vehicle is mg, where m is the mass of the vehicle and g is the acceleration due gravity (9.8 m s^{-2}). This acts vertically downwards.

The forces operating on the vehicle and ground, in a lateral or vertical direction, are pictured in Fig. 9.8.

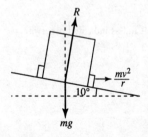

Figure 9.8 A vehicle rounding a banked bend in the road. R is the reactive force of the ground on the vehicle. The vehicle provides a force of mg, the weight of the vehicle, operating vertically downwards. The vehicle needs a sideways force of mv^2/r in order to maintain the locally circular motion.

BOOLEAN ALGEBRA

10.1 INTRODUCTION

Boolean algebra can be thought of as the study of the set $\{0,1\}$ with the operations $+$ (or), . (and) and $^-$ (not). It is particularly important because of its use in the design of logic circuits. Usually a high voltage represents TRUE (or 1), and a low voltage represents FALSE (or 0). The operation of OR $(+)$ is then performed on two voltage inputs using an OR gate and AND (.) using an AND gate, and NOT is performed using a NOT gate. This very simple algebra is very powerful as it forms the basis of computer hardware.

You will probably have noticed that the operations of $\wedge$ (AND), $\vee$ (OR) and $\neg$ (NOT) used in Chapter 9 for propositions are very similar to the operations $\cap$ (AND), $\cup$ (OR) and $'$ (NOT) (complement) used for sets. This connection is not surprising as membership of a set, A, could be defined using a statements like '3 is a member of A', which is either TRUE or FALSE. In simplifying logic circuits use is made of the different interpretations that can be put upon the operations and variables. We can use truth tables, borrowed from the theory of propositions, as given in Chapter 9, or we can use Venn diagrams, borrowed from set theory, as give in Chapter 7.

The first thing we shall examine in this chapter is what we mean by an algebra and why we are able to skip between these various interpretations. Then we look at implementing and minimizing logic circuits.

10.2 ALGEBRA

Before we look at Boolean algebra we shall take a look at some ideas about algebra:

(a) What is an algebra?
(b) What is an operation?
(c) What do we mean by the properties (or laws or axioms) of an algebra?

An algebra is a set with operations defined on it. In Chapter 1 we looked at the algebra of real numbers and defined an operation as a way of combining two numbers to give a single number. We could therefore define an operation as a way of combining two elements of the set to result in another element of the set.

Example 10.1 The set of real numbers, $\mathbb{R}$, has the operations '+' and '.', e.g. $3+5=8$ and $3.4=12$, and we could combine any two numbers in this way and we would always get another real number.

Example 10.2 Consider the set of sets in some universal set $\mathscr{E}$, e.g.

$$\mathscr{E} = \{a,b,c,d,e\}$$

$$A = \{a,d\}, \ B = \{a,b,c\}$$

Then $A \cap B = \{a\}$ and $A \cup B = \{a,b,c,d\}$. The operations of $\cap$ and $\cup$ also result in another set contained in $\mathscr{E}$.

In both of these examples the operations are *binary* operations because they use *two* inputs to give one output.

There is another sort of operation which is important, called a *unary* operation, because it only has *one* input to give one output. Consider Example 10.2: $A' = \{b,c,e\}$ gives the complement of A. This is a unary operation, as only one input, A, was needed to define the output A'.

If we can find a rule which is always true for an algebra, then that is called a property, law or axiom of that algebra. For example, $(3+5)+4 = 3+(5+4)$ is an application of the associative law of addition, which can be expressed in general in the following way for the set of real numbers:

$$\forall a,b,c \in \mathbb{R}, \quad (a+b)+c = a+(b+c)$$

If we can list all the properties of a particular algebra then we can give that algebra a name. For instance, the real numbers with the operations of '+' and '.' form a *field*.

10.3 BOOLEAN ALGEBRAS

Sets as a Boolean algebra

The sets contained in some universal set $\mathscr{E}$ display a number of properties which can be shown using Venn diagrams.

Example 10.3 Show, using Venn diagrams, that, for any three sets A, B, C in some universal set $\mathscr{E}$, $A \cap (B \cup C) = (A \cap B) \cup (A \cap C)$.

SOLUTION This can be shown to be true by drawing a Venn diagram of the left-hand side of the expression and a Venn diagram of the right-hand side of the expression. Operations are performed in the order indicated by the brackets, and the result of each operation is given a different shading. This is done in Figs 10.1(a) and (b). The region shaded in Fig. 10.1(a), representing $A \cap (B \cup C)$, is the same as that representing $(A \cap B) \cup (A \cap C)$ in Fig. 10.1(b), hence showing that $A \cap (B \cup C) = (A \cap B) \cup (A \cap C)$.

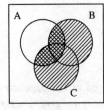
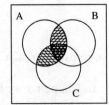

(a) $B \cup C$ ▨
$A \cap (B \cup C)$ ▩

(b) $A \cap B$ ▨
$A \cap C$ ▨
$(A \cap B) \cup (A \cap C)$ ≡

Figure 10.1 (a) A Venn diagram of $A \cap (B \cup C)$. (b) A Venn diagram of $(A \cap B) \cup (A \cap C)$.

In the same way other properties can be shown to be true. A full list of the properties gives, for every A, B, C$\subseteq\mathscr{E}$:

(B1)	$A\cup A=A$	$A\cap A=A$	idempotent
(B2)	$A\cup(B\cup C)=(A\cup B)\cup C$	$A\cap(B\cap C)=(A\cap B)\cap C$	associative
(B3)	$A\cup B=B\cup A$	$A\cap B=B\cap A$	commutative
(B4)	$A\cup(A\cap B)=A$	$A\cap(A\cup B)=A$	absorption
(B5)	$A\cup(B\cap C)=(A\cup B)\cap(A\cup C)$	$A\cap(B\cup C)=(A\cap B)\cup(A\cap C)$	distributive laws
(B6)	$A\cup\mathscr{E}=\mathscr{E}$	$A\cap\varnothing=\varnothing$	bound laws
(B7)	$A\cup\varnothing=A$	$A\cap\mathscr{E}=A$	identity laws
(B8)	$A\cup A'=\mathscr{E}$	$A\cap A'=\varnothing$	complement laws
(B9)	$\varnothing'=\mathscr{E}$	$\mathscr{E}'=\varnothing$	0 and 1 laws
(B10)	$(A\cup B)'=A'\cap B'$	$(A\cap B)'=A'\cup B'$	De Morgan's laws

Notice that all the laws come in pairs (called duals). A dual of a rule is given by replacing $\cup$ by $\cap$ and $\varnothing$ by $\mathscr{E}$ and vice versa.

Propositions

We looked at propositions in Chapter 9. Propositions can be given a value of either TRUE (T) or FALSE (F). Examples of propositions are $3=5$, which is false, and $2<3$, which is true. The logical operators of $\wedge$ (AND), $\vee$ (OR) and $\neg$ (NOT) are defined using truth tables, which we repeat in Table 10.1.

Table 10.1 The truth tables defining the logical operators

p	q	$p\wedge q$		p	q	$p\vee q$		p	$\neg p$
T	T	T		T	T	T		T	F
T	F	F		T	F	T		F	T
F	T	F		F	T	T			
F	F	F		F	F	F			

Properties of propositions and their operations can be shown using truth tables.

Example 10.4 Show, using truth tables, that for any propositions p, q, r,

$$(p\wedge q)\wedge r=p\wedge(q\wedge r)$$

SOLUTION The truth tables are given in Table 10.2. Note that there are eight lines in the truth table in order to represent all the possible states (T, F) for the three variables p, q and r. As each can be either TRUE or FALSE, then in total there are $2^3=8$ possibilities. To find $(p\wedge q)\wedge r$, $p\wedge q$ is performed first and the result of that is ANDed with r. To find $p\wedge(q\wedge r)$, then $q\wedge r$ is performed first and p is ANDed with the result. As the resulting columns are equal we can conclude that

$$(p\wedge q)\wedge r=p\wedge(q\wedge r)$$

Table 10.2 A truth table to show $(p \wedge q) \wedge r = p \wedge (q \wedge r)$

p	q	r	$p \wedge q$	$(p \wedge q) \wedge r$	$q \wedge r$	$p \wedge (q \wedge r)$
T	T	T	T	T	T	T
T	T	F	T	F	F	F
T	F	T	F	F	F	F
T	F	F	F	F	F	F
F	T	T	F	F	T	F
F	T	F	F	F	F	F
F	F	T	F	F	F	F
F	F	F	F	F	F	F

Example 10.5 Show that for any two propositions p,q:

$$\neg(p \wedge q) \Leftrightarrow (\neg p) \vee (\neg p)$$

SOLUTION The truth table is given in Table 10.3.

Table 10.3 A truth table to show $\neg(p \wedge q) \Leftrightarrow (\neg p) \vee (\neg q)$. The fourth column gives the truth values of $\neg(p \wedge q)$ and the seventh column gives the truth value of $(\neg p) \vee (\neg p)$. As the two columns are the same we can conclude that $\neg(p \wedge q) = (\neg p) \vee (\neg q)$

p	q	$p \wedge q$	$\neg(p \wedge q)$	$\neg p$	$\neg q$	$\neg p \vee \neg q$
T	T	T	F	F	F	F
T	F	F	T	F	T	T
F	T	F	T	T	F	T
F	F	F	T	T	T	T

It turns out that all the properties we listed for sets are also true for propositions. We list them again: for any three propositions p, q, r

(B1)	$p \vee p \Leftrightarrow p$	$p \wedge p \Leftrightarrow p$	idempotent
(B2)	$p \vee (q \vee r) \Leftrightarrow (p \vee q) \vee r$	$p \wedge (q \wedge r) \Leftrightarrow (p \wedge q) \wedge r$	associative
(B3)	$p \vee q \Leftrightarrow q \vee p$	$p \wedge q \Leftrightarrow q \wedge p$	commutative
(B4)	$p \vee (p \wedge q) \Leftrightarrow p$	$p \wedge (p \vee q) \Leftrightarrow p$	absorption
(B5)	$p \vee (q \wedge r) \Leftrightarrow (p \vee q) \wedge (p \vee r)$	$p \wedge (q \vee r) \Leftrightarrow (p \wedge q) \vee (p \wedge r)$	distributive laws
(B6)	$p \vee T \Leftrightarrow T$	$p \wedge F \Leftrightarrow F$	bound laws
(B7)	$p \vee F \Leftrightarrow p$	$p \wedge T \Leftrightarrow p$	identity laws
(B8)	$p \vee \neg p \Leftrightarrow T$	$p \wedge \neg p \Leftrightarrow F$	complement laws
(B9)	$\neg F \Leftrightarrow T$	$\neg T \Leftrightarrow F$	0 and 1 laws
(B10)	$\neg(p \vee q) \Leftrightarrow \neg p \wedge \neg q$	$\neg(p \wedge q) \Leftrightarrow \neg p \vee \neg q$	De Morgan's laws

Notice again that all the laws are duals of each other. A dual of a rule is given by replacing $\vee$ by $\wedge$ and F by T, and vice versa.

The Boolean set {0,1}

The simplest Boolean algebra is that defined on the set $\{0,1\}$. The operations on this set are AND (.), OR (+), and NOT ($^-$). The operations can be defined using truth tables as in Table 10.1, shown again in Table 10.4. This time notice that the first two are usually ordered in order to mimic binary counting, starting with 0 0, then 0 1, then 1 0 and then 1 1. This is merely a convention, and the rows may be ordered any way you like. a and b are two variables which may take the values 0 or 1.

Table 10.4 The operations of AND (.), OR (+) and NOT ($^-$) defined for any variables a, b taken from the Boolean set {0,1}

a	b	$a.b$	a	b	$a+b$	a	$\bar{a}$
0	0	0	0	0	0	0	1
0	1	0	0	1	1	1	0
1	0	0	1	0	1		
1	1	1	1	1	1		

This now looks far more like arithmetic. However, beware, because although the operation AND behaves like multiplication, $0.0=0$, $0.1=0$, $1.0=0$ and $1.1=1$, as in 'ordinary' arithmetic, the operation OR behaves differently: $1+1=1$.

All the laws as given for sets and for propositions hold again and they can be listed as follows. For any three variables a, b, $c \in \{0,1\}$:

(B1)	$a+a=a$	$a.a=a$	idempotent
(B2)	$a+(b+c)=(a+b)+c$	$a.(b.c)=(a.b).c$	associative
(B3)	$a+b=b+a$	$a.b=b.a$	commutative
(B4)	$a+(a.b)=a$	$a.(a+b)=a$	absorption
(B5)	$a+(b.c)=(a+b).(a+c)$	$a.(b+c)=(a.b)+(a.c)$	distributive laws
(B6)	$a+1=1$	$a.0=0$	bound laws
(B7)	$a+0=a$	$a.1=a$	identity laws
(B8)	$a+\bar{a}=1$	$a.\bar{a}=0$	complement laws
(B9)	$\bar{0}=1$	$\bar{1}=0$	0 and 1 laws
(B10)	$\overline{(a+b)}=\bar{a}.\bar{b}$	$\overline{(a.b)}=\bar{a}+\bar{b}$	De Morgan's laws

We often leave out the '.' so that 'ab' means '$a.b$'. We also adopt the convention that '.' takes priority over '+', and hence miss out some of the brackets.

Example 10.6 Evaluate the following, where '+', '.' and $^-$ are Boolean operators:

(a) $1.1.0+\bar{0}.1$

(b) $(1.\bar{1})+1$

(c) $(\bar{1}+1).0+(1+1).0$

SOLUTION
(a) We use the convention that '.' is performed first: $1.1.0 + \bar{0}.1 = 0 + 1.1 = 0 + 1 = 1$
(b) $(1.\bar{1}) + 1 = (1.0) + 1 = 0 + 1 = 1$
(c) $(1 + 1).0 + (1 + 1).0 = 1.0 + 1.0 = 0 + 0 = 0$

The algebraic laws can be used to simplify a Boolean expression.

Example 10.7 Simplify

$$abc + \bar{a}bc + b\bar{c}$$

SOLUTION

$$abc + \bar{a}bc + b\bar{c}$$

$$= (a + \bar{a})bc + b\bar{c} \qquad \text{(using a distributive law)}$$

$$= 1.bc + b\bar{c} \qquad \text{(using a complement law)}$$

$$= bc + b\bar{c} \qquad \text{(using an identity law)}$$

$$= b(c + \bar{c}) \qquad \text{(by one of the distributive laws)}$$

$$= b \qquad \text{(using a complement and identity law)}$$

Although it is possible to simplify in this way it can be quite difficult to spot the best way to perform the simplification, and hence there are special techniques used in the design of digital circuits which are more efficient.

10.4 DIGITAL CIRCUITS

Switching circuits form the basis of computer hardware. Usually a high voltage represents TRUE (or 1) while a low voltage represents FALSE (or 0). Digital circuits can be represented using letters for each input.

There are three basic gates which combine inputs, representing the operators NOT (¯), AND (.) and OR (+). These are shown in Fig. 10.2.

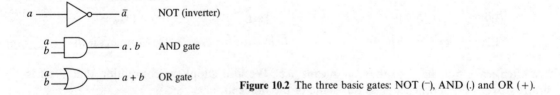

Figure 10.2 The three basic gates: NOT (¯), AND (.) and OR (+).

Other gates

Other common gates used in the design of digital circuits are the NAND gate $(\overline{ab})$, i.e. not(ab), the NOR gate $(\overline{a+b})$, i.e. not$(a+b)$ and the EXOR gate $(a \oplus b)$, i.e. exclusive or: $a \oplus b = a\bar{b} + \bar{a}b$. These gates are shown in Fig. 10.3.

$a \over b$ ⟶ $\overline{a.b}$ NAND gate

$a \over b$ ⟶ $\overline{a+b}$ NOR gate

$a \over b$ ⟶ $a \oplus b$ EXOR gate $a \oplus b = a\overline{b} + \overline{a}b$

Figure 10.3 Three other common gates: NAND ($\overline{ab}$), NOR ($\overline{a+b}$) and EXOR ($a \oplus b = a\overline{b} + \overline{a}b$).

Implementing a logic circuit

First, we need to simplify the expression. Each letter represents an input that can be on or off (1 or 0). The operations between inputs are represented by the gates. The output from the circuit represents the entire Boolean expression.

> **Example 10.8** Implement $ab\overline{c} + a\overline{b} + a\overline{c}$.

SOLUTION We can use absorption to write this as $a\overline{b} + a\overline{c}$, and this can be implemented as in Fig. 10.4 using AND, OR and NOT gates. Alternatively, we can use the distributive and De Morgan's laws to write the expression as $a\overline{b} + \overline{c}a = a(\overline{b} + \overline{c}) = a\overline{bc}$ which can be implemented using an AND gate and a NAND gate.

(a)

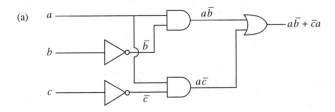

(b)

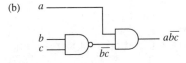

Figure 10.4 (a) An implementation of $ab\overline{c} + a\overline{b} + a\overline{c} = a\overline{b} + a\overline{c}$. (b) An alternative implementation using $a\overline{b} + a\overline{c} = a\overline{bc}$.

Minimization and Karnaugh maps

It is clear that there are numbers of possible implementations of the same logic circuit. However, in order to use fewer components in building the circuit it is important to be able to minimize the Boolean expression. There are several methods for doing this. A popular method is to use a Karnaugh map. Before using a Karnaugh map, the Boolean expression must be written in the form of a 'sum of products'. To do this we may either use some of the algebraic rules or it may be simpler to produce a truth table and then copy the 0s and 1s into the Karnaugh map. Example 10.9 is initially in the sum of product form and Example 10.10 uses a truth table to find the Karnaugh map.

> **Example 10.9** Minimize $ab + \overline{a}b + a\overline{b}$ using a Karnaugh map and draw the implementation of the resulting expression as a logic circuit.

SOLUTION Draw a Karnaugh map as in Fig. 10.5(a). If there are two variables in the expression then there are $2^2 = 4$ squares in the Karnaugh map. Figure 10.5(b) shows a Karnaugh map

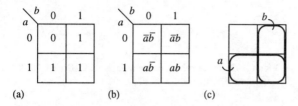

Figure 10.5 (a) A two-variable Karnaugh map representing $ab+\bar{a}b+a\bar{b}$. (b) A two-variable Karnaugh map with all the boxes labelled. (c) A Karnaugh map is like a Venn diagram. The second row represents the set a and the second column represents the set b.

with the squares labelled term by term. Figure 10.5(c) shows that the map is like a Venn diagram of the sets a and b. In Figure 10.5(a) we put a 0 or 1 in the square depending on whether that term is present in our expression. Adjacent 1s indicate that we can simplify the expression. Figure 10.6 indicates how we go about the minimization. We draw a line around any two adjacent 1s and write down the term representing that section of the map. We are able to encircle the second row, representing a, and the second column, representing b. As all the 1s have now been included we know that $a+b$ is a minimization of the expression. Notice that it does not matter if one of the squares with a 1 in it has been included twice, but we must not leave any out. The implementation of $a+b$ is drawn in Fig. 10.7.

Figure 10.6 A two-variable Karnaugh map representing $ab+\bar{a}b+a\bar{b}$.

Figure 10.7 An implementation of $a+b$.

Example 10.10 Minimize $c(b+\overline{(ab)})+\bar{c}ab$ and draw the implementation of the resulting expression as a logic circuit.

SOLUTION First we need to find the expression as a sum of products. This can be done by finding the truth table and then copying the result into the Karnaugh map. The truth table is found in Table 10.5. Notice that we calculate various parts of the expression and build up to the final expression. With practice the expression can be calculated directly; for instance, when $a=0$, $b=0$ and $c=0$ then $c(b+\overline{(ab)})+\bar{c}ab=0(0+\overline{(0.0)})+\bar{0}.0.0=0(0+1)+1.0=0$.

Table 10.5 A truth table to find $c(b+\overline{(ab)})+\bar{c}ab$

a	b	c	ab	$\bar{c}$	$\overline{ab}$	$\bar{c}ab$	$b+\overline{(ab)}$	$c(b+\overline{(ab)})$	$c(b+\overline{(ab)})+\bar{c}ab$
0	0	0	0	1	1	0	1	0	0
0	0	1	0	0	1	0	1	1	1
0	1	0	0	1	1	0	1	0	0
0	1	1	0	0	1	0	1	1	1
1	0	0	0	1	1	0	1	0	0
1	0	1	0	0	1	0	1	1	1
1	1	0	1	1	0	1	1	0	1
1	1	1	1	0	0	0	1	1	1

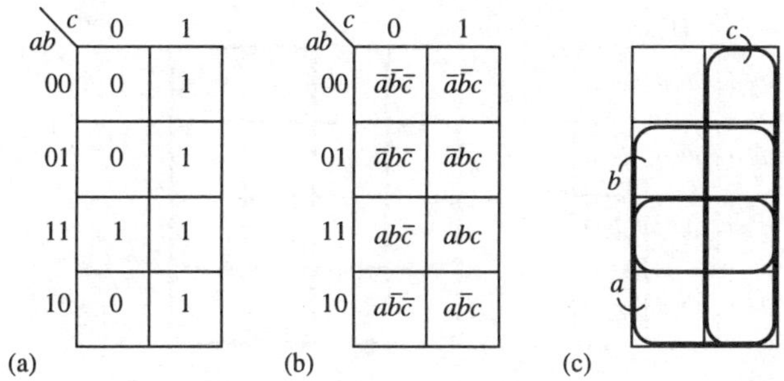

(a)

ab \ c	0	1
00	0	1
01	0	1
11	1	1
10	0	1

(b)

ab \ c	0	1
00	$\bar{a}\bar{b}\bar{c}$	$\bar{a}\bar{b}c$
01	$\bar{a}b\bar{c}$	$\bar{a}bc$
11	$ab\bar{c}$	abc
10	$a\bar{b}\bar{c}$	$a\bar{b}c$

(c)

Figure 10.8 (a) A three-variable Karnaugh map representing $c(b+\overline{(ab)})+\bar{c}ab$. (b) A three-variable Karnaugh map with all the boxes labelled. (c) A Karnaugh map is like a Venn diagram. The third and fourth rows represent the set a and the second and third rows represent the set b. c is represented by the second column.

Draw a Karnaugh map as in Fig. 10.8(a) and copy in the expression values as found in Table 10.5. There are three variables in the expression, so therefore there are $2^3 = 8$ squares in the Karnaugh map. Figure 10.8(b) shows a Karnaugh map with the squares labelled term by term. Figure 10.8(c) shows the Venn diagram equivalence with sets a, b and c. In Fig. 10.8(a) we put a 0 or 1 in the square depending on whether that term is present, as given in the truth table in Table 10.5.

Adjacent 1s indicate that we can simplify the expression. Figure 10.9 indicates how we go about the minimization. We draw a line around any four adjacent 1s and write down the term representing that section of the map. The second column represents c and has been encircled. Then we look for any two adjacent 1s. We are able to encircle the third row, representing ab. As all the 1s have now been included we know that $c + ab$ is a minimization of the expression. An implementation of $c + ab$ is drawn in Fig. 10.10.

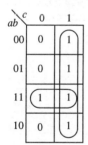

Figure 10.9 A three-variable Karnaugh map representing $c(b+\overline{(ab)})+\bar{c}ab$.

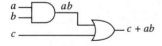

Figure 10.10 An implementation of $c + ab$.

Example 10.11 Minimize $ab\bar{c} + \bar{a}bd + abc\bar{d} + a\bar{b}\bar{c}d + abc$ using a Karnaugh map and draw the implementation of the resulting expression as a logic circuit.

SOLUTION Draw a Karnaugh map as in Fig. 10.11(a). There are four variables in the expression, so there are $2^4 = 16$ squares in the Karnaugh map. Figure 10.11(b) shows a Karnaugh map with the squares labelled term by term. Figure 10.11(c) shows the Venn

(a)

ab \ cd	00	01	11	10
00	0	0	0	0
01	0	1	1	0
11	1	1	1	1
10	0	1	0	0

(b)

ab \ cd	00	01	11	10
00	$\bar{a}\bar{b}\bar{c}\bar{d}$	$\bar{a}\bar{b}\bar{c}d$	$\bar{a}\bar{b}cd$	$\bar{a}\bar{b}c\bar{d}$
01	$\bar{a}b\bar{c}\bar{d}$	$\bar{a}b\bar{c}d$	$\bar{a}bcd$	$\bar{a}bc\bar{d}$
11	$ab\bar{c}\bar{d}$	$ab\bar{c}d$	$abcd$	$abc\bar{d}$
10	$a\bar{b}\bar{c}\bar{d}$	$a\bar{b}\bar{c}d$	$a\bar{b}cd$	$a\bar{b}c\bar{d}$

(c) *(Karnaugh map drawn as a Venn diagram, with labels d, c across the top and b, a down the left side.)*

Figure 10.11 (a) A four-variable Karnaugh map representing $ab\bar{c}+\bar{a}bd+abc\bar{d}+a\bar{b}\bar{c}d+abc$. (b) A four-variable Karnaugh map with all the squares labelled. (c) A Karnaugh map is like a Venn diagram. The third and fourth rows represent the set a and the second and third rows represent the set b. c is represented by the third and fourth columns and d by the second and third columns.

diagram equivalence with sets a, b, c and d. In Fig. 10.11(a) we put a 0 or 1 in the square depending on whether that term is present in our expression. However, the term $ab\bar{c}$ involves only three of the four variables. In this case it must occupy two squares. As d could be either 0 or 1 for '$ab\bar{c}$' to be true, we fill in the squares for $ab\bar{c}d$ and $ab\bar{c}\bar{d}$. The number of squares to be filled with a 1 to represent a certain product is 2^m, where m is the number of missing variables in the expression. In this case $ab\bar{c}$ has no d term in it so the number of squares representing it is 2^1.

Adjacent 1s indicate that we can simplify the expression. Figure 10.12 indicates how we go about the minimization. We draw a line around any eight adjacent 1s, of which there are none. Next we look for any four adjacent 1s and write down the term representing that section of the map. The third row represents ab and has been encircled. The middle four squares represent bd and have been encircled. Then we look for any two adjacent 1s. The bottom two squares of the second column represent $a\bar{c}d$. As all the 1s have now been included we know that $ab+bd+a\bar{c}d$ is a minimization of the expression. This is implemented in Fig. 10.13.

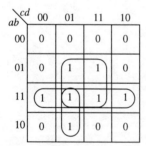

Figure 10.12 A four-variable Karnaugh map representing $ab\bar{c}+\bar{a}bd+abc\bar{d}+a\bar{b}\bar{c}d+abc$.

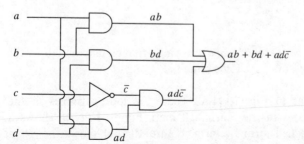

Figure 10.13 An implementation of $ab+bd+a\bar{c}d$.

Example 10.12 To display the digits 0 to 9 a seven segment LED (light emitting diode) display may be used as shown in Fig. 10.14. The various states may be represented using a four-variable digital circuit. The logic control signals for the lamp drivers are given by the truth table given in Table 10.6. The X indicates a 'don't care' condition in the truth table. The column for the segment labelled p can be copied into a Karnaugh map as given in Fig. 10.15. Wherever a 1 appears in the truth table representation for p there is a 1 copied to the Karnaugh map. Similarly the 0s and the 'don't care' crosses are copied. Minimize the Boolean expression for p using the Karnaugh map.

Figure 10.14 A seven-segment LED (light emitting diode) display. The number 1 requires the segments labelled q and r to light up and the other segments to be off.

(a)

ab\\cd	00	01	11	10
00	1	0	1	1
01	0	1	1	0
11	X	X	X	X
10	1	1	X	X

(b)

ab\\cd	00	01	11	10
00	1	0	1	1
01	0	1	1	0
11	X	X	X	X
10	1	1	X	X

Figure 10.15 (a) A Karnaugh map for the segment labelled P in Fig. 10.14. This has been copied from the truth table given in Table 10.5. (b) A minimization of the Karnaugh map. The 'don't care' X's may be treated as 1's if it is convenient but they can also be treated as 0's.

Table 10.6 A truth table giving the logic control signals for the lamp drivers for the LED segments pictured in Fig. 10.14

Digit displayed	Circuit inputs				Segments						
	a	b	c	d	p	q	r	s	t	u	v
0	0	0	0	0	1	1	1	1	1	1	0
1	0	0	0	1	0	1	1	0	0	0	0
2	0	0	1	0	1	1	0	1	1	0	1
3	0	0	1	1	1	1	1	1	0	0	1
4	0	1	0	0	0	1	1	0	0	1	1
5	0	1	0	1	1	0	1	1	0	1	1
6	0	1	1	0	0	0	1	1	1	1	1
7	0	1	1	1	1	1	1	0	0	0	0
8	1	0	0	0	1	1	1	1	1	1	1
9	1	0	0	1	1	1	1	1	0	1	1
—	1	0	1	0	X	X	X	X	X	X	X
—	1	0	1	1	X	X	X	X	X	X	X
—	1	1	0	0	X	X	X	X	X	X	X
—	1	1	0	1	X	X	X	X	X	X	X
—	1	1	1	0	X	X	X	X	X	X	X
—	1	1	1	1	X	X	X	X	X	X	X

SOLUTION The minimization is represented in Fig. 10.15(b). We first look for any eight adjacent squares with a 1 or an X in them. The bottom two rows are encircled, giving the term a. Now we look for groups of four. The central four squares represent bd and the third column represents cd. Finally, we can count the four corner squares as adjacent. This is because two squares may be considered as adjacent if they are located symmetrically with respect to any of the lines which divide the Karnaugh map into equal halves, quarters or eighths. This means that squares that could be curled round to meet each other, as if the Karnaugh map were drawn on a cylinder, are considered adjacent, as also are the four corner squares. Here the four corner squares represent the term $\bar{b}\,\bar{d}$. Hence the minimization for p gives $p = a + cd + bd + \bar{b}\,\bar{d}$.

10.5 SUMMARY

1. An algebra is a set with operations defined on it. A *binary* operation is a way of combining *two* elements of the set to result in another element of the set. A unary operation has only one input element producing one output.
2. A Boolean algebra has the operations of AND, OR and NOT defined on it and obeys the set of laws given in Sec. 10.3 as (B1) to (B10). Examples of a Boolean algebra are the set of sets in some universal set $\mathscr{E}$, with the operations of $\cap$, $\cup$ and $'$; the set of propositions with the operations of $\wedge$, $\vee$ and $\neg$; and the set $\{0,1\}$ with the operations '.', '+' and '‾'.
3. Logic circuits can be represented as Boolean expressions. Usually a high voltage is represented by 1 or TRUE and a low voltage by 0 or FALSE. There are three basic gates to represent the operators AND (.), OR (+) and NOT (‾).
4. A Boolean expression may be minimized by first expressing it as a sum of products and then using a Karnaugh map to combine terms.

10.6 EXERCISES

10.1 Show the following properties of sets using Venn diagrams:

(a) $A \cup (A \cap B) = A$ (b) $A \cup (B \cap C) = (A \cup B) \cap (A \cup C)$

10.2 $p = $ 'It rained yesterday'
$q = $ 'I used an umbrella yesterday'

(a) Construct English sentences to express the following ($\wedge \Leftrightarrow$ 'and', $\vee \Leftrightarrow$ 'or', $\neg \Leftrightarrow$ 'not'):

   (i) $p \wedge q$ (ii) $p \vee q$ (iii) $\neg p \vee q$
   (iv) $p \wedge \neg q$ (v) $\neg (p \wedge q)$ (vi) $\neg p \vee \neg q$

(b) Given that p is true and q is false, what is the truth value of each part of section (a)?

10.3 Show the following properties of propositions using truth tables:

(a) $p \vee (p \wedge q) \Leftrightarrow p$ (b) $\neg (p \vee q) \Leftrightarrow \neg p \wedge \neg q$

10.4 Using Venn diagrams or truth tables find simpler expressions for the following:

(a) $ab + a\bar{b}$ (b) $(ab)(ac)$
(c) $a + abc$ (d) $\overline{(ab)}a$

10.5 (a) Draw implementations of the following as logic circuits:

     (i) $\bar{a}b + a\bar{b}$ (ii) $a + bc$
     (iii) $\bar{a} + ab$ (iv) $\bar{a}\bar{b}c$

(b) If $a = 1$, $b = 0$ and $c = 1$, what is the value of each of the expressions in section (a)?

10.6 Minimize the following expressions and draw their logic circuits:

(a) $ab\bar{c} + a\bar{b} + abc$ (b) $\overline{abc}(a + b + c) + a$
(c) $\overline{(\bar{a} + c)}(a + b) + ab$ (d) $(a + b)(a + d) + ab\bar{c} + ab\bar{d} + abcd$

10.7 Obtain a Boolean expression for the logic networks shown in Fig. 10.16.

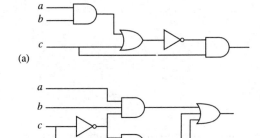

(a)

(b)   **Figure 10.16** Logic networks for Exercise 10.7.

10.8 Consider the LED segment labelled r in Fig. 10.14, given in the text. Follow the method given in Example 10.12 to find a minimized expression for r and draw its logic network.

ELEVEN

TRIGONOMETRIC FUNCTIONS AND WAVES

11.1 INTRODUCTION

Waves occur naturally in a number of situations: the movement of disturbed water, the passage of sound through the air, the vibrations of a plucked string. If the movement of a particular particle is plotted against time, we get the distinctive wave shape, called a sinusoid. The mathematical expression of a wave is found by using the trigonometric functions, sine and cosine. In Chapter 6 we looked at right-angled triangles and defined the trigonometric ratios. The maximum angle in a right-angled triangle is $90°$, so to find the trigonometric functions $\sin(t)$, $\cos(t)$ and $\tan(t)$, where t can be extended over the real numbers, we need a new way of defining them. This we do by using a rotating rod. Usually the function will be used to relate, for instance, the height of the rod to time. Therefore it does not always make sense to think of the input to the cosine and sine functions as being an angle. This problem is overcome by using a new measure for the angle called the radian, which easily relates the angle to the distance travelled by the tip of the rotating rod.

Waves may interfere with each other, as for instance on a plucked string, where the disturbance bounces off the ends, producing a standing wave. Amplitude modulation of, for instance, radio waves, works by the superposition of a message on a higher signal frequency. These situations require an understanding of what happens when two or more cosine or sine functions are added, subtracted or multiplied, and hence we also study trigonometric identities.

11.2 TRIGONOMETRIC FUNCTIONS AND RADIANS

Consider a rotating rod of length 1. Imagine, for instance, that it is a position marked on a bicycle tyre at the tip of one of the spokes, as the bicycle travels along. The distance travelled by the tip of the rod in one complete revolution is 2π (the circumference of the circle of radius 1). The height of the rod (measured from the centre of the wheel), y, can be plotted against the distance travelled by the tip, as in Fig. 11.1. Similarly, the position to the right or left of the origin, x, can be plotted against the distance travelled by the tip of the rod, as in Fig. 11.2. Figure 11.1 defines the function $y = \sin(t)$ and Fig. 11.2 defines the function $x = \cos(t)$.

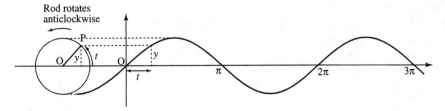

Figure 11.1 The function $y = \sin(t)$, where t is the distance travelled by the tip of a rotating rod of length 1 unit and y is the height of the rod.

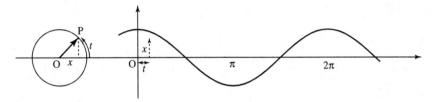

Figure 11.2 The function $x = \cos(t)$, where t is the distance travelled by the tip of a rotating rod of length 1 unit and x is the position to the right or left of the origin.

This definition of the trigonometric function is very similar to that used for the ratios in the triangle, if the hypotenuse is of length 1 unit. The definitions become the same for angles up to a right angle if radians are used as a measure of the angle in the triangle instead of degrees.

Instead of 360° making a complete revolution, 2π radians make a complete revolution. Some examples of degree to radian conversion are given in Fig. 11.3.

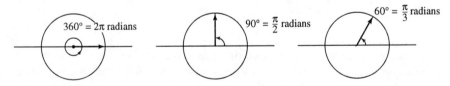

Figure 11.3 (a) $360° = 2\pi$ radians. (b) $90° = \pi/2$ radians. (c) $60° = \pi/3$ radians.

To convert degrees to radians use the fact that 360° is the same as 2π radians, or equivalently that 180° is the same as π radians. Hence, to convert degrees to radians multiply by $\pi/180$, and to convert radians to degrees multiply by $180/\pi$.

Remember that π is approximately 3.1416, so these conversions can be expressed approximately as: to convert degrees to radians multiply by 0.017 45 (that is, $1° \simeq 0.017\,45$ radians), and to convert radians to degrees multiply by 57.3 (that is, 1 radian $\simeq 57.3°$).

Example 11.1
(a) Express 45° in radians.
        Multiply 45 by $\pi/180$, giving $\pi/4 \simeq 0.785$. Hence $45° \simeq 0.785$ radians.
(b) Express 17° in radians.
        Multiply 17 by $\pi/180$, giving $17\pi/180 \simeq 0.297$. Hence $17° \simeq 0.297$ radians.
(c) Express 120° in radians.
        Multiply 120 by $\pi/180$ giving $2\pi/3 \simeq 2.094$. Hence $120° \simeq 2.094$ radians.

(d) Express 2 radians in degrees.

Multiply 2 by $180/\pi$, giving 114.6. Hence 2 radians $\simeq 114.6°$.

(e) Express $5\pi/6$ radians in degrees.

Multiply $5\pi/6$ by $180/\pi$, giving 150. Hence $5\pi/6$ radians $= 150°$.

(f) Express 0.5 radians in degrees.

Multiply 0.5 by $180/\pi$, giving 28.6. Hence 0.5 radians $\simeq 28.6°$.

The trigonometric functions can also be defined using a rotating rod of length r, as in Fig. 11.4. The function values are given by:

$$\cos(\alpha) = \frac{x}{r}, \qquad \sin(\alpha) = \frac{y}{r}, \qquad \tan(\alpha) = \frac{y}{x} = \frac{\sin(\alpha)}{\cos(\alpha)}$$

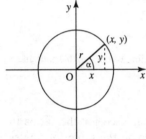

Figure 11.4 The trigonometric functions defined in terms of a rotating rod of length r.

Also:

$$\sec(\alpha) = \frac{1}{\cos(\alpha)}, \qquad \operatorname{cosec}(\alpha) = \frac{1}{\sin(\alpha)}, \qquad \cotan(\alpha) = \frac{1}{\tan(\alpha)}$$

where α is measured in radians (one complete revolution is 2π radians).

Notice that the definitions are exactly the same as given in Chapter 6 for the trigonometric ratios, where r is the hypotenuse and x and y are the adjacent and opposite sides to the angle, except that x and y can now take both positive and negative values and the angles can be as big as we like or negative (if the rod rotates clockwise):

$$\cos(\alpha) = \frac{\text{adjacent}}{\text{hypotenuse}}$$

$$\sin(\alpha) = \frac{\text{opposite}}{\text{hypotenuse}}$$

$$\tan(\alpha) = \frac{\text{opposite}}{\text{adjacent}}$$

etc.

To get the correct function values from the calculator the calculator should be in 'radian' mode. However, by custom engineers often use degrees, so we will use the convention that if the 'units' are not specified then radians must be used, and for the input to be in degrees it must be expressly marked, e.g. $\cos(30°)$.

Important relationship between the sine and the cosine

From Pythagoras's theorem, looking at the diagram in Fig. 11.4 we have $x^2 + y^2 = r^2$. Dividing both sides by r^2 we get:

$$\frac{x^2}{r^2} + \frac{y^2}{r^2} = 1$$

and using the definitions of $\sin(\alpha) = y/r$ and $\cos(\alpha) = x/r$ we get

$$(\cos(\alpha))^2 + (\sin(\alpha))^2 = 1$$

and this is written in shorthand as

$$\cos^2(\alpha) + \sin^2(\alpha) = 1$$

were $\cos^2(\alpha)$ means $(\cos(\alpha))^2$.

11.3 GRAPHS AND IMPORTANT PROPERTIES

We can now draw graphs of the functions for all input values t as in Figs 11.5, 11.6 and 11.7.

These are all important examples of periodic functions. To show that the $\cos(t)$ or $\sin(t)$ function is periodic, translate the graph to the left ot right by 2π. The resulting graph will fit exactly on top of the original untranslated graph. 2π is called the fundamental period, as translating by 4π, 6π, 8π etc. also results in the graph fitting exactly on top of the original function. The fundamental period is therefore defined as the smallest period that has this property, and all other periods are multiples of the fundamental period. This periodic property can be expressed using a letter, n, to represent any integer, giving

$$\sin(t + 2\pi n) = \sin(t)$$

$$\cos(t + 2\pi n) = \cos(t)$$

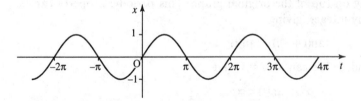

Figure 11.5 The graph of $y = \sin(t)$, where t can take any value. Notice that the function repeats itself every 2π. This shows that the function is periodic with period 2π. Notice also that the value of $\sin(t)$ is never more than 1 and never less than -1. The function is odd as $\sin(-t) = -\sin(t)$.

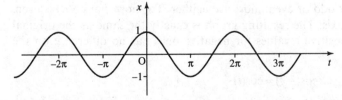

Figure 11.6 The graph of $x = \cos(t)$, where t can take any value. Notice that the function repeats itself every 2π. This shows that the function is periodic with period 2π. Notice also that the value of $\cos(t)$ is never more than 1 and never less than -1. The function is even as $\cos(-t) = \cos(t)$.

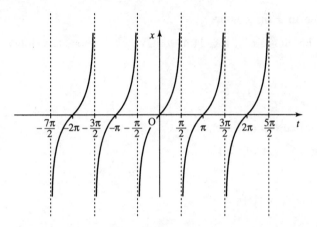

Figure 11.7 The graph of $z = \tan(t)$, where t can take any value except odd multiples of $\pi/2$ (for instance $\tan(t)$ is not defined for $t = \pi/2$, $3\pi/2$, $5\pi/2$). Notice that the function repeats itself every π. This shows that the function is periodic with period π. The function values extend from $-\infty$ to ∞, i.e. the range of $\tan(t)$ is all the real numbers. The function is odd as $\tan(-t) = -\tan(t)$.

Thus adding or subtracting any multiple of 2π to or from the value of t gives the same value of the functions $x = \cos(t)$, $y = \sin(t)$.

The other important thing to remember about $\cos(t)$ and $\sin(t)$ is that although the domain of the functions is all the real numbers, the function values themselves lie between -1 and $+1$:

$$-1 \leqslant \cos(t) \leqslant 1$$

$$-1 \leqslant \sin(t) \leqslant 1$$

We say that the functions are bounded by -1 and $+1$ or, in other words, the range of the cosine and sine functions is $[-1,1]$.

$z = \tan(t)$ has fundamental period π. If the graph is translated by π to the left or right then the resulting graph will fit exactly on top of the original graph. This periodic property can be expressed using n to represent any integer, giving

$$\tan(t + \pi n) = \tan(t)$$

The values of $\tan(t)$ are not bounded. We can also say that

$$-\infty < \tan(t) < \infty$$

Symmetry

Other important properties of these functions are the symmetries of the functions. $\cos(t)$ is an even function and $\sin(t)$ and $\tan(t)$ are odd. Unlike the terms odd and even when used to describe numbers, not all functions are either odd or even: most are neither. To show that $\cos(t)$ is even, reflect the graph along the vertical axis. The resulting graph is exactly the same as the original graph. This shows that swapping positive t values for negative ones has no difference on the function values, that is

$$\cos(-t) = \cos(t)$$

Other examples of even functions were given in Chapter 8 and the general property of even functions was given there as $f(t) = f(-t)$. The functions sine and tangent are odd. If they are reflected along the vertical axis then the resulting graph is an upside down version of the original.

This shows that swapping positive t values for negative ones gives the negative of the original function. This property can be expressed by

$$\sin(-t) = -\sin(t)$$

$$\tan(-t) = -\tan(t)$$

For a general function $y = f(t)$ the property of being odd can be expressed by $f(-t) = -f(t)$.

The relationships between the sine and cosine

Take the graph of $\sin(t)$ and translate it to the left by $90°$ or $\pi/2$ and we get the graph of $\cos(t)$. Equivalently, take the graph of $\cos(t)$ and translate it to the right by $\pi/2$ and we get the graph of $\sin(t)$. Using the ideas of translating functions given in Chapter 8 we get

$$\sin(t + \pi/2) = \cos(t)$$

$$\cos(t - \pi/2) = \sin(t)$$

Other relationships can be shown using triangles as in Fig. 11.8, giving $\cos(\alpha - \pi/2) = \sin(\alpha)$ and $\sin(\pi/2 - \alpha) = \cos(\alpha)$.

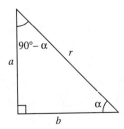

Figure 11.8 $\sin(\alpha) = a/r$ and $\cos(90° - \alpha) = a/r$. Then $\cos(90° - \alpha) = \sin(\alpha)$. As the cosine is an even function, $\cos(90° - \alpha) = \cos(-(90° - \alpha)) = \cos(\alpha - 90°)$, which confirms that $\cos(\alpha - 90°) = \sin(\alpha)$.
$\cos(\alpha) = b/r$ and $\sin(90° - \alpha) = b/r$, so $\cos(\alpha) = \sin(90° - \alpha)$.

From Pythagoras we also have that

$$a^2 + b^2 = r^2$$

Dividing both sides by r^2 we get

$$\frac{a^2}{r^2} + \frac{b^2}{r^2} = 1$$

and using the definitions of $\sin(\alpha) = a/r$ and $\cos(\alpha) = b/r$ we get

$$\cos^2(\alpha) + \sin^2(\alpha) = 1$$

and rearranging this we have $\cos^2(\alpha) = 1 - \sin^2(\alpha)$ or $\sin^2(\alpha) = 1 - \cos^2(\alpha)$.

Example 11.2 Given $\sin(A) = 0.5$ and $0 \leqslant A \leqslant 90°$, use trigonometric identities to find:

(a) $\cos(A)$
(b) $\sin(90° - A)$
(c) $\cos(90° - A)$

SOLUTION

(a) Using $\cos^2(A)=1-\sin^2(A)$ and $\sin(A)=0.5$

$$\Rightarrow\cos^2(A)=1-(0.5)^2=0.75$$
$$\Leftrightarrow\cos(A)\simeq\pm0.866$$

As A is between $0°$ and $90°$ then the cosine must be positive giving $\cos(A)\simeq0.866$.

(b) As $\sin(90°-A)=\cos(A)$ then $\sin(90°-A)\simeq0.866$.

(c) As $\cos(90°-A)=\sin(A)$ then $\cos(90°-A)=0.5$.

The functions $A\cos(at+b)+B$ and $A\sin(at+b)+B$

The graphs of these functions can be found by using the ideas of Chapter 8 for graph sketching.

Example 11.3 Sketch the graph of y against t, where

$$y=2\cos\left(2t+\frac{2\pi}{3}\right)$$

The stages in sketching this graph are shown in Fig. 11.9.

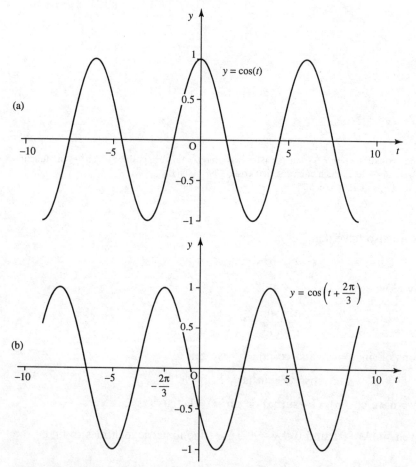

Figure 11.9 Sketching the graph of $y=2\cos(2t+2\pi/3)$. (a) Start with $y=\cos(t)$. (b) Shift to the left by $2\pi/3$ to give $y=\cos(t+2\pi/3)$. (c) Squash the graph in the t-axis to give $y=\cos(2t+2\pi/3)$. (d) Stretch the graph in the y-axis, giving $y=2\cos(2t+2\pi/3)$.

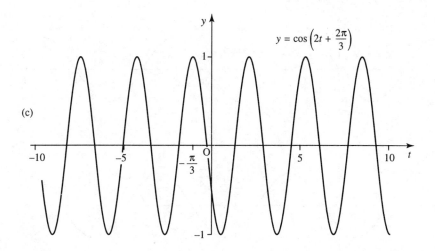

(c)

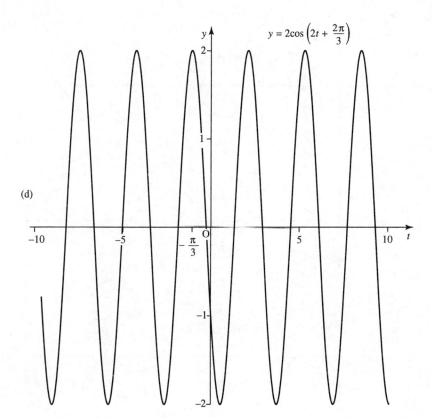

(d)

Figure 11.9 continued

Example 11.4 Sketch the graph of z against q, where

$$z = \tfrac{1}{2} \sin\left(\pi q - \frac{\pi}{4}\right) - \tfrac{1}{2}$$

The stages in sketching this graph are given in Fig. 11.10.

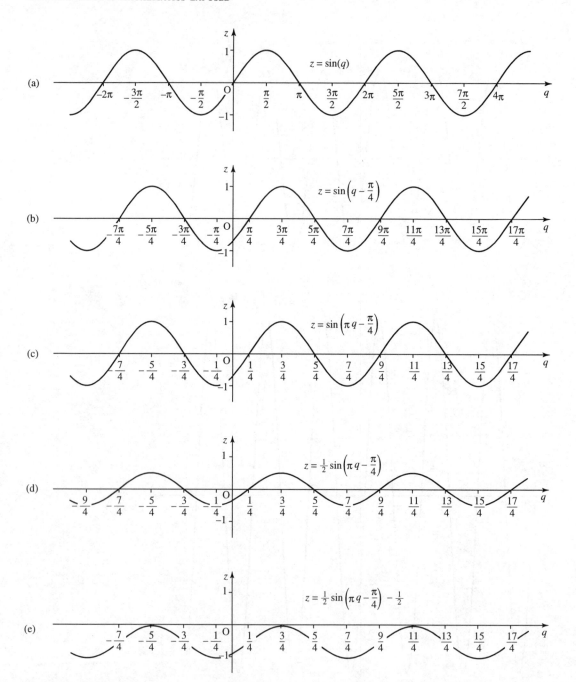

Figure 11.10 Sketching the graph of $z = \frac{1}{2}\sin(\pi q - \pi/4) - \frac{1}{2}$. (a) Start with $z = \sin(q)$. (b) Shift to the right by $\pi/4$ to give $z = \sin(q - \pi/4)$. (c) Squash the graph in the a-axis to give $z = \sin(\pi q - \pi/4)$. (d) Squash the graph in the z-axis, giving $z = \frac{1}{2}\sin(\pi q - \pi/4)$. (e) Translate in the z-direction by 1 to get $z = \frac{1}{2}\sin(\pi q - \pi/4) - \frac{1}{2}$.

Amplitude, fundamental period, phase and cycle rate

Figure 11.11 shows some examples of functions $y = A\cos(ax + b)$, and in each case the amplitude, phase, fundamental period and cycle rate has been found. In Fig. 11.11(a), $y = \frac{1}{2}\cos(5\pi x + \pi/2)$ is drawn, and has a peak value of 0.5 and a trough value of -0.5. Therefore the amplitude is

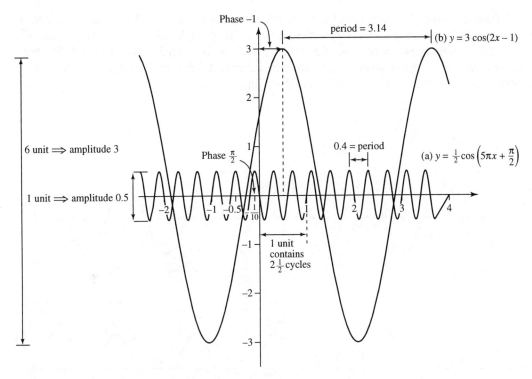

Figure 11.11 (a) $y=\frac{1}{2}\cos(5\pi x+\pi/2)$. (b) $y=3\cos(2x-1)$.

half the difference: $(0.5-(-0.5))/2=0.5$. The period is the minimum amount the graph needs to be shifted to the left or right (excluding no shift) in order to fit over the original graph. In this case, the period is 0.4. The phase is found by finding the proportion of the cycle that the graph has been shifted to the left or right. In this case the proportion of shift is $\frac{1}{4}$. Now multiply that by a standard cycle length of 2π to give the phase angle of $\pi/2$. The cycle rate is the number of cycles in unit length given by the reciprocal of the period: $1/0.4=2.5$.

Curve (b), $y=3\cos(2x-1)$, has a peak value of 3 and a trough value of -3. Therefore the amplitude is half the difference: $(3-(-3))/2=3$. The period is the minimum amount the graph needs to be shifted to the left or right (excluding no shift) in order to fit over the original graph. In this case the period is π. The phase is found by finding the proportion of the cycle that the graph has been shifted to the left. In this case, the proportion of shift is $-0.5/\pi$. Now multiply that by a standard cycle length of 2π to give the phase angle of -1. The cycle rate is the number of cycles in unit length, given by the reciprocal of the period: $1/\pi\simeq0.32$.

We can generalize from these examples to say that for the function $y=A\cos(ax+b)$, A positive, we have the following: the **amplitude** is half the difference between the function values at the peak and the trough of the wave, and in the case where $y=A\cos(ax+b)$ is given by A.

The **fundamental period**, P, or **cycle length** is the smallest non-zero distance that the graph can be shifted to the right or left so that it lies on top of the original graph. This can be found by looking for two consecutive values where the function takes its maximum value, that is when the cosine takes the value 1. Using the fact that $\cos(0)=1$ and $\cos(2\pi)=1$, then

$$\cos(ax+b) \text{ becomes } \cos(0) \text{ when } ax+b=0 \Leftrightarrow x=-b/a$$
$$\cos(ax+b) \text{ becomes } \cos(2\pi) \text{ when } ax+b=2\pi \Leftrightarrow x=2\pi/a-b/a$$

and the difference between them is $2\pi/a$, giving the fundamental period of the function $\cos(ax+b)$ as $2\pi/a$.

The **phase** is given by the number b in the expression $A \cos(ax + b)$. The phase is related to the amount the function $A \cos(ax + b)$ is shifted to the left or right with respect to the function $A \cos(ax)$. It expresses the proportion of a standard cycle (maximum 2π) that the graph has been shifted by, and therefore a phase can always be expressed between 0 and 2π or more often between $-\pi$ and π. Various phase shifts are given in Fig. 11.12.

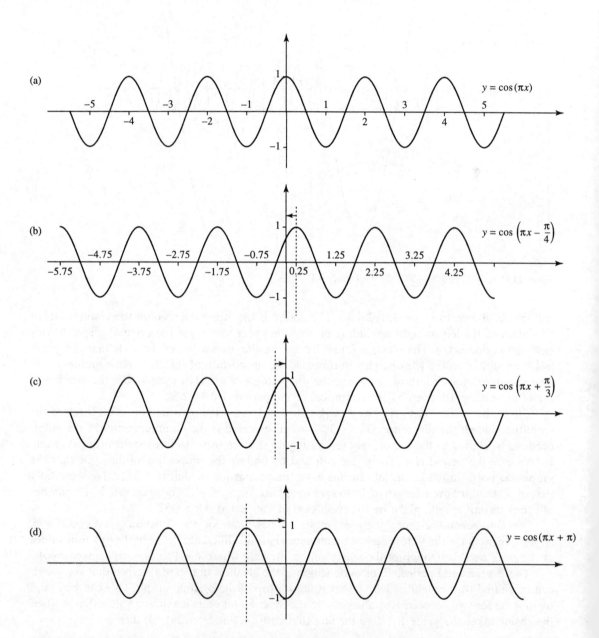

Figure 11.12 Examples of phase shifting. (a) A graph $y = \cos(\pi x)$. (b) $y = \cos(\pi x - \pi/4)$ has phase of $-\pi/4$ and is shifted by $\frac{1}{8}$ of a cycle (given by the proportion that the phase, $-\pi/4$, represents of a standard cycle of 2π). (c) $y = \cos(\pi x + \pi/3)$ has phase of $\pi/3$ and is shifted by $\frac{1}{6}$ of a cycle (given by the proportion that the phase, $\pi/3$, represents of a standard cycle of 2π). (d) $y = \cos(\pi x + \pi)$ has phase of π and is shifted by $\frac{1}{2}$ of a cycle (given by the proportion that the phase, π, represents of a standard cycle of 2π).

The cycle rate, or frequency, is the number of cycles in one unit. This can be found by considering its relation to the length of the cycle. The longer the cycle the fewer cycles there will be in one unit.

If the length of one cycle is P (the fundamental period) then there is 1 cycle in P units and $1/P$ cycles in 1 unit.

The cycle rate is the reciprocal of the fundamental period. As for the function $y = A \cos(ax + b)$, the fundamental period is $P = 2\pi/a$ then the number of cycles is $1/P = a/2\pi$. Examples are given in Fig. 11.13.

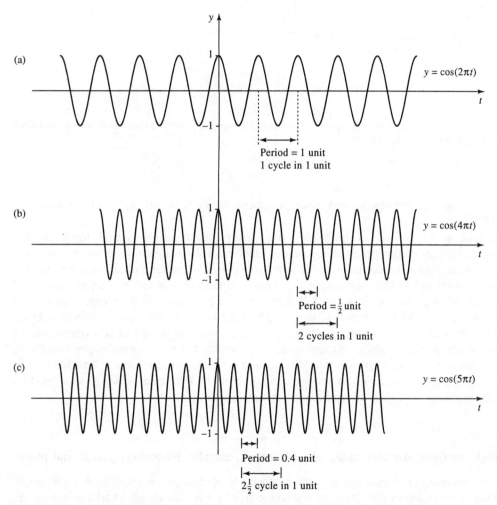

Figure 11.13 The relationship between cycle length (fundamental period) and the number of cycles in 1 unit. (a) $y = \cos(2\pi t)$ has cycle length $2\pi/2\pi = 1$ and therefore 1 cycle in 1 unit. (b) $y = \cos(4\pi t)$ has cycle length $2\pi/4\pi = \frac{1}{2}$ and therefore 2 cycles in 1 unit. (c) $y = \cos(5\pi t)$ has cycle length $2\pi/5 = 0.4$ and therefore $2\frac{1}{2}$ cycles in 1 unit.

11.4 WAVE FUNCTIONS OF TIME AND DISTANCE

A wave allows energy to be transferred from one point to another without any particles of the medium moving between the two points. Water waves move along the surface of a pond in response to a child rythmically splashing a hand in the water. The child's boat floating in the

Figure 11.14 A wave created by rythmically splashing a hand at the edge of a pond. The child's boat bobs up and down without moving in the direction of the wave.

path of the wave merely bobs up and down without moving in the direction of the wave (see Fig. 11.14).

If we look at the position of the boat as the wave passes, it moves up and down with the height expressed against time giving a sinusoidal function. This is then a wave function of time, and in the expression $y = A \cos(\omega t + \phi)$ the letter A represents the amplitude, ϕ represents the phase and ω is related to the wave frequency. This is explained in detail in the next section. If we take a snapshot picture of the surface of the water at a particular point in time then we will also get a wave shape where we now have a graph of the height of the water expressed against the distance from the wave's origin. In this case, where $y = A \cos(kx + \phi)$, A still represents the amplitude, ϕ the phase but the coefficient of x, k, is now related to the wavelength. Ideally we want an expression that can give the height, y, at any position x at any time t. This function is called the progressive wave function and we can combine the two ideas of waves as a function of time and distance to obtain an expression for this function.

Sinusoidal functions of time: amplitude, frequency, angular frequency, period and phase

Waves that represent a displacement from a central fixed position varying with time can be represented by an expression such as $y = A \cos(\omega t + \phi)$ or $y = A \sin(\omega t + \phi)$, where t is in seconds. Examples include an alternating voltage measured across a particular circuit element or the position of the centre of an ear drum as it vibrates in response to a pure sound wave. As we saw in the previous section, A represents the wave **amplitude**. ω is called the **angular frequency**; as it gives the number of cycles in 2π, it is measured in radians per second. The number of cycles in 1 s is called the **frequency** $= f \, (= \omega/2\pi)$ and is measured in hertz (Hz). ϕ is the **phase** and the cycle length is $2\pi/\omega$ seconds. In the case of a function of time the cycle length is called the **periodic time** or just the **period** and we often use the greek letter, τ (tau), to represent this where $\tau = 2\pi/\omega$ seconds. Then we have that $y = A \cos(\omega t + \phi)$ can be rewritten as

$$y = A \cos(2\pi f t + \phi)$$

using the frequency. As $f = 1/\tau$, this can be written as

$$y = A \cos\left(\frac{2\pi}{\tau} t + \phi\right)$$

Example 11.5
(a) For $y = 3 \cos(t + 1)$, find the amplitude, frequency, period, angular frequency and phase, where t is expressed in seconds.

Compare $y = 3 \cos(t + 1)$ with $y = A \cos(\omega t + \phi)$. We can see that the angular frequency $\omega = 1$, the phase $\phi = 1$ and the amplitude $A = 3$.

As $f = \omega/2\pi$, we have $f = 1/2\pi$, and the period $\tau = 1/f = 1/(1/2\pi) = 2\pi$ seconds.

(b) For $V = 12 \cos(314t + 1.6)$, find the amplitude, frequency, period, angular frequency and phase, where t is expressed in seconds.

Compare $V = 12 \cos(314t + 1.6)$ with $V = A \cos(\omega t + \phi)$. We can see that the angular frequency $\omega = 314$, the phase $\phi = 1.6$ and the amplitude $A = 12$.

As $f = \omega/2\pi$, we have $f = 314/2\pi \simeq 50$ Hz, and the period $\tau = 1/f = 1/50 = 0.02$ s.

Sinusoidal functions of distance: amplitude, cycle rate, wavelength and phase

Waves that give the displacement from a central fixed position of various different points at a fixed moment in time can be represented by an expression such as $y = A \cos(kx + \phi)$ or $y = A \sin(kx + \phi)$, where x is in metres. Examples include the position of a vibrating string at a particular moment or the surface of a pond in response to a disturbance. As we saw in the previous section, A represents the wave amplitude. k is called the **wavenumber** and represents the number of cycles in 2π. The spatial frequency gives the number of cycles in 1 metre $(= k/2\pi)$. The cycle length is called the **wavelength** and we often use the greek letter, λ, to represent this. The phase is ϕ. The expression for y can be rewritten, using the wavelength, as

$$y = A \cos\left(\frac{2\pi}{\lambda} t + \phi\right)$$

Example 11.6
(a) For $y = 4 \cos(x + 0.5)$, find the amplitude, wavelength, wavenumber, spatial frequency and phase, where x is expressed in metres.

Compare $y = 4 \cos(x + 0.5)$ with $y = A \cos(kx + \phi)$. We can see that the wavenumber $k = 1$, the phase $\phi = 0.5$ and the amplitude $A = 4$.

As spatial frequency $= k/2\pi$, this gives $1/2\pi$ wavelengths per metre, and the wavelength $\lambda = 2\pi/k = 2\pi/1 = 2\pi$ m.

(b) For $y = 2 \sin(2\pi x)$, find the amplitude, wavelength, wavenumber, spatial frequency and phase, where x is expressed in metres.

Using $\sin(\theta) = \cos(\theta - \pi/2)$, we get $2 \sin(2\pi x) = 2 \cos(2\pi x - \pi/2)$. Compare $y = 2 \cos(2\pi x - \pi/2)$ with $y = A \cos(kx + \phi)$. We can see that the wavenumber $k = 2\pi$, the phase $\phi = -\pi/2$ and the amplitude $A = 2$. As spatial frequency $= k/2\pi$ this gives 1 wavelength per metre, and the wavelength $\lambda = 2\pi/k = 2\pi/2\pi = 1$ m.

Waves in time and space

The two expressions for a wave function of time and space can be combined as

$$y = A \cos(\omega t - kx)$$

and this is called a progressive wave equation. The $-$ sign is used to give a wavefront travelling from left to right and t should be taken as positive with $\omega t \geqslant kx$.

Notice that if we look at the movement of a particular point by fixing x then we replace x by x_0, and this just gives a function of time, $y = A \cos(\omega t + \phi)$, where $\phi = -kx_0$.

If we look at the wave at a single moment in time then we fix time and replace t by t_0, and this just gives a function of distance x, $y = A \cos(kx + \phi)$, where $\phi = -\omega t_0$.

Waves are of two basic types. Mechanical waves need a medium through which to travel, e.g. sound waves, water waves and seismic waves. Electromagnetic waves can travel through a vacuum, e.g. light rays and X-rays. In all cases where they can be expressed as a progressive or travelling wave, the frequency, wavelength etc. can be found from the expression of the wave in the same way.

Figure 11.15 shows three snapshot pictures of the progressive wave $y = \cos(15t - 3x)$ at $t = 0$, $t = 2$ and $t = 5$. This wave has angular frequency $\omega = 15$ and wavenumber $k = 3$, and therefore the frequency f is $2\pi/15$ and the wavelength $\lambda = 2\pi/3$. By considering the amount the wavefront moves in a period of time we are able to find the wave velocity.

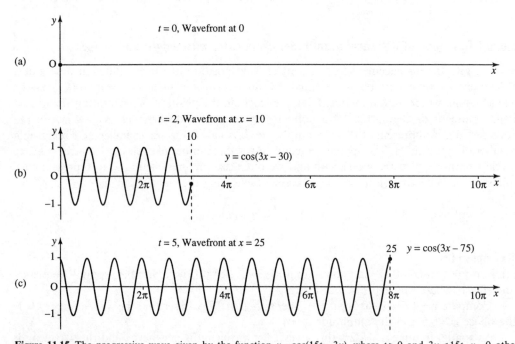

Figure 11.15 The progressive wave given by the function $y = \cos(15t - 3x)$, where $t > 0$ and $3x < 15t$, $y = 0$ otherwise.
(a) No wave at $t = 0$.
(b) $t = 2$ gives $y = \cos(30 - 3x)$
$\qquad = \cos(3x - 30)$ for $3x < 30$, i.e. $x < 10$
(c) $t = 5$ gives $y = \cos(75 - 3x)$
$\qquad = \cos(3x - 75)$ for $3x < 75$, i.e. $x < 25$
Notice that the wavefront has moved 25 m in 5 s, giving a velocity of $25/5 = 5$ m s^{-1}.

Velocity of a progressive wave

The progressive wave $y = A \cos(\omega t - kx)$ vibrates f times per second and the length of each cycle, the wavelength, is λ. In this case the wavefront must move through a distance of λf metres per second, and hence the velocity v is given by

$$v = f\lambda$$

where $f = \omega/2\pi$ and $\lambda = 2\pi/k$ then $v = \omega/k$.

Example 11.7 A wave is propagated from a central position as in Fig. 11.16 and is given by the function $y = 2\cos(6.28t - 1.57r)$, where $t > 0$ and $1.57r \leqslant 6.28t$. Find the frequency, periodic time, spatial frequency, wavenumber and wavelength. The wave is pictured for $t = 5$ in Fig. 11.16.

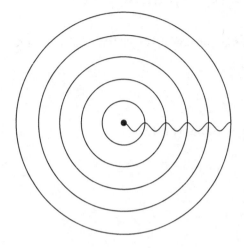

Figure 11.16 $y = 2\cos(6.28t - 1.57r)$, where $1.57r \leqslant 6.28t$ when $t = 5$, giving $y = 2\cos(31.4 - 1.57r)$, $r < 20$. The concentric circles represent the peak amplitudes of the wave. The wavefront has moved to $r = 20$ at $t = 5$, giving a wave velocity of $20/5 = 4 \text{ m s}^{-1}$.

SOLUTION Comparing $y = 2\cos(6.28t - 1.57r)$ with $y = A\cos(\omega t - kr)$ gives $A = 2$, angular frequency $\omega = 6.28$ and wave number $k = 1.57$. Hence frequency $f = \omega/2\pi \Rightarrow f = 6.28/2\pi \simeq 1$ Hz, periodic time $\tau = 1/f = 1$ s, spatial frequency $= k/2\pi = 1.57/2\pi \simeq \frac{1}{4}$, wavelength $\lambda = 2\pi/k = 2\pi/1.57 \simeq 4$ m and velocity $= f\lambda = 1 \times 4 = 4 \text{ m s}^{-1}$.

Measuring amplitudes: decibels

In Chapter 8 we looked at sound decay in a room and found that the expression was exponential and could be expressed by using a power of 10. Because of this property of sound decay, and decay of other wave forms, and also because of the need to have a unit which can be used easily to express relatively small quantities, decibels are often used to represent wave amplitudes. In this case the measurement is always in relation to some reference level.

Sound pressure is parallel in electronics to the voltage. The sound pressure level is measured in decibels and is defined as $20\log_{10}(p/p_0)$, where p is the actual sound pressure and p_0 the reference pressure in N m^{-2}. The reference pressure used is approximately the threshold of audibility for sound at 1000 Hz, and is given by

$$p_0 = 2 \times 10^{-5} \text{ N m}^{-2}$$

Voltage, measured in decibels, is given by $20\log_{10}(V/V_0)$.

Sound intensity is parallel to power in a circuit. The sound intensity level $= 10\log_{10}(I/I_0)$, where I is the sound intensity and I_0 is the sound intensity at the threshold of audibility: $I_0 = 10^{-12} \text{ W m}^{-2}$.

Because the reference points used for the measurement of the amplitude of sound are the same whether measuring the sound pressure level or the sound intensity level, measurement of either will give the same result on the same wave.

Example 11.8 The sound generated by a car has intensity 2×10^{-5} W m^{-2}. Find the sound intensity level and sound pressure level.

SOLUTION The sound intensity level is

$$10 \log_{10}\left(\frac{2 \times 10^{-5}}{10^{-12}}\right) = 10 \log_{10}(2 \times 10^7) = 70 \log_{10}(2) \simeq 21.1 \text{ dB}$$

As this is the same as the sound pressure level, the sound pressure level $= 21.1$ dB.

Example 11.9 An amplifier outputs 5 W when the input power is 0.002 W. Calculate the power gain.

SOLUTION The power gain is given by

$$10 \log_{10}\left(\frac{5}{0.002}\right) = 10 \log_{10}(2500) \simeq 34 \text{ dB}$$

11.5 TRIGONOMETRIC IDENTITIES

Compound angle identities

It can often be useful to write an expression for, for instance, $\cos(A + B)$ in terms of trigonometric ratios for A and B. A common mistake is to assume that $\cos(A + B) = \cos(A) + \cos(B)$ but this can easily be disproved. Take as an example $A = 45°$ and $B = 45°$, then

$$\cos(A + B) = \cos(45° + 45°) = \cos(90°) = 0$$

$$\cos(A) + \cos(B) = \cos(45°) + \cos(45°) \simeq 0.707 + 0.707 = 1.414$$

showing that

$$\cos(A + B) = \cos(A) + \cos(B) \text{ is FALSE}$$

The correct expression is

$$\cos(A + B) = \cos(A) \cos(B) - \sin(A) \sin(B)$$

Other compound angle identities are as follows:

$$\sin(A + B) = \sin(A) \cos(B) + \cos(A) \sin(B)$$

$$\tan(A + B) = \frac{\tan(A) + \tan(B)}{1 - \tan(A) \tan(B)}$$

There are various ways of showing these to be true. In Fig. 11.17 we show that $\sin(A + B) = \sin(A) \cos(B) + \cos(A) \sin(B)$ by using a geometrical argument. Draw two triangles YZW and YWX so that *$\angle A$ and $\angle B$ are adjacent angles and the two triangles are right-angled, as shown. Draw the lines XX' and YY' so that they form right angles to each other, as shown.* Notice that $\angle YXY'$ is also $\angle A$. From $\triangle WX'X$, $\sin(A + B) = XX'/XW$, and as $X'Y'YZ$ is a rectangle then $X'Y' = ZY$. So

$$\sin(A + B) = \frac{XY' + ZY}{XW} = \frac{XY'}{XW} + \frac{ZY}{XW} = \frac{ZY}{XW} + \frac{XY'}{XW}$$

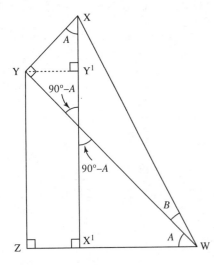

Figure 11.17 $\sin(A+B)=\sin(A)\cos(B)+\cos(A)\sin(B)$.

As WY/WY = 1 and XY/XY = 1,

$$\sin(A+B)=\frac{ZY}{XW}\frac{WY}{WY}+\frac{XY'}{XW}\frac{XY}{XY}$$

$$=\frac{ZY}{WY}\frac{WY}{XW}+\frac{XY'}{XY}\frac{XY}{XW}$$

Looking at the triangles containing these sides we can see that this gives $\sin(A+B)=\sin(A)\cos(B)+\cos(A)\sin(B)$.

A similar argument can be used for $\cos(A+B)$ and $\tan(A+B)$ is usually found by using the expressions for $\sin(A+B)$, $\cos(A+B)$ and the definition of the tangent.

$$\tan(A+B)=\frac{\sin(A+B)}{\cos(A+B)}=\frac{\sin(A)\cos(B)+\cos(A)\sin(B)}{\cos(A)\cos(B)-\sin(A)\sin(B)}$$

Divide the top and bottom lines by $\cos(A)\cos(B)$, giving

$$\tan(A+B)=\frac{\dfrac{\sin(A)\cos(B)}{\cos(A)\cos(B)}+\dfrac{\cos(A)\sin(B)}{\cos(A)\cos(B)}}{\dfrac{\cos(A)\cos(B)}{\cos(A)\cos(B)}-\dfrac{\sin(A)\sin(B)}{\cos(A)\cos(B)}}$$

$$\tan(A+B)=\frac{\tan(A)+\tan(B)}{1-\tan(A)\tan(B)}$$

From these three identities for $\sin(A+B)$, $\cos(A+B)$ and $\tan(A+B)$ we can obtain many other expressions. A list of important trigonometric identities is given in Table 11.2.

Example 11.10 Using $\cos(2A)=\cos^2(A)-\sin^2(A)$ and $\cos^2(A)+\sin^2(A)=1$, show that $\cos^2(A)=\frac{1}{2}(\cos^2(A)+1)$.

SOLUTION From $\cos^2(A)+\sin^2(A)=1$, we find $\sin^2(A)=1-\cos^2(A)$ (subtracting $\cos^2(A)$ from

Table 11.2 Summary of important trigonometric identities

$\cos(A \pm B) = \cos(A)\cos(B) \mp \sin(A)\sin(B)$

$\sin(A \pm B) = \sin(A)\cos(B) \pm \cos(A)\sin(B)$

$\tan(A \pm B) = \dfrac{\tan(A) \pm \tan(B)}{1 \mp \tan(A)\tan(B)}$

$\sin(X) + \sin(Y) = 2\sin\frac{1}{2}(X+Y)\cos\frac{1}{2}(X-Y)$

$\sin(X) - \sin(Y) = 2\cos\frac{1}{2}(X+Y)\sin\frac{1}{2}(X-Y)$

$\cos(X) + \cos(Y) = 2\cos\frac{1}{2}(X+Y)\cos\frac{1}{2}(X-Y)$

$\cos(X) - \cos(Y) = -2\sin\frac{1}{2}(X+Y)\sin\frac{1}{2}(X-Y)$

$\sin(2A) = 2\sin(A)\cos(A)$

$\cos(2A) = \cos^2(A) - \sin^2(A)$

$\tan(2A) = \dfrac{2\tan(A)}{1 - \tan^2(A)}$

$\cos(2A) = 2\cos^2(A) - 1$

$\cos(2A) = 1 - 2\sin^2(A)$

$\cos^2(A) + \sin^2(A) = 1$

$\cos^2(A) = \frac{1}{2}(\cos(2A) + 1)$

$\sin^2(A) = \frac{1}{2}(1 - \cos(2A))$

$\cos(A - (\pi/2)) = \sin(A)$

$\sin(A + (\pi/2)) = \cos(A)$

---

both sides). Substitute this into $\cos(2A) = \cos^2(A) - \sin^2(A)$:

$$\cos(2A) = \cos^2(A) - (1 - \cos^2(A))$$

$$\Leftrightarrow \cos(2A) = \cos^2(A) - (1 + \cos^2(A))$$

$$\Leftrightarrow \cos(2A) = 2\cos^2(A) - 1$$

$$\Leftrightarrow \cos(2A) + 1 = 2\cos^2(A) \qquad \text{(adding 1 to both sides)}$$

$$\Leftrightarrow \tfrac{1}{2}(\cos(2A) + 1) = \cos^2(A) \qquad \text{(dividing by 2)}$$

Hence $\cos^2(A) = \frac{1}{2}(\cos^2(A) + 1)$.

Example 11.11 From

$$\cos(A \pm B) = \cos(A)\cos(B) \mp \sin(A)\sin(B)$$

$$\sin(A \pm B) = \sin(A)\cos(B) \pm \cos(A)\sin(B)$$

show that

$$\sin(X) + \sin(Y) = 2\sin\tfrac{1}{2}(X+Y)\cos\tfrac{1}{2}(X-Y)$$

SOLUTION Using

$$\sin(A + B) = \sin(A)\cos(B) + \cos(A)\sin(B) \tag{11.1}$$

and

$$sin(A - B) = sin(A) cos(B) - cos(A) sin(B) \tag{11.2}$$

and set

$$X = A + B \tag{11.3}$$

and

$$Y = A - B \tag{11.4}$$

Using Eqs (11.3) and (11.4) we can solve for A and B. Add Eq. (11.3) and Eq. (11.4), giving $X + Y = 2A \Leftrightarrow A = (X + Y)/2$. Subtract Eq. (11.3) from Eq. (11.4), giving

$$X - Y = A + B - (A - B) \Leftrightarrow X - Y = 2B \Leftrightarrow B = \frac{X - Y}{2}$$

Add Eq. (11.1) and Eq. (11.2) to give

$$sin(A + B) + sin(A - B) = sin(A) cos(B) + cos(A) sin(B) + sin(A) cos(B) - cos(A) sin(B)$$

$$\Leftrightarrow sin(A + B) + sin(A - B) = 2 sin(A) cos(B)$$

Substitute for A and B, giving

$$sin(X) + sin(Y) = 2 sin\frac{(X + Y)}{2} cos\frac{(X - Y)}{2}$$

Example 11.12 Given that $cos(60°) = \frac{1}{2}$, find $cos(30°)$.

SOLUTION Using $cos^2(A) = \frac{1}{2}(cos(2A) + 1)$ and putting $A = 30°$ then

$$cos^2(30°) = \frac{1}{2}(cos(60°) - 1) = \frac{1}{2}(\frac{1}{2} + 1) = \frac{1}{2}(\frac{3}{2}) = \frac{3}{4}$$

$$\Rightarrow cos(30°) = \pm \sqrt{\frac{3}{2}}$$

From knowledge of the graph of the cosine we know that $cos(30°) > 0$, so $cos(30°) = \sqrt{\frac{3}{2}}$.

Example 11.13 Using $sin(90°) = 1$ and $cos(90°) = 0$, find $sin(45°)$.

SOLUTION Using $sin^2(A) = \frac{1}{2}(1 - cos(2A))$ and putting $A = 45°$ and $2A = 90°$:

$$sin^2(45°) = \frac{1}{2}(1 - cos(90°)) = \frac{1}{2} \quad \text{(as } cos(90°) = 0)$$

$$\Leftrightarrow sin(45°) = \pm\sqrt{\frac{1}{2}}$$

From our knowledge of the graph of the sine function we know that $sin(45°) > 0$; hence

$$sin(45°) = \sqrt{\frac{1}{2}}$$

11.6 SUPERPOSITION

The principle of superposition of waves states that the effect of a number of waves can be found by summing the disturbances that would have been produced by the individual waves separately. This behaviour is quite different from that of travelling particles, which will bump into each other, thereby altering the velocities of both.

The idea of superposition is used to explain the behaviour of:

1. Stationary waves formed by two wave trains of the same amplitude and frequency travelling at the same speed in opposite directions.
2. Interference of coherent waves from identical sources.
3. Two wave trains of close frequency travelling at the same speed in opposite directions, causing beats.
4. Diffraction effects.

We look at some examples of these applications.

Standing waves

Suppose that a wave is created by plucking a string of a musical instrument; then when the wave reaches the end of the string it is reflected back. The reflected wave will have the same frequency as the initial wave but a different phase, and will be travelling in the opposite direction.

The sum of the incident and reflected waves forms a standing wave. An example is shown in Fig. 11.18. Figure 11.18(a) shows the incident wave in a string at some instant in time. Its phase is $18°$. Beyond the barrier is shown the hypothetical continuation of the wave as if there were no barrier. The reflected wave is found by turning this section upside down and reflecting it, as shown in (b). The reflected wave has phase $180° -$ phase of the incident wave $= 180° - 18° = 162°$. In Fig. 11.18(c), the sum of the incident and reflected waves produces a standing wave. The maximum and minimum values on this are called antinodes and the zero values are called nodes. As the string is fixed at both ends there must be nodes at the ends.

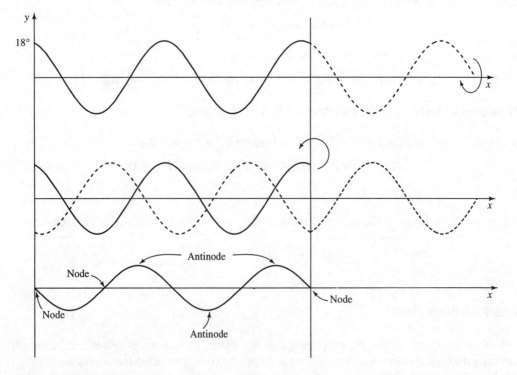

Figure 11.18 (a) Incident wave in a string at some instant in time. (b) The reflected wave. (c) The sum of the incident and reflected waves.

At different moments in time the phase of the incident wave will be different. This changes the amplitude of the standing wave but does not change the positions of the nodes or antinodes (for a given frequency of wave). Only waves whose wavelengths exactly divide into $2l$ (twice the length of the string) can exist on the string because their amplitude must be 0 at the two end points. Each possible wavelength defines a mode of vibration of the string. $\lambda = 2l$ is called the fundamental mode and is shown in Fig. 11.19.

Figure 11.19 The fundamental mode for a standing wave in a string of length l has wavelength $2l$, so that half a cycle fits into the length of the string. This is the longest wavelength possible.

The standing wave can be explained using

$$\cos(X) + \cos(Y) = 2 \cos \tfrac{1}{2}(X + Y) \cos \tfrac{1}{2}(X - Y)$$

The example given in Fig. 11.18 has a wavelength of 4, giving wavenumber $360/4 = 90$. The incident wave (phase $18°$) is $y = \cos(90°x + 18°)$ and the reflected wave is $y = \cos(90°x + 162°)$. Summing these gives

$$\cos(90°x + 18°) + \cos(90°x + 162°)$$

$$= 2 \cos(\tfrac{1}{2}(90°x + 18° + 90°x + 162°)) \cos(\tfrac{1}{2}(90°x + 18° - (90°x + 162°)))$$

$$= 2 \cos(90°x + 90°) \cos(72°)$$

As a $90°$ phase-shifted version of a cosine gives a sine, this gives $-2 \cos(72°) \sin(90°x)$.

We see that the result is a sine wave of the same spatial frequency as the incident and reflected wave, but with an amplitude of $2 \cos 72° \simeq 0.62$.

This result can also be found for a general situation, now expressing the phases etc. in radians.

The initial wave is $\cos(kx + \delta)$ and the reflected wave is $\cos(kx + \pi - \delta)$. The standing wave is given by summing these, giving

$$\cos(kx + \delta) + \cos(kx + \pi - \delta)$$

$$= 2 \cos\left(\frac{kx + \delta + kx + \pi - \delta}{2}\right) \cos\left(\frac{kx + \delta - (kx + \pi - \delta)}{2}\right)$$

$$= 2 \cos\left(kx + \frac{\pi}{2}\right) \cos\left(\delta - \frac{\pi}{2}\right)$$

The standing wave has the same spatial frequency as the original waves. As $\cos(kx + (\pi/2)) = -\sin(kx)$ and $\cos(\delta - (\pi/2)) = \sin(\delta)$ this becomes $-2 \sin(kx) \sin(\delta)$.

So the instantaneous amplitude of the standing wave is $2 \sin(\delta)$, where δ is the phase of the incident wave.

11.7 INVERSE TRIGONOMETRIC FUNCTIONS

From the graphs of the trigonometric functions, $y = \sin(x)$, $y = \cos(x)$ and $y = \tan(x)$, we notice that for any one value of y there are several possible values of x. This means that there are no inverse functions if all input values for x are allowed. However, we can see on a calculator that there is a function listed above the sine button and marked as $\sin^{-1}$, so is it in fact the inverse function?

Try the following with the calculator in degree mode. Enter 60 and press sin, then press $\sin^{-1}$. This is shown in Table 11.3(a). The same process is repeated for 120° and for −120°. However, for the latter two cases the inverse function does not work.

Table 11.3 Sin and $\sin^{-1}$ on a calculator

(a) $\sin^{-1}$	sin	(b) $\sin^{-1}$	sin	(c) $\sin^{-1}$	sin
60°	→ 0.8660	120°	→ 0.8660	−120°	→ −0.8660
60°	← 0.8660	60°	← 0.8660	−60°	← −0.8660

We can restrict the range of values allowed into sin(x) to the range −90° to +90° and then $\sin^{-1}(x)$ is a true inverse. The inverse function of $y=\sin(x)$ is defined as $f(x)=\sin^{-1}(x)$ (often written as arcsin(x) to avoid confusion with 1/sin(x)). It is only the inverse function if the domain of the sine function is limited to $-\pi/2 \leqslant x \leqslant \pi/2$. Thus $\sin^{-1}(\sin(x))=x$ if x lies within the limits given above and $\sin(\sin^{-1}(x))=x$ if $-1 \leqslant x \leqslant 1$.

The graph of $y=\sin^{-1}(x)$ is given in Fig. 11.20.

$f(x)=\cos^{-1} x$ is the inverse of $y=\cos x$ if the domain of $\cos x$ is limited to $0 \leqslant x \leqslant \pi$. Thus $\cos^{-1}(\cos x)=x$ if x is limited to the interval above and $\cos(\cos^{-1} x)=x$ if $-1 \leqslant x \leqslant 1$.

The graph of $y=\cos^{-1} x$ is given in Fig. 11.21.

$f(x)=\tan^{-1}(x)$ is the inverse of $y=\tan(x)$ if the domain of $\tan(x)$ is limited to $-\pi/2 < x < \pi/2$. Thus $\tan^{-1}(\tan(x))=x$ if x is limited as above, and $\tan(\tan^{-1}(x))=x$ for all x.

The graph of $y=\tan^{-1}(x)$ is given in Fig. 11.22.

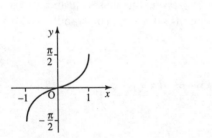

Figure 11.20 Graph of $y=\sin^{-1}(x)$.

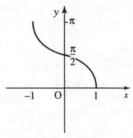

Figure 11.21 Graph of $y=\cos^{-1}(x)$.

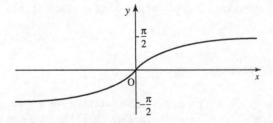

Figure 11.22 Graph of $y=\tan^{-1}(x)$.

11.8 SOLVING EQUATIONS

The solutions to the equations $\sin(x)=a$, $\cos(x)=a$ and $\tan(x)=a$ are shown in Figs 11.23, 11.24 and 11.25. Where the lines $y=a$ cross the sine, cosine or tangent graph gives the solutions to the equations. In Fig. 11.23, solutions to $\sin(x)=a$ are give by values of x where the line $y=a$ crosses the graph $y=\sin(x)$. Notice that there are two solutions in every cycle. The first solution is $\sin^{-1}(a)$ and the next is given by $\pi-\sin^{-1}(a)$. Solutions in the other cycles can be found by adding a multiple of 2π to these two solutions.

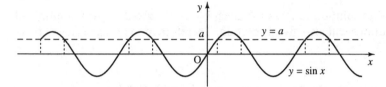

Figure 11.23 Solutions of $\sin(x)=a$.

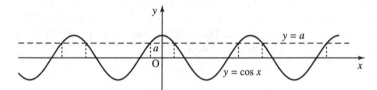

Figure 11.24 Solutions of $\cos(x)=a$.

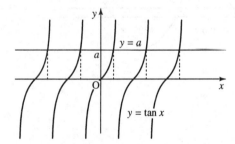

Figure 11.25 Solutions of $\tan(x)=a$.

In Fig. 11.24, solutions to $\cos(x)=a$ are given by values of x where the line $y=a$ crosses the graph $y=\cos(x)$. Notice that there are two solutions in every cycle. The two solutions in $[-\pi,\pi]$ are $\cos^{-1}(a)$ and $-\cos^{-1}(a)$. Other solutions can be found by adding a multiple of 2π to these two solutions.

In Fig. 11.25, solutions to $\tan(x)=a$ are given by values of x where the line $y=a$ crosses the graph $y=\tan(x)$. Notice that there is one solution in every cycle. The solution in $[0,\pi]$ is $\tan^{-1}(a)$. Solutions in the other cycles can be found by adding a multiple of π to this solution.

Example 11.14 Find solutions to $\sin(x)=0.5$ in the range $[2\pi,4\pi]$.

SOLUTION From the graph of $y=\sin(x)$ and the line $y=0.5$ in Fig. 11.26 the solutions can be worked out from where the two lines cross. The solution nearest $x=0$ is given by

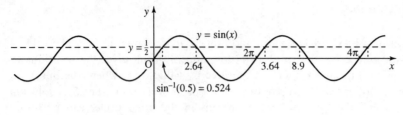

Figure 11.26 Solutions of $y = \sin(x)$ and $y = 0.5$.

$\sin^{-1}(0.5) \simeq 0.524$. The other solution in $[0, 2\pi]$ is given by $\pi - 0.524 \simeq 2.62$. Any multiple of 2π added on to these solutions will also give a solution. Therefore in the range $[2\pi, 4\pi]$ there are solutions 3.64 and 8.9.

Example 11.15 Find solutions to $\cos(x) = 0.3$ in the range $[-480°, 480°]$.

SOLUTION From the graph of $y = \cos(x)$ and the line $y = 0.3$ in Fig. 11.27 the solutions can be worked out from where the two lines cross. The solution nearest $x = 0$ is given by $\cos^{-1}(0.3) = 73°$. The other solution in $[0°, 360°]$ is given by $-73°$. Any multiple of $360°$ added on to these solutions will also give a solution. Therefore in the range $[-480°, 480°]$ there are solutions $-433°$, $-287°$, $-73°$, $73°$, $287°$ and $433°$.

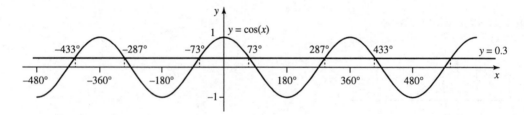

Figure 11.27 Solutions of $y = \cos(x)$ and $y = 0.3$.

Example 11.16 Find solutions to $\tan(x) = 0.1$ in the range $[360°, 540°]$.

SOLUTION From the graph of $y = \tan(x)$ and the line $y = 0.1$ on Fig. 11.28 the solutions can be worked out from where the two lines cross. The solution nearest $x = 0$ is given by $\tan^{-1}(0.1) \simeq 6°$. Any multiple of $180°$ added on to this solution will also give a solution. Therefore in the range $[360°, 540°]$ there is one solution $366°$.

11.9 SUMMARY

1. Trigonometric functions can be defined using a rotating rod of length 1. The sine is given by plotting the height of the tip of the rod against the distance travelled. The cosine is given by plotting the position that the tip of the rod is to the left or right of the origin against the distance travelled. Hence if the tip of the rod is at point (x, y) and the tip has travelled a

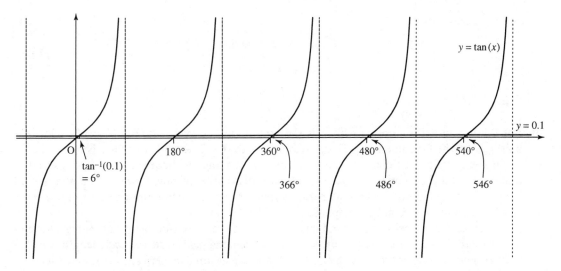

Figure 11.28 Solutions of $y = \tan(x)$ and $y = 0.1$.

distance of t units then $\sin(t) = y$, $\cos(t) = x$ and the tangent is given by

$$\tan(t) = \frac{\sin(t)}{\cos(t)} = \frac{y}{x}$$

2. If angles are measured in radians then this definition is the same for angles between 0 and $\pi/2$ as that given by defining the cosine, sine and tangent from the sides of a triangle (of hypotenuse 1) as in Chapter 6.

There are 2π radians in a complete revolution (360°) and therefore

$$\pi \text{ radians} = 180°$$

$$1 \text{ radian} = \frac{180°}{\pi}$$

$$1° = \frac{\pi}{180} \text{ radians}$$

The trigonometric ratios can now be defined using a rotating rod of length r and the angle, α, made by the rod with the x-axis. Then, if the tip of the rod is at point (x,y)

$$\sin(\alpha) = \frac{y}{r} \qquad \cos(\alpha) = \frac{x}{r} \qquad \tan(\alpha) = \frac{\sin(\alpha)}{\cos(\alpha)} = \frac{x}{y}$$

α is expressed in radians, although engineers often use degrees. If degrees are intended then they must be expressly marked.

3. $\operatorname{Sin}(t)$ and $\tan(t)$ are odd functions, $\cos(t)$ is even. This can be expressed by:

$$\sin(-t) = -\sin(t)$$

$$\cos(-t) = \cos(t)$$

$$\tan(-t) = -\tan(t)$$

$\operatorname{Sin}(t)$ and $\cos(t)$ are periodic with period 2π and $\tan(t)$ is periodic with period π. This can be expressed by

$$\sin(t + 2\pi n) = \sin(t)$$

$$\cos(t + 2\pi n) = \cos(t)$$

$$\tan(t + \pi n) = \tan(t) \qquad \text{where } n \in \mathbb{Z}.$$

4. For the function $y = A \cos(ax + b)$, A is the amplitude, the cycle rate (number of cycles in 1 unit)$= a/2\pi$, the fundamental period, or cycle length, $P = 2\pi/a$ and the phase is b.

   For a function of time, $y = A \cos(\omega t + \phi)$, ω is the angular frequency and is measured in radians/second. The number of cycles in one second is the frequency, $f = \omega/2\pi$ and is measured in Hz. The cycle length is called the periodic time or period (often represented by τ), $\tau = 2\pi/\omega$, and is measured in seconds. The phase is ϕ.

   For a function of distance, $y = A \cos(kx + \phi)$, k is the wave number, the number of cycles in one metre is the spatial frequency, $= k/2\pi$ wavelengths per metre, the cycle length is called the wavelength (often represented by λ), $= 2\pi/k$ and is measured in metres, and the phase is ϕ.

   The function $y = A \cos(\omega t - kx)$, with $t > 0$ and $\omega t \geqslant kx$ is called the progressive wave equation and has velocity $v = \lambda f$ metres per second.

5. Wave amplitudes are often measured on a logarithmic scale using decibels.

6. There are many trigonometric identities, summarized in Table 11.2. Some of the more fundamental ones are:

$$\cos(A \pm B) = \cos(A) \cos(B) \mp \sin(A) \sin(B)$$

$$\sin(A \pm B) = \sin(A) \cos(B) \pm \cos(A) \sin(B)$$

$$\tan(A) = \frac{\sin(A)}{\cos(A)}$$

$$\cos^2(A) + \sin^2(A) = 1$$

from which others can be derived.

7. The principle of superposition of waves gives that the effect of a number of waves can be found by summing the disturbances that would have been produced by the waves separately. One application of this is the case of stationary waves, which can be explained mathematically using trigonometric identities.

8. $\sin^{-1}(x)$ is the inverse function of $\sin(x)$ if the domain of $\sin(x)$ is limited to $-\pi/2 \leqslant x \leqslant \pi/2$, in which case $\sin^{-1}(\sin(x)) = x$.

   $\cos^{-1}(x)$ is the inverse function of $\cos(x)$ if the domain of $\cos(x)$ is limited to $0 \leqslant x \leqslant \pi$, in which case $\cos^{-1}(\cos(x)) = x$.

   $\tan^{-1}(x)$ is the inverse function of $\tan(x)$ if the domain of $\tan(x)$ is limited to $-\pi/2 < x < \pi/2$, in which case $\tan^{-1}(\tan(x)) = x$.

9. Inverse trigonometric functions are used in solving trigonometric equations. There are many solutions to trigonometric equations and graphs can be used to help see where the solutions lie.

   $\sin(x) = a$ has one solution given by $x = \sin^{-1}(a)$ and another given by $\pi - \sin^{-1}(a)$. Other solutions can be found by adding or subtracting a multiple of 2π.

   $\cos(x) = a$ has one solution given by $x = \cos^{-1}(a)$ and another given by $-\cos^{-1}(a)$. Other solutions can be found by adding or subtracting a multiple of 2π.

   $\tan(x) = a$ has one solution given by $x = \tan^{-1}(a)$. Other solutions can be found by adding or subtracting a multiple of π.

11.10 EXERCISES

11.1 Without using a calculator express the following angles in degrees (remember that π radians $= 180°$, $\pi \simeq 3.142$):

(a) $\dfrac{2\pi}{3}$ radians   (b) 4π radians   (c) $\dfrac{3\pi}{5}$ radians

(d) 6.284 radians   (e) 1.571 radians

    Check your answers using a calculator.

11.2 Without using a calculator, express the following angles in radians.

(a) 45°   (b) 135°   (c) 10°   (d) 150°

    Check your answers using a calculator.

11.3 Given that $\cos(\pi/3) = \tfrac{1}{2}$, without using a calculator find:

(a) $\sin\left(\dfrac{\pi}{3}\right)$ (b) $\tan\left(\dfrac{\pi}{3}\right)$ (c) $\cos\left(\dfrac{2\pi}{3}\right)$

(d) $\sin\left(\dfrac{7\pi}{3}\right)$ (e) $\tan\left(\dfrac{4\pi}{3}\right)$

    Check your answers using a calculator.

11.4 By considering transformations of the graphs of sin(x), cos(x) and tan(x), sketch the graphs of the following:

(a) $y = \sin\left(x + \dfrac{\pi}{4}\right)$ (b) $y = \tan\left(x - \dfrac{\pi}{2}\right)$ (c) $y = 3\sin(x)$

(d) $y = \tfrac{1}{2}\cos(x)$ (e) $y = \sin(\pi x)$ (f) $y = 2\sin\left(\tfrac{1}{2}x + \dfrac{\pi}{6}\right)$

(g) $y = \sin(x) + 3)$ (h) $y = -\cos(x)$

11.5 From the graphs in Fig. 11.29 find the phase, amplitude, period (cycle length) and number of cycles in one unit.

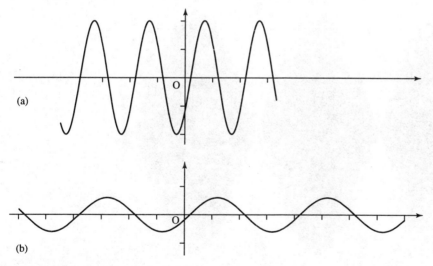

(a)

(b)

Figure 11.29 Graphs for Exercise 11.5.

11.6 For the following functions of time, find the amplitude, period, angular frequency and phase.

(a) $y = 3\cos\left(4t + \dfrac{\pi}{2}\right)$ (b) $V = \sin(377t + 0.4)$

(c) $p = 40\cos(3000t - 0.8)$

11.7 For the following functions of distance, x, find the amplitude, wavelength, spatial frequency and wavenumber.

(a) $y = 0.5 \cos\left(2x - \dfrac{\pi}{2}\right)$ (b) $y = 2 \cos(72x + 0.33)$

(c) $y = 52 \sin(80x)$

11.8 Given a progressive wave with $t > 0$:

$$y = \begin{cases} 3 \cos(2t - 5x) & \text{for } 5x < 2t \\ 0 & \text{otherwise} \end{cases}$$

(a) sketch the waves for $t = 1$, $t = 5$ and $t = 10$.
(b) sketch the wave as a function of time for

   (i) $x = 2$ $(t > 5)$ (ii) $x = 4$ $(t > 10)$

(c) find the wave velocity and use your graphs to justify it.

11.9 A pneumatic drill produces a sound pressure of 6 N m^{-2}. Given that the reference pressure is 2×10^{-5} N m^{-2}, find the sound pressure level in decibels.

11.10 The reference level on a voltmeter is set as 0.775 V. Calculate the reading in decibels when the voltage reading is 0.4V.

11.11 Show, using trigonometric identities, that

(a) $\cos(X + \delta) - \cos(X - \delta) = -2 \sin(\delta) \sin(X)$
(b) $\sin(X + \delta) + \sin(X - \delta) = 2 \sin(X) \cos(\delta)$

11.12 Two wave trains have very close frequencies and can be expressed by the sinusoids $y = 2 \sin(6.14t)$ and $y = 2 \sin(6.19t)$. Their sum is sketched in Fig. 11.30. Use the expression for the summation of two sines to find the beat frequency (the number of times the magnitude of the amplitude envelope reaches a maximum each second).

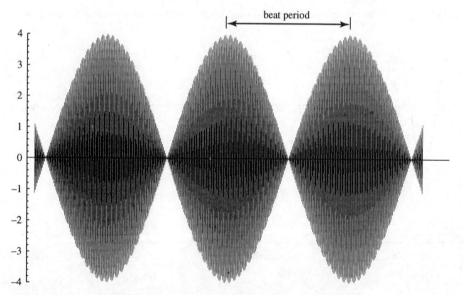

Figure 11.30 Two sinusoids $y = 2 \sin(6.14t)$ and $2 \sin(6.19t)$ for Exercise 11.12.

11.13 A single frequency of 200 Hz (message signal) is amplitude modulated with a carrier frequency of 2 MHz. Express the message signal as $m = a \cos(\omega_1 t)$ and the carrier as $c = b \cos(\omega_2 t)$ and assuming that the modulation gives the product $mc = ab \cos(\omega_1 t) \cos(\omega_2 t)$ use trigonometric identities to show that the modulated signal can be represented by the sum of two frequencies at

$$2 \times 10^6 \pm 200 \text{ Hz}$$

11.14 (a) Give the wavelengths of three modes of vibration on a string of length 0.75 m.

(b) The velocity v is approximately given by $v = \sqrt{(T/m)}$, where T is the tension and m is the mass per unit length of the string. Given that $T = 2200$ N and $m = 0.005$ kg m^{-1}, find the frequency of the fundamental mode.

11.15 Use $a\cos(\omega t + \delta) = a\cos(\omega t)\cos(\delta) - a\sin(\omega t)\sin(\delta)$ to find c and d in the expression $2\cos(3t + \pi/3) = c\cos(3t) + d\sin(3t)$.

11.16 Express as single sines or cosines:

(a) $\sin(43°)\cos(61°) + \cos(43°)\sin(61°)$
(b) $\sin(22°)\cos(18°) - \cos(22°)\cos(18°)$
(c) $\cos(63°)\cos(11°) + \sin(63°)\sin(11°)$
(d) $\sin(41°)\sin(22°) - \cos(41°)\cos(22°)$
(e) $\sin(2x)\cos(x) + \cos(2x)\sin(x)$
(f) $\cos^2(x) - \sin^2(x)$

11.17 Express $\cos(x + y + z)$ in terms of the sines and cosines of x, y and z.

11.18 Find all the solutions to the following equations in the interval $[0, 6\pi]$:

(a) $\sin(x) = -\frac{1}{2}$ (b) $\tan(x) = \frac{1}{3}$ (c) $\cos(x) = -0.8$
(d) $\sin^2(x) = 0.25$ (e) $\cos(x) + 2\cos^2(x) = 0$
(f) $\sin^2(x) - 2 = 2$

TWELVE
DIFFERENTIATION

12.1 INTRODUCTION

We have used functions to expression relationships between variables. For instance, in an electrical circuit the current and the voltage through a resistor can be related by $V = IR$. Here the one physical relationship can be found by simply substituting in the formula the known value of the other. Another common relationship between physical quantities is that one quantity is the rate of change of another: for instance, speed is defined as rate of change of distance with respect to time. It is simple to find the average speed of a moving object if we know how far it has travelled in a certain length of time – this is given by:

$$\frac{\text{Distance travelled}}{\text{Time taken}}$$

This does not give an idea of the speed at any particular instant. If I am travelling by coach from London to Birmingham the journey takes about 2.75 h to go about 110 miles. This means that the average speed is about 40 m.p.h. However, I know that the coach by no means travels at a constant speed. Through central London it travels at around 12 m.p.h. and on the motorway around 70 m.p.h. Is there a way that we can estimate the speed at any particular moment as accurately as possible, armed only with a milometer, to give the distance travelled, and a stopwatch? I can measure the distance travelled in a 10 s interval and find that it is 0.2 miles. That means that the average speed is 0.2/10 miles per second = 0.02 miles per second = 0.02 × 60 × 60 miles per hour = 72 m.p.h. This gives a pretty good idea of the speed at any moment within that 10 s interval, as the length of time is small enough that the speed probably has not changed too much. The smaller the period of time over which we take the measurement, the more accurately we should be able to estimate the speed at any one instant. The speed is found by looking at the ratio of the distance travelled over the time taken where the time interval is taken as small as possible. This is an approximation to the instantaneous rate of change of distance with respect to time.

The rate of change of one quantity with respect to another is called its derivative. If we know the expression defining the function then we are able to find its derivative. The techniques used for differentiating are described in this chapter.

Many physical quantities are related through differentiation. Some of these are current and charge, acceleration and velocity, force and work done, momentum and force, and power and energy. Applications of differentiation are therefore very widespread in all areas of engineering.

12.2 THE AVERAGE RATE OF CHANGE AND THE GRADIENT OF A CHORD

A ball is thrown from the ground and after t seconds is at a distance s metres from the ground, where

$$s = 20t - 5t^2$$

Find:

(a) The average velocity of the ball between $t=1$ s and $t=1.1$ s
(b) The average velocity between $t=1$ and $t=1.01$ s
(c) The average velocity between $t=1$ and $t=1.001$ s
(d) The average velocity between $t=1$ and $t=1.0001$ s

Guess the velocity at $t=1$.

The average velocity is given by the distance moved divided by the time taken. This can be represented by

$$\text{average velocity} = \frac{\text{change in distance}}{\text{change in time}} = \frac{\delta s}{\delta t} = \frac{s(t_2) - s(t_1)}{t_2 - t_1}$$

where t_2 and t_1 are the times between which we are finding the average. δs ('delta s') is used to represent a change in s, and δt ('delta t') is used to represent a change in t. The average velocity, $\delta s/\delta t =$ 'delta s over delta t'.

So, to solve this problem we can use a table, as given in Table 12.1.

Table 12.1 The average velocities over various intervals are calculated using a table

t_1	t_2	$s(t_1) = 20t_1 - 5t_1^2$	$s(t_2) = 20t_2 - 5t_2^2$	$t_2 - t_1$	$s(t_2) - s(t_1)$	Average velocity $= \dfrac{\delta s}{\delta t}$
1	1.1	15	15.95	0.1	0.95	9.5
1	1.01	15	15.099 5	0.01	0.995	9.95
1	1.001	15	15.009 995	0.001	0.009 995	9.995
1	1.000 1	15	15.000 999 95	0.000 1	$9.999\,5 \times 10^{-4}$	9.999 5
						$\vdots$
		Velocity at $t=1$				10

The graph of this function is given in Fig. 12.1. If the line is drawn which joins two points lying on the graph of the function then this is called a chord. We found that when $t=1$ then $s=15$, so the point $(1,15)$ lies on the graph. When $t=1.1$, $s=15.95$, so we also mark the point $(1.1,15.95)$. The triangle containing the chord joining these two points has height $\delta s =$ change in $s = 15.95 - 15 = 0.95$, and base length $\delta t =$ change in $t = 1.1 - 1.0 = 0.1$. This means that the gradient $= \delta s/\delta t = 0.95/0.1 = 9.5$. Another chord can be drawn from $t=1$ to $t=1.01$. When $t=1.01$, $s=15.0995$, so we mark the point $(1.01,15.0995)$. The triangle containing the chord joining $(1,15)$ to $(1.01,15.0995)$ has height $\delta s =$ change in $s = 15.0995 - 15 = 0.0995$, and base length $\delta t =$ change in $t = 1.01 - 1.0 = 0.1$. This means that the gradient $= \delta s/\delta t = 0.0995/0.01 = 9.95$.

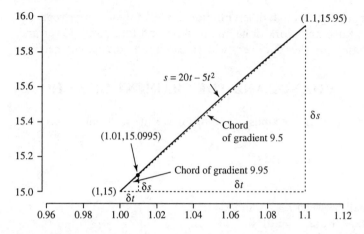

Figure 12.1 Part of the graph of $x = 20t - 5t^2$. The chord joining (1,15) to (1.1,15.95) has gradient $\delta s/\delta t = 9.5$. The chord joining (1,15) to (1.01,15.0995) has gradient 9.95.

As the ends of the chord are put nearer together, the gradient of the chord gives a very good approximation to the instantaneous rate of change. Unfortunately, the chord becomes so small we can hardly see it! To get round this problem we can extend the line at either end. So a chord between two points that are very close together appears to only just touch the function curve. A line which just touches at one point is called a tangent. As the two points on the chord approach each other, the line of the chord approaches the tangent. Therefore the gradient of the chord must also give a good approximation to the gradient of the tangent.

We can guess from Table 12.1 that the instantaneous velocity at $t = 1$ is 10 m s^{-1}. Although the length of time over which we take the average gets smaller and smaller, i.e. tends towards zero, the average velocity does not get nearer to zero, but instead approaches the value of the instantaneous velocity.

The instantaneous velocity is represented by ds/dt, the derivative of s with respect to t, and can be defined using

$$\text{instantaneous velocity} = \frac{\mathrm{d}s}{\mathrm{d}t} = \lim_{\delta t \to 0} \frac{\delta s}{\delta t}$$

This is read as 'ds by dt is the limit as delta t tends to 0 of delta s over delta t'. Note that ds/dt is read as 'ds by dt' (not 'ds over dt') because the line between the ds and the dt does not mean 'divided by'. However, because it does represent a rate of change it usually 'works' to treat ds/dt like a fractional expression. This is because we can always approximate the instantaneous rate of change by the average rate of change (which is a fraction):

$$\frac{\mathrm{d}s}{\mathrm{d}y} \simeq \frac{\delta s}{\delta t} \qquad \text{for small } \delta t$$

i.e. 'ds by dt is approximately delta s over delta t for small delta t'.

$\delta s/\delta t$ represents the gradient of a chord and ds/dt represents the gradient of the tangent. If $\delta s/\delta t$ is used as an approximation to ds/dt then we are using the chord to approximate to the tangent.

12.3 THE DERIVATIVE FUNCTION

We saw in the previous section that if the ends of the chord are put closer and closer together the gradient of the chord approaches the gradient of a tangent. The tangent to a curve is a line which only touches the curve at one point. The gradient of the tangent is also more simply referred to as the slope of the curve at that point. If the slope of the curve is found for every point on the curve then we get the derivative function.

The derivative of a function $y = f(x)$ is defined as

$$\frac{dy}{dx} = \lim_{\delta x \to 0} \frac{\delta y}{\delta x}$$

provided that this limit exists. As δy is the change in y and $y = f(x)$ then, at the points x and $x + \delta x$, y has values $f(x)$ and $f(x + \delta x)$, so the increase in y is given by $\delta y = f(x + \delta x) - f(x)$. We have

$$\frac{dy}{dx} = \lim_{\delta x \to 0} \frac{\delta y}{\delta x} = \lim_{\delta x \to 0} \frac{f(x + \delta x) - f(x)}{\delta x}$$

The derivative of $y = f(x)$, dy/dx, 'dy by dx', can also be represented as $f'(x)$ (read as 'f dashed of x').

$f'(a)$ is the gradient of the tangent to the curve $f(x)$ at the point $x = a$. This is found by finding the gradient of the chord between two points at $x = a + \delta x$ and $x = a$ and taking the limit as δx tends to zero.

The gradient of a chord gives the average rate of change of a function over an interval, and the gradient of the tangent – the derivative – gives the instantaneous rate of change of the function at a point. These definitions are pictured in Fig. 12.2. Derivative functions can be found by evaluating the limit shown in Fig. 12.2. This is called differentiating from first principles.

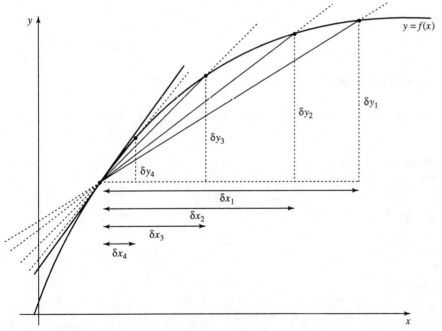

Figure 12.2 The gradient of the chord $\delta y/\delta x$ approaches the gradient of the tangent (or the slope of the curve) as the ends of the chord get closer together (i.e. δx tends to zero). This is written as

$$\frac{dy}{dx} = \lim_{\delta x \to 0} \frac{\delta y}{\delta x}$$

read as 'dy by dx equals the limit, as the change in x (delta x) tends to zero, of the change in y (delta y) divided by the change in x (delta x)'.

Example 12.1 If $y = x^2$, find dy/dx using

$$\frac{dy}{dx} = \lim_{\delta x \to 0} \frac{\delta y}{\delta x}$$

SOLUTION Consider a small change so that x goes from x to $x + \delta x$. As $y = x^2$ we can find the function value as $x + \delta x$ by replacing x by $x + \delta x$, giving $y + \delta y = (x + \delta x)^2$. Therefore, the change in y, δy, is given by

$$(x + \delta x)^2 - x^2 = x^2 + 2x\,\delta x + \delta x^2 - x^2 = 2x\,\delta x + (\delta x)^2$$

Therefore

$$\frac{\delta y}{\delta x} = \frac{2x\,\delta x + (\delta x)^2}{\delta x}$$

As long as δx does not actually equal 0 we can divide the top and bottom line by δx, giving

$$\frac{\delta y}{\delta x} = 2x + \delta x$$

and therefore

$$\frac{dy}{dx} = \lim_{\delta x \to 0} 2x + \delta x = 2x$$

This has shown that

$$y = x^2 \Rightarrow \frac{dy}{dx} = 2x$$

12.4 SOME COMMON DERIVATIVES

We begin by listing derivatives of some simple functions, as given in Table 12.2.

Table 12.2 The derivatives of some simple functions

$f(x)$	$f'(x)$
C	0
x^n	nx^{n-1}
$\cos(x)$	$-\sin(x)$
$\sin(x)$	$\cos(x)$
$\tan(x)$	$\sec^2(x) = \dfrac{1}{\cos^2(x)}$

We can also express the lines of this table using d/dx as an operator, giving for instance

$$\frac{d}{dx}(x^n) = nx^{n-1}$$

This can be read as 'the derivative of x^n is nx^{n-1}' or 'd by dx of x^n is nx^{n-1}'.

To see the validity of a couple of these, refer to Fig. 12.3 for the derivative of a constant, C, and to Fig. 12.4 for the derivative of $\cos(x)$.

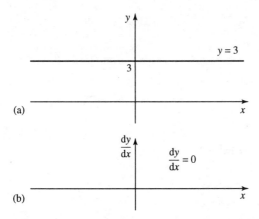

Figure 12.3 (a) The derivative of a constant, for instance $y = 3$. The slope at any point is zero (the line has zero gradient everywhere). (b) The graph of the derivative is given as $dy/dx = 0$.

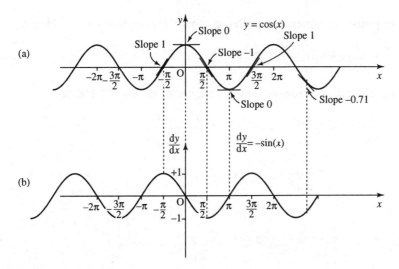

Figure 12.4 (a) The graph of $y = \cos(x)$ with a few tangents marked. Notice that when travelling from left to right, when we are going uphill the slope (and therefore the derivative) must be positive, and when going downhill the derivative must be negative. At the top of the hills and the bottom of the troughs the slope is 0. Joining up the points on the bottom graph gives something like an upside-down sine wave. (b) The graph of the derivative $dy/dx = -\sin(x)$.

Example 12.2 Differentiate

(a) x (b) x^5 (c) $\dfrac{1}{x^3}$

(d) $\sqrt{x}$ (e) $\dfrac{1}{\sqrt{x^3}}$

SOLUTION To differentiate a power of x we must first write the expression in the form x^n where n is some number. We can then use the fact that

$$\frac{d}{dx}(x^n) = nx^{n-1}$$

from Table 12.2.

(a) $x = x^1$; therefore $n = 1$. Substituting $n = 1$ in

$$\frac{d}{dx}(x^n) = nx^{n-1}$$

gives

$$\frac{d}{dx}(x^1) = 1x^{1-1} = 1x^0 = 1$$

as $x^0 = 1$. Hence

$$\frac{d}{dx}(x) = 1$$

(b)

$$\frac{d}{dx}(x^5) = 5x^{5-1} = 5x^4$$

(c) $1/x^3 = x^{-3}$ (using the properties of negative powers given in Chapter 4), so $n = -3$. Hence

$$\frac{d}{dx}(x^{-3}) = (-3)x^{-3-1} = -3x^{-4} = \frac{-3}{x^4}$$

(d) $\sqrt{x} = x^{1/2}$ (using the properties of roots given in Chapter 4), so $n = 1/2$. Hence

$$\frac{d}{dx}(x^{1/2}) = \tfrac{1}{2}x^{1/2-1} = \tfrac{1}{2}x^{-1/2} = \tfrac{1}{2}\frac{1}{x^{1/2}} = \frac{1}{2\sqrt{x}}$$

(e)

$$\frac{1}{\sqrt{x^3}} = \frac{1}{x^{3/2}} = x^{-3/2}$$

so $n = -3/2$. Hence

$$\frac{d}{dx}(x^{-3/2}) = -\tfrac{3}{2}x^{-3/2-1} = -\tfrac{3}{2}x^{-5/2} = -\tfrac{3}{2}\frac{1}{x^{5/2}} = -\tfrac{3}{2}\frac{1}{\sqrt{x^5}}$$

Example 12.3 The energy stored in a stretched spring of extension x metres is found to be $E = x^2$ joules (Fig. 12.5). The force exerted by hanging a weight on the spring is given by mg, where g is the acceleration due to gravity $g \simeq 10$ m s^{-1} and m is the mass. Given that the spring is extended by 0.5 m and that $F = dE/dx$ find the mass hanging on the spring.

SOLUTION Find the expression for the force by differentiating $E = x^2$:

$$F = \frac{dE}{dx} = 2x$$

As $x = 0.5$, $F = 2(0.5) = 1$ N.

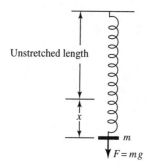

Figure 12.5 A spring has a weight hanging from it of unknown mass m. The spring extends by an amount x and the energy stored in the spring is known to be $E=x^2$. The force due to gravity is $F=mg$, where g is the acceleration due to gravity.

Now, $F=mg$, and as $g \simeq 10$ then

$$1 = m \times 10 \qquad m = \tfrac{1}{10} = 0.1 \text{ kg}$$

The mass on the spring is 0.1 kg.

12.5 FINDING THE DERIVATIVE OF COMBINATIONS OF FUNCTIONS

To find the derivative of functions which are the sum, difference, quotient, product or composite of any of the functions given in Table 12.2 we use that table and the rules given in this section.

Derivatives of $af(x)$ where a is a constant

$$\frac{\mathrm{d}}{\mathrm{d}x}(af(x)) = af'(x)$$

This is only true if a is a constant, not if it is a function of x.

Example 12.4 Differentiate $y = 2x^3$. Notice that this is a constant, 2, multiplied by x^3. The derivative of x^3 is found by looking at Table 12.2. The derivative of x^n is given by nx^{n-1}. In this case $n = 3$, so

$$\frac{\mathrm{d}}{\mathrm{d}x}(x^3) = 3x^2$$

and hence

$$\frac{\mathrm{d}y}{\mathrm{d}x} = 2(3x^2) = 6x^2$$

Derivatives of a sum (or a difference) of functions

If y can be written as the sum of two functions, i.e. $y = u + v$, where u and v are functions of x, then:

$$\frac{\mathrm{d}y}{\mathrm{d}x} = \frac{\mathrm{d}u}{\mathrm{d}x} + \frac{\mathrm{d}y}{\mathrm{d}x}$$

Example 12.5 Differentiate $y = \sin(t) + \sqrt{t}$.

SOLUTION

$$y = \sin(t) + \sqrt{t}$$

$$\Leftrightarrow y = \sin(t) + t^{1/2}$$

To differentiate a sum, differentiate each part:

$$\frac{dy}{dt} = \cos(t) + \tfrac{1}{2}t^{-1/2}$$

$$\frac{dy}{dt} = \cos(t) + \frac{1}{2\sqrt{t}}$$

Derivatives of composite functions (function of a function)

If $y = f(x)$ is a composite function, so that we can write $y = h(u)$, where $u = g(x)$, then

$$\frac{dy}{dx} = \frac{dy}{du}\frac{du}{dx}$$

This is called the chain rule.

Example 12.6 Differentiate $y = \sin(2x)$.

SOLUTION We can substitute $u = 2x$, giving $y = \sin(u)$. Then

$$\frac{du}{dx} = 2 \qquad \text{and} \qquad \frac{dy}{du} = \cos(u)$$

Therefore

$$\frac{dy}{dx} = \frac{dy}{du}\frac{du}{dx} = 2\cos(u)$$

Finally, resubstitute for u, giving

$$\frac{dy}{dx} = 2\cos(2x)$$

Note that we always need to make substitutions so that our function is the composite of the simple functions that we know how to differentiate i.e. x^n, $\sin(x)$, $\cos(x)$, $\tan(x)$ (or a constant times these functions, or the sum of these functions). One simple way to guess the required substitution is to look for a bracket.

Example 12.7 Differentiate $y = (5x - 2)^3$.

SOLUTION Substitute $u = 5x - 2$ (the function in the bracket) so that $y = u^3$. Then

$$\frac{du}{dx} = 5 \qquad \text{and} \qquad \frac{dy}{du} = 3u^2$$

Therefore

$$\frac{dy}{dx} = \frac{dy}{du}\frac{du}{dx}$$

gives

$$\frac{dy}{dx} = 5(3u^3) = 15(5x - 2)^2$$

(resubstituting $u = 5x - 2$).

Example 12.8 Differentiate $y = \cos(2x^2 + 3)$.

SOLUTION Substitute $u = 2x^2 + 3$, giving $y = \cos(u)$. Then

$$\frac{du}{dx} = 4x \qquad \text{and} \qquad \frac{dy}{du} = -\sin(u)$$

Therefore

$$\frac{dy}{dx} = \frac{dy}{du}\frac{du}{dx} = -4x \sin(u) = -4x \sin(2x^2 + 3)$$

Because of the widespread use of the chain rule, it is useful to be able to differentiate a composite function 'in your head'. This is a technique that comes with practice (like mental arithmetic).

Example 12.9 Differentiate

$$V = \frac{1}{t + 1}$$

SOLUTION Rewrite

$$V = \frac{1}{t + 1}$$

as $V = (t + 1)^{-1}$, and think of this as $V = (\ )^{-1}$ where $(\ ) = t + 1$. Now differentiate V with respect to $(\ )$ and multiply by the derivative of $(\ )$ with respect to t. That is,

$$\frac{dV}{dt} = \frac{dV}{d(\ )}\frac{d(\ )}{dt}$$

where $(\ )$ can be any expression. So

$$\frac{dV}{dt} = (-1(t + 1)^{-2})\frac{d}{dt}(t + 1)$$

$$\frac{dV}{dt} = (-1(t + 1)^{-2})1$$

$$\frac{dV}{dt} = -\frac{1}{(t + 1)^2}$$

Example 12.10 Differentiate

$$y = \frac{1}{(3t^2 + 2t)^2}$$

SOLUTION Rewrite

$$y = \frac{1}{(3t^2 + 2t)^2}$$

as $y = (3t^2 + 2t)^{-2}$, and think of this as $y = (\ \ )^{-2}$ where $(\ \ ) = 3t^2 + 2t$. Now differentiate y with respect to $(\ \ )$ and multiply by the derivative of $(\ \ )$ with respect to t. That is,

$$\frac{dy}{dt} = \frac{dy}{d(\ \ )} \frac{d(\ \ )}{dt}$$

where $(\ \ )$ can be any expression. So

$$\frac{dy}{dt} = (-2(3t^2 + 2t)^{-3}) \frac{d}{dt}(3t^2 + 2t)$$

$$\Rightarrow \frac{dy}{dt} = (-2(3t^2 + 2t)^{-3})(6t + 2)$$

$$\Leftrightarrow \frac{dy}{dt} = -\frac{12t + 4}{(3t^2 + 2t)^3}$$

Derivatives of inverse trigonometric functions

By the definition of the inverse function we know that $y = \sin^{-1}(x) \Leftrightarrow \sin(y) = x$, where $-1 \leqslant x \leqslant 1$, and therefore the derivatives are related. We can find the derivative of $\sin^{-1}(x)$ as in the following example.

Example 12.11 Given $y = \sin^{-1}(x)$ and

$$\frac{d}{dx}(\sin(x)) = \cos(x)$$

find dy/dx.

SOLUTION As $y = \sin^{-1}(x)$, then, by the definition of the inverse (assuming x is limited to $[-1,1]$, $\sin(y) = x$. The left-hand side of this is a function of y, which we can call w. Hence $w = \sin(y)$, and by the chain rule

$$\frac{dw}{dx} = \frac{dw}{dy} \frac{dy}{dx}$$

the equation $\sin(y) = x$ becomes $w = x$.

Differentiating $w = x$ with respect to x gives $dw/dx = 1$. Differentiating $w = \sin(y)$ with respect to y we get

$$\frac{dw}{dy} = \cos(y)$$

Hence

$$\frac{dw}{dx} = \frac{dw}{dy} \frac{dy}{dx}$$

becomes

$$1 = \cos(y)\frac{dy}{dx}$$

Dividing both sides by $\cos(y)$ (if $\cos(y) \neq 0$) gives

$$\frac{dy}{dx} = \frac{1}{\cos(y)}$$

This is an expression for the derivative we want to find, but it is a function of y instead of x. Use $\sin(y) = x$ and the trigonometric identity $\cos^2(y) = 1 - \sin^2(y)$, giving $\cos(y) = \sqrt{1 - \sin^2(y)}$ (for $-\pi/2 \leqslant y \leqslant \pi/2$). As $\sin(y) = x$ then $\cos(y) = \sqrt{1 - x^2}$, giving

$$\frac{dy}{dx} = \frac{1}{\sqrt{1 - x^2}}$$

The same method can be used to find the derivatives of $\cos^{-1}(x)$ and $\tan^{-1}(x)$, and we can now add these functions to the list, giving a new table of standard derivatives, as in Table 12.3.

Table 12.3 The derivatives of some simple functions

$f(x)$	$f'(x)$
C	0
x^n	nx^{n-1}
$\cos(x)$	$-\sin(x)$
$\sin(x)$	$\cos(x)$
$\tan(x)$	$\sec^2(x)$
$\sin^{-1}(x)$	$\dfrac{1}{\sqrt{1-x^2}}$
$\cos^{-1}(x)$	$\dfrac{-1}{\sqrt{1-x^2}}$
$\tan^{-1}(x)$	$\dfrac{1}{1+x^2}$

Derivatives of a product of two functions

If y can be written as the product of two functions, so that

$$y = uv$$

where u and v are functions of x, then

$$\frac{dy}{dx} = u\frac{dv}{dx} + v\frac{du}{dx}$$

Example 12.12 Find the derivative of $y = 5x\sin(x)$.

SOLUTION $y = uv$, where $u = 5x$ and $v = \sin(x)$. Hence

$$\frac{du}{dx} = 5 \quad \text{and} \quad \frac{dv}{dx} = \cos(x)$$

Using the derivative of a product formula, then:

$$\frac{dy}{dx} = 5x \, \cos(x) + 5 \, \sin(x)$$

Example 12.13 Find the derivative of $y = \sin(t) \cos(3t)$.

SOLUTION $y = uv$, where $u = \sin(t)$ and $v = \cos(3t)$. Hence

$$\frac{du}{dt} = \cos(t) \qquad \text{and} \qquad \frac{dv}{dt} = -3 \, \sin(3t)$$

Using the derivative of a product formula, then:

$$\frac{dy}{dt} = \cos(t) \, \cos(3t) - 3 \, \sin(t) \, \sin(3t)$$

Derivatives of a quotient of two functions

If y can be written as the quotient of two functions, so that $y = u/v$, where u and v are functions of x, then

$$\frac{dy}{dx} = \frac{v(du/dx) - u(dv/dx)}{v^2}$$

Example 12.14 Find the derivative of

$$y = \frac{\sin(3x)}{x+1}$$

SOLUTION We have $u = \sin(3x)$ and $v = x + 1$, so

$$\frac{du}{dx} = 3 \, \cos(3x) \qquad \text{and} \qquad \frac{dv}{dx} = 1$$

Hence

$$\frac{dy}{dx} = \frac{(x+1)(3 \, \cos(3x)) - (\sin(3x))1}{(x+1)^2}$$

$$= \frac{3(x+1) \, \cos(3x) - \sin(3x)}{(x+1)^2}$$

Example 12.15 Find the derivative of

$$z = \frac{12t}{1+t^3}$$

SOLUTION Setting $u = 12t$ and $v = 1 + t^3$, we have

$$\frac{du}{dt} = 12 \qquad \text{and} \qquad \frac{dv}{dt} = 3t^2$$

Hence

$$\frac{dz}{dt} = \frac{(1+t^3)12 - (12t)(3t^2)}{(1+t^3)^2} = \frac{12 + 12t^3 - 36t^3}{(1+t^3)^2} = \frac{12 - 24t^3}{(1+t^3)^2}$$

12.6 APPLICATIONS OF DIFFERENTIATION

As we mentioned in the introduction to this chapter, many physical quantities important in engineering are related by the derivative, as described below.

Mechanics

$v = \dfrac{dx}{dt}$ v = velocity, x = distance, t = time

$a = \dfrac{dv}{dt}$ a = acceleration, v = velocity, t = time

$F = \dfrac{dW}{dx}$ F = force, W = work done (or energy used), x = distance moved in the direction of the force

$F = \dfrac{dp}{dt}$ F = force, p = momentum, t = time

$P = \dfrac{dW}{dt}$ P = power, W = work done (or energy used), t = time

$\dfrac{dE}{dv} = p$ E = kinetic energy, v = velocity, p = momentum

Gases

$\dfrac{dW}{dV} = p$ p = pressure, W = work done under isothermal expansion, V = volume

Circuits

$I = \dfrac{dQ}{dt}$ I = current, Q = charge, t = time

$V = L\dfrac{dI}{dt}$ V = voltage drop across an inductor, L = inductance, I = current, t = time

Electrostatics

$E = -\dfrac{dV}{dx}$ V = potential, E = electric field, x = distance

Example 12.16 A ball is thrown in the air so that the height of the ball is found to be $s = 3t - 5t^2$. Find

(a) The ball's initial velocity when first thrown into the air.

(b) The time when it returns to the ground.

(c) Its final velocity as it hits the ground.

SOLUTION

(a)
$$v = \frac{ds}{dt} = 3t - 5t^2 \Rightarrow \frac{ds}{dt} = 3 - 10t$$

The ball is initially thrown into the air when $t = 0$, so

$$\frac{ds}{dt} = 3 - 10(0) = 3 \text{ m s}^{-1}$$

(b) The ball returns to the ground on the second occasion, so that the distance travelled $s = 0$. The time is given by solving the equation for t when $s = 0$:

$$0 = 3t - 5t^2$$

$$\Leftrightarrow t(3 - 5t) = 0$$

$$\Leftrightarrow t = 0 \text{ or } 3 - 5t = 0$$

$$\Leftrightarrow t = 0 \text{ or } t = 3/5$$

Therefore it must return to the ground when $t = \frac{3}{5} = 0.6$ s.

(c) When $t = 0.6$, using $v = ds/dt = 3 - 10t$,

$$v = 3 - 10(0.6) = 3 - 6 = -3 \text{ m s}^{-1}$$

Therefore the velocity as it hits the ground is -3 m s^{-1}.

Example 12.17 A rocket is moving with a velocity of $v = 4t^2 + 10\,000$ m s^{-1} over a brief period of time while leaving the Earth's atmosphere. Find its acceleration after 2 s.

SOLUTION Use $a = dv/dt$ as $v = 4t^2 + 10\,000$. Then $a = 8t$, and at $t = 2$ this gives $a = 16$, so the acceleration after 2 s is 16 m s^{-2}.

Example 12.18 The potential due to a point charge Q at a position r from the charge is given by

$$V = \frac{Q}{4\pi\varepsilon_0 r}$$

where ε_0, the permittivity of free space, $\simeq 8.85 \times 10^{-12}$ F m^{-1} and $\pi \simeq 3.14$.

Given that $Q = 1$ C, find the electric field strength at a distance of 5 m using $E = -dV/dr$.

SOLUTION

$$V = \frac{Q}{4\pi\varepsilon_0 r}$$

Substitute for ε_0 and π and use $Q = 1$, so that

$$V = \frac{1}{4 \times 3.14 \times 8.85 \times 10^{-12} r} \simeq \frac{9 \times 10^9}{r} = 9 \times 10^9 r^{-1}$$

Now,

$$E = -\frac{dV}{dr} = -9 \times 10^9(-r^{-2}) = 9 \times 10^9 r^{-2}$$

When $r = 5m$,

$$E = \frac{9 \times 10^9}{25} = 3.6 \times 10^8 \text{ V m}^{-1}$$

12.7 SUMMARY

1. The average rate of change of a function over a certain interval is the same as the gradient of the chord drawn on the graph of the function. This gradient of a function, $y = f(x)$, is given by

$$\frac{\delta y}{\delta x} = \frac{\text{change in } y}{\text{change in } x}$$

2. If the chord is very short then the gradient of the chord is approximately the gradient of the tangent to the graph at a particular spot, i.e. the slope of the graph at that point. The slope of the graph gives the instantaneous rate of change of the function with respect to its independent variable, known as its derivative. This is represented by dy/dx. Then we have the definition

$$\frac{dy}{dx} = \lim_{\delta x \to 0} \frac{\delta y}{\delta x} = \lim_{\delta x \to 0} \frac{f(x + \delta x) - f(x)}{\delta x}$$

This is read as 'dy by dx is the limit, as delta x tends to 0, of delta y over delta x'.

The derivative of $y = f(x)$, dy/dx ('dy by dx') can also be represented by $f'(x)$ (read as 'f dashed of x').

3. Derivatives of simple functions are given in Table 12.3. Rules are used to differentiate combinations of these functions. These are:

Product with a constant

$$\frac{d}{dx}(af(x)) = af'(x)$$

Sum
If $y = u + v$ then

$$\frac{dy}{dx} = \frac{du}{dx} + \frac{dv}{dx}$$

Composite function (function of a function), called the chain rule
If $y = f(x)$, where $y = h(u)$ and $u = g(x)$ then

$$\frac{dy}{dx} = \frac{dy}{du} \frac{du}{dx}$$

Product
If $y = uv$ then

$$\frac{dy}{dx} = v\frac{du}{dx} + u\frac{dv}{dx}$$

Quotient
If $y = u/v$ then

$$\frac{dy}{dx} = \frac{v(du/dx) - u(dv/dx)}{v^2}$$

4. There are many applications of differentiation in all areas of engineering, some of which are listed in Sec. 12.6.

12.8 EXERCISES

12.1 A car is travelling such that its distance, s (metres) from its starting position after a time t (seconds) is

$$s = \tfrac{1}{15}t^3 + 2t \qquad 0 \leqslant t < 10$$

$$s = 22(t - 10) + 86.67 \qquad t \geqslant 10$$

(a) What is its average velocity in the first 10 s?
(b) Give the velocity as a function of time.
(c) What is the instantaneous velocity when (i) $t = 5$, (ii) $t = 10$, (iii) $t = 15$?
(d) What is the average acceleration for the first 10 s?
(e) What is the average acceleration between $t = 10$ and $t = 15$?
(f) Give the acceleration as a function of time.
(g) What is the instantaneous acceleration when (i) $t = 5$, (ii) $t = 10$, (iii) $t = 15$?

12.2 Differentiate the following:

(1) $3x^2 + 6x - 12$ (2) $x^{1/2} - x^{-1/2}$ (3) $\sqrt{2x^3} - \dfrac{5}{6x^2}$

(4) $\sin(3x^3 + x)$ (5) $2\cos(6x - 2)$ (6) $\tan(x^2)$

(7) $\dfrac{1}{2x - 3}$ (8) $(4x - 5)^6$ (9) $\dfrac{1}{\sqrt{x^2 - 1}}$

(10) $\sin^{-1}(5 - 2x)$ (11) $\tan\left(\dfrac{1}{x}\right)$ (12) $\sqrt{x^2 + 2}$

(13) $(x + 4)^{-3/2}$ (14) $\sin^2(x)$ (15) $5\cos^3(x)$

(16) $\dfrac{1}{\sin^3(x)}$ (17) $\cos^2(5x)$ (18) $x^3\sqrt{x + 1}$

(19) $5x\cos(x)$ (20) $6x^2\sin(x)$ (21) $(3x + 1)\tan(5x)$

(22) $x^3\cos^{-1}(x)$ (23) $\dfrac{x^3}{\cos(x)}$ (24) $\dfrac{1}{\sin^2(x)}$

(25) $\dfrac{\sin(x)}{2x + 10}$ (26) $\dfrac{x^2}{\tan(x)}$ (27) $\dfrac{3x^2}{\sqrt{x - 1}}$

(28) $\dfrac{5x^2 - 1}{5x^2 + 1}$ (29) $\dfrac{(x - 1)\cos(x)}{x^2 - 1}$ (30) $\sin^{-1}(x^2)$

(31) $x^2\sqrt{(x - 1)}\sin(x)$ (32) $\cos^2(x^2)$ (33) $\tan^2(\sqrt{5x - 1})$

12.3 A current i is travelling through a single-turn loop of radius 1 m. A four-turn search coil of effective area 0.03 m^2 is placed inside the loop. The magnetic flux linking the search coil is given by

$$\Phi = \mu_0 \frac{iA}{2r} \text{ Wb}$$

where r (m) is the radius of the current carrying loop, A (m^2) is the area of the search coil and μ_0 is the permeability of free space $= 4\pi \times 10^{-7}$ H m^{-1}. Find the e.m.f. induced in the search coil, given by $\mathscr{E} = -N(\mathrm{d}\Phi/\mathrm{d}t)$, where N is the number of turns in the search coil, and the current is given by $i = 20\sin(20\pi t) + 50\sin(30\pi t)$.

THIRTEEN
INTEGRATION

13.1 INTRODUCTION

In Chapter 12 we saw that many physical quantities are related by one being the rate of change, the derivative, of the other. It follows that there must be a way of expressing the 'inverse' relationship. This is called integration. Velocity is the rate of change of distance with time, distance is the integral of velocity with respect to time.

Unfortunately, there are two issues which complicate the simple idea that 'integration is the inverse of differentiation'. First, we find that there are many different functions which are the integral of the same function. Luckily, these functions only differ from each other by a constant. To find all the possible integrals of a function we can find any one of them and add on some constant, called the constant of integration. This type of integral is called the indefinite integral. As it is not satisfactory to have an unknown constant left in the solution to a problem, we employ some other information to find its value. Once the unknown constant is replaced by some value to fit a certain problem, then we have the particular integral.

The second problem with integration is that most functions cannot be integrated exactly, even apparently simple functions like

$$\frac{\sin(x)}{x}$$

For this reason, numerical methods of integration are particularly important. These methods all depend on understanding the idea of integration as the area under a graph. The definite integral of a function, $y = f(x)$, is the integral between two values of x, and therefore gives a number (not a function of x) as a result. There is no uncertainty and hence it is called the definite integral.

This chapter is concerned with methods of finding the integral, the formulae that can be used for finding exact integrals and also with numerical integration. We also look more closely at the definitions of definite and indefinite integrals and at applications of integration.

13.2 INTEGRATION

Integration is the inverse process to differentiation. Consider the following examples (where C is a constant):

Function	→ derivative
Integral	← function
$x^2 + C$	$2x$
$\sin(x) + C$	$\cos(x)$

The derivative of $x^2 + C$ with respect to x is $2x$; therefore the integral of $2x$ with respect to x is $x^2 + C$. This can be written

$$\frac{\mathrm{d}}{\mathrm{d}x}(x^2 + C) = 2x$$

$$\Leftrightarrow \int 2x \, \mathrm{d}x = x^2 + C$$

$\int 2x \, \mathrm{d}x$ is read 'the integral of $2x$ with respect to x'. The $\int$ notation is like an 's', representing a sum, and its origin will be explained more when we look at the definite integral.

The second example gives

$$\frac{\mathrm{d}}{\mathrm{d}x}(\sin(x) + C) = \cos(x)$$

$$\Leftrightarrow \int \cos(x) \, \mathrm{d}x = \sin(x) + C$$

The derivative of $\sin(x) + C$ with respect to x is $\cos(x)$, which is the same as saying that the integral of $\cos(x)$ with respect to x is $\sin(x) + C$.

Constant of integration

Because the derivative of a constant is zero it is not possible to determine the exact integral simply through using inverse differentiation.

For instance, the derivatives of $x^2 + 1$, $x^2 - 2$ and $x^2 + 1000$ all give $2x$. Therefore we express the integral of $2x$ as $x^2 + C$, where C is some constant called the constant of integration.

To find the value that the constant of integration should take in the solution of a particular problem we use some other known information. Supposing we know that a ball has velocity $v = 20 - 10t$. We want to find the distance travelled in time t and we also know that the ball was thrown from the ground, which is at distance 0. We can work out the distance travelled by doing 'inverse differentiation', giving the distance $s = 20t - 5t^2 + C$, where C is the constant of integration. As we also know that $s = 0$ when $t = 0$, we can substitute these values to give $0 = C$, and hence the solution is that $s = 20t - 5t^2$.

In solving this problem we used the fact that $v = \mathrm{d}s/\mathrm{d}t$, and therefore we know that $\mathrm{d}s/\mathrm{d}t = 20 - 10t$. This is called a differential equation, because it is an equation and contains an expression including a derivative. This is the one sort of differential equation which can be solved directly by integrating. Some other sorts of differential equation are solved in Chapters 14, 16 and 20.

In this case, the solution $s = 20t - 5t^2 + C$ represents all possible solutions of the differential equation and is therefore called the general solution. If a value of C is found to solve a given problem, then this is the particular solution.

Example 13.1 Find y such that $dy/dx = 3x^2$ given that y is 5 when $x = 0$.

SOLUTION We know that x^3 differentiated gives $3x^2$, so

$$\frac{dy}{dx} = 3x^2 \Leftrightarrow y = \int 3x^2 \, dx$$

$$\Leftrightarrow y = x^3 + C$$

where C is some constant. This is the general solution to the differential equation. To find the particular solution for this example, use the fact that y is 5 when x is 0. Substitute in $y = x^3 + C$, giving $5 = 0 + C$, so $C = 5$. This gives the particular solution as $y = x^3 + 5$.

13.3 FINDING INTEGRALS

To find the table of standard integrals we take Table 12.3 for differentiation, swap the columns, rewrite a couple of the entries in a more convenient form and add on the constant of integration. This gives Table 13.1.

Table 13.1 The table of standard integrals

$f(x)$	$\int f(x) \, dx$
1	$x + C$
$x^n \; (n \neq -1)$	$\dfrac{x^{n+1}}{n+1} + C$
$\sin(x)$	$-\cos(x) + C$
$\cos(x)$	$\sin(x) + C$
$\sec^2(x)$	$\tan(x) + C$
$\dfrac{1}{\sqrt{1-x^2}}$	$\sin^{-1}(x) + C$
$\dfrac{-1}{\sqrt{1-x^2}}$	$\cos^{-1}(x) + C$
$\dfrac{1}{1+x^2}$	$\tan^{-1}(x) + C$

As integration is 'anti-differentiation', we can spot the integral in the standard cases, i.e. those listed in Table 13.1.

Example 13.2 (a) Find $\int x^3 \, dx$.
From Table 13.1

$$\int x^n \, dx = \frac{x^{n+1}}{n+1} + C$$

where $n \neq -1$. Here $n = 3$, so

$$\int x^3 \, dx = \frac{x^{3+1}}{3+1} + C = \frac{x^4}{4} + C$$

Check:
Differentiate

$$\frac{x^4}{4} + C$$

giving

$$\frac{4x^3}{4} = x^3$$

which is the original expression that we integrated, hence showing that we integrated correctly.
(b) Find

$$\int \frac{1}{x^2 + 1} \, dx$$

From the table,

$$\int \frac{1}{1+x^2} \, dx = \tan^{-1}(x) + C$$

Check:
Differentiate $\tan^{-1}(x) + C$, giving

$$\frac{1}{1+x^2}$$

(c) Find $\int x^{-1/2} \, dx$.
   From the table,

$$\int x^n \, dx = \frac{x^{n+1}}{n+1} + C$$

where $n \neq -1$; in this case $n = -1/2$, so

$$\int x^{-1/2} \, dx = \frac{x^{-1/2+1}}{-1/2+1} + C = \frac{x^{1/2}}{1/2} + C = 2x^{1/2} + C$$

Check:
Differentiate $2x^{1/2} + C$, giving

$$2(\tfrac{1}{2})x^{1/2-1} = x^{-1/2}$$

(d) Find $\int 1 \, dx$.
   From the table, $\int 1 \, dx = x + C$.

Check:
Differentiate $x + C$, giving 1.

We can also find integrals of some combinations of the functions listed in Table 13.1. To do this we need to use rules similar to those for differentiation. However, because when integrating we are working 'backwards', the rules are not so simple as those used to perform differentiation, and furthermore they will not always give a method that will work in finding the desired integral.

Integration of sums and $af(x)$

We can use the facts that

$$\int (f(x)+g(x))\,dx = \int f(x)\,dx + \int g(x)\,dx$$

and also that

$$\int af(x)\,dx = a\int f(x)\,dx$$

Example 13.3
(a) $\int 3x^2+2x-1\,dx = x^3+x^2-x+C$
Check:

$$\frac{d}{dx}(x^3+x^2-x+C)=3x^2+2x-1$$

(b) $\int 3\sin(x)+\cos(x)\,dx = -3\cos(x)+\sin(x)+C$
Check:

$$\frac{d}{dx}(-3\cos(x)+\sin(x)+C)=3\sin(x)+\cos(x)$$

(c)
$$\int \frac{1}{\sqrt{1-x^2}}-\frac{2}{1+x^2}\,dx = \sin^{-1}(x)-2\tan^{-1}(x)+C$$

Check:

$$\frac{d}{dx}(\sin^{-1}(x)-2\tan^{-1}(x)+C)=\frac{1}{\sqrt{1-x^2}}-\frac{2}{1+x^2}$$

Changing the variable of integration

In Chapter 12 we looked at differentiating composite functions. If $y=f(x)$, where we can make a substitution for y in terms of u, i.e. $y=g(u)$, with $u=h(x)$, then

$$\frac{dy}{dx}=\frac{dy}{du}\frac{du}{dx}$$

We can use this to integrate in very special cases by making a substitution for a new variable. The idea is to rewrite the integral so that we end up with one of the functions in Table 13.1. To see when this might work as a method of integration we begin by looking at differentiating a composite function.

Consider the derivative of

$$y=(3x+2)^3$$

We differentiate this using the chain rule, giving

$$\frac{dy}{dx} = 3(3x+2)^2 \frac{d}{dx}(3x+2) = 3(3x+2)^2 3$$

As integration is backwards differentiation, then

$$\int 3(3x+2)^2 3 \, dx = (3x+3)^3 + C$$

Suppose that we had started with the problem of finding the following integral:

$$\int 3(3x+2)^2 3 \, dx$$

If we could spot that the expression to be integrated comes about from differentiating using the chain rule, then we would be able to perform the integration. We can substitute $u = 3x + 2$, giving $du/dx = 3$, and the integral becomes

$$\int 3u^2 \frac{du}{dx} \, dx$$

We then use the 'trick' of replacing $(du/dx) \, dx$ by du, giving

$$\int 3u^2 \, du$$

As the expression to be integrated only involves the variable u we can perform the integration, and we get

$$\int 3u^2 \, du = u^3 + C$$

Substituting again for $u = 3x + 2$ we get the integral as

$$\int 3(3x+2)^2 3 \, dx = (3x+2)^3 + C$$

We used the trick of replacing '$(du/dx) \, dx$' by du. This can be justified in the following argument.

By the definition of the integral as inverse differentiation, if y is differentiated with respect to x and then integrated with respect to x we will get back to y, give or take a constant. This is expressed by

$$\int \frac{dy}{dx} \, dx = y + C \tag{13.1}$$

If y is a composite function that can be written in terms of the variable u, then

$$\frac{dy}{dx} = \frac{dy}{du} \frac{du}{dx}$$

Substituting the chain rule for dy/dx into Eq. (13.1) gives

$$\int \frac{dy}{du} \frac{du}{dx} \, dx = y + C \tag{13.2}$$

If y is a function of u then we could just differentiate with respect to u and then integrate again and we will get back to the same expression, give or take a constant. That is,

$$\int \frac{dy}{du} \, du = y + C \tag{13.3}$$

Considering Eqs (13.2) and (13.3) together, then

$$\int \frac{dy}{du} \frac{du}{dx} \, dx = \int \frac{dy}{du} \, du$$

so that we can represent this result symbolically by

$$\frac{du}{dx} \, dx = du$$

In practice, we make a substitution for u and change the variable of integration by finding du/dx and substituting

$$dx = du \bigg/ \frac{du}{dx}$$

Example 13.4 Find the integral

$$\int -(4-2x)^3 \, dx$$

SOLUTION Make the substitution $u = 4 - 2x$. Then $du/dx = -2$, so $du = -2 \, dx$ and $dx = -du/2$. The integral becomes

$$\int -u^3 \left(-\frac{du}{2} \right) = \int \frac{u^3}{2} \, du = \frac{u^4}{8} + C$$

Re-substitute for $u = 4 - 2x$, giving

$$\int -(4-2x)^3 \, dx = \tfrac{1}{8}(4-2x)^4 + C$$

Check:
Differentiate the result:

$$\frac{d}{dx}(\tfrac{1}{8}(4-2x)^4 + C) = \tfrac{1}{8}4(4-2x)^3 \frac{d}{dx}(4-2x)$$

$$= \tfrac{1}{8}4(4-2x)^3(-2)$$

$$= -(4-2x)^3$$

As this is the original expression that we integrated, this has shown that our result was correct.

When using this method, to find a good thing to substitute look for something in a bracket, or an 'implied' bracket. Such substitutions will not always lead to an expression which it is possible to integrate. However, if the integral is of the form $\int f(u) \, dx$, where u is a linear function of x or if the integral is of the form

$$\int f(u) \frac{du}{dx} \, dx$$

then a substitution will work providing $f(u)$ is a function with a known integral (i.e. a function listed in Table 13.1).

Integrations of the form $\int f(ax+b)\,dx$

For the integral $\int f(ax+b)\,dx$, make the substitution

$$u = ax + b$$

Example 13.5 Find $\int \sin(3x+2)\,dx$.

SOLUTION Substitute $u = 3x+2$. Then

$$\frac{du}{dx} = 3 \Rightarrow du = 3\,dx \Rightarrow dx = \frac{du}{3}$$

and the integral becomes

$$\int \sin(u)\frac{du}{3} = -\frac{\cos(u)}{3} + C$$

Re-substitute $u = 3x+2$, giving

$$\int \sin(3x+2)\,dx = -\frac{\cos(3x+2)}{3} + C$$

Check:

$$\frac{d}{dx}\left(-\frac{\cos(3x+2)}{3} + C\right) = \frac{\sin(3x+2)}{3}\frac{d}{dx}(3x+2) = \frac{3\sin(3x+2)}{3} = \sin(3x+2)$$

Example 13.6 Integrate

$$\frac{1}{\sqrt{1-(3-x)^2}}$$

with respect to x.

SOLUTION Notice that this is very similar to the expression which integrates to $\sin^{-1}(x)$ or $\cos^{-1}(x)$. We substitute for the expression in the bracket $u = 3-x$, giving

$$\frac{du}{dx} = -1 \Rightarrow dx = -du$$

The integral becomes

$$\int \frac{1}{\sqrt{1-u^2}}(-du) = \int \frac{-du}{\sqrt{1-u^2}}$$

From Table 13.1 this integrates to give

$$\cos^{-1}(u) + C$$

Re-substituting $u = 3-x$ gives

$$\int \frac{1}{\sqrt{1-(3-x)^2}}\,dx = \cos^{-1}(3-x) + C$$

Check

$$\frac{d}{dx}(\cos^{-1}(3-x)+C)=\frac{-1}{\sqrt{1-(3-x)^2}}\frac{d}{dx}(3-x)$$

$$=\frac{1}{\sqrt{1-(3-x)^2}}$$

Integrals of the form $\int f(u)(du/dx)\,dx$

Make the substitution for $u(x)$. This type of integration will often work when the expression to be integrated is of the form of a product. One of the terms will be a composite function. This term could well involve an expression in brackets, in which case we will probably substitute the expression in the bracket for a new variable u. The integral should simplify, provided the other part of the product is of the form du/dx.

Example 13.7 Find $\int x\sin(x^2)\,dx$.

SOLUTION Substitute

$$u=x^2\Rightarrow\frac{du}{dx}=2x\Rightarrow du=2x\,dx\Rightarrow dx=\frac{du}{2x}$$

giving

$$\int x\sin(x^2)\,dx=\int x\sin(u)\frac{du}{2x}=\int\tfrac{1}{2}\sin(u)\,du$$

$$=-\tfrac{1}{2}\cos(u)+C$$

As $u=x^2$, we have

$$\int x\sin(x^2)\,dx=-\tfrac{1}{2}\cos(x^2)+C$$

Check:

$$\frac{d}{dx}(-\tfrac{1}{2}\cos(x^2)+C)=\tfrac{1}{2}\sin(x^2)\frac{d}{dx}(x^2)=\tfrac{1}{2}\sin(x^2)(2x)$$

$$=x\sin(x^2)$$

Example 13.8 Find

$$\int\frac{3x}{(x^2+3)^4}\,dx$$

SOLUTION Substitute $u=x^2+3$. Then

$$\frac{du}{dx}=2x\Rightarrow du=2x\,dx\Rightarrow dx=\frac{du}{2x}$$

The integral becomes

$$\int\frac{3x\,du}{u^4\,2x}=\int\frac{3}{2}u^{-4}\,du$$

which can be integrated, giving

$$\frac{3}{2}\frac{u^{-4+1}}{(-4+1)}+C=-\tfrac{1}{2}u^{-3}+C$$

Re-substituting for $u=x^2+3$ gives

$$\int\frac{3x}{(x^2+3)^4}\,dx=-\tfrac{1}{2}(x^2+3)^{-3}+C=-\tfrac{1}{2}\frac{1}{(x^2+3)^3}+C$$

Check:

$$\frac{d}{dx}(-\tfrac{1}{2}(x^2+3)^{-3}+C)=-\tfrac{1}{2}(-3)(x^2+3)^{-4}\frac{d}{dx}(x^2+3)$$

using the function of a function rule. Hence

$$-\tfrac{1}{2}(-3)(x^2+3)^{-4}(2x)=3x(x^2+3)^{-4}=\frac{3x}{(x^2+3)^4}$$

Example 13.9 Find

$$\int\cos^2(x)\,\sin(x)\,dx$$

SOLUTION This can be rewritten as $\int(\cos(x))^2\,\sin(x)\,dx$. Substitute $u=\cos(x)$; then

$$\frac{du}{dx}=-\sin(x)\Rightarrow dx=\frac{-du}{\sin(x)}$$

The integral becomes

$$\int-u^2\,\sin(x)\frac{du}{\sin(x)}=\int-u^2\,du$$

Integrating gives

$$\frac{-u^3}{3}+C$$

Re-substitute for u, giving

$$\int(\cos(x))^2\,\sin(x)\,dx=-\tfrac{1}{3}\cos^3(x)+C$$

Check:
Differentiate

$$-\tfrac{1}{3}\cos^3(x)+C=-\tfrac{1}{3}(\cos(x))^3+C$$

giving

$$-\tfrac{3}{3}(\cos(x))^2(-\sin(x))=\cos^2(x)\,\sin(x)$$

This method of integration will only work when the integral is of the form

$$\int f(u)\frac{du}{dx}\,dx$$

that is, there is a function of a function multiplied by the derivative of the substituted variable, or where the substituted variable is a linear function.

Sometimes you may want to try to perform this method of integration and discover that it fails to work. In this case another method must be used.

Example 13.10 Find

$$\int \frac{x^2}{(x^2+1)^2}\,dx$$

Substitute $u = x^2 + 1$; then

$$\frac{du}{dx}=2x \Rightarrow dx=\frac{du}{2x}$$

The integral becomes

$$\int \frac{x^2}{u^2}\frac{du}{dx}=\int \frac{x}{2u^2}\,du$$

This substitution has not worked. We are no nearer being able to perform the integration. There is still a term in x involved in the integral, so we are not able to perform an integration with respect to u only.

In some of these cases integration by parts may be used.

Integrations by parts

This can be useful for integrating some products, e.g. $\int x \sin x\,dx$. The formula is derived from the formula for differentiation of a product:

$$\frac{d}{dx}(uv)=\frac{du}{dx}v+u\frac{dv}{dx}$$

$$\Leftrightarrow u\frac{dv}{dx}=\frac{d}{dx}(uv)-\frac{du}{dx}v \qquad \text{(subtracting } (du/dx)v \text{ from both sides)}$$

$$\Rightarrow \int u\frac{dv}{dx}\,dx=uv-\int v\frac{du}{dx}\,dx \qquad \text{(integrating both sides)}$$

As we found before, $(du/dx)\,dx$ can be replaced by du, so $(dv/dx)\,dx$ can be replaced by dv, and this gives a compact way of remembering the formula:

$$\int u\,dv=uv-\int v\,du$$

To use the formula, we need to make a wise choice as to which term is u (which we then need to differentiate to find du) and which term is dv (which we then need to integrate to find v). Note that the second term, $v\,du$, must be easy to integrate.

Example 13.11 Find

$$\int x \sin x \, dx$$

SOLUTION Use $u=x$ and $dv=\sin(x)\,dx$. Then

$$\frac{du}{dx}=1 \quad \text{and} \quad v=\int \sin(x)\,dx=-\cos(x)$$

Substitute in $\int u\,dv=uv-\int v\,du$ to give

$$\int x \sin(x)\,dx=-x\cos(x)-\int -\cos(x).1\,dx$$

$$=-x\cos(x)+\sin(x)+C$$

Check:

$$\frac{d}{dx}(-x\cos(x)+\sin(x)+C)=-\cos(x)+x\sin(x)+\cos(x)$$

$$=x\sin(x)$$

We can now solve the problem that we tried to solve using a substitution, but failed.

Example 13.12 Find

$$\int \frac{x^2}{(x^2+1)^2}\,dx$$

SOLUTION We can spot that if we write this as

$$\int x\frac{x}{(x^2+1)^2}\,dx$$

then the second term in the product can be integrated. We set

$$u=x \quad \text{and} \quad dv=\frac{x\,dx}{(x^2+1)^2}=x(x^2+1)^{-2}\,dx$$

Then

$$du=dx \quad \text{and} \quad v=-\tfrac{1}{2}(x^2+1)^{-1}$$

(To find v we have performed the integration $\int x(x^2+1)^{-2}\,dx=-\tfrac{1}{2}(x^2+1)^{-1}$. Check this result by substituting for x^2+1.) Substitute in $\int u\,dv=uv-\int v\,du$ to give

$$\int x\frac{x}{(x^2+1)^2}\,dx=-\frac{x}{2}(x^2+1)^{-1}-\int \frac{-(x^2+1)^{-1}}{2}\,dx$$

$$=\frac{-x}{2(x^2+1)}+\tfrac{1}{2}\int \frac{dx}{(x^2+1)}$$

Note that the remaining integral is a standard integral given in Table 13.1 as $\tan^{-1}(x)$, so the integral becomes

$$\int \frac{x^2}{(x^2+1)^2}\,dx = \frac{-x}{2(x^2+1)} + \tfrac{1}{2}\tan^{-1}(x) + C$$

Check:

$$\frac{d}{dx}\left(\frac{-x}{2(x^2+1)} + \tfrac{1}{2}\tan^{-1}(x) + C\right) = \frac{d}{dx}\left(\frac{-x}{2}(x^2+1)^{-1} + \tfrac{1}{2}\tan^{-1}(x) + C\right)$$

$$= -\tfrac{1}{2}(x^2+1)^{-1} + \frac{x}{2}(2x)(x^2+1)^{-2} + \tfrac{1}{2}\frac{1}{x^2+1}$$

$$= \frac{x^2}{(x^2+1)^2}$$

Integrating using Trigonometric Identities

There are many possible ways of using trigonometric identities in order to perform integration. We shall just look at examples of how to deal with powers of trigonometric functions. For even powers of a trigonometric function then the double angle formula may be used. For odd powers of a trigonometric function, a method involving the substitution

$$\cos^2(x) = 1 - \sin^2(x) \qquad \text{or} \qquad \sin^2(x) = 1 - \cos^2(x)$$

is used.

Example 13.13 Find $\int \sin^2(x)\,dx$.

SOLUTION As $\cos(2x) = 1 - 2\sin^2(x)$, then

$$\sin^2(x) = \frac{1 - \cos(2x)}{2}$$

The integral becomes

$$\int \sin^2(x)\,dx = \int \frac{1 - \cos(2x)}{2}\,dx = \tfrac{1}{2}x - \tfrac{1}{4}\sin(2x) + C$$

Check:

$$\frac{d}{dx}(\tfrac{1}{2}x - \tfrac{1}{4}\sin(2x) + C) = \tfrac{1}{2} - \tfrac{1}{4}.2\cos(2x) = \tfrac{1}{2}(1 - \cos(2x))$$

$$= \sin^2(x) \qquad \text{(from the double angle formula)}$$

Example 13.14 Find $\int \cos^3(x)\,dx$.

SOLUTION As $\cos^2(x) + \sin^2(x) = 1$, then $\cos^2(x) = 1 - \sin^2(x)$. Hence

$$\int \cos^3(x)\,dx = \int \cos(x)\cos^2(x)\,dx$$

$$= \int \cos(x)(1 - \sin^2(x))\,dx$$

$$= \int \cos(x) - \cos(x) \sin^2(x) \, dx$$

$$= \int \cos(x) \, dx - \int \cos(x) \sin^2(x) \, dx$$

The second part of this integral is of the form

$$\int f(u) \frac{du}{dx} \, dx$$

Substitute $u = \sin(x)$; then

$$du = \cos(x) \, dx \Rightarrow dx = \frac{du}{\cos(x)}$$

Hence

$$\int \cos(x) \sin^2(x) \, dx = \int \cos(x) u^2 \frac{du}{\cos(x)} = \int u^2 \, du = \frac{u^3}{3} + C$$

Re-substituting $u = \sin(x)$ gives

$$\int \cos(x) \sin^2(x) \, dx = \frac{\sin^3(x)}{3} + C$$

Therefore

$$\int \cos^3(x) \, dx = \int \cos(x) \, dx - \cos(x) \sin^2(x) \, dx$$

$$= \sin(x) - \frac{\sin^3(x)}{3} + C$$

Check:

$$\frac{d}{dx} \left(\sin(x) - \frac{\sin^3(x)}{3} + C \right) = \cos(x) - \tfrac{3}{3} \sin^2(x) \cos(x)$$

$$= \cos(x)(1 - \sin^2(x))$$

$$= \cos(x) \cos^2(x)$$

$$= \cos^3(x)$$

13.4 APPLICATIONS OF INTEGRATION

We listed in Sec. 12.7 the applications of differentiation that are important in engineering. Here we list the equivalent relationships using integration.

Mechanics

$x = \int v \, dt$ $v = $ velocity, $x = $ distance, $t = $ time

$v = \int a \, dt$ $a = $ acceleration, $v = $ velocity, $t = $ time

$W = \int F \, dx$ $F = $ force, $W = $ work done (or energy used), $x = $ distance moved in the direction of the force

$p = \int F\, dt$ F = force, p = momentum, t = time

$W = \int P\, dt$ P = power, W = work done (or energy used), t = time

$p = \int E\, dv$ E = kinetic energy, v = velocity, p = momentum

Gases

$p = \int W\, dV$ p = pressure, W = work done under isothermal expansion, V = volume

Electrical circuits

$Q = \int I\, dt$ I = current, Q = charge, t = time.

$I = \dfrac{1}{L}\int V\, dt$ V = voltage drop across an inductor, L = inductance, I = current, t = time

Electrostatics

$V = -\int E\, dx$ v = potential, E = electric field, x = distance

Example 13.15 A car moving with a velocity of 12 m s^{-1} accelerated uniformly for 10 s at 1 m s^{-2} and then kept a constant velocity. Calculate:

(a) The distance travelled during the acceleration.
(b) The velocity reached after 20 m.
(c) The time taken to travel 100 m from the time that the acceleration first started.

SOLUTION Take as time 0 the time when the car begins to accelerate. From $t=0$ to $t=10$ the acceleration is 1 m s^{-1}. Hence

$$\frac{dv}{dt} = 1$$

Therefore

$$v = \int 1\, dt = t + C$$

for $0 \leqslant t \leqslant 10$ giving $v = t + C$.

To find the constant C we need to use other information given in the problem. We know that at $t=0$ the velocity is 12 m s^{-1}. Substituting this into $v = t + C$ gives

$$12 = 0 + C \Leftrightarrow C = 12$$

so $v = t + 12$.

For $t > 10$ the velocity is constant; therefore $v = 10 + 12 = 22$ m s^{-1} for $t \geqslant 10$.

The velocity function is therefore

$$v = \begin{cases} t + 12 & 0 \leqslant t \leqslant 10 \\ 22 & t > 10 \end{cases}$$

To find the distance travelled we need to integrate one more time:

$$v = \frac{dx}{dt} = t + 12 \qquad \text{for} \qquad 0 \leqslant t \leqslant 10$$

$$x = t^2 + 12t + C \qquad \text{for} \qquad 0 \leqslant t \leqslant 10$$

To find the value of C, consider that the distance travelled is 0 at $t=0$. Hence $C=0$, and therefore

$$x=t^2+12t \qquad \text{for } 0\leqslant t\leqslant 10$$

For $t>10$ we have a different expression for the velocity:

$$v=\frac{dx}{dt}=22 \qquad \text{for} \qquad t>10$$

$$x=22t+C \qquad \text{for} \qquad t>10$$

To find the value of C in this expression we need some information about the distance travelled, for instance, at $t=10$. Using $x=t^2+12t$ we get that at $t=10$, $x=100+120=220$ m. Substituting this into $x=22t+C$ gives $220=22\times 10+C$, which gives $C=0$, so $x=22t$ for $t>10$.

The function for x is therefore

$$x=\begin{cases} t^2+12t & \text{for} & 0\leqslant t\leqslant 10 \\ 22t & \text{for} & t>10 \end{cases}$$

As we now have expressions for v and for x we are in a position to answer the questions.

(a) The distance travelled during the acceleration is the distance after 10 s:

$$x=(10)^2+12(10)=220 \text{ m}$$

(b) To find the velocity reached after 20 m we need first to find the time taken to travel 20 m:

$$x=20 \Rightarrow 20=t^2+12t$$

$$t^2+12t-20=0$$

Using the formula for solving a quadratic equation:

$$at^2+bt+c=0 \Leftrightarrow t=\frac{-b\pm\sqrt{b^2-4ac}}{2a}$$

$$t=\frac{-12\pm\sqrt{144+80}}{2}=\frac{-12\pm\sqrt{224}}{2}$$

$$t\simeq -13.5 \text{ or } t\simeq 1.5$$

As t cannot be negative, $t\simeq 1.5$ s.

Substituting the value for t into the expression for v gives

$$v=t+12=1.5+12=13.5 \text{ m s}^{-1}$$

So the velocity after 20 m is 13.5 m s^{-1}.

(c) The time taken to travel 100 m from when the acceleration first started can be found from $x=100$:

$$100=t^2+12t \Leftrightarrow t^2+12t-100=0$$

Using the quadratic formula gives

$$t=\frac{-12\pm\sqrt{144+400}}{2}\simeq\frac{-12\pm 23.32}{2}$$

$$t\simeq -17.66 \text{ or } t\simeq 5.66$$

The time taken to travel 100 m is 5.66 s.

Example 13.16 The current across a 1 μF capacitor from time $t=3$ ms to $t=4$ ms is given by $i(t)=-2t$. Find the voltage across the capacitor during that period of time, given that $V=-0.1$ V when $t=3$ ms.

SOLUTION For a capacitor, $V=Q/C$, where $Q=\int I\,dt$, and C is the capacitance, V the voltage drop across the capacitor, Q the charge and I the current. So

$$V=\frac{1}{1\times 10^{-6}}\int -2t\,dt$$

$$=10^6(-t^2)+C$$

When $t=3$ ms, $V=0.1$, so $V=0.1$ when $t=3\times 10^{-3}$:

$$0.1=10^6(-10^{-6}\times 9)+C\Leftrightarrow C=0.1+9\Leftrightarrow C=9.1$$

So $V=-10^6t^2+9.1$

13.5 THE DEFINITE INTEGRAL

The definite integral from $x=a$ to $x=b$ is defined as the area under the curve between those two points. In the graph in Fig. 13.1, the area under the graph has been approximated by dividing it into rectangles. The height of each is the value of y, and if each rectangle is the same width, δx, then the area of the rectangle is $y\,\delta x$.

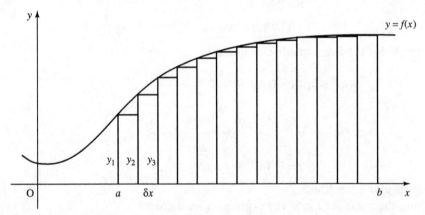

Figure 13.1 A graph of $y=f(x)$ and the area under the graph from $x=a$ to $x=b$. This is approximated by splitting the area into strips of width δx.

If the rectangle is very thin, then y will not vary very much over its width and the area can reasonably be approximated as the sum of all these rectangles.

The symbol for a sum is Σ (the capital Greek letter sigma).

The area under the graph is approximately

$$A=y_1\,\delta x+y_2\,\delta x+y_3\,\delta x+y_4\,\delta x+\ldots=\sum_{x=a}^{x=b-\delta x} y\,\delta x$$

We would assume that if δx is made smaller then the approximation to the exact area would improve. An example is given for the function $y=x$ in Fig. 13.2. Between the values of 1 and 2 we divide the area into strips, firstly of width 0.1, then 0.01, then 0.001.

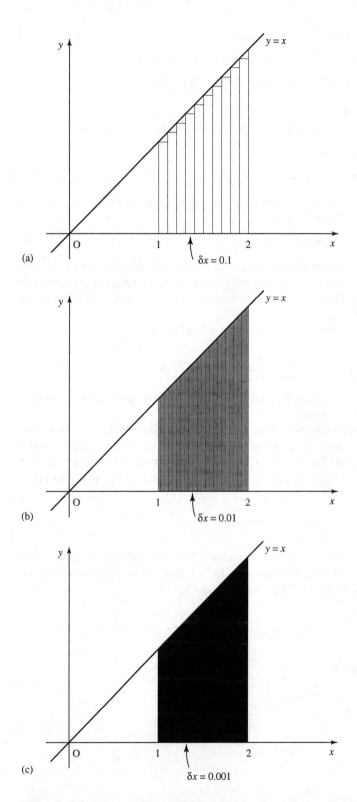

Figure 13.2 The area under the graph $y=x$ between $x=1$ and $x=2$. (a) Dividing into strips of width 0.1 gives 1.45. (d) Dividing into strips of width 0.01 gives 1.495. (c) Dividing into strips of width 0.001 gives 1.4995.

When $\delta x = 0.1$ the calculation gives

$$1 \times 0.1 + 1.1 \times 0.1 + 1.2 \times 0.1 + 1.3 \times 0.1 + 1.4 \times 0.1 + 1.5 \times 0.1$$
$$+ 1.6 \times 0.1 + 1.7 \times 0.1 + 1.8 \times 0.1 + 1.9 \times 0.1 = 1.45$$

When $\delta x = 0.01$ the calculation gives

$$1 \times 0.001 + 1.01 \times 0.01 + 1.02 \times 0.01 + \ldots + 1.98 \times 0.01 + 1.99 \times 0.01 = 1.495$$

When $\delta x = 0.001$ the calculation gives

$$1 \times 0.001 + 1.001 \times 0.001 + 1.002 \times 0.001 + \ldots + 1.998 \times 0.001 + 1.999 \times 0.001 = 1.4995$$

The exact answer is given by the area of the trapezoid which is equal to the average length of the parallel sides multiplied by the width, giving

$$0.5(1 + 2) \times 1 = 1.5$$

We can see that the smaller the strips the nearer the area approximates to the exact area of 1.5. Therefore, as the width of the strips gets smaller and smaller, there is a better approximation to the area, and we say that in the limit, as the width tends to zero, we have the exact area, which is called the definite integral.

The area under the curve $y = f(x)$ between $x = a$ and $x = b$ is found as

$$\int_a^b y \, dx = \lim_{\delta x \to 0} \sum_{x=a}^{x=b-\delta x} y \, \delta x$$

which is read as 'The definite integral of y from $x = a$ to $x = b$ equals the limit as δx tends to 0 of the sum of y times δx for all x from $x = a$ to $x = b - \delta x$'.

This is the definition of the definite integral, which gives a number as its result, not a function.

We need to show that our two ways of defining integration (the indefinite integral as the inverse process of differentiation and the definite integral as the area under the curve) are consistent. To do this, consider an integral of y from some starting point, a, up to any point x. Then the area A is

$$A = \int_a^x y \, dx$$

Notice that as we move the final point x then A will change. Now consider moving the final value by a small amount δx. This will increase the area by δA, and δA is approximately the area of a rectangle of height y and width δx. This is shown in Fig. 13.3.

So we have $\delta A \simeq y \, \delta x$, that is

$$\frac{\delta A}{\delta x} \simeq y$$

Taking the limit as δx tends to 0 gives

$$y = \frac{dA}{dx}$$

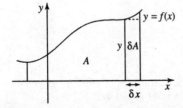

Figure 13.3 The area A is given by the definite integral $\int_a^x y \, dx$, and the increase in the area, $\delta A \simeq y \, \delta x$.

This shows that finding the area under the graph does in fact give a function which when differentiated gives back the function of the original graph, i.e. the area function gives the 'inverse' of differentiation.

The area function, A, is not unique because different functions will be found by moving the position of the starting point for the area. However, in each case dA/dx will be the original function.

This is illustrated for the area under $y = \frac{1}{2}t$ in Fig. 13.4, where there are two area functions, one starting from $t = -1$ and the other from $t = 0$.

The first area function is

$$A = \frac{t^2}{4} - \frac{1}{4}$$

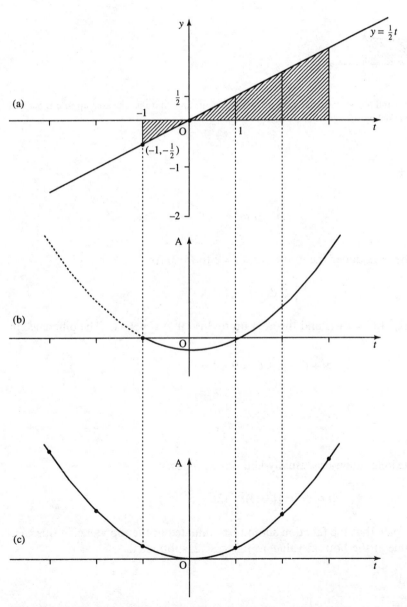

Figure 13.4 (a) The graph of $y = t/2$. (b) The area under the graph of $y = t/2$ starting from $t = -1$. (c) The area under the graph of $y = t/2$ starting from $t = 0$.

and the second is

$$A = \frac{t^2}{4}$$

The definite integral, the area under a particular section of the graph, can be found, as in Fig. 13.5, by subtracting areas.

In practice we do not need to worry about the starting value for finding the area. The effect of any constant of integration will cancel out.

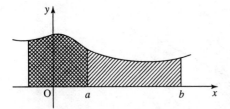

Figure 13.5 The area between a and b is given by the area up to b minus the area up to a. The area up to a is marked by ✕. The area up to b is marked by ///.

Example 13.17 Find

$$\int_2^3 2t \, dt$$

SOLUTION This is the area under the graph from $t = 2$ to $t = 3$. As

$$\int 2t \, dt = t^2 + C$$

the area up to 2 is $(2)^2 + C = 4 + C$ and the area up to 3 is $(3)^2 + C = 9 + C$. The difference in the areas is

$$9 + C - (4 + C) = 9 - 4 = 5$$

Therefore

$$\int_2^3 2t = 5$$

The working of a definite integral is usually laid out as follows:

$$\int_2^3 2t \, dt = [t^2]_2^3 = (3)^2 - (2)^2 = 5$$

The square brackets indicate that the function should be evaluated at the top value, in this case 3, and then have the value at the bottom value, in this case 2, subtracted.

Example 13.18 Find

$$\int_{-1}^1 3x^2 + 2x - 1 \, dx$$

SOLUTION

$$\int_{-1}^{1} 3x^2 + 2x - 1 \, dx = [x^3 + x^2 - x]_{-1}^{1}$$

$$= (1^3 + 1^2 - 1) - ((-1)^3 + (-1)^2 - (-1))$$
$$= 1 - (-1 + 1 + 1)$$
$$= 1 - 1 = 0$$

Example 13.19 Find

$$\int_{0}^{\pi/6} \sin(3x + 2) \, dx$$

SOLUTION

$$\int_{0}^{\pi/6} \sin(3x + 2) \, dx = [-\tfrac{1}{3}\cos(3x + 2)]_{0}^{\pi/6}$$

$$= -\tfrac{1}{3}\cos\left(3\frac{\pi}{6} + 2\right) - (-\tfrac{1}{3}\cos(2))$$

$$= -\tfrac{1}{3}\cos\left(\frac{\pi}{2} + 2\right) + \tfrac{1}{3}\cos(2) \simeq 0.1644$$

Example 13.20 Find the shaded area in Fig. 13.6, where $y = -x^2 + 6x - 5$.

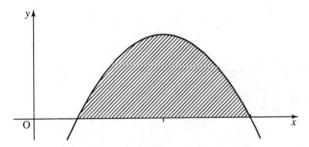

Figure 13.6 The shaded area is bound by the graph of $y = -x^2 + 6x - 5$ and the x-axis.

SOLUTION First we find where the curve crosses the x-axis, i.e. when $y = 0$:
$$0 = -x^2 + 6x - 5 \Leftrightarrow x^2 - 6x + 5 = 0$$
$$\Leftrightarrow (x - 5)(x - 1) = 0 \Leftrightarrow x = 5 \quad \text{or} \quad x = 1$$

This has given the limits of the integration. Now we integrate:

$$\int_{1}^{5} -x^2 + 6x - 5 \, dx = \left[\frac{-x^3}{3} + \frac{6x^2}{2} - 5x\right]_{1}^{5}$$

$$= -\frac{(5)^3}{3} + 3(5)^2 - 5(5) - \left(-\frac{(1)^3}{3} + 3(1)^2 - 5(1)\right)$$

$$= -\tfrac{125}{3} + 75 - 25 + \tfrac{1}{3} - 3 + 5 = 10\tfrac{2}{3}$$

Therefore the shaded area is $10\tfrac{2}{3}$ units^2.

Finding the area when the integral is negative

The integral can be negative if the curve is below the x-axis, as in Fig. 13.7, where the area under the curve $y=\sin(x)$ from $x=\pi$ to $x=3\pi/2$ is illustrated.

$$\int_{\pi}^{3\pi/2} \sin(x)\,dx = [-\cos(x)]_{\pi}^{3\pi/2} = -\cos\left(\frac{3\pi}{2}\right) + \cos(\pi) = -1$$

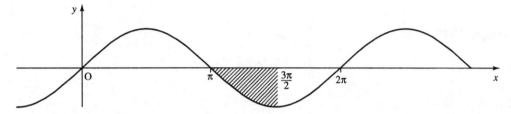

Figure 13.7 The area under the curve given by $\int_{\pi}^{3\pi/2} \sin(x)\,dx$.

The integral is negative because the values of y are negative in that region. In the case where all of that portion of the curve is below the x-axis, to find the area we just take the modulus. Therefore the shaded area $A=1$.

This is important because negative and positive areas can cancel out, giving an integral of 0. In Fig. 13.8, the area under the curve $y=\sin(x)$ from $x=0$ to $x=2\pi$ is pictured. The area under the curve has a positive part from 0 to π and an equal negative part from π to 2π.

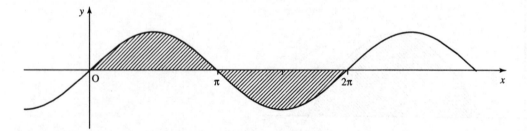

Figure 13.8 The area under the graph $y=\sin(x)$ from $x=0$ to 2π.

The following gives an integral of 0:

$$\int_{0}^{2\pi} \sin(x)\,dx = [-\cos(x)]_{0}^{2\pi} = -\cos(2\pi) - (-\cos(0)) = -1 - (-1) = 0$$

To stop the positive and negative parts of the integration cancelling we find the total shaded area in two stages:

$$\int_{0}^{\pi} \sin(x)\,dx = [-\cos(x)]_{0}^{\pi} = 2$$

and

$$\int_{\pi}^{2\pi} \sin(x)\,dx = [-\cos(x)]_{\pi}^{2\pi} = -2$$

So the total area is $2+|-2|=4$.

We have seen that if we wish to find the area bounded by a curve which crosses the x-axis, then we must find where it crosses the x-axis first and perform the integration in stages.

Example 13.21 Find the area bounded by the curve $y = x^2 - x$ and the x-axis and the lines $x = -1$ and $x = 1$.

SOLUTION First we find if the curve crosses the x-axis.

$$x^2 - x = 0 \Leftrightarrow x(x - 1) = 0 \Leftrightarrow x = 0 \quad \text{or} \quad x = 1$$

The sketch of the graph with the required area shaded is given in Fig. 13.9. Therefore the area is the sum of A_1 and A_2.

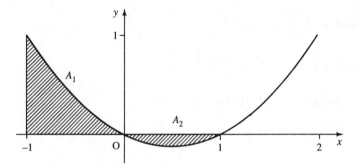

Figure 13.9 Sketch of $y = x(x-1)$, with the area bounded by the x-axis and $x = -1$ (nd $x = 1$ marked. The area above the x-axis is marked as A_1 and the area below the x-axis is marked as A_2.

We find A_1 by integrating from -1 to 0:

$$\int_{-1}^{0} x^2 - x \, dx = \left[\frac{x^3}{3} - \frac{x^2}{2} \right]_{-1}^{0} = 0 - \left(\frac{(-1)^3}{3} - \frac{(-1)^2}{2} \right) = \frac{1}{3} + \frac{1}{2} = \frac{5}{6}$$

Therefore $A_1 = \frac{5}{6}$.

We find A_2 by integrating from 0 to 1 and taking the modulus:

$$\int_{0}^{1} x^2 - x \, dx = \left[\frac{x^3}{3} - \frac{x^2}{2} \right]_{0}^{1} = \frac{1}{3} - \frac{1}{2} = -\frac{1}{6}$$

Therefore $A_2 = \frac{1}{6}$.

Then the total area is $A_1 + A_2 = \frac{5}{6} + \frac{1}{6} = 1$.

13.6 THE MEAN VALUE AND r.m.s. VALUE

The mean value of a function is the value it would have if it were constant over the range but with the same area under the graph (i.e. with the same integral) (see Fig. 13.10).

The formula for the mean value is

$$M = \frac{1}{b-a} \int_{a}^{b} y \, dx$$

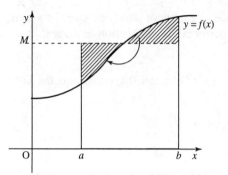

Figure 13.10 The mean value of a function is the value it would take if it were constant over the range but with the same integral.

Example 13.22 Find the mean value of $i(t) = 20 + 2\sin(\pi t)$ for $t = 0$ to 0.5.

SOLUTION Using the formula with $a = 0$, $b = 0.5$ gives

$$M = \frac{1}{0.5 - 0} \int_0^{0.5} 20 + 2\sin(\pi t) \, dt$$

$$2\left[20t - \frac{2}{\pi}\cos(\pi t) \right]_0^{0.5} = 2\left(10 - 0 - \left(0 - \frac{2}{\pi}(1) \right) \right)$$

$$\approx 21.27$$

The root mean squared (r.m.s.) value

The 'root mean squared value' (r.m.s. value) means the square root of the mean value of the square of y. The formula for the r.m.s. value of y between $x = a$ and $x = b$ is

$$\text{r.m.s. } (y) = \sqrt{\frac{1}{(b-a)} \int_a^b y^2 \, dx}$$

The advantage of the r.m.s. value is that as all the values for y are squared they are positive, so the r.m.s. value will not give 0 unless we are considering the zero function. If the function represents the voltage then the r.m.s. value can be used to calculate the average power in the signal. In contrast, the mean value gives zero if calculated for the sine or cosine over a complete cycle, giving no additional useful information.

Example 13.23 Find the r.m.s. value of $y = x^2 - 3$ between $x = 1$ and $x = 3$.

SOLUTION

$$(\text{r.m.s.}(y))^2 = \frac{1}{3-1} \int_1^3 (x^2 - 3)^2 \, dx = \frac{1}{2} \int_1^3 x^4 - 6x^2 + 9 \, dx$$

$$= \frac{1}{2}\left[\frac{x^5}{5} - \frac{6x^3}{3} + 9x \right]_1^3$$

$$= \frac{1}{2}((\frac{243}{5} - 54 + 27) - (\frac{1}{5} - 2 + 9)) = 7.2$$

Then the r.m.s. value is $\sqrt{7.2} \approx 2.683$.

13.7 NUMERICAL METHODS OF INTEGRATION

Many problems may be difficult to solve analytically. In such cases, numerical methods may be used. This is often necessary in order to perform integrations.

The following integrals could not be solved by the methods of integration we have met so far:

$$\int_2^3 \frac{\sin(x)}{x^2+1}\,dx$$

$$\int_{-3}^2 2^{x^2}\,dx$$

Numerical methods can usually only give an approximate answer.

General method

We wish to approximate the integral

$$\int_a^b f(x)\,dx$$

Formulae for numerical integration are obtained by considering the area under the graph and splitting the area into strips, as in Fig. 13.11. The area of the strips can be approximated using the trapezoidal rule or Simpson's rule. In each case, we assume that the thickness of each strip is h and that there are N strips, so that

$$h = \frac{(b-a)}{N}$$

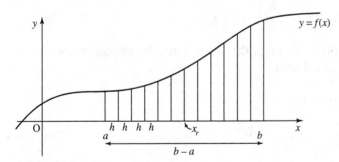

Figure 13.11 Numerical integration is performed by splitting the area into strips of width h. The area of the strips is approximated using the trapezoidal rule or Simpson's rule.

Numerical methods are obviously to be used with a computer or possibly a programmable calculator. However, it is a good idea to be able to check some simple numerical results, which needs some understanding of the algorithms used.

The trapezoidal rule

The strips are approximated to trapeziums with parallel sides of length y_{r-1} and y_r, as in Fig. 13.12. The area of each strip is

$$\frac{h}{2}(y_{r-1}+y_r)$$

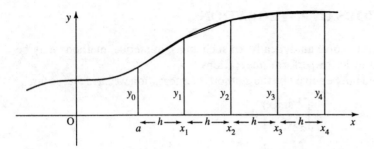

Figure 13.12 The trapezoidal rule is found by approximating each of the strips as a trapezium.

The formula becomes

$$A = h\left(\frac{y_0}{2} + y_1 + y_2 + \ldots + \frac{y_N}{2}\right)$$

where

$$x_r = a + rh$$

$$y_r = f(x_r)$$

and

$$N = \frac{(b-a)}{h}$$

A computer program would more likely use the equivalent recurrence relation, where A_r is the area up to the rth strip (at $x = x_r$):

$$A_r = A_{r-1} + \frac{h}{2}(y_{r-1} + y_r)$$

for $r = 1$ to N and $A_0 = 0$. This is simply stating that the area is found by adding on the area of one strip at a time to the previously found area.

Example 13.24 We wish to approximate

$$\int_1^3 x^2 \, dx$$

The limits of the integration are 1 and 3, so $a = 1$ and $b = 3$. We choose a step size of 0.5, and therefore

$$N = \frac{b-a}{h} = \frac{3-1}{0.5} = 4$$

Using $x_r = a + rh$ and $y_r = f(x_r)$, which in this case gives $y_r = x_r^2$, we get

$$x_0 = 1 \qquad y_0 = (1)^2 = 1$$

$$x_1 = 1.5 \qquad y_1 = (1.5)^2 = 2.25$$

$$x_2 = 2 \qquad y_2 = (2)^2 = 4$$

$$x_3 = 2.5 \qquad y_3 = (2.5)^2 = 6.25$$

$$x_4 = 3 \qquad y_4 = (3)^2 = 9$$

Using the formula for the trapezoidal rule:

$$A = h\left(\frac{y_0}{2} + y_1 + y_2 + \dots + \frac{y_N}{2}\right)$$

we get

$$A = 0.5(0.5 + 2.25 + 4 + 6.25 + 4.5) = 8.75$$

Hence by the trapezoidal rule:

$$\int_1^3 x^2 \, dx \simeq 8.75$$

Simpson's rule

For Simpson's rule the area of each strip is approximated by drawing a parabola through three adjacent points (see Fig. 13.13). Notice that the number of strips must be even.

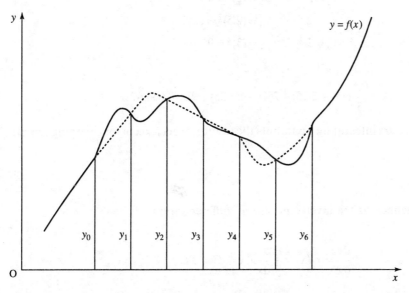

Figure 13.13 Simpson's rule is found by approximating the areas of the strips by drawing a parabola through 3 adjacent points. To do this the total number of strips must be even.

The area of the strips in this case is not obvious, as in the case of the trapezoidal rule. Three strips together have an area of

$$\frac{h}{3}(y_{2n-2} + 4y_{2n-1} + y_{2n})$$

where $r = 2n$. The formula then becomes

$$A = \frac{h}{3}(y_0 + 4y_1 + 2y_2 + 4y_3 + 2y_4 + \dots + 2y_{N-2} + 4y_{N-1} + y_N)$$

where $x_r = a + rh$, $y_r = f(x_r)$ and $N = (b-a)/h$, as before.

Again, a computer program would more likely use a recurrence relation to define the area:

$$A_{2n} = A_{2(n-1)} + \frac{h}{3}(y_{2n-2} + 4y_{2n-1} + y_{2n})$$

where $n = 1, \ldots, N/2$ and $A_0 = 0$.

Example 13.25 Find $\int_1^3 x^2 \, dx$ using Simpson's rule with $h = 0.5$.

SOLUTION From the limits of the integral we find that $a = 1$ and $b = 3$. Then

$$N = \frac{(b-a)}{h} = \frac{3-1}{0.5} = 4$$

Using $x_r = a + rh$ and $y_r = f(x_r)$, which in this case gives $y_r = x_r^2$, we get

$$
\begin{aligned}
x_0 &= 1 & y_0 &= (1)^2 = 2 \\
x_1 &= 1.5 & y_1 &= (1.5)^2 = 2.25 \\
x_2 &= 2 & y_2 &= (2)^2 = 4 \\
x_3 &= 2.5 & y_3 &= (2.5)^2 = 6.25 \\
x_4 &= 3 & y_4 &= (3)^2 = 9
\end{aligned}
$$

Hence

$$A = \frac{0.5}{3}(1 + 4(2.25) + 2(4) + 4(6.25) + 9) \simeq 8.666\,67$$

In this case, as we are integrating a parabola the result is exact (except for rounding errors).

13.8 SUMMARY

1. Integration can be defined as the inverse process of differentiation.
   If $y = f(x)$, then

$$\frac{dy}{dx} = f'(x) \Leftrightarrow \int \frac{dy}{dx} \, dx = f(x) + C$$

or equivalently

$$\int \frac{dy}{dx} \, dx = y + C$$

This is called indefinite integration, and C is the constant of integration.
2. A table of standard integrals can be found as in Table 13.1 by swapping the columns of Table 12.3, rearranging them in a more convenient form and adding the constant of integration.
3. Integrals of combinations of the functions given in Table 13.1 cannot always be found, but some methods can be tried as follows:

(a) Substitute $u = ax + b$ to find $\int f(ax + b) \, dx$
(b) Substitute when the integral is of the form

$$\int f(u) \frac{du}{dx} \, dx$$

(c) Use integration by parts when the integral is of the form

$$\int u \, dv \text{ using the formula } \int u \, dv = uv - \int v \, du$$

(d) Use trigonometric identities to integrate powers of $\cos(x)$ and $\sin(x)$.

4. Integration has many applications, some of which are listed in Sec. 12.4.
5. The definite integral, from $x=a$ to $x=b$, is defined as the area under the curve between those two values. This is written as $\int_a^b f(x) \, dx$.
6. The mean value of a function is the value the function would have if it were constant over the range but with the same area under the graph.

   The mean value from $x=a$ to $x=b$ of y is

$$M = \frac{1}{b-a} \int_a^b y \, dx$$

7. The root mean squared value (r.m.s. value) is the square root of the mean value of the square of y. The r.m.s. value from $x=a$ to $x=b$ is given by

$$\text{r.m.s.}(y) = \sqrt{\frac{1}{b-a} \int_a^b y^2 \, dx}$$

8. Two methods of numerical integration are using the trapezoidal rule and Simpson's rule, where $A \simeq \int f(x) \, dx$.

Trapezoidal rule

$$A = h\left(\frac{y_0}{2} + y_1 + y_2 + \ldots + \frac{y_N}{2}\right)$$

Simpson's rule

$$A = \frac{h}{3}(y_0 + 4y_1 + 2y_2 + 4y_3 + 2y_4 + \ldots + 2y_{N-2} + 4y_{N-1} + y_N)$$

   In both cases, $x_r = a + rh$ and $N = (b-a)/h$. For Simpson's rule, N must be even. h is called the step size and N is the number of steps.

13.9 EXERCISES

13.1 Find the following integrals

(a) $\int x^3 + x^2 \, dx$

(b) $\int 2\sin(x) + \sec^2(x) \, dx$

(c) $\int \frac{1}{x^3} \, dx$

(d) $\int 1 + x^2 + 3x^3 \, dx$

(e) $\int 1 - 5x \, dx$

(f) $\int \cos(2-4x) \, dx$

(g) $\int \sqrt{2x-1} \, dx$

(h) $\int \frac{1}{\sqrt{x+2}} \, dx$

(i) $\int x(x^2-4)^3 \, dx$

(j) $\int x\sqrt{1+x^2} \, dx$

(k) $\int \frac{\cos(x)}{(1+\sin(x))^2} \, dx$

(l) $\int (x^2+x-6)(2x+1) \, dx$

(m) $\int \frac{4x^2}{(x^3-7)^2} \, dx$

(n) $\int_1^2 x \cos(x) \, dx$

(o) $\int x^2 \cos(x) \, dx$

(p) $\int_2^4 x\sqrt{x-1} \, dx$

(q) $\int x(2x-3)^4 \, dx$

(r) $\int_0^{\pi/2} \sin^5(x) \, dx$

(s) $\int \cos^4(x) \, dx$

(t) $\int \sin(3x) \cos(5x) \, dx$

13.2 Given that $v = ds/dt = 3 - t$, find s in terms of t if $s = 5$ when $t = 0$. What is the value of s when $t = 2$?

13.3 Find the equation of the curve with the gradient $dy/dt = -5$ which passes through the origin.

13.4 A curve $y = f(x)$ passes through the point $(0,1)$ and its gradient at any point is $1 - 2x^2$. Find the function.

13.5 The voltage across an inductor of inductance 3 H is measured as $V = 2\sin(2t - \pi/6)$. The current at $t = 0$ is 0. Given that $V = L\,di/dt$, find the current after 10 s.

13.6 The velocity of a spring is found to be $v = 6\sin(3\pi t)$. Assuming that the spring is perfect, so that

$$v = -\frac{1}{k}\frac{dF}{dt}$$

where k is the spring constant (known to be 0.5), v the velocity and F the force operating on the spring, find the force, given that it is 0 N initially.

13.7 The magnitude of the magnetic flux density at the mid-point of the axis of a solenoid, as in Fig. 13.14 can be found by the integral

$$B = \int_{\beta_1}^{\beta_2} \frac{\mu_0 nI}{2} \sin(\beta)\,d\beta$$

where μ_0 is the permeability of free space ($\simeq 4\pi \times 10^{-7}$ H m^{-1}), n is the number of turns and I is the current. If the solenoid is so long that $\beta_1 \simeq 0$ and $\beta_2 \simeq \pi$, show that $B = \mu_0 nI$.

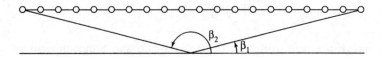

Figure 13.14 Field on the axis of a solenoid for Exercise 13.26.

13.8 Find the area under the curve $y = x + x^2$ between the lines $x = 1$ and $x = 4$.

13.9 Find the area bounded by the x-axis and the portion of the curve $y = 2(x - 1)(x - 4)$ which lies below it.

13.10 Find the total area bounded by the curve $y = 2x - x^2$, the x-axis and the lines $x = -1$ and $x = 1$.

13.11 Find the mean value of $i(t) = 5 - \cos(t/2)$ for $t = 0$ to $t = 5$.

13.12 Calculate the r.m.s. value of $i = 3\cos(50\pi t)$ between $t = 0$ and $t = 0.01$.

13.13 Approximate

$$\int_0^1 \frac{\sin(x)}{x}\,dx$$

(a) Using the trapezoidal rule with $h = 0.2$ and $h = 0.1$.
(b) Using Simpson's rule with $h = 0.5$ and $h = 0.25$.

13.14 Find an approximate value of

$$A = \int_1^3 \frac{dx}{x}$$

(a) by the trapezoidal rule with $N = 6$.
(b) By Simpson's rule with $N = 6$.

13.15 Approximate $\int_0^1 x^5\,dx$ using Simpson's rule with $N = 10$.

FOURTEEN

THE EXPONENTIAL FUNCTION

14.1 INTRODUCTION

In Example 8.11 we looked at an acoustical absorption problem. We found that after a single note on a trombone had been played, the sound intensity decayed according to the expression $I = 10^{-11t}$. This is a decaying exponential function. Many other physical situations involve decay or growth in an exponential fashion, for instance population growth or the decay of charge on a discharging capacitor. The functions $y = a^t$, where a can be any positive number, are called exponential functions. In this chapter we shall look at how they describe this particular type of growth or decay, when the growth or decay is proportional to the current size of the 'population', y. This situation can be described by a differential equation of the form $dy/dt = ky$. The special case where $k = 1$, giving the equation $dy/dt = y$, leads us to define the number $e \simeq 2.718\,281\,8$ and the function $y = e^t$ which is called *the* exponential function. The inverse function is $y = \log_e(t)$, which is also referred to as $y = \ln(t)$ and called the natural or Napierian logarithm.

The function $y = e^t$ is neither even nor odd; however, it is possible to split any real function into an even part and an odd part, and in this case we find that this gives the hyperbolic functions. The hyperbolic functions have properties that are surprisingly similar to the properties of the trigonometric functions, and hence have similar names, $\cosh(t)$, $\sinh(t)$ (the hyperbolic sine and hyperbolic cosine), from which we can define also $\tanh(t)$, the hyperbolic tangent. We also look at differentiation and integration problems involving the exponential and hyperbolic functions.

14.2 EXPONENTIAL GROWTH AND DECAY

Supposing, following some deed of heroism, the police offered you the choice of the following rewards:

(a) Tomorrow you receive 1p, the following day 2p, then 4p, 8p and so on. Each day the amount doubles, which continues for the next month.
(b) Tomorrow you receive £2, the following day £4, then £6, £8 and so on, so that each day you receive £2 more than the day before. Again you receive payments on every day for the next month.

If, although not motivated by personal greed, you wish to receive the highest possible reward (in order, presumably, to donate the amount to charity) which reward should you accept? Option (b) superficially appears to be the best because at least it starts off with enough money to buy a sandwich. However, a closer look reveals that if you choose option (a), then on the last day (assuming there are 31 days in the month) you receive in excess of £10 000 000, with the total reward exceeding £20 million. However, option (b) only reaps £62 on the final day, with a total reward of only £992.

Both options are examples of growth. Option (b) gives a constant growth rate, of £2. If y_n is the amount received on day n, the way that this grows could be expressed by $y_{n+1} = y_n + 2$. This is expressing the fact that the amount received on day $n+1$ is £2 more than the amount received on day n. This also means that the change each day in the amount received is 2, which can be expressed as $\Delta y_n = 2$. $\Delta y_n = y_{n+1} - y_n$, and represents the change in y from day n to day $n+1$.

Option (a) is an example of exponential growth (or geometric growth). The amount received each day is proportional to the amount received the day before, in this case twice as much. y_n can be expressed by $y_{n+1} = 2y_n$. The change in y is equal to the value of y itself: $\Delta y_n = y_n$.

Because of the nature of exponential growth it is unlikely that you would be given such an attractive reward as option (a) represents. However, exponential growth is not beyond the reach of the everyday person, as savings accounts offer this opportunity. Unfortunately, the amount you receive does not increase as quickly as doubling each day, but it is based on how much you have already in the bank, and hence it is exponential.

Suppose you opened an account which paid annual interest of 6 per cent and the annual rate of inflation was 3 per cent. Then the real rate of growth is approximately 3 per cent per annum. If y_n is the value of the amount you have in the bank after n years, then $y_{n+1} = 1.03 y_n$. We can also express this by saying that the interest received each year, that is the change in y_n, Δy_n, is 3 per cent of y_n, that is $\Delta y_n = 0.03 y_n$, where $\Delta y_n = y_{n+1} - y_n$. If the rate of interest remains constant then if you deposit £1 tomorrow then your descendants, in only 500 years time, will receive an amount worth over £2.5 million in real terms.

The models of growth that we have discussed so far give examples of recurrence relations, also called difference equations. Their solutions are not difficult to find. For instance, if $y_{n+1} = 2y_n$, and we know that on day one we received 1p, i.e. $y_1 = 0.01$, then we can substitute $n = 1, 2, 3 \dots$ (as we did in Section 7.4) to find values of the function giving 0.01, 0.02, 0.04, 0.08, 0.16, 0.32 \dots. Clearly there is a power of 2 involved in the expression for y_n, so we can guess that $y = 0.01 \, (2^{n-1})$. By checking a few values of n we can confirm that this is indeed the amount received each day. When we deposit £1 in the bank at a real rate of growth of 3 per cent we get the recurrence relation $y_{n+1} = 1.03 y_n$, where y_n is the present-day value of the amount in the savings account after n years. Substituting a few values, beginning with $y_0 = 1$ (the amount we initially deposit), then we get 1, 1.03, 1.06, 1.09, 1.13, 1.16, 1.19 \dots (to the nearest penny). Each time we multiply the amount by 1.03, so in the solution for y there must be a power of 1.03. We can guess the solution as $y_n = (1.03)^n$. By checking a few values of n we can confirm that this is in fact the amount in the bank after n years.

The models we have looked at so far are discrete models. In the case of the money in the bank the increase occurs at the end of each year. However, if we consider population growth, for instance, then it is not possible to say that the population grows at the end of a certain period: the growth could happen at any moment of time. In this case, providing the population is large enough, it is easier to model the situation continuously, using a differential equation. Such models take the form of $dy/dt = ky$. dy/dt is the rate of growth, if k is positive, or the rate of decay, if k is negative. The equation states that the rate of growth or decay of a population of size y is proportional to the size of the population.

Example 14.1 A malfunctioning fridge maintains a temperature of 6 °C, which allows a population of bacteria to reproduce such that, on average, each bacterium divides every

20 minutes. Assuming no bacteria die in the time under consideration, find a differential equation to describe the population growth.

SOLUTION If the population at time t is given by p, then the rate of change of the population is dp/dt. The increase in the population is such that it approximately doubles every 20 minutes, that is it increases by p in 20×60 s. That gives the rate of increase as $p/20 \times 60$ per second. Hence the differential equation describing the population is

$$\frac{dp}{dt} = \frac{p}{1200}$$

Example 14.2 A capacitor in an RC circuit, has been charged to a charge of Q_0. The voltage source has been removed and the circuit closed, as in Fig. 14.1. Find a differential equation which describes the rate of discharge of the capacitor if $C = 0.001$ μF and $R = 10$ MΩ.

Figure 14.1 A closed RC circuit.

SOLUTION The voltage across a capacitor is given by Q/C, where C is the capacitance and Q is the charge on the capacitor. The voltage across the resistor is given by Ohm's law as IR. From Kirchhoff's voltage law, the sum of the voltage drops in the circuit must be 0; therefore, as the circuit is closed, we get voltage across the resistor + voltage across the capacitor = 0,

$$\Rightarrow IR + \frac{Q}{C} = 0$$

By definition, the current is the rate of change of charge with respect to time, that is $I = dQ/dt$, giving the differential equation

$$R\frac{dQ}{dt} + \frac{Q}{C} = 0$$

We can rearrange this equation:

$$R\frac{dQ}{dt} + \frac{Q}{C} = 0$$

$$\Leftrightarrow R\frac{dQ}{dt} = -\frac{Q}{C}$$

$$\Leftrightarrow \frac{dQ}{dt} = -\frac{Q}{RC}$$

We can see that this is an equation for exponential decay. The rate of change of the charge on the capacitor is proportional to the remaining charge at any point in time, with a constant of proportionality given by $1/RC$. In this case, as $R = 10$ MΩ and $C = 0.001$ μF, then

$$\frac{dQ}{dt} = -100Q$$

Example 14.3 Radioactivity is the emission of α- or β-particles and γ-rays due to the disintegration of the nuclei of atoms. The rate of disintegration is proportional to the number of atoms at any point in time, and the constant of proportionality is called the radioactivity decay constant. The radioactive decay constant for radium B is approximately $4.3 \times 10^{-4}\,\text{s}^{-1}$. Give a differential equation which describes the decay of the number of particles N in a piece of radium B.

SOLUTION If the number of particles at time t is N, then the rate of change is dN/dt. The decay is proportional to the number of atoms, and we are given that the constant of proportionality is $4.3 \times 10^{-4}\,\text{s}^{-1}$, so therefore we have

$$\frac{dN}{dt} = -4.3 \times 10^{-4}\,N$$

as the equation which describes the decay.

Example 14.4 An object is heated so that its temperature is 400 K and the temperature of its surroundings is 300 K and then it is left to cool. Newton's law of cooling states that the rate of heat loss is proportional to the excess temperature over the surroundings. Furthermore, if m is the mass of the object and c is its specific heat capacity, then the rate of change of heat is proportional to the rate of fall of temperature of the body, and is given by $dQ/dt = -mc(d\phi/dt)$, where Q is the heat in the body and ϕ is its temperature.
 Find a differential equation for the temperature which describes the way the body cools.

SOLUTION Newton's law of cooling gives

$$\frac{dQ}{dt} = A(\phi - \phi_s)$$

where A is some constant of proportionality, Q is the heat in the body, ϕ is its temperature and ϕ_s is the temperature of its surroundings. As we also know that

$$\frac{dQ}{dt} = -mc\frac{d\phi}{dt}$$

this can be substituted in our first equation, giving

$$-mc\frac{d\phi}{dt} = A(\phi - \phi_s)$$

$$\Leftrightarrow \frac{d\phi}{dt} = -\frac{A}{mc}(\phi - \phi_s)$$

A/mc can be replaced by a constant k, giving

$$\frac{d\phi}{dt} = -k(\phi - \phi_s)$$

In this case, the temperature of the surroundings is known to be 300 K so the equation describing the rate of change of temperature is

$$\frac{d\phi}{dt} = -k(\phi - 300)$$

We have established that there are a number of important physical situations that can be described by the equation $dy/dt = ky$. The rate of change of y is proportional to its value. We would like to solve this equation, i.e. find y explicitly as a function of t. In Chapter 13, we solved simple differential equations, such as $dy/dt = 3t$ by integrating both sides with respect to t, e.g.

$$\frac{dy}{dt} = 3t \Leftrightarrow y = \int 3t \, dt \Leftrightarrow y = \frac{3t^2}{2} + C$$

However, we cannot solve the equation $dy/dt = ky$ in this way because the right-hand side is a function of y, not of t. If we integrate both sides with respect to t we get

$$\frac{dy}{dt} = ky \Leftrightarrow y = \int ky \, dt$$

Although this is true we are no nearer solving for y, as we need to know y as a function of t in order to find $\int ky \, dt$.

When we solved equations in Chapter 3 we said that one method that was to guess a solution and substitute that value for the unknown into the equation to see if it gave a true statement. This would be a very long method to use unless we are able to make an informed guess. We can use this method with this differential equation as we know from our experience with problems involving discrete growth that a solution should involve an exponential function of the form $y = a^t$. The main problem is to find the value of a which will go with any particular equation. To do this we begin by looking for an exponential function which would solve the equation $dy/dt = y$, that is we want to find the function whose derivative is equal to itself.

14.3 THE EXPONENTIAL FUNCTION $y = e^t$

Figures 14.2(a) and 14.3(a) give graphs of $y = 2^t$ and $y = 3^t$ which are two exponential functions. We can sketch their derivative functions by drawing tangents to the graph and measuring the gradient of the tangent at various different points. The derivative functions are pictured in Fig. 14.2(b), dy/dt where $y = 2^t$, and in Fig. 14.3(b), dy/dt where $y = 3^t$.

We can see that for these exponential functions, the derivative has the same shape as the original function, but has been scaled in the y direction, i.e. multiplied by a constant, k, so that $dy/dt = ky$, as we expected.

$$\frac{d}{dt}(2^t) = (C)(2^t)$$

and

$$\frac{d}{dt}(3^t) = (D)(3^t)$$

where C and D are constants. We can see from the graphs that $C < 1$ and $D > 1$, that is that the derivative of 2^t gives a squashed version of the original graph and the derivative of $3t$ gives a stretched version of the original graph.

It would seem reasonable that there would be a number somewhere between 2 and 3, which we can call e, which has the property that the derivative of e^t is exactly the same as the original graph; that is, that

$$\frac{d}{dt}(e^t) = e^t$$

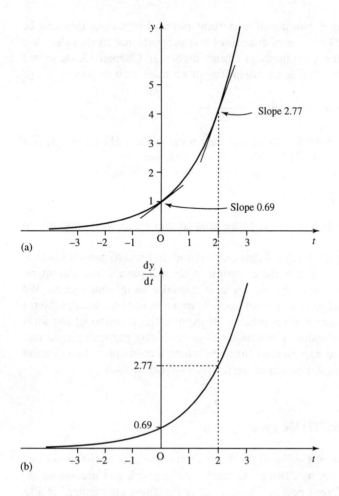

Figure 14.2 (a) The graph of $y = 2^t$ with some tangents marked. (b) The graph of the derivative (the gradient of the tangent at any point on $y = 2^t$ plotted against t).

There are various methods for finding the value of e and an investigation into finding e to one decimal place is given in the Appendix. e is an irrational number, which means that it cannot be written exactly as a fraction (or as a decimal). Its value to five decimal places is e = 2.718 28.

The function e^t is often referred to as $\exp(t)$. e^t and its derivative are pictured in Fig. 14.4.

By definition of the logarithm (as given in Chapter 4) we know that the inverse function to e^t is $\log_e(t)$ (log, base e, of t). This is often represented by the shorthand of $\ln(t)$, and called the natural or Napierian logarithm.

We are now able to solve the differential equation $dy/dt = y$, as we know that one solution is $y = e^t$ because the derivative of $y = e^t$ is e^t. When we discussed differential equations in Chapter 13 we noticed that there was an arbitrary constant which was involved in the solution of a differential equation. In this case the constant represents the initial size of the population, or the initial charge or the initial number of atoms or the initial temperature. The general solution to $dy/dt = y$ is $y = y_0 e^t$, where y_0 is the value of y at time $t = 0$. We can show that this is in fact the general solution by substituting into the differential equation.

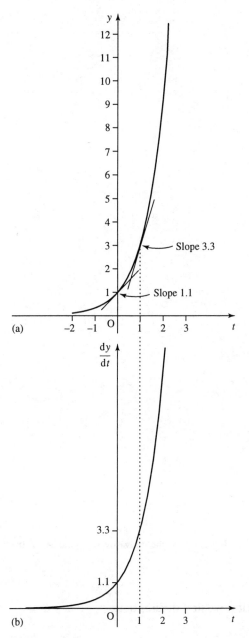

Figure 14.3 (a) The graph of $y=3^t$ with some tangents marked. (b) The graph of the derivative (the gradient of the tangent at any point on $y=3^t$ plotted against t).

Example 14.5

(a) Show that any function of the form $y=y_0e^t$, where y_0 is a constant, is a solution to the equation

$$\frac{dy}{dt}=y$$

(b) Show that in the function $y=y_0e^t$ then $y=y_0$ when $t=0$.

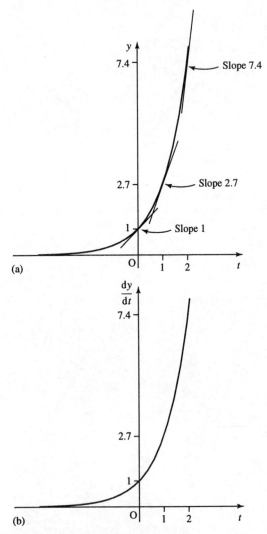

Figure 14.4 (a) The graph of $y = e^t$ with some tangents marked. (b) The graph of the derivative (the gradient of the tangent at any point on e^t plotted against t).

SOLUTION

(a) To show that any function $y = y_0 e^t$ is a solution we first differentiate:

$$\frac{d}{dt}(y_0 e^t) = y_0 e^t$$

(as y_0 is a constant and $d(e^t)/dt = e^t$). Substitute for dy/dt and for y into the differential equation and we get $y_0 e^t = y_0 e^t$, which is a true statement for all t. Hence the solutions to $dy/dt = y$ are $y = y_0 e^t$.

(b) Substitute $t = 0$ in the function $y = y_0 e^t$ and we get $y = y_0 e^0$. As any number raised to the power of 0 is 1, we have $y = y_0$. Hence y_0 is the value of y at $t = 0$.

Using the function of a function rule we can find the derivative of e^{kt}, where k is some constant, and show that this function can be used to solve differential equations of the form $dy/dt = ky$.

The derivative of e^{kt}

To find the derivative of $y=e^{kt}$ where k is a constant substitute $u=kt$ so that $y=e^u$:

$$\frac{du}{dt}=k \quad \text{and} \quad \frac{dy}{du}=e^u$$

therefore, using the chain rule,

$$\frac{dy}{dt}=\frac{dy}{du}\frac{du}{dt}=ke^u=ke^{kt}$$

(resubstituting $u=kt$). Therefore

$$\frac{d}{dt}(e^{kt})=ke^{kt}$$

Notice that if we substitute y for e^{kt} into

$$\frac{d}{dt}(e^{kt})=ke^{kt}$$

we get $dy/dt=ky$, which was the differential equation we set out to solve for our growth or decay problems. This tells us that one solution to the equation $dy/dt=ky$ is $y=e^{kt}$. The general solution must involve a constant, so we try $y=y_0e^{kt}$, where y_0 is the initial size of the population, or initial temperature etc.

Example 14.6
(a) Show that any function of the form $y=y_0e^{kt}$, where y_0 is a constant, is a solution to the equation

$$\frac{dy}{dt}=ky$$

(b) Show that in the function $y=y_0e^{kt}$, $y=y_0$ when $t=0$.

SOLUTION
(a) To show that any function $y=y_0e^{kt}$ is a solution we first differentiate:

$$\frac{d}{dt}(y_0e^{kt})=y_0ke^{kt}$$

as y_0 is a constant and $d(e^{kt})/dt=ke^{kt}$. Substitute for dy/dt and for y in the differential equation and we get

$$y_0ke^{kt}=ky_0e^{kt}, \text{ which is a true statement for all values of } t.$$

Hence the solutions to $dy/dt=ky$ are $y=y_0e^{kt}$.
(b) Substitute $t=0$ in the function $y=y_0e^t$ and we get $y=y_0e^0$. As any number raised to the power of 0 is 1, we have $y=y_0$. Hence y_0 is the value of y at $t=0$.

Example 14.7 Solve the differential equation given in Example 14.2, describing the discharge of a capacitor in a closed RC circuit with $R=10$ MΩ and $C=0.001$ μF:

$$\frac{dQ}{dt}=-100Q$$

Find a particular solution given that at $t=0$ the voltage drop across the capacitor was 10 V.

SOLUTION We have discovered that the solution to a differential equation of the form $dy/dt = ky$ is given by $y = y_0 e^{kt}$, where y_0 is the initial value of y. Comparing $dy/dt = ky$ with $dQ/dt = -100Q$, replacing y by Q and k by -100 gives the solution:

$$Q = Q_0 e^{-100t}$$

To find the value of Q_0 we need to find the value of Q when $t = 0$. We are told that the initial value of the voltage across the capacitor was 10 V and we know that the voltage drop across a capacitor is Q/C. Therefore we have

$$\frac{Q_0}{0.001 \times 10^6} = 10 \Leftrightarrow Q_0 10 \times .001 \times 10^6$$

$$\Leftrightarrow Q_0 = 10\,000$$

Therefore the equation which describes the charge as the capacitor discharges is

$$Q = 10\,000 e^{-100t}$$

The derivative of a^t

The derivative of $y = 2^t$ can now be found by observing that $2 = e^{(\ln(2))}$. Therefore

$$y = 2^t = (e^{\ln(2)})^t = e^{\ln(2)t}$$

This is of the form e^{kt} with $k = \ln(2)$. As

$$\frac{d}{dt}(e^{kt}) = ke^{kt}$$

then

$$\frac{d}{dt}(e^{\ln(2)t}) = \ln(2)e^{t\ln(2)}$$

Using again the fact that $e^{\ln(2)t} = (e^{\ln(2)})^t = 2^t$ we get

$$\frac{d}{dt}(2^t) = \frac{d}{dt}(e^{\ln(2)t}) = \ln(2)e^{\ln(2)t} = \ln(2)2^t$$

that is

$$\frac{d}{dt}(2^t) = \ln(2)2^t$$

Compare this result to that which we found by sketching the derivative of $y = 2^t$ in Fig. 14.2(b). We said that the derivative graph was a squashed version of the original graph. This result tells us that the scaling factor is $\ln(2) \simeq 0.693$, which confirms our observation that the scaling factor is $C < 1$.

Using the same argument for any exponential function $y = a^t$ we find that $dy/dt = \ln(a)a^t$.

In finding these results we have used the fact that an exponential function to whatever base, a, can be written as e^{kt}, where $k = \ln(a)$.

The derivative of $y = \ln(x)$

$y = \ln(x)$ is the inverse function of $f(x) = e^x$, and therefore we can find the derivative in a manner similar to that used to find the derivatives of the inverse trigonometric functions.

$$y = \ln(x) \quad \text{where } x > 0$$
$$\Leftrightarrow e^y = e^{\ln(x)} \quad \text{(take the exponential of both sides)}$$
$$\Leftrightarrow e^y = x \quad \text{(as exp is the inverse function to ln, } e^{\ln(x)} = x)$$

We wish to differentiate both sides with respect to x, but the left-hand side is a function of y, so we use the chain rule, setting $w = e^y$. We know that $dw/dy = e^y$, so the equation $e^y = x$ becomes $w = x$. Differentiating both sides of $w = x$ with respect to x gives $dw/dx = 1$, where

$$\frac{dw}{dx} = \frac{dw}{dy}\frac{dy}{dx}$$

from the chain rule. So

$$e^y \frac{dy}{dx} = 1$$

and re-substituting $x = e^y$ we get

$$x\frac{dy}{dx} = 1 \Leftrightarrow \frac{dy}{dx} = \frac{1}{x} \quad \text{(we can divide by } x \text{ as } x > 0)$$

Hence

$$\frac{d}{dx}(\ln x) = \frac{1}{x}$$

The derivative of the log, whatever the base, can be found using the change of base rule for logarithms, as given in Chapter 4. We can write

$$\log_a(x) = \frac{\ln(x)}{\ln(a)}$$

Therefore

$$\frac{d}{dx}(\log_a(x)) = \frac{d}{dx}\left(\frac{\ln(x)}{\ln(a)}\right) = \frac{1}{\ln(a)x}$$

14.4 THE HYPERBOLIC FUNCTIONS

Any function defined for both positive and negative values of x can be written as the sum of an even function and an odd function. That is, for any function $y = f(x)$ we can write

$$f(x) = f_e(x) + f_o(x)$$

where

$$f_e(x) = \frac{f(x) + f(-x)}{2}$$

and

$$f_o(x) = \frac{f(x) - f(-x)}{2}$$

The even and odd parts of the function e^x are given the names of hyperbolic cosine and hyperbolic sine. The names of the functions are usually shortened to cosh(x) (read as 'cosh of x') and sinh(x) (read as 'shine of x').

$$e^x = \cosh(x) + \sinh(x)$$

and

$$\cosh(x) = \frac{e^x + e^{-x}}{2}$$

$$\sinh(x) = \frac{e^x - e^{-x}}{2}$$

They are called the hyperbolic sine and cosine because they bear the same sort of relationship to the hyperbola as the sine and cosine do to the circle. When we introduced the trigonometric functions in Chapter 11 we used a rotating rod of length r. The horizontal and vertical position of the tip of the rod as it travels around the circle defines the cosine and sine function. A point (x,y) on the circle can be defined using $x = r\cos(\alpha)$, $y = r\sin(\alpha)$. These are called parametric equations for the circle, and α is the parameter. If the parameter is eliminated then we get the equation of the circle $x^2/r^2 + y^2/r^2 = 1$. This is pictured in Fig. 14.5(a). Any point on a hyperbola can similarly be defined in terms of a parameter, α, and then we get $x = a\cosh(\alpha)$, $y = b\sinh(\alpha)$.

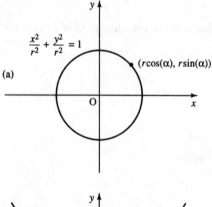

(a)

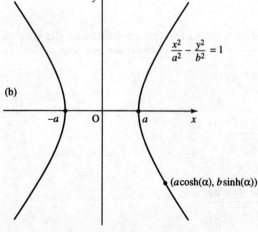

(b)

Figure 14.5 (a) $x = r\cos(\alpha)$, $y = r\sin(\alpha)$ defines a point on the circle $x^2/r^2 + y^2/r^2 = 1$. (b) $x = a\cosh(\alpha)$ and $y = b\sinh(\alpha)$ defines a point on the hyperbola $x^2/a^2 - y^2/b^2 = 1$.

If the parameter is eliminated from the equations we get the equation for the hyperbola as

$$\frac{x^2}{a^2} - \frac{y^2}{b^2} = 1$$

Figure 14.5(b) has a graph of the hyperbola.
    The function $y = \tanh(x)$ is defined, similarly to $\tan(x)$, as

$$\tanh(x) = \frac{\sinh(x)}{\cosh(x)}$$

and the reciprocals of these three main functions may be defined as

$$\operatorname{cosech}(x) = \frac{1}{\sinh(x)} \qquad \text{(the hyperbolic cosecant)}$$

$$\operatorname{sech}(x) = \frac{1}{\cosh(x)} \qquad \text{(the hyperbolic secant)}$$

$$\coth(x) = \frac{1}{\tanh(x)} \qquad \text{(the hyperbolic cotangent)}$$

The graphs of $\cosh(x)$, $\sinh(x)$ and $\tanh(x)$ are shown in Fig. 14.6.

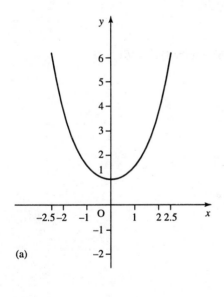

(a)

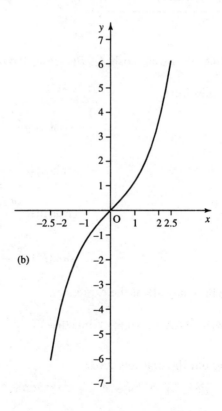

(b)

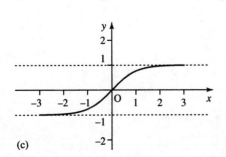

(c)

Figure 14.6 (a) The graph of $y = \cosh(x)$. (b) The graph of $y = \sinh(x)$. (c) The graph of $y = \tanh(x)$.

Hyperbolic identities

The hyperbolic identities are similar to those for trigonometric functions. A list of the more important ones is given in Table 14.1.

Table 14.1 Summary of important hyperbolic identities

$$\cosh(x) = \frac{e^x + e^{-x}}{2}$$

$$\sinh(x) = \frac{e^x - e^{-x}}{2}$$

$$\tanh(x) = \frac{\sinh(x)}{\cosh(x)} = \frac{e^x - e^{-x}}{e^x + e^{-x}}$$

$$\cosh(x) + \sinh(x) = e^x$$

$$\cosh(x) - \sinh(x) = e^{-x}$$

$$\cosh(A \pm B) = \cosh(A)\cosh(B) \pm \sinh(A)\sinh(B)$$

$$\sinh(A \pm B) = \sinh(A)\cosh(B) \pm \cosh(A)\sinh(B)$$

$$\tanh(A \pm B) = \frac{\tanh(A) \pm \tanh(B)}{1 \pm \tanh(A)\tanh(B)}$$

Example 14.8 Show that $\cosh(A + B) = \cosh(A)\cosh(B) + \sinh(A)\sinh(B)$.

SOLUTION Substitute

$$\cosh(A) = \frac{e^A + e^{-A}}{2}$$

$$\sinh(A) = \frac{e^A - e^{-A}}{2}$$

$$\cosh(B) = \frac{e^B + e^{-B}}{2}$$

$$\sinh(B) = \frac{e^B - e^{-B}}{2}$$

into the right-hand side of the expression:

$$\cosh(A)\cosh(B) + \sinh(A)\sinh(B) = \frac{(e^A + e^{-A})(e^B + e^{-B})}{2 \quad 2} + \frac{(e^A - e^{-A})(e^B - e^{-B})}{2 \quad 2}$$

Multiplying out the brackets gives

$$\tfrac{1}{4}(e^{A+B} + e^{A-B} + e^{-A+B} + e^{-(A+B)} + (e^{A+B} - e^{A-B} - e^{-A+B} + e^{-(A+B)}))$$

Simplifying then gives

$$\tfrac{1}{4}(2e^{A+B} + 2e^{-(A+B)}) = \tfrac{1}{2}(e^{A+B} + e^{-(A+B)})$$

which is the definition of $\cosh(A + B)$.

We have shown that the right-hand side of the expression is equal to the left-hand side, and therefore

$$\cosh(A+B)=\cosh(A)\cosh(B)+\sinh(A)\sinh(B)$$

Inverse hyperbolic functions

The graphs of the inverse hyperbolic functions, $\sinh^{-1}(x)$, $\cosh^{-1}(x)$ and $\tanh^{-1}(x)$, are given in Fig. 14.7.

As $\cosh(x)$ is not a one-to-one function, it has no true inverse. However, if we limit x to zero or positive values only, then $\cosh^{-1}(x)$ is indeed the inverse function, and $\cosh^{-1}(\cosh(x))=x$. The $\sinh^{-1}(x)$ function is defined for all values of x, but $\cosh^{-1}(x)$ is defined for $x \geqslant 1$ only and $\tanh^{-1}(x)$ is defined for $-1<x<1$.

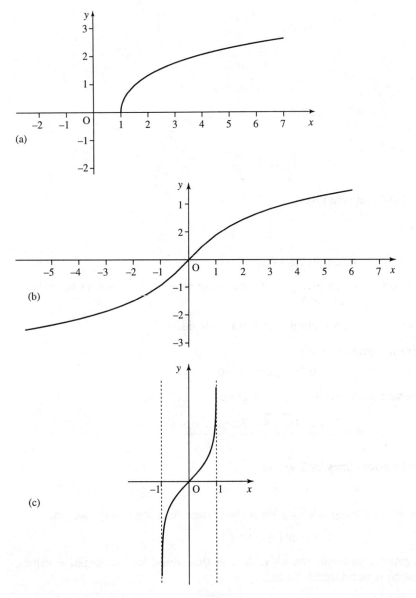

Figure 14.7 (a) The graph of $y=\cosh^{-1}(x)$, $x \geqslant 1$. (b) The graph of $y=\sinh^{-1}(x)$. (c) The graph of $\tanh^{-1}(x)$, $-1 \leqslant x \leqslant 1$.

As the hyperbolic functions are defined in terms of the exponential function we might suspect that the inverse would be defined in terms of the logarithm. The logarithmic equivalences are:

$$\sinh^{-1}(x) = \ln(x + \sqrt{x^2 + 1}) \qquad \text{all } x$$

$$\cosh^{-1}(x) = \ln(x + \sqrt{x^2 - 1}) \qquad x \geqslant 1$$

$$\tanh^{-1}(x) = \frac{1}{2}\ln\left(\frac{1+x}{1-x}\right) \qquad -1 < x < 1$$

Example 14.9 Show that $\sinh^{-1}(x) = \ln(x + \sqrt{x^2 + 1})$ using the definitions

$$y = \sinh^{-1}(x) \Leftrightarrow \sinh(y) = x$$

and

$$\sinh(y) = \frac{e^y - e^{-y}}{2}$$

SOLUTION

$$y = \sinh^{-1}(x)$$

$$\Leftrightarrow \sinh(y) = x$$

Using

$$\sinh(y) = \frac{e^y - e^{-y}}{2}$$

to substitute on the left-hand side gives

$$\frac{e^y - e^{-y}}{2} = x$$

$$\Leftrightarrow e^y - e^{-y} = 2x \qquad \text{(multiplying by 2)}$$

$$\Leftrightarrow e^{2y} - 1 = 2xe^y \qquad \text{(multiplying by } e^y \text{ and using properties of powers to write } e^y.e^y = e^{2y})$$

$$\Leftrightarrow e^{2y} - 2xe^y - 1 = 0 \qquad \text{(subtracting } 2xe^y \text{ from both sides)}$$

This is now a quadratic equation in e^y:

$$(e^y)^2 - 2xe^y - 1 = 0$$

Using the formula where $a = 1$, $b = -2x$, $c = -1$ gives

$$e^y = \frac{2x \pm \sqrt{4x^2 + 4}}{2} = \frac{2x \pm 2\sqrt{x^2 + 1}}{2}$$

Dividing the top and bottom lines by 2 gives

$$e^y = x \pm \sqrt{x^2 + 1}$$

Taking ln of both sides and using $\ln(e^y) = y$ (ln is the inverse function of exp) we get:

$$y = \ln(x \pm \sqrt{x^2 + 1})$$

We discount the negative sign inside the logarithm, as this would lead to negative values, for which the logarithm is not defined. So finally

$$\sinh^{-1}(x) = \ln(x + \sqrt{x^2 + 1})$$

Calculations

The hyperbolic and inverse hyperbolic functions are sometimes not given on a calculator. To calculate a hyperbolic function, use the definitions

$$\cosh(x) = \frac{e^x + e^{-x}}{2}$$

$$\sinh(x) = \frac{e^x - e^{-x}}{2}$$

$$\tanh(x) = \frac{\sinh(x)}{\cosh(x)} = \frac{e^x - e^{-x}}{e^x + e^{-x}}$$

To calculate the inverse hyperbolic functions use their logarithmic equivalences.

> **Example 14.10** Calculate the following, and where possible use the appropriate inverse function to check your result:
>
> (a) $\sinh(1.444)$ (b) $\tanh^{-1}(-0.5)$ (c) $\cosh(-1)$
> (b) $\cosh^{-1}(3)$ (e) $\cosh^{-1}(0)$

SOLUTION
(a) From the definition,

$$\sinh(1.444) = \tfrac{1}{2}(e^{1.444} - e^{-1.444})$$

$$\simeq 2.000\,815\,2$$

$$= 2.001 \text{ to 4 s.f.}$$

Check:
Use the inverse function of sinh, i.e. find $\sinh^{-1}(2.000\,815\,2)$. From the logarithmic equivalence,

$$\sinh^{-1}(2.000\,815\,2) = \ln(2.000\,815\,2 + \sqrt{2.000\,815\,2^2 + 1})$$

$$= 1.444$$

As this is the original number input to the sinh function, we have found $\sinh^{-1}(\sinh(1.444)) = 1.444$, which confirms the accuracy of our calculation.
(b) To calculate $\tanh^{-1}(-0.5)$, use the logarithmic equivalence, giving

$$\tanh^{-1}(0.5) = \tfrac{1}{2}\ln\left(\frac{1+(-0.5)}{1-(-0.5)}\right) = \tfrac{1}{2}\ln(\tfrac{1}{3})$$

$$\simeq -0.549\,306\,1$$

$$= -0.5493 \text{ to 4 s.f.}$$

Check:
Use the inverse function of $\tanh^{-1}$, i.e. find

$$\tanh(-0.549\,306\,1) = \frac{e^{-0.549\,306\,1} - e^{0.549\,306\,1}}{e^{-0.549\,306\,1} + e^{0.549\,306\,1}}$$

$$\simeq -0.5$$

This is the original number input to the $\tanh^{-1}$ function, and this confirms the accuracy of our calculation.

$$(c) \ \cosh(-1) = \frac{e^{-1} + e^1}{2} \simeq 1.543\,080\,6$$

$$= 1.543 \text{ to 4 s.f.}$$

Check:
Use the inverse function of cosh, i.e. $\cosh^{-1}$:

$$\cosh^{-1}(1.543\,080\,6) = \ln(1.543\,080\,6 + \sqrt{1.543\,080\,6^2 - 1})$$

$$= 1$$

This is not the number that we first started with, which was -1. However, we know that $\cosh^{-1}(x)$ is only a true inverse of $\cosh(x)$ if the domain of $\cosh(x)$ is limited to positive values. We did not expect the inverse to 'work' in this case, where we started with a negative value.

(d) To calculate $\cosh^{-1}(3)$ use

$$\cosh^{-1}(3) = \ln(3 + \sqrt{3^2 - 1})$$

$$= \ln(3 + \sqrt{8})$$

$$\simeq 1.762\,747\,2$$

$$= 1.763 \text{ to 4 s.f.}$$

Check:
The inverse function to $\cosh^{-1}$ is cosh, so we find

$$\cosh(1.762\,747\,2) = \frac{e^{1.762\,747\,2} + e^{-1.762\,747\,2}}{2}$$

$$= 3$$

This confirms the accuracy of our calculation, as we have shown $\cosh(\cosh^{-1}(3)) = 3$.

(e) Using the logarithmic definition of $\cosh^{-1}$ leads to an attempt to take the square root of a negative number. This confirms that $\cosh^{-1}(0)$ is not defined in $\mathbb{R}$.

Derivatives

Derivatives of the hyperbolic functions can be found by reverting to their definitions in terms of the exponential function.

Example 14.11 Show that

$$\frac{d}{dx}(\sinh(x)) = \cosh(x)$$

SOLUTION As

$$\sinh(x) = \frac{e^x - e^{-x}}{2}$$

then

$$\frac{d}{dx}(\sinh(x)) = \frac{d}{dx}\frac{(e^x - e^{-x})}{2} = \frac{e^x - (-1)e^{-x}}{2} = \frac{e^x + e^{-x}}{2}$$

$$= \cosh(x)$$

Therefore

$$\frac{d}{dx}(\sinh(x)) = \cosh(x)$$

In a way similar to Example 14.6 we can find

$$\frac{d}{dx}(\cosh(x)) = \sinh(x)$$

and

$$\frac{d}{dx}(\tanh(x)) = \operatorname{sech}^2(x)$$

The derivatives of the inverse hyperbolic functions can be found using the same method as given for the derivatives of the inverse trigonometric functions (in Chapter 11), giving

$$\frac{d}{dx}(\sinh^{-1}(x)) = \frac{1}{\sqrt{1+x^2}}$$

$$\frac{d}{dx}(\cosh^{-1}(x)) = \frac{1}{\sqrt{x^2-1}}$$

$$\frac{d}{dx}(\tanh^{-1}(x)) = \frac{1}{1-x^2}$$

14.5 MORE DIFFERENTIATION AND INTEGRATION

We are now able to add the functions $y = e^x$ and $y = \ln(x)$, $y = a^x$, $y = \log_a(x)$ and the hyperbolic and inverse hyperbolic functions to the list of functions in our table of derivatives giving, Table 14.2. By swapping the columns and rearranging some of the terms in a more convenient fashion, and adding the constant of integration, we get the table of integrals as given in Table 14.3.

The methods of differentiation and integration of combined functions, discussed in Chapters 12 and 13, can equally be applied to exponential and logarithmic functions.

Example 14.12 Find derivatives of the following:

(a) $y = e^{-2t^2 + 3}$ (b) $x = e^{-t}\cos(3t)$ (c) $y = \frac{\sinh(x)}{x}(x \neq 0)$

SOLUTION (a) To differentiate $y = e^{-2t^2 + 3}$ using the function of a function rule, think of this as $y = e^{(\ )}$ ('$y = e$ to the bracket'). Now differentiate y with respect to ( ) and multiply by

Table 14.2 The derivatives of some simple functions

$f(x)$	$f'(x)$
C	0
x^n	nx^{n-1}
$\cos(x)$	$-\sin(x)$
$\sin(x)$	$\cos(x)$
$\tan(x)$	$\sec^2(x)$
$\sin^{-1}(x)$	$\dfrac{1}{\sqrt{1-x^2}}$
$\cos^{-1}(x)$	$\dfrac{-1}{\sqrt{1-x^2}}$
$\tan^{-1}(x)$	$\dfrac{1}{1+x^2}$
e^x	e^x
a^x	$(\ln(a))a^x$
$\ln(x)$	$\dfrac{1}{x}$
$\log_a(x)$	$\dfrac{1}{\ln(a)x}$
$\cosh(x)$	$\sinh(x)$
$\sinh(x)$	$\cosh(x)$
$\tanh(x)$	$\mathrm{sech}^2(x)$
$\sinh^{-1}(x)$	$\dfrac{1}{\sqrt{1+x^2}}$
$\cosh^{-1}(x)$	$\dfrac{1}{\sqrt{x^2-1}}$
$\tanh^{-1}(x)$	$\dfrac{1}{1-x^2}$

Table 14.3 A table of standard integrals

$f(x)$	$\int f(x)\,dx$
1	$x+C$
$x^n \ (n \neq -1)$	$\dfrac{x^{n+1}}{n+1}+C$
$\sin(x)$	$-\cos(x)+C$
$\cos(x)$	$\sin(x)+C$
$\sec^2(x)$	$\tan(x)+C$
$\dfrac{1}{\sqrt{1-x^2}}$	$\sin^{-1}(x)+C$
$\dfrac{-1}{\sqrt{1-x^2}}$	$\cos^{-1}(x)+C$
$\dfrac{1}{1+x^2}$	$\tan^{-1}(x)+C$
e^x	e^x+C
a^x	$\dfrac{a^x}{\ln(a)}+C$
$\dfrac{1}{x}$	$\ln(x)+C$
$\cosh(x)$	$\sinh(x)+C$
$\sinh(x)$	$\cosh(x)+C$
$\mathrm{sech}^2(x)$	$\tanh(x)+C$
$\dfrac{1}{\sqrt{1+x^2}}$	$\sinh^{-1}(x)+C$
$\dfrac{1}{\sqrt{x^2-1}}$	$\cosh^{-1}(x)+C$
$\dfrac{1}{1-x^2}$	$\tanh^{-1}(x)+C$

the derivative of ( ) with respect to t. That is, use

$$\frac{dy}{dt} = \frac{dy}{d(\ )}\frac{d(\ )}{dt}$$

where ( ) represents the expression in the bracket.

$$\frac{dy}{dt} = e^{(-2t^2+3)}\frac{d}{dt}(-2t^2+3) = e^{(-2t^2+3)}(-4t)$$

$$= -4te^{(-2t^2+3)}$$

(b) To find the derivative of $x = e^{-t}\cos(3t)$, write $x = uv$ so $u = e^{-t}$ and $v = \cos(3t)$; then

$$\frac{du}{dt} = -e^{-t} \quad \text{and} \quad \frac{dv}{dt} = -3\sin(3t)$$

where we have used the chain rule to find both of these derivatives.
    Now use the product rule

$$\frac{dx}{dt} = -e^{-t}\cos(3t) - e^{-t}3\sin(3t) = -e^{-t}\cos(3t) - 3e^{-t}\sin(3t)$$

(c) To find the derivative of

$$y = \frac{\sinh(x)}{x}$$

we use the formula for the quotient of two functions, where $y = u/v$, $u = \sinh(x)$, $v = x$ and

$$\frac{dy}{dx} = \frac{(du/dx)v - u(dv/dx)}{v^2}$$

Hence we get

$$\frac{d}{dx}\left(\frac{\sinh(x)}{x}\right) = \frac{\cosh(x).x - \sinh(x).1}{x^2}$$

$$= \frac{x\cosh(x) - \sinh(x)}{x^2}$$

Example 14.13 Find the following integrals:

(a) $\int xe^{x^2+2}\,dx$ (b) $\int \sinh(t)\cosh^2(t)\,dt$

(c) $\int xe^x\,dx$ (d) $\int_1^2 \ln(x)\,dx$

(e) $\int \dfrac{3x^2+2x}{x^3+x^2+2}\,dx$

SOLUTION

(a) $\int xe^{x^2+2}\,dx$

Here we have a function of a function, $e^{(x^2+2)}$ multiplied by a term that is something like the derivative of the term in the bracket.

Try a substitution, $u = x^2 + 2$:

$$\Rightarrow \frac{du}{dx} = 2x \Rightarrow dx = \frac{du}{2x}$$

then

$$\int xe^{x^2+2}\,dx = \int xe^u \frac{du}{2x} = \int \tfrac{1}{2}e^u\,du$$

$$= \tfrac{1}{2}e^u + C$$

Re-substituting $u = x^2 + 2$ gives

$$\int xe^{x^2+2}\,dx = \tfrac{1}{2}e^{x^2+2} + C$$

(b) To find $\int \sinh(t)\cosh^2(t)\,dt$ we remember that $\cosh^2(t) = (\cosh(t))^2$, so

$$\int \sinh(t)\cosh^2(t)\,dt = \int \sinh(t)(\cosh(t))^2\,dt$$

Sinh(t) is the derivative of the function in the bracket, $\cosh(t)$, so a substitution, $u = \cosh(t)$, should work:

$$u = \cosh(t) \Rightarrow \frac{du}{dt} = \sinh(t)$$

$$\Rightarrow dt = \frac{du}{\sinh(t)}$$

$$\int \sinh(t)(\cosh(t))^2 \, dt = \int \sinh(t) u^2 \frac{du}{\sinh(t)}$$

$$= \int u^2 \, du$$

$$= \frac{u^3}{3} + C$$

Re-substituting $u = \cosh(t)$ gives

$$\int \sinh(t) \cosh^2(t) \, dt = \frac{\cosh^3(t)}{3} + C$$

(c) $\int x e^x \, dx$

Use integration by parts, $\int u \, dv = uv - \int v \, du$, and choose $u = x$ and $dv = e^x \, dx$, giving

$$\frac{du}{dx} = 1 \qquad \text{and} \qquad v = \int e^x \, dx = e^x$$

Then

$$\int x e^x \, dx = x e^x - \int e^x \, dx$$

$$= x e^x - e^x + C$$

(d) $\int_1^2 \ln(x) \, dx$

Write $\ln(x) = 1.\ln(x)$ and use integration by parts with $u = \ln(x)$ and $dv = 1.dx$:

$$\Rightarrow du = \frac{dx}{x} \qquad \text{and} \qquad v = \int 1.dx = x$$

$$\int_1^2 \ln(x) \, dx = [x \ln(x)]_1^2 - \int_1^2 x \frac{1}{x} \, dx$$

$$= 2 \ln(2) - 1 \ln(1) - \int_1^2 1 \, dx$$

$$= 2 \ln(2) - [x]_1^2$$

$$= 2 \ln(2) - (2 - 1) \approx 0.3863 \text{ to 4 s.f.}$$

(e) We rewrite

$$\int \frac{3x^2 + 2x}{x^3 + x^2 + 2} \, dx = \int (3x^2 + 2x)(x^3 + x^2 + 2)^{-1} \, dx$$

Notice that there are two brackets. To decide what to substitute we notice that

$$\frac{d}{dx}(x^3 + x^2 + 2) = 3x^2 + 2x$$

so it should work to substitute $u = x^3 + x^2 + 2$:

$$\Rightarrow \frac{du}{dx} = 3x^2 + 2x \Rightarrow dx = \frac{du}{3x^2 + 2x}$$

The integral becomes

$$\int(3x^2+2x)u^{-1}\frac{du}{3x^2+2x}=\int\frac{du}{u}$$

$$=\ln(u)+C$$

Re-substituting for u gives

$$\int\frac{3x^2+2x}{x^3+x^2+2}\,dx=\ln(x^3+x^2+2)+C$$

Integration using partial fractions

The fact that expressions like $1/(3x+2)$ can be integrated using a substitution which results in an integral of the form

$$\int\frac{1}{u}\,du=\ln(u)+C$$

is exploited when we perform the integration of fractional expressions like

$$\frac{2x-1}{(x-3)(x+1)}$$

We first rewrite the function to be integrated using partial fractions.

Example 14.14 Find

(a) $\displaystyle\int\frac{2x-1}{(x-3)(x+1)}\,dx$

(b) $\displaystyle\int\frac{x^2}{(x+2)(2x-1)^2}\,dx$

SOLUTION

(a) $\displaystyle\int\frac{2x-1}{(x-3)(x+1)}\,dx$

Rewrite the expression using partial fractions. We need to find A and B so that

$$\frac{2x-1}{(x-3)(x+1)}=\frac{A}{x-3}+\frac{B}{x+1}$$

where this should be true for all values of x.

Multiply by $(x-3)(x+1)$, giving $2x-1=A(x+1)+B(x-3)$. This is an identity, so we can substitute values for x:

substitute $x=-1$ giving $-3=B(-4)\Leftrightarrow B=3/4$

substitute $x=3$ giving $5=A(4)$ $\Leftrightarrow A=5/4$

Hence

$$\frac{2x-1}{(x-3)(x+1)}=\frac{5}{4(x-3)}+\frac{3}{4(x+1)}$$

So

$$\int \frac{2x-1}{(x-3)(x+1)}\,dx = \int \frac{5}{4(x-3)} + \frac{3}{4(x+1)}\,dx$$

As $(x-3)$ and $(x+1)$ are linear functions of x we can find each part of this integral using substitutions of $u=x-3$ and $u=x+1$ giving

$$\int \frac{5}{4(x-3)} + \frac{3}{4(x+1)}\,dx = \tfrac{5}{4}\ln(x-3) + \tfrac{3}{4}\ln(x+1) + C$$

so

$$\int \frac{2x-1}{(x-3)(x+1)}\,dx = \tfrac{5}{4}\ln(x-3) + \tfrac{3}{4}\ln(x+1) + C$$

Check:

$$\frac{d}{dx}(\tfrac{5}{4}\ln(x-3) + \tfrac{3}{4}\ln(x+1) + C) = \frac{5}{4(x-3)} + \frac{3}{4(x+1)}$$

Writing this over a common denominator gives

$$\frac{d}{dx}(5\ln(x-3) + 3\ln(x+1) + C) = \frac{20(x+1) + 12(x-3)}{16(x-3)(x+1)}$$

$$= \frac{20x + 20 + 12x - 36}{16(x-3)(x+1)}$$

$$= \frac{32x - 16}{16(x-3)(x+1)}$$

$$= \frac{2x-1}{(x-3)(x+1)}$$

(b) $\int \dfrac{x^2}{(x+2)(2x-1)^2}\,dx$

Again we can use partial fractions. Because of the repeated factor in the denominator we use both a linear and a squared term in that factor.

We need to find A, B and C so that

$$\frac{x^2}{(x+2)(2x-1)^2} = \frac{A}{x+2} + \frac{B}{2x-1} + \frac{C}{(2x-1)^2}$$

where this should be true for all values of x.

Multiply by $(x+2)(2x-1)^2$, giving

$$x^2 = A(2x-1)^2 + B(2x-1)(x+2) + C(x+2)$$

This is an identity, so we can substitute values for x:

substitute $x=\tfrac{1}{2}$ giving $0.25 = C(2.5) \Leftrightarrow C = 0.1$

substitute $x=-2$ giving $4 \quad = A(-5)^2 \Leftrightarrow A = 4/25 = 0.16$

substitute $x=0$ giving $0 \quad = A + B(-1)(2) + C(2)$

Using the fact that $A=0.16$ and $C=0.1$ we get

$$0=0.16-2B+0.2$$

$$\Leftrightarrow 0=0.36-2B$$

$$\Leftrightarrow 2B=0.36$$

$$\Leftrightarrow B=0.18$$

Then we have

$$\int \frac{x^2 \, dx}{(x+2)(2x-1)^2} = \int \frac{0.16}{x+2} + \frac{0.18}{2x-1} + \frac{0.1}{(2x-1)^2} \, dx$$

$$=0.16 \ln(x+2) + \frac{0.18}{2} \ln(2x-1) - \frac{0.1}{2}(2x-1)^{-1} + C$$

$$=0.16 \ln(x+2) + 0.09 \ln(2x-1) - \frac{0.05}{2x-1} + C$$

Check:

$$\frac{d}{dx}\left(0.16 \ln(x+2) + 0.09 \ln(2x-1) - \frac{0.05}{2x-1} + C\right)$$

$$=\frac{d}{dx}(0.16 \ln(x+2) + 0.09 \ln(2x-1) - 0.05(2x-1)^{-1} + C)$$

$$=\frac{0.16}{x+2} + \frac{0.09}{2x-1}(2) + 0.05(2)(2x-1)^{-2}$$

$$=\frac{0.16}{x+2} + \frac{0.18}{2x-1} + \frac{0.1}{(2x-1)^2}$$

Writing this over a common denominator gives

$$\frac{0.16(2x-1)^2 + 0.18(2x-1)(x+2) + 0.1(x+2)}{(x+2)(2x-1)^2}$$

$$=\frac{0.16(4x^2-4x+1) + 0.18(2x^2+3x-2) + 0.1x+0.2}{(x+2)(2x-1)^2}$$

$$=\frac{0.64x^2-0.64x+0.16+0.36x^2+0.54x-0.36+0.1x+0.2}{(x+2)(2x-1)^2}$$

$$=\frac{x^2}{(x+2)(2x-1)^2}$$

14.6 SUMMARY

1. Many physical situations involve exponential growth or decay where the rate of change of y is proportional to its current value.
2. All exponential functions, $y=a^t$, are such that $dy/dt = ky$; that is, the derivative of an exponential function is also an exponential function scaled by a factor k.

3. The exponential function $y = e^t$ has the property that $dy/dt = y$; that is, its derivative is equal to the original function:

$$\frac{d}{dt}(e^t) = e^t \quad \text{where } e \simeq 2.718\,28.$$

The inverse function to e^t is $\log_e(t)$, which is abbreviated to $\ln(t)$. This is called the natural or Napierian logarithm.

4. The general solution to $dy/dt = ky$ is $y = y_0 e^{kt}$ where y_0 is the value of y at $t = 0$.

5.
$$\frac{d}{dt}(a^t) = \ln(a)a^t$$

and

$$\frac{d}{dt}(\log_a(t)) = \frac{1}{\ln(a)t}$$

6. The hyperbolic cosine (cosh) and hyperbolic sine (sinh) are the even and odd parts of the exponential function:

$$e^x = \cosh(x) + \sinh(x)$$

$$\cosh(x) = \frac{e^x + e^{-x}}{2}$$

$$\sinh(x) = \frac{e^x - e^{-x}}{2}$$

These functions get the name hyperbolic because of their relationship to a hyperbola. The hyperbolic tangent is defined by

$$\tanh(x) = \frac{\sinh(x)}{\cosh(x)} = \frac{e^x - e^{-x}}{e^x + e^{-x}}$$

There are various hyperbolic identities, which are similar to the trigonometric identities, and they are given in Table 14.1.

7. The inverse hyperbolic functions $\cosh^{-1}(x)$ $(x \geqslant 1)$, $\sinh^{-1}(x)$ and $\tanh^{-1}(x)$ $(-1 < x < 1)$ have the following logarithmic identities:

$$\sinh^{-1}(x) = \ln(x + \sqrt{x^2 + 1}) \quad \text{all } x \in \mathbb{R}$$

$$\cosh^{-1}(x) = \ln(x + \sqrt{x^2 - 1}) \quad x \geqslant 1$$

$$\tanh^{-1}(x) = \frac{1}{2}\ln\left(\frac{1+x}{1-x}\right) \quad -1 < x < 1$$

$\text{Cosh}^{-1}(x)$ is the inverse of $\cosh(x)$ if the domain of $\cosh(x)$ is limited to the positive values of x.

8. Adding the derivatives and integrals of the exponential, ln, hyperbolic and inverse hyperbolic functions to the tables of standard derivatives and integrals gives Tables 14.2 and 14.3.

9. Partial fractions can be used to integrate fractional functions such as

$$\frac{x+1}{(x-1)(x+2)}$$

14.7 EXERCISES

14.1 Using

$$\frac{d}{dt}(e^t) = e^t$$

show that the function $2e^{3t}$ is a solution to the differential equation

$$\frac{dy}{dt} = 3y$$

14.2 Assuming $p = p_0 e^{kt}$, find p_0 and k such that

$$\frac{dp}{dt} = \frac{p}{1200}$$

and $p = 1$ when $t = 0$.

14.3 Assuming $N = N_0 e^{kt}$, find N_0 and k such that

$$\frac{dN}{dt} = -4.3 \times 10^{-4} N$$

and $N = 5 \times 10^6$ at $t = 0$.

14.4 Assuming $\phi = Ae^{kt} + 300$ find A and k such that

$$\frac{d\phi}{dt} = -0.1(\phi - 300)$$

and $\phi = 400$ when $t = 0$.

14.5 Using the definitions

$$\cosh(x) = \frac{e^x + e^{-x}}{2}$$

and

$$\sinh(x) = \frac{e^x - e^{-x}}{2}$$

show that

(a) $\cosh^2(x) - \sinh^2(x) = 1$
(b) $\sinh(x - y) = \sinh(x)\cosh(y) - \cosh(x)\sinh(y)$

14.6 Using $y = \tanh^{-1}(x) \Leftrightarrow \tanh(y) = x$, where $-1 < x < 1$ and

$$\tanh(y) = \frac{e^y - e^{-y}}{e^y + e^{-y}}$$

show that

$$\tanh^{-1}(x) = \tfrac{1}{2}\ln\left(\frac{1+x}{1-x}\right)$$

where $-1 < x < 1$.

14.7 Calculate the following and where possible use the appropriate inverse functions to check your result:

(a) $\cosh(2.1)$ (b) $\tanh(3)$ (c) $\sinh^{-1}(0.6)$
(d) $\tanh^{-1}(1.5)$ (e) $\cosh^{-1}(-1.5)$

14.8 Differentiate the following:

(a) $z = e^{t^2 - 2}$ (b) $x = e^{-t}\cosh(2t)$

(c) $\dfrac{x^2-1}{\sinh(x)}$ (d) $\ln(x^3-3x)$

(e) $\log_2(2x)$ (f) a^{4t}

(g) $2^t t^2$ (h) $\dfrac{1}{(e^{t-1})^2}$

14.9 Find the following integrals:

(a) $\int e^{4t-3}\,dt$ (b) $\displaystyle\int_2^3 \dfrac{dt}{4t-1}$

(c) $\int x \sinh(2x^2)\,dx$ (d) $\int x \ln(x)\,dx$

(e) $\int_0^1 e^x x^2\,dx$ (f) $\displaystyle\int \dfrac{\sinh(t)}{\cosh(t)}\,dt$

(g) $\displaystyle\int \dfrac{2(x-1)}{x^2-2x-4}\,dx$ (h) $\displaystyle\int \dfrac{t+1}{(t-3)(t-1)}\,dt$

(i) $\displaystyle\int_2^4 \dfrac{-t\,dt}{t^2(t-1)}$

14.10 The charge on a discharging capacitor in an RC circuit decays according to the expression

$$Q = 0.001 e^{-10t}$$

Find an expression for the current using $I = dQ/dt$, and find after how long the current is half of its initial value.

14.11 A charging capacitor in an RC circuit with a d.c. voltage of 5 V charges according to the expression

$$q = 0.005(1 - e^{-0.5t})$$

Given that the current $i = dq/dt$, calculate the current

(a) when $t = 0$
(b) after 10 s
(c) after 20 s

FIFTEEN

VECTORS

15.1 INTRODUCTION

Many things can be represented by a simple number, for instance time, distance and mass: these are then called scalar quantities. Others, however, are better represented by both their size (or magnitude) and a direction. Some of these are velocity, acceleration and force. These quantities are called vector quantities because they are represented by vectors.

A simple example of a vector is one which describes displacement. Supposing someone is standing in a room with floor tiles, as in Fig. 15.1. Then moving from one position to another can be described by the number of tiles to the right and the number of tiles towards the top of the page.

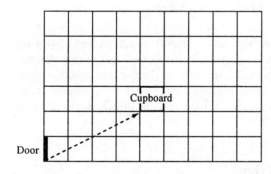

Figure 15.1 A tiled room. To reach the cupboard from the door we need to move four tiles to the right and two tiles towards the top of the page. This can be represented by the vector (4,2).

In the example, to move from the door to the cupboard can be represented by (4,2). This vector consists of two numbers, where the order of the numbers is important. Moving (4,2) results in a different final position from that if we move (2,4). The magnitude of the displacement can be found by drawing a straight line from the starting position to the final position and measuring the length. From Pythagoras (Chapter 6), this can be found as $\sqrt{4^2 + 2^2} = \sqrt{20} \simeq 4.47$. The direction can be described by an angle, for instance the angle made to the wall with the window on it.

This example shows that a two-dimensional vector (2d) can be used to represent movement on a flat surface. A two-dimensional vector is two numbers, where the order of the numbers is important.

If the room in Fig. 15.1 also had wall tiles then we could represent a position above the floor by a number of tiles toward the ceiling. This three-dimensional (3d) vector can be represented by three numbers. It can also be represented by a distance travelled and the direction (the angle made to the floor and the angle made to the wall).

Velocity is an example of a vector quantity. This can be described by two things, the speed, which is the rate of change of distance travelled with respect to time, and also the direction in which it is travelling. Similarly force can be described by the size, or magnitude, of the force and also the direction in which it operates.

Vectors have their own rules for addition and subtraction. If two forces of equal magnitude operate on one object then the net effect will depend on the direction of the forces. If the forces operate in opposite directions they could balance each other out, like two tug-of-war teams in a stalemate struggle. Alternatively they could operate in the same direction or partially in the same direction and cause the object to have an acceleration.

For the examples of vectors given so far, the maximum dimension of the vector is three as there are only three spatial dimensions. However, there are many examples when vectors of higher dimension are useful. For instance, a path through the network given in Fig. 15.2 can be represented by a list of 1s and 0s to indicate whether each of the edges is included in the path. A path from S to T can be represented by a vector, for instance:

$$a \quad b \quad c \quad d \quad e \quad f \quad g \quad h$$

$$0 \quad 1 \quad 0 \quad 0 \quad 1 \quad 0 \quad 0 \quad 0 \quad \text{represents the path } be$$

$$0 \quad 0 \quad 1 \quad 0 \quad 0 \quad 1 \quad 1 \quad 0 \quad \text{represents the path } cfg$$

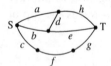

Figure 15.2 A network consisting of sides a, b, c, d, e, f, g and h.

Although there are many other types of vector we shall concentrate on vectors of two or three dimensions used to represent physical quantities in space. Many of the ideas in this chapter are only true for geometrical vectors of two and three dimensions. As three-dimensional vectors can only be correctly represented by making a three-dimensional model it is important to concentrate on understanding two-dimensional vectors, as they can be drawn on a piece of paper, allowing results to be checked easily.

15.2 VECTORS AND VECTOR QUANTITIES

A vector is a string of numbers, e.g.

$$(1, 2, -1)$$

$$(0, 1)$$

$$(3, -4, 2, -6, 8)$$

$$(2.6, 9, -1.2, 0.3)$$

The length of the string is called the dimension of the vector. For the examples above, the dimensions are 3, 2, 5 and 4, respectively. The commas can be left out, so the examples above can be written

$$(1 \quad 2 \quad -1)$$

$$(0 \quad 1)$$

$$(3 \quad -4 \quad 2 \quad -6 \quad 8)$$

$$(2.6 \quad 9 \quad -1.2 \quad 0.3)$$

Vectors may also be written as columns, giving

$$\begin{pmatrix} 1 \\ 2 \\ -1 \end{pmatrix} \begin{pmatrix} 1 \\ 0 \end{pmatrix} \begin{pmatrix} 3 \\ -4 \\ 2 \\ -6 \\ 8 \end{pmatrix} \begin{pmatrix} 2.6 \\ 9 \\ -1.2 \\ 0.3 \end{pmatrix}$$

Whether vectors are written as columns or rows only becomes important when we look at matrices (Chapter 19). However, the order of the numbers in the vector is important: (0,1) is a different vector from (1,0).

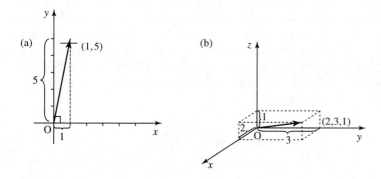

A

a

B **Figure 15.3** A vector is drawn in a diagram as a line segment with an arrow to indicate its direction.

We shall mainly deal with two- or three-dimensional vectors. Vectors are represented on a diagram by a line segment with an arrow, as in Fig. 15.3. In printed material, vectors can be represented by bold letters: **a**. They are also represented by $\underline{a}$ or $\vec{a}$ or $\overrightarrow{AB}$.

In the rest of this chapter we shall assume that we are dealing with two- or three-dimensional vectors represented in rectangular form, also called Cartesian form. This means that the numbers in the vectors correspond to the x, y, z values for a set of rectangular axes, as shown in Fig. 15.4. This assumption is important for many of the geometrical interpretations presented here.

Figure 15.4 (a) A two-dimensional rectangular set of axes and the vector (1,5). The axes are at right angles and the numbers in the vector give the x, y translation it represents. (b) A three-dimensional rectangular set of axes and the vector (2,3,1). The axes are at right angles and the numbers in the vector correspond to the x, y, z translation it represents.

Position vectors and translation vectors

Vectors can represent points in a plane, as in Fig. 15.5(a) or points in space, as in Fig. 15.5(b). These are called position vectors. They can be thought of as representing a translation from the origin.

Vectors can also represent a translation which can be applied to figures. In Fig. 15.6, a four-sided figure ABCD has been translated through the vector (2,3).

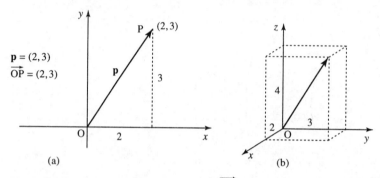

(a)   (b)

Figure 15.5 (a) The position vector $\mathbf{p}=(2,3)$ or $\overrightarrow{OP}=(2,3)$ is used to represent a point in the plane. The point can be found by translating from the origin by 2 in the x-direction followed by 3 in the y-direction; hence $\mathbf{p}=(2,3)$. (b) The position vector $\mathbf{p}=(2,3,4)$ is used to represent a point in space. The point can be found by translating from the origin by 2 in the x-direction, followed by 3 in the y-direction, followed by 4 in the z-direction; hence $\mathbf{p}=(2,3,4)$.

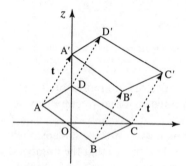

Figure 15.6 (a) The vector $\mathbf{t}=(2,3)$ is used to represent a translation of the figure ABCD. Each of the points defining the figure has been translated by (2,3).

Vector quantities

Vectors can represent physical quantities that have both a magnitude and a direction. In Fig. 15.7 there is an example of the forces acting on a body that is being pulled up a slope. By using vectors and vector addition, the resultant force acting on the body can be found and therefore the direction in which the body will travel can be found, together with the size of the acceleration. Other quantities with both magnitude and direction are velocity, acceleration and moment. Quantities that only have magnitude and no direction are called scalar quantities and can be represented using a number, e.g. mass and length.

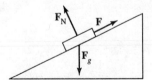

Figure 15.7 The object is being pulled up the slope using a force $\mathbf{F}$ which has a direction parallel to the slope of the hill. There is also a force due to gravity $\mathbf{F}_g$ acting vertically downwards and a force at right angles to the plane, $\mathbf{F}_N$.

15.3 ADDITION AND SUBTRACTION OF VECTORS

Addition

To add two vectors, add corresponding elements of the vectors.

Example 15.1

$$\mathbf{a} = (2,3) \quad \text{and} \quad \mathbf{b} = (4,2)$$

$$\mathbf{a} + \mathbf{b} = (2,3) + (4,2) = (2+4, 3+2) = (6,5)$$

$$\mathbf{c} = (1,3,1.5) \quad \text{and} \quad \mathbf{d} = (5,-2,1)$$

$$\mathbf{c} + \mathbf{d} = (1,3,1.5) + (5,-2,1) = (1+5, 3+(-2), 1.5+1) = (6,1,2.5)$$

If the vectors are represented in the plane then the vector sum can be found using the parallelogram law, as in Fig. 15.8. The resultant or vector sum of **a** and **b** is found by drawing vector **a** and then drawing vector **b** from the tip of vector **a** which gives the point C. Then **a** + **b** can be found by drawing a line starting at O to the point C. If we imagine walking from O to A, along vector **a**, and then from A to C along vector **b**, this has the same effect as walking direct from O to C along vector **c**. We can also use the parallelogram to show that **a** + **b** = **b** + **a**. To find **b** + **a**, start with vector **b** and draw vector **a** from the tip of vector **b**; this also gives the point C. Then if we walk from O to B along vector **b** and then from B to C along vector **a**, this has the same effect as walking along the other two sides of the parallelogram or walking direct from O to C. Hence

$$\mathbf{a} + \mathbf{b} = \mathbf{b} + \mathbf{a} = \mathbf{c}$$

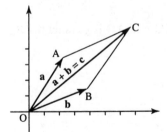

Figure 15.8 The resultant or vector sum of **a** and **b**.

Subtraction

To subtract one vector from another, subtract corresponding elements of the vectors.

Example 15.2

$$\mathbf{a} = (2,3) \quad \text{and} \quad \mathbf{b} = (4,2)$$

$$\mathbf{a} - \mathbf{b} = (2,3) - (4,2) = (2-4, 3-2) = (-2,1)$$

Using a vector diagram we can perform vector subtraction in two ways. Draw **a** and −**b** and add as before, or simply draw vectors **a** and **b** from the same point. The line joining the tip of **b** to the tip of **a** gives the vector **a** − **b**. These methods are explained in Fig. 15.9.

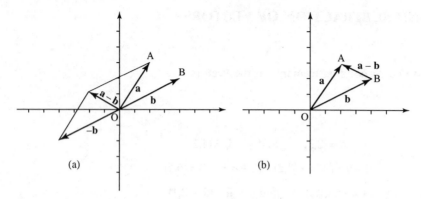

Figure 15.9 (a) To find $\mathbf{a}-\mathbf{b}$ by using addition, draw $\mathbf{a}$ and $\mathbf{b}$. Then $-\mathbf{b}$ is the vector in the opposite direction. Then add $\mathbf{a}$ and $-\mathbf{b}$ by drawing a parallelogram as in Fig. 15.8. (b) Use the triangle OAB. $\overrightarrow{BA}$ gives the vector $\mathbf{a}-\mathbf{b}$. To see this, imagine walking directly from B to A. This is the same as walking from B to O, which is backwards along $\mathbf{b}$ and therefore is the vector $-\mathbf{b}$, and then along $\overrightarrow{OA}$ which is the vector $\mathbf{a}$. Hence $\overrightarrow{BA} = -\mathbf{b}+\mathbf{a}=\mathbf{a}-\mathbf{b}$.

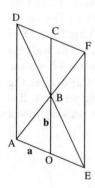

Figure 15.10 Using vectors. See Example 15.3.

Example 15.3 In Fig. 15.10, $\overrightarrow{OA}=\mathbf{a}$, $\overrightarrow{OB}=\mathbf{b}$, OB=BC, OA=EO and AD is parallel to OC and EF. Write the following vectors in terms of $\mathbf{a}$ and $\mathbf{b}$:

(a) $\overrightarrow{OE}$　(b) $\overrightarrow{OC}$　(c) $\overrightarrow{BA}$　(d) $\overrightarrow{AB}$
(e) $\overrightarrow{AD}$　(f) $\overrightarrow{BE}$　(g) $\overrightarrow{BF}$

SOLUTION

(a) $\overrightarrow{OE}$ is the same length as $\mathbf{a}$ but in the opposite direction; therefore $\overrightarrow{OE} = -\mathbf{a}$.

(b) $\overrightarrow{OC}$ is in the same direction as $\mathbf{b}$, but twice the length; therefore $\overrightarrow{OC}=2\mathbf{b}$.

(c) $\overrightarrow{BA}$ is in a triangle with $\mathbf{a}$ and $\mathbf{b}$. To get from B to A we would walk in the reverse direction along $\mathbf{b}$ and then along $\mathbf{a}$:

$$\overrightarrow{BA} = -\mathbf{b}+\mathbf{a}=\mathbf{a}-\mathbf{b}$$

(d) $\overrightarrow{AB} = -\overrightarrow{BA} = -(\mathbf{a}-\mathbf{b})=\mathbf{b}-\mathbf{a}$.

(e) $\overrightarrow{AD}$ is parallel to $\overrightarrow{OC}$ in the same direction. As $\overrightarrow{OC}=2\mathbf{b}$ then $\overrightarrow{AD}=2\mathbf{b}$.

(f) To find $\overrightarrow{BE}$ we need to know $\overrightarrow{OE}$. $\overrightarrow{OE}$ is the same length as $\mathbf{a}$ in the opposite direction; therefore $\overrightarrow{OE} = -\mathbf{a}$. To get from B to E we could go from B to O $(-\mathbf{b})$ and then from O to E $(-\mathbf{a})$; therefore $\overrightarrow{BE} = -\mathbf{b}-\mathbf{a}$.

(g) $\overrightarrow{BF}$ is the same length as $\overrightarrow{AB}$ and in the same direction. Therefore $\overrightarrow{BF} =\overrightarrow{AB} =\mathbf{b}-\mathbf{a}$.

15.4 MAGNITUDE AND DIRECTION OF A 2D VECTOR – POLAR COORDINATES

We have already noted that a vector has magnitude and direction. A two-dimensional vector can be represented by its length (also called magnitude or modulus), r (or $|\mathbf{r}|$), and its angle to the x-axis, also called its argument, θ. If the vector is (x,y), then $r^2 = x^2 + y^2$, from Pythagoras's theorem.

The angle θ is given by $\tan^{-1}(y/x)$ if x is positive, or by $\tan^{-1}(y/x) + \pi$ if x is negative. Hence $\mathbf{r} = (r,\theta)$ in polar coordinates, and can also be written as $r\underline{/\theta}$, so it is clear that the second number represents the angle.

As it is usual to give the angle between $-\pi$ and $+\pi$, it may be necessary to subtract 2π from the angle given by this formula. (As 2π is a complete rotation this will make no difference to the position of the vector.)

Example 15.4 Find the magnitude and direction of:

(a) $(2,3)$ (b) $(-1,-4)$ (c) $(1,-2.2)$ (d) $(-2,5.6)$

SOLUTION To perform these conversions to polar form it is a good idea to draw a diagram of the vector in order to be able to check that the angle is of the correct size. Figure 15.11 shows the diagrams for each part of the example.

(a) $\mathbf{r} = (2,3)$ has magnitude $\sqrt{2^2 + 3^2} \approx 3.606$, and the angle is given by $\tan^{-1}(3/2) \approx 0.983$. Therefore in polar coordinates $\mathbf{r}$ is $3.606 \underline{/0.983}$.

(b) $\mathbf{r} = (-1,-4)$ has magnitude $\mathbf{r} = \sqrt{(-1)^2 + (-4)^2} \approx 4.123$, and the angle is given by $\tan^{-1}(-4/-1) + \pi \approx 1.326 + 3.142 \approx 4.467$. As this angle is bigger than π subtract 2π (a complete revolution) to give -1.816. Therefore in polar coordinates $\mathbf{r} = 4.123 \underline{/-1.816}$. Note that the angle is between $-\pi$ and $-\pi/2$, meaning that the vector must lie in the third quadrant, which we can see is correct from the diagram.

(c) $\mathbf{r} = (1,-2.2)$ has magnitude $r = \sqrt{1^2 + (-2.2)^2} \approx 2.416$, and the angle is given by $\tan^{-1}(-2.2/1) = 1.144$. Therefore in polar coordinates $\mathbf{r} = 2.416 \underline{/-1.144}$. Note that the angle is between $-\pi/2$ and 0, meaning that the vector must lie in the fourth quadrant.

(d) $\mathbf{r} = (-2,5.6)$ has magnitude $r = \sqrt{(-2)^2 + (5.6)^2} \approx 5.946$ and the angle is given by $\tan^{-1}(5.6/-2) + \pi \approx 1.914$. Therefore in polar coordinates $\mathbf{r} = 5.946 \underline{/1.914}$. Note that the angle is between $\pi/2$ and π, meaning that the vector must lie in the second quadrant.

Many calculators have a rectangular to polar conversion facility. Look this up on the instructions with your calculator and check the results. Remember that to get the result in radians you should first put your calculator into radian mode.

Conversion from polar coordinates to rectangular coordinates

If a vector is given by its length and angle to the x-axis, i.e. $\mathbf{r} = r\underline{/\theta}$ then

$$x = r \cos(\theta)$$

$$y = r \sin(\theta)$$

Hence in rectangular coordinates $\mathbf{r} = (r \cos(\theta), r \sin(\theta))$. This result can easily be found from the triangle, as shown in Fig. 15.12; examples are given in Fig. 15.13.

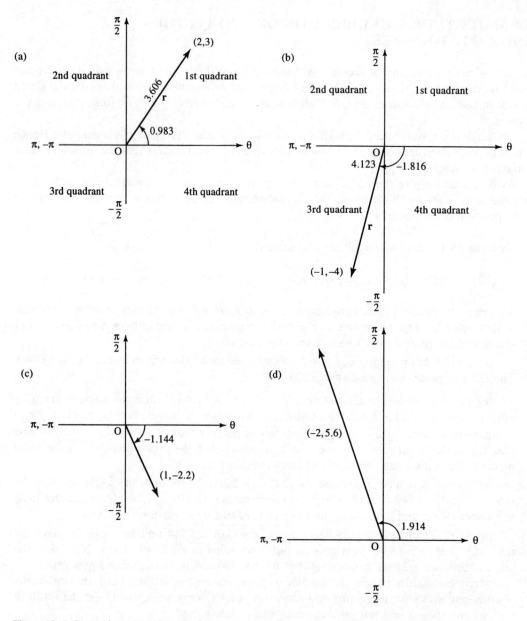

Figure 15.11 Converting vectors to polar form. (a) $\mathbf{r} = (2,3)$. (b) $\mathbf{r} = (-1,-4)$. (c) $\mathbf{r} = (1,-2.2)$. (d) $\mathbf{r} = (-2,5.6)$.

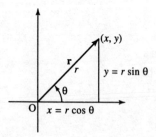

Figure 15.12 If a vector $\mathbf{r}$ is known in polar coordinates, $\mathbf{r} = r\underline{/\theta}$, then from the triangle $\cos(\theta) = x/r \Leftrightarrow x = r\cos(\theta)$ and $\sin(\theta) = y/r \Leftrightarrow y = r\sin(\theta)$.

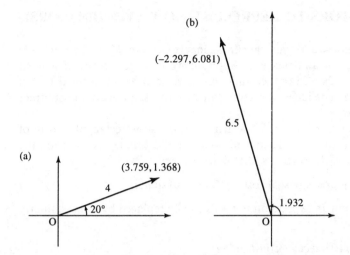

Figure 15.13 (a) $4\underline{/20°}$ in rectangular coordinates is given by $x=4\cos(20°)\simeq3.759$ and $y=4\sin(20°)\simeq1.368$. Therefore the vector is $(3.759,1.368)$. (b) $6.5\underline{/1.932}$ in rectangular coordinates is given by $x=6.5\cos(1.932)\simeq-2.297$ and $y=6.5\sin 1.932\simeq6.081$. Therefore the vector is $(-2.297,6.081)$.

Adding two vectors expressed in polar coordinates

To add two vectors expressed in polar coordinates, first, express them in rectangular coordinates, then find the sum and convert back into polar coordinates.

Example 15.5 Find $2\underline{/20°}+4\underline{/50°}$.

SOLUTION Using $x=r\cos(\theta)$ and $y=r\sin(\theta)$ we can express the two vectors in rectangular coordinates, giving

$$2\underline{/20°}\simeq(1.879,0.684)$$

$$4\underline{/50°}\simeq(2.5711,3.046)$$

Therefore $2\underline{/20°}+4\underline{/50°}\simeq(1.879,0.684)+(2.571,3.064)=(4.45,3.748)$.

Finally, this can be represented in polar coordinates by using $r=\sqrt{x^2+y^2}$ and $\theta=\tan^{-1}(y/x)$ ($+\pi$ if x is negative), giving $(4.55,3.748)\simeq5.818\underline{/40.106°}$.

Example 15.6 Find $4\underline{/1}+2\underline{/-1.6}$.

SOLUTION Using $x=r\cos(\theta)$ and $y=r\sin(\theta)$ we can express the two vectors in rectangular coordinates, giving

$$4\underline{/1}\simeq(2.161,3.366)$$

$$2\underline{/-1.6}\simeq(-0.058,-1.999)$$

Therefore $4\underline{/1}+2\underline{/-1.6}\simeq(2.161,3.366)+(-0.058,-1.999)=(2.103,1.367)$.

Finally, this can be represented in polar coordinates by using $r=\sqrt{x^2+y^2}$ and $\theta=\tan^{-1}(y/x)$ ($+\pi$ if x is negative), giving $(2.103,1.367)\simeq2.508\underline{/0.576}$.

15.5 APPLICATION OF VECTORS TO REPRESENTING WAVES (PHASORS)

In Sec. 11.3 we found the amplitude, phase and cycle rate (frequency) of a wave. $f(t) = A \cos(\omega t + \phi)$ has amplitude A, angular frequency ω and phase ϕ. Suppose we consider waves of a fixed frequency (say 50 Hz, giving $\omega = 50 \times 2\pi \simeq 314$); then different waves can be represented by the amplitude and phase, giving $y = A/\phi$. The ideas of vectors can then be used to add and subtract waves and find their combined effect.

If a wave can be represented in polar form by A/ϕ then what does the rectangular form of the vector represent? We find that if the wave is split into cosine and sine terms by using the trigonometric identity $\cos(A + B) = \cos(A) \cos(B) - \sin(A) \sin(B)$ we get

$$f(t) = A \cos(\omega t + \phi) = A \cos(\phi) \cos(\omega t) - A \sin(\phi) \sin(\omega t)$$

As $A \cos(\phi)$ is a constant, not involving an expression in t, this can be replaced by c and similarly $A \sin(\phi)$ can be replaced by d, giving

$$f(t) = c \cos(\omega t) - d \sin(\omega t)$$

where $c = A \cos(\phi)$ and $d = A \sin(\phi)$.

So the vector (c,d) used to represent a wave represents the function $f(t) = c \cos(\omega t) - d \sin(\omega t)$, and if expressed in polar form, A/ϕ, it represents the equivalent expression

$$f(t) = A \cos(\omega t + \phi)$$

Example 15.7 Express the following as a single cosine term and hence give the amplitude and phase of the resultant function:

$$x = 3 \cos(2t) + 2 \sin(2t)$$

SOLUTION Comparing $x = 3 \cos(2t) + 2 \sin(2t)$ with the expression $f(t) = c \cos(\omega t) - d \sin(\omega t)$ gives $c = 3$, $d = -2$ and $\omega = 2$. Expressing the vector $(3, -2)$ in polar form gives $3.605/-0.588$, and hence $x = 3.605 \cos(2t - 0.588)$, giving the amplitude as 3.605 and phase as -0.588.

Check:
Expand $\quad x = 3.605 \cos(2t - 0.588) \quad$ using $\quad \cos(A - B) = \cos(A) \cos(B) + \sin(A) \sin(B)$:
$3.605 \cos(2t - 0.588) = 3.605 \cos(2t) \cos(0.588) + 3.605 \sin(2t) \sin(0.588) = 3 \cos(2t) + 2 \sin(2t)$
which is the original expression.

Example 15.8 Express the following as a single cosine term and hence give the magnitude and phase of the resultant function:

$$y = -2 \cos(t) - 4 \sin(t)$$

SOLUTION Comparing $y = -2 \cos(t) - 4 \sin(t)$ with the expression $f(t) = c \cos(\omega t) - d \sin(\omega t)$ gives $c = -2$, $d = 4$ and $\omega = 1$. Expressing the vector $(-2,4)$ in polar form gives $4.472/2.034$, and hence $y = 4.472 \cos(t + 2.034)$.

Check:
Expand $\quad y = 4.472 \cos(t + 2.034) \quad$ using $\quad \cos(A + B) = \cos(A) \cos(B) - \sin(A) \sin(B)$:
$4.472 \cos(t + 2.034) = 4.472 \cos(t) \cos(2.034) - 4.472 \sin(t) \sin(2.034) = -2 \cos(t) - 4 \sin(t)$.

Example 15.9 Express $x = 3 \cos(20t + 5)$ as the sum of cosine and sine terms.

SOLUTION Representing x as the phasor $3\underline{/5}$ with angular frequency 20, then $3\underline{/5}$ converts to rectangular form as the vector $(0.851, -2.877)$ and this now gives the values of (c,d) in the expression $f(t) = c \cos(\omega t) - d \sin(\omega t)$, giving $x = 0.851 \cos(20t) + 2.877 \sin(20t)$.

Example 15.10 Find the resultant wave found from combining the following into one term:

$$f(t) = 3 \cos(314t + 0.5) + 2 \cos(314t + 0.9)$$

SOLUTION As both terms are of the same angular frequency, 314 rad s^{-1}, we can express the two component parts by their amplitude and phase and then add the two vectors, giving $3\underline{/0.5} + 2\underline{/0.9}$. Expressing this in rectangular form gives $(2.633, 1.438) + (1.243, 1.567) = (3.876, 3.005)$. Finally, expressing this again in polar form gives $4.904\underline{/0.659}$, so the resultant expression is $f(t) = 4.904 \cos(314t + 0.659)$.

This method is a shorthand version of writing out all the trigonometric identities. It is even quicker if you use the polar–rectangular and rectangular–polar conversion facility on a calculator.

15.6 MULTIPLICATION OF A VECTOR BY A SCALAR AND UNIT VECTORS

Multiplying a vector by a scalar has the effect of changing the length without affecting the direction. Each number in the vector is multiplied by the scalar.

Example 15.11 If $a = (1, -2)$ then

$$3a = 3(1, -2) = (3 \times 1, 3 \times (-2)) = (3, -6)$$

$$0.5a = 0.5(1, -2) = (0.5 \times 1, 0.5 \times (-2)) = (0.5, -1)$$

This is pictured in Fig. 15.14.

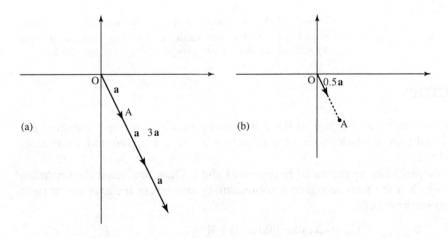

Figure 15.14 The vector $(1, -2)$: (a) multiplied by 3; (b) multiplied by 0.5.

Unit vectors

Unit vectors have length 1. They are often represented by vectors with a cap on them: $\hat{r}$. Hence $\hat{r}$ means the unit vector in the same direction as r.

To find the unit vector in the same direction as r, divide r by its length.

In Section 15.5 we found the length of a two-dimensional vector, (x,y) is $\sqrt{x^2+y^2}$ similarly it can be shown that in three dimensions (x,y,z) has length $\sqrt{x^2+y^2+z^2}$.

Example 15.12 Find unit vectors in the direction of the following vectors:

(a) $(1,-1)$ (b) $(3,4)$ (c) $(0.5,1,0.2)$

SOLUTION

(a) Find the length of $(1,-1)$, given by $\sqrt{x^2+y^2}=\sqrt{1^2+(-1)^2}\simeq1.414$. Therefore the unit vector is

$$\frac{1}{1.414}(1,-1)\simeq(0.707,-0.707)$$

(b) Find the length of $(3,4)$, given by $\sqrt{x^2+y^2}=\sqrt{3^2+4^2}=5$. Therefore the unit vector is $\frac{1}{5}(3,4)=(0.6,0.8)$ (see Fig. 15.15).

(c) Find the length of $(0.5,1,0.2)$, given by $\sqrt{x^2+y^2+z^2}=\sqrt{(0.5)^2+1^2+0.2^2}\simeq1.136$. Therefore the unit vector is

$$\frac{1}{1.136}(0.5,1,0.2)\simeq(0.44,0.88,1.176)$$

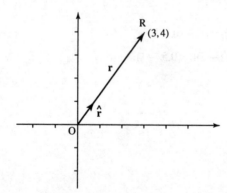

Figure 15.15 The vector $r=(3,4)$ has modulus, or length, $r=\sqrt{3^2+4^2}=5$. The unit vector in the same direction is found by dividing the vector r by its length, giving $\hat{r}=\frac{1}{5}(3,4)=(0.6,0.8)$.

15.7 BASIS VECTORS

Vectors in a plane are made up of a part in the x-direction and a part in the y-direction e.g. $(2,3)=(2,0)+(0,3)$. i and j are used to represent unit vectors in the x-direction and y-direction, i.e. $i=(1,0)$ and $j=(0,1)$.

Any vector in the plane can be expressed in terms of i and j. These are called the Cartesian unit basis vectors, which is the name given to a coordinate system where the axes are at right angles to each other (orthogonal).

$$(2,3)=2(1,0)+3(0,1)=2i+3j$$

See Fig. 15.16.

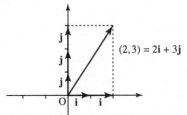

Figure 15.16 Any vector in a plane can be expressed in terms of the vectors **i** and **j**. For instance, $(2,3) = 2(1,0) + 3(0,1) = 2\mathbf{i} + 3\mathbf{j}$.

The unit vector in the z-direction is often given the symbol **k**, and in three dimensions, using rectangular axes,

$$\mathbf{i} = (1,0,0)$$
$$\mathbf{j} = (0,1,0)$$
$$\mathbf{k} = (0,0,1)$$

then

$$(5, -1,2) = (5,0,0) + (0, -1,0) + (0,0,2)$$
$$= 5(1,0,0) + (-1)(0,1,0) + 2(0,0,1)$$
$$= 5\mathbf{i} - \mathbf{j} + 2\mathbf{k}$$

The vectors **i** and **j** form a basis set because all two-dimensional geometrical vectors can be expressed in terms of them. Similarly, all three-dimensional vectors can be expressed in terms of **i, j** and **k**. There are many other sets of vectors that can be used as a basis set: for instance, if we were in a room shaped like a parallelogram we could express any position in the room by moving parallel to one of the sides and then parallel to the other sides. This is pictured in Fig. 15.17. Other basis sets are not as useful for interpreting spatial vectors, as they do not give the same geometrical results. For instance, the interpretation of the scalar product, given in the next section, relies on the fact that we use Cartesian basis vectors.

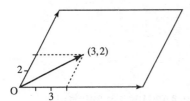

Figure 15.17 A room shaped as a parallelogram. Any position in the room can be found by moving parallel to the sides. The basis vectors used are not at right angles.

15.8 PRODUCTS OF VECTORS

There are two products of vectors that are commonly used: the scalar product, which results in a scalar, and the vector or cross product, which gives a vector as the result. However, both of these products are irreversible: they have no inverse operation. In other words, it is not possible to divide by a vector.

Scalar product

The scalar product of two vectors is defined by

$$\mathbf{a} \cdot \mathbf{b} = (a_1, a_2) \cdot (b_1, b_2) = a_1 b_1 + a_2 b_2$$

Notice that the scalar product results in a simple number as the result.

Example 15.13 Find the scalar product of (2,3) and (1,2).

SOLUTION Using the definition:

$$(2,3)\cdot(1,2)=(2)(1)+(3)(2)=2+6=8.$$

Interpretation of the scalar product

The scalar product of **a** and **b** is related to the length of the vectors in the following way:

$$\mathbf{a}\cdot\mathbf{b}=ab\ \cos(\theta)$$

where θ is the angle between the two vectors, a is the magnitude of **a** and b is the magnitude of **b** (see Fig. 15.18).

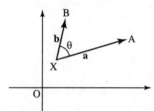

Figure 15.18 The scalar product $\mathbf{a}\cdot\mathbf{b}=ab\ \cos(\theta)$, where θ is the angle between the two vectors.

The magnitude of a vector is the square root of the dot product with itself, hence $\mathbf{a}\cdot\mathbf{a}=a^2$.
The scalar product can be used to find the angle between two vectors. It can also be used to find the length of a vector and can be used to test if two vectors are at right angles (orthogonal).

Example 15.14 Find the angle between $(1,-1)$ and $(3,2)$.

SOLUTION If $\mathbf{a}=(1,-1)$ and $\mathbf{b}=(3,2)$ then $\mathbf{a}\cdot\mathbf{b}=(1,-1)\cdot(3,2)=(1)(3)+(-1)(2)=3-2=1$.
We now use the relationship

$$\mathbf{a}\cdot\mathbf{b}=ab\ \cos(\theta)$$

$$\Leftrightarrow\cos(\theta)=\frac{\mathbf{a}\cdot\mathbf{b}}{ab}$$

to find the angle between the vectors. We find the magnitude of **a** and the magnitude of **b**:

$$a=\sqrt{1+(-1)^2}\simeq1.414$$

and

$$b=\sqrt{3^2+2^2}\simeq3.606$$

Hence

$$\cos(\theta)=\frac{\mathbf{a}\cdot\mathbf{b}}{ab}$$

becomes

$$\cos(\theta)=\frac{1}{1.414\times3.606}\simeq0.196$$

giving

$$\theta = \cos^{-1}(0.196) \simeq 1.373 \text{ radians}$$

Example 15.15 Show that $(2, -1)$ and $(-0.5, 1)$ are at right angles.

SOLUTION If two vectors are at an angle θ, with $\cos(\theta) = 0$, then $\theta = \pm 90°$, so the vectors are at right angles. Hence if we find that $\mathbf{a} \cdot \mathbf{b} = 0$ this shows that $\mathbf{a}$ and $\mathbf{b}$ are at right angles (as long as one of the vectors is not the null vector $(0, 0)$).

In this case the scalar product gives

$$(2, -1) \cdot (-0.5, -1) = 2(-0.5) + (-1)(-1) = -1 + 1 = 0$$

As the scalar product of the two vectors is 0 then the angle between them is $90°$, so they are at right angles.

Example 15.16 Show that $(1, -1)$ and $(-2, -2)$ are at right angles.

SOLUTION

$$(1, -1) \cdot (-2, -2) = (1)(-2) + (-1)(-2) = -2 + 2 = 0$$

Hence they are at right angles.

Vector (or cross) product

The vector product of $\mathbf{a}$ and $\mathbf{b}$ is defined by $\mathbf{a} \times \mathbf{b} = ab \sin(\theta)\hat{\mathbf{n}}$, where $\hat{\mathbf{n}}$ is the unit vector normal to the plane of $\mathbf{a}$ and $\mathbf{b}$ and θ is the angle between $\mathbf{a}$ and $\mathbf{b}$.

If $\mathbf{a}$ and $\mathbf{b}$ are vectors which lie in the x, y plane, then the vector product will be a vector normal to that plane, i.e. wholly in the z-direction. It can be found from the following expression:

$$(a_1, a_2, 0) \times (b_1, b_2, 0) = (0, 0, a_1 b_2 - a_2 b_1) = (a_1 b_2 - a_2 b_1)\mathbf{k}$$

where $\mathbf{k}$ is the unit vector in the z-direction.

Example 15.17

$$(1, 2, 0) \times (-3, -1, 0) = (0, 0, (1)(-1) - (2)(-3)) = (0, 0, 5)$$

This is pictured in Fig. 15.19.

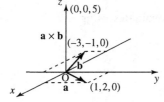

Figure 15.19 The vector product of two vectors lying in the x, y plane: $(1, 2, 0) \times (-3, -1, 0) = (0, 0, 5)$.

Application of the vector product

The vector product can be used to find the area of a parallelogram with sides OA and OB (see Fig. 15.20). The area of a parallelogram is given by $ab \sin(\theta)$ where a and b are the lengths of the sides and θ is the angle between them. Consider vectors $\mathbf{a}$ and $\mathbf{b}$ representing the sides of the parallelogram. We know that $\mathbf{a} \times \mathbf{b} = ab \sin(\theta)\hat{\mathbf{n}}$, where $\hat{\mathbf{n}}$ is a unit vector normal to the plane

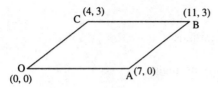

Figure 15.20 The magnitude of the vector product of two vectors gives the area of the parallelogram formed by the vectors. The area of OABC is given by $|(4,3) \times (7,0)| = |(4.0) - (3.7)| = |0 - 21| = 21$ square units.

of **a** and **b**. As **n̂** is a unit vector it has a magnitude of 1 so $|\mathbf{a} \times \mathbf{b}| = ab \sin(\theta)$, which is exactly the same as the formula for the area of the parallelogram. Therefore the area of the parallelogram can be found by taking the magnitude of the vector product of the vectors which define the sides of the parallelogram.

As $\sin(\theta) = 0$ when $\theta = 0$ or $\theta = 180°$, the vector product can also be used to test for parallel vectors (vectors pointing in the same direction or in exactly opposing directions). Again, we only need to consider the magnitude of the vector product.

Example 15.18 Show that the vectors $(0.2, -5)$ and $(-1, 25)$ are parallel.

SOLUTION

$$|(0.2, -5) \times (-1, 25)| = |(0.2 \times 25) - (-5 \times -1)| = |5 - 5| = 0$$

As we know that $|\mathbf{a} \times \mathbf{b}| = ab \sin(\theta)$, then as a and b are of non-zero length $|\mathbf{a} \times \mathbf{b}| = 0 \Leftrightarrow \sin(\theta) = 0$. This shows that the vectors are parallel.

15.9 SUMMARY

1. A vector is a string of numbers where the length of the string is called the dimension of the vector.
2. Vectors are used to represent points on a plane or in space, translations, and physical quantities that have both magnitude and direction (called vector quantities).
3. The vector sum is found by adding corresponding elements of the vectors or, from a diagram, by using a parallelogram.
4. To subtract vectors, subtract corresponding elements of the vectors. A triangle may be used to perform vector subtraction in a diagram.
5. Two-dimensional vectors $\mathbf{r} = (x, y)$ can be expressed in polar coordinates using

$$r = \sqrt{x^2 + y^2}$$

and

$$\theta = \tan^{-1}(y/x) \qquad (+\pi \text{ if } x \text{ is negative})$$

so that $(x, y) = r\underline{/\theta}$.

r or $|\mathbf{r}|$ is the magnitude, or length, of the vector and θ is the angle that the vector makes to the x-axis, also called its argument.

To convert from polar to rectangular coordinates use

$$x = r \cos(\theta) \qquad \text{and} \qquad y = r \sin(\theta)$$

To add vectors given in polar form they must first be converted to rectangular form.

6. Waves of a fixed frequency can be represented by phasors giving the amplitude and phase. $f(t) = A \cos(\omega t + \phi)$ can be represented by its amplitude and phase $A/\underline{\phi}$. Converting this vector to rectangular form gives (c,d), where

$$f(t) = c \cos(\omega t) - d \sin(\omega t)$$

Using ideas of conversion from polar to rectangular form and vector addition, waves of the same frequency can easily be combined.

7. Unit vectors have length 1. To find the unit vector in the same direction as a vector $\mathbf{r}$, divide the vector by its length:

$$\hat{\mathbf{r}} = \frac{\mathbf{r}}{r}$$

8. Any vectors in the plane can be represented in terms of $\mathbf{i} = (1,0)$ and $\mathbf{j} = (0,1)$, and in three dimensions by $\mathbf{i} = (1,0,0)$, $\mathbf{j} = (0,1,0)$ and $\mathbf{k} = (0,0,1)$. These are the Cartesian unit basis vectors and they are at right angles to each other.

9. Where $\mathbf{a} = (a_1, a_2)$ and $\mathbf{b} = (b_1, b_2)$ are two vectors, then the scalar product is given by $\mathbf{a} \cdot \mathbf{b} = a_1 b_1 + a_2 b_2$ and $\mathbf{a} \cdot \mathbf{b} = ab \cos(\theta)$ where a and b are the magnitudes of the vectors $\mathbf{a}$ and $\mathbf{b}$ and θ is the angle between them. The scalar product can be used to find the angle between two vectors.

10. The vector product is given by $\mathbf{a} \times \mathbf{b} = ab \sin(\theta)\hat{\mathbf{n}}$, where a and b are the magnitudes of the vectors $\mathbf{a}$ and $\mathbf{b}$, θ is the angle between them and $\hat{\mathbf{n}}$ is a unit vector normal to the plane of $\mathbf{a}$ and $\mathbf{b}$. If $\mathbf{a}$ and $\mathbf{b}$ are vectors in the x,y plane, i.e. $\mathbf{a} = (a_1, a_2, 0)$ and $\mathbf{b} = (b_1, b_2, 0)$, we have

$$(a_1, a_2, 0) \times (b_1, b_2, 0) = (0, 0, a_1 b_2 - a_2 b_1) = (a_1 b_2 - a_2 b_1)\mathbf{k}$$

where $\mathbf{k}$ is the unit vector in the z-direction.

The magnitude of the vector product can be used to find the area of a parallelogram.

15.10 EXERCISES

15.1 In the diagram in Fig. 15.21, $\overrightarrow{OA} = \mathbf{a}$ and $\overrightarrow{OB} = \mathbf{b}$, $OA = BC = OD$, $OB = AC = EO$, and EOB and DOA are straight lines. Write the following in terms of $\mathbf{a}$ and $\mathbf{b}$:

(a) $\overrightarrow{AB}$ (b) $\overrightarrow{BA}$ (c) $\overrightarrow{OC}$ (d) $\overrightarrow{OE}$
(e) $\overrightarrow{OD}$ (f) $\overrightarrow{ED}$ (g) $\overrightarrow{DE}$ (h) $\overrightarrow{DA}$
(i) $\overrightarrow{BE}$ (j) $\overrightarrow{EA}$

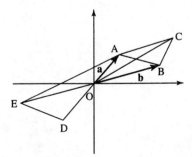

Figure 15.21 Vectors for Exercise 15.1.

15.2 Given $\mathbf{a}=(1,3)$, $\mathbf{b}=(-1,2)$, $\mathbf{c}=(3,6,2)$ and $\mathbf{d}=(6,4,-1)$, find the following:

(a) $\mathbf{a}+\mathbf{b}$ (b) $\mathbf{a}-\mathbf{b}$ (c) $\mathbf{b}-\mathbf{a}$ (d) $-\mathbf{b}+\mathbf{a}$
(e) $2\mathbf{b}$ (f) $\mathbf{a}+2\mathbf{b}$ (g) $3\mathbf{a}-\mathbf{b}$ (h) $\mathbf{c}-\mathbf{d}$
(i) $10\mathbf{c}$ (j) $\mathbf{c}+6\mathbf{d}$ (k) $6\mathbf{c}-\mathbf{d}$

15.3 Express the following in polar form, $r\underline{/\theta}$, where r is the length of the vector and θ its angle to the x-axis:

(a) $(1,3)$ (b) $(3,-1)$ (c) $(-1,-3)$ (d) $(5,-6)$

15.4 Express the following vectors $r\underline{/\theta}$, where r is the modulus of the vector and θ the angle to the x-axis, in rectangular form. The angle is expressed in radians:

(a) $5\underline{/\pi}$ (b) $1\underline{/-\pi}$ (c) $\frac{1}{2}\underline{/-\pi/4}$ (d) $3\underline{/\pi/3}$

15.5 Express the following as sums of cosine and sine terms in ωt:

(a) $f(t)=3\cos(3t-2)$ (b) $f(t)=10\cos(20t+5)$

15.6 Express the following as single cosine terms:

(a) $f(t)=4\cos(10t)-3\sin(10t)$ (b) $g(t)=-2\cos(157t)+10\sin(157t)$

15.7 Express the following as a single wave:

(a) $6\cos(2t-3)+10\cos(2t+2)$ (b) $\cos(t-\pi/2)+\cos(t+\pi/2)$ (c) $2\cos(628t-1.57)-6\cos(628t)$

15.8 Find the unit vectors in the same direction as the following:

(a) $(6,8)$ (b) $(5,12)$ (c) $(5,-12)$ (d) $(1,1)$
(e) $(3,2)$ (f) $(2,0)$ (g) $(0,-3)$ (h) $(2,4,4)$
(i) $(1,-1,2)$ (j) $(0.5,0,-0.5)$

15.9 Express the following vectors in terms of $\mathbf{i}=(1,0)$ and $\mathbf{j}=(0,1)$ or in terms of $\mathbf{i}=(0,0,1)$, $\mathbf{j}=(0,1,0)$, $\mathbf{k}=(0,0,1)$ for three-dimensional vectors:

(a) $(5,2)$ (b) $(-1,-2)$ (c) $(-6,2)$ (d) $(-1,2,-3)$ (e) $(0.2,-1.6,3.3)$

15.10 Find the following scalar products:

(a) $(1,-2){\cdot}(3,3)$ (b) $(9,2){\cdot}(-1,6)$ (c) $(6,-1){\cdot}(-1,-3)$

15.11 Find the angle between the following pairs of vectors:

(a) $(1,-2)$ and $(5,1)$ (b) $(6,-1)$ and $(1,6)$ (c) $(2,-1)$ and $(4,9)$

15.12 Show that the following pairs of vectors are at right angles to each other:

(a) $(2,1)$ and $(-1,2)$ (b) $(-6,3)$ and $(1,2)$
(c) $(0.5,-2)$ and $(4,1)$

15.13 Show that the following pairs of vectors are parallel:

(a) $(-3,1)$ and $(1.5,-0.5)$ (b) $(6,3)$ and (18.9)

15.14 Find the area of the parallelogram OABC, where two adjacent sides are:

(a) $OA=(1,-1)$ and $OC=(5,2)$
(b) $OA=(4,-1)$ and $OC=(2,2)$
(c) $OA=(-3,1)$ and $OC=(2,3)$

SIXTEEN
COMPLEX NUMBERS

16.1 INTRODUCTION

In the previous chapter we have shown that a single frequency wave can be represented by a phasor. We begin this chapter with a brief look at linear system theory. Such systems, when the input is a single frequency wave, produce an output at the same frequency, which may be phase-shifted with a scaled amplitude. Using complex numbers the system can be represented by a number which multiplies the input phasor, having the effect of rotating the phasor and scaling the amplitude. We can define j as the number which rotates the phasor by $\pi/2$ without changing the amplitude. If this multiplication is repeated, hence rotating the phasor by $\pi/2 + \pi/2 = \pi$, then the system output will be inverted. In this way we can get the fundamental definition $j^2 = -1$. j is clearly not a real number, as any real number squared is positive. j is called an imaginary number.

The introduction of imaginary numbers allows any quadratic equation to be solved. In previous chapters we said that the equation $ax^2 + bx + c = 0$ had no solutions when the formula leads to an attempt to take the square root of a negative number. The introduction of the number j makes square roots of negative numbers possible, and in these cases the equation has complex roots. A complex number, z, has a real and imaginary part, $z = x + jy$ where x is the real part and y is the imaginary part. We saw in Chapter 1 that real numbers are represented by points on a number line. Complex numbers need a whole plane to represent them.

We shall look at operations involving complex numbers, the conversion between polar and Cartesian (rectangular) form and the application of complex numbers to alternating current theory.

By looking at the problem of motion in a circle we show the equivalence between polar and exponential form of complex numbers and represent a wave in complex exponential form. We can also obtain formulae for the sine and cosine in terms of complex exponentials, and we solve complex equations $z^n = c$, where c is a complex number.

16.2 PHASOR ROTATION BY $\pi/2$

In system theory, a system is represented, as in Fig. 16.1, as a box with an input and an output. We think about the system after it has been in operation for a length of time, so any initial switching effects have disappeared.

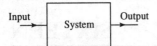

Figure 16.1 A system is characterized by a box with an input and output.

Of particular importance are systems which, when the input is a single-frequency wave, produce an output at the same frequency that can be characterized by a phase shift and a change of amplitude of the wave. Examples of such systems are electrical circuits which are made up of lumped elements, i.e. resistors, capacitors and inductors. Here the input and output are voltages. Such a system is shown in Fig. 16.2(a). Equivalent mechanical systems are made up of masses, springs and dampers, and the input and output are the external force applied and the tension in the spring. An example of such a mechanical system is shown in Fig. 16.2(b).

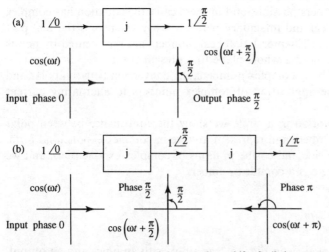

Figure 16.2 (a) An electrical system made up of resistors, capacitors and inductors with voltage as input and output. (b) A mechanical system made up of masses, springs and dampers. The input is the external force and the output is tension in the spring.

We saw in Chapter 15 that a single-frequency wave can be characterized by its amplitude and phase, and these can be represented by vectors, called phasors. The advantage of complex numbers is that a phasor can be treated as a number and the system can be represented also by a number multiplying the input phasor.

Consider a single-frequency input of 0 phase and amplitude 1. If there is a system which has the effect of simply shifting the phase by $\pi/2$, then we represent this by the imaginary number j. So $1\underline{/0} \times j = 1\underline{/\pi/2}$. This system is shown in Fig. 16.3(a). Supposing now we consider a system

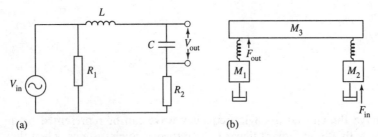

Figure 16.3 (a) A system which produces a phase shift of $\pi/2$, i.e. rotates a phasor by $\pi/2$. This may be represented as a multiplication by j. (b) A system consisting of two subsystems, both of which produce a phase shift of $\pi/2$, giving a combined shift of π. As a phase shift of π inverts a wave, i.e. $\cos(\omega t + \pi) = -\cos(\omega t)$, this is equivalent to multiplication by -1. Hence $j \times j = -1$.

which can be broken into two components, both of which shift the phase by $\pi/2$ as shown in Fig. 16.3(b). The combined effect of the two systems is to multiply the input by $j \times j$. The final output wave, shifted now by π, is $\cos(\omega t + \pi) = -\cos(\omega t)$ so it is -1 times the initial input. For this to be so, $j \times j = -1$. This is the central definition for complex numbers:

$$j \times j = -1$$

meaning that $j = \sqrt{-1}$, where j is an operator which rotates a phasor by $\pi/2$.

We shall return to these linear time-invariant systems in Chapter 20.

16.3 COMPLEX NUMBERS AND OPERATIONS

Complex numbers allow us to find solutions to all quadratic equations. Equations like $x^2 + 4 = 0$ do not have real roots because

$$x^2 + 4 = 0 \Leftrightarrow x^2 = -4$$

and there is no real number which if squared will give -4.

If we introduce new numbers by using $j = \sqrt{-1}$, then a solution to $x^2 + 4 = 0$ is $x = j2$. $j2$ is an imaginary number. To check that $j2$ is in fact a solution to $x^2 + 4 = 0$, substitute it into the equation $x^2 + 4 = 0$, giving

$$(j2)^2 + 4 = 0$$

$$\Leftrightarrow j^2(2)^2 + 4 = 0$$

$$\Leftrightarrow (-1)(4) + 4 = 0 \qquad (\text{using } j^2 = -1)$$

$$\Leftrightarrow 0 = 0$$

which is true. Therefore $x = j2$ is a solution. In order to solve all possible quadratic equations we need to use complex numbers, that is numbers that have both real and imaginary parts.

Mathematicians often use i instead of j to represent $\sqrt{-1}$. However, j is used in engineering work to avoid confusion with the symbol for current.

Real and imaginary parts and the complex plane

A complex number, z, can be written as the sum of its real and imaginary parts:

$$z = a + jb$$

where a and b are real numbers.

The real part of z is a ($\text{Re}(z) = a$).

The imaginary part of z is b ($\text{Im}(z) = b$).

Complex numbers can be represented in the complex plane (often called an Argand diagram) as the points (x, y) where

$$z = x + jy$$

e.g. $z = 1 - j2$ is shown in Fig. 16.4. The methods used for visualizing and adding and subtracting complex numbers are the same as those used for two-dimensional vectors in Chapter 15.

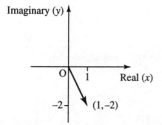

Figure 16.4 The number $z = 1 - j2$. The real part is plotted along the x-axis and the imaginary part along the y-axis.

Equality of two complex numbers

Two complex numbers can only be equal if their real parts are equal and their imaginary parts are equal.

Example 16.1 If $a - 2 + jb = 6 + j2$, where a and b are known to be real numbers, find a and b.

SOLUTION

$$a - 2 + jb = 6 + j2$$

We know that a and b are real, so

$$a - 2 = 6 \qquad \text{(real parts must be equal)}$$

$$\Leftrightarrow a = 8$$

$$b = 2 \qquad \text{(imaginary parts must be equal)}$$

Check by substituting $a = 8$ and $b = 2$ into

$$a - 2 + jb = 6 + j2$$

giving

$$8 - 2 + j2 = 6 + j2$$

$$\Leftrightarrow 6 + j2 = 6 + j2$$

which is correct.

Addition of complex numbers

To add complex numbers, add the real parts and the imaginary parts.

Example 16.2 Given $z_1 = 3 + j4$ and $z_2 = 1 - j2$, find $z_1 + z_2$.

SOLUTION

$$z_1 + z_2 = 3 + j4 + 1 - j2 = (3 + 1) + j(4 - 2) = 4 + j2$$

On the Argand diagram the numbers add like vectors by the parallelogram law, as in Fig. 16.5.

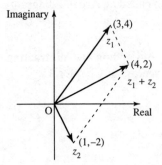

Figure 16.5 Adding two complex numbers using the parallelogram law.

Subtraction of complex numbers

To subtract complex numbers we subtract the real and imaginary parts.

Example 16.3 Given $z_1 = 3 + j4$ and $z_2 = 1 - j2$, find $z_1 - z_2$.

SOLUTION

$$z_1 - z_2 = 3 + j4 - (1 - j2) = 3 - 1 + j(4 - - 2) = 2 + j6$$

On the Argand diagram, reverse the vector z_2 to give $-z_2$ and then add z_1 and $(-z_2)$, as in Fig. 16.6.

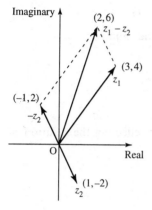

Figure 16.6 To find $z_1 - z_2$ on an Argand diagram, reverse vector z_2 to give $-z_2$ and then add, giving $z_1 + -z_2$.

Multiplication of complex numbers

To multiply complex numbers, multiply out the brackets, as for any other expression, and remember that $j^2 = -1$.

Example 16.4 Given $z_1 = 3 + j4$ and $z_2 = 1 - j2$, find $z_1 z_2$.

SOLUTION

$$z_1 z_2 = (3 + j4)(1 - j2) = 3 + j4 + 3(-j2) + (j4)(-j2)$$
$$= 3 + j4 - j6 - j^2 8$$
$$= 3 - j2 + 8 \quad (\text{using } j^2 = -1)$$
$$= 11 - j2$$

Example 16.5 Find $(4 - j2)(8 - j)$.

SOLUTION Multiplying as before gives

$$(4 - j2)(8 - j) = 32 - j16 - j4 + (-j2)(-j)$$
$$= 32 - j20 + j^2 2$$

Using $j^2 = -1$ gives $32 - j20 - 2 = 30 - j20$.

The complex conjugate

The complex conjugate of a number, $z = x + jy$, is the number with equal real part and the imaginary part negated. This is represented by z^*:

$$z^* = x - jy$$

A number multiplied by its conjugate is always real and positive (or zero). For example,

$$z = 3 + j4 \Rightarrow z^* = 3 - j4$$

$$zz^* = (3 + j4)(3 - j4) = (3)(3) + (j4)3 + 3(-j4) + (j4)(-j4)$$

$$= 9 + j12 - j12 - j^2 16$$

$$= 9 - j^2 16$$

$$= 9 + 16 = 25 \qquad (\text{using } j^2 = -1)$$

Note that the conjugate of the conjugate takes you back to the original number:

$$z = 3 + j4$$

$$z^* = 3 - j4$$

$$z^{**} = 3 + j4 = z$$

The conjugate of a number can be found on an Argand diagram by reflecting the position of the number in the real axis (see Fig. 16.7).

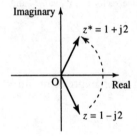

Figure 16.7 The complex conjugate of a number can be found by reflecting the number in the real axis. The diagram shows $1 - j2$ and its conjugate $1 + j2$.

Example 16.6 Find complex conjugates of the following and show that zz^* is real and positive, or zero, in each case:

(a) $2 - j5$ (b) $-4 + j2$ (c) -5 (d) $j6$
(e) $a + jb$, where a and b are real

SOLUTION
(a) The conjugate of $2 - j5$ is

$$(2 - j5)^* = 2 + j5$$

Hence

$$zz^* = (2 - j5)(2 + j5) = (2)(2) + (-j5)2 + 2(j5) + (-j5)(j5)$$

$$= 4 - j10 + j10 - j^2 25$$

$$= 4 - j^2 25 \qquad (\text{using } j^2 = -1)$$

$$= 4 + 25 = 29$$

which is real and positive. We have shown that $2 - j5$ multiplied by its conjugate $2 + j5$ gives a real, positive number.

(b) $(-4+j2)* = -4-j2$

$$(-4+j2)(-4-j2) = (-4)(-4)+(j2)(-4)+(-4)(-j2)+(j2)(-j2)$$
$$= 16-j8+j8-j^24$$
$$= 16-j^24 \quad \text{(using } j^2 = -1)$$
$$= 16+4=20$$

which is real and positive.

(c) -5 is a real number and therefore its complex conjugate is the same: -5. $(-5)* = -5$, and $(-5)(-5) = 25$, which is real and positive.

(d) $(j6)* = -j6$

$$(j6)(-j6) = -j^236 = 36$$

which is real and positive.

(e) $(a+jb)* = a-jb$

$$(a+jb)(a-jb) = (a)(a)+(jb)(a)+(a)(-jb)+(jb)(-jb)$$
$$= a^2+jab-jab-j^2b^2 = a^2-j^2b^2$$

Using $j^2 = -1$ this gives a^2+b^2. As a and b are real, this must be a real number. Also, we know that the square of a real number is greater than or equal to 0. So a^2+b^2 is real, and it is positive if $a \neq 0$, $b \neq 0$ or zero if both a and b are zero.

It is a good idea to remember this last result that $a+jb$ multiplied by its conjugate, $a-jb$, gives a^2+b^2. That is, any number multiplied by its complex conjugate gives the sum of the square of the real part and the square of the imaginary part. This is the same as the value of the modulus of z squared, i.e.

$$zz* = |z|^2$$

Division of complex numbers

To divide complex numbers we use the fact that a number multiplied by its conjugate is real to transform the bottom line of the fraction to a real number. If we multiply the bottom line by its complex conjugate we must also multiply the top line in order not to change the value of the number.

Example 16.7 Given $z_1 = 3+j4$ and $z_2 = 1-j2$, find z_1/z_2.

$$\frac{z_1}{z_2} = \frac{3+j4}{1-j2} = \frac{(3+j4)(1+j2)}{(1-j2)(1+j2)}$$

Here we have multiplied the top and bottom lines by z_2^* to make the bottom line entirely real. Hence we get

$$\frac{-5+j10}{1+2^2} = \frac{-5+j10}{5} = \frac{-5}{5}+\frac{j10}{5} = -1+j2$$

Example 16.8 Find

$$\frac{-3+j2}{10+j5}$$

in the form $x+jy$.

SOLUTION Multiply the top and bottom lines by the complex conjugate of $10+j5$ to make the bottom line real:

$$\frac{-3+j2}{10+j5} = \frac{(-3+j2)(10-j5)}{(10+j5)(10-j5)}$$

$$= \frac{(-3)(10)+(j2)(10)+(-3)(-j5)+(j2)(-j5)}{10^2+5^2}$$

$$= \frac{-30+j20+j15-j^2 10}{125} = \frac{-30+j20+j15+10}{125} = \frac{-20+j35}{125}$$

$$= \frac{-20}{125} + \frac{j35}{125} = -0.16+j0.28$$

16.4 SOLUTION OF QUADRATIC EQUATIONS

In Chapter 2 we looked at solutions of $ax^2+bx+c=0$, where a, b and c are real numbers, and said that the solutions are given by the formula

$$x = \frac{-b \pm \sqrt{b^2-4ac}}{2a}$$

We discovered that there are no real solutions if $b^2-4ac<0$ because we would need to take the square root of a negative number. We can now find complex solutions in this case by using $j=\sqrt{-1}$.

Example 16.9 Solve the following where $x \in \mathbb{C}$, the set of complex numbers:

(a) $x^2+3x+5=0$ (b) $x^2-x+1=0$
(c) $4x^2-2x+1=0$ (d) $4x^2+1=0$

SOLUTION

(a) To solve $x^2+3x+5=0$, compare it with $ax^2+bx+c=0$, giving $a=1$, $b=3$ and $c=5$. Substitute these values in

$$x = \frac{-b \pm \sqrt{b^2-4ac}}{2a}$$

giving

$$x = \frac{-3 \pm \sqrt{3^2-4(1)(5)}}{2(1)} = \frac{-3 \pm \sqrt{-11}}{2}$$

We write $-11=(-1).11$, so

$$\sqrt{-11} = \sqrt{-1}\sqrt{11} \simeq j3.317 \qquad (\text{using } j=\sqrt{-1})$$

Therefore

$$x \simeq \frac{-3 \pm j3.317}{2} \Rightarrow x \simeq -1.5+j1.658 \vee x \simeq 1.5-j1.658$$

(b) To solve $x^2 - x + 1 = 0$, compare it with $ax^2 + bx + c = 0$, giving $a = 1$, $b = -1$ and $c = 1$. Substitute in

$$x = \frac{-b \pm \sqrt{b^2 - 4ac}}{2a}$$

giving

$$x = \frac{1 \pm \sqrt{1^2 - 4(1)(1)}}{2(1)} = \frac{1 \pm \sqrt{-3}}{2}$$

We write $-3 = (-1).3$ so

$$\sqrt{-3} = \sqrt{-1}\sqrt{3} \simeq j1.732 \qquad (\text{using } j = \sqrt{-1})$$

Therefore

$$x \simeq \frac{1 \pm j1.732}{2} \Leftrightarrow x \simeq 0.5 + j0.866 \vee x \simeq 0.5 - j0.866$$

(c) To solve $4x^2 - 2x + 1 = 0$, compare it with $ax^2 + bx + c = 0$, giving $a = 4$, $b = -2$ and $c = 1$. Substitute in

$$x = \frac{-b \pm \sqrt{b^2 - 4ac}}{2a}$$

giving

$$x = \frac{-(-2) \pm \sqrt{(-2)^2 - 4(4)(1)}}{2(4)} = \frac{2 \pm \sqrt{-12}}{8}$$

We write $-12 = (-1).12$, so

$$\sqrt{-12} = \sqrt{-1}\sqrt{12} \simeq j3.4641 \qquad (\text{using } j = \sqrt{-1})$$

Therefore

$$x \simeq \frac{2 \pm j3.4641}{8} \Leftrightarrow x \simeq 0.25 + j0.433 \vee x \simeq 0.25 - j0.433$$

(d) To solve $4x^2 + 1 = 0$, we could use the formula (as in the other cases), but it is quicker to do the following:

$$4x^2 + 1 \Leftrightarrow 4x^2 = -1 \qquad \text{(subtracting 1 from both sides)}$$

$$\Leftrightarrow x^2 = \frac{-1}{4} \qquad \text{(dividing both sides by 4)}$$

$$\Leftrightarrow x = \pm\sqrt{-0.25} \qquad \text{(taking the square root of both sides)}$$

As $\sqrt{-0.25} = \sqrt{-1}\sqrt{0.25} = j0.5$, we get $x = \pm j0.5 \Leftrightarrow x = j0.5 \vee x = -j0.5$.

If $ax^2 + bx + c = 0$ and the coefficients a, b and c are all real numbers then we find that the two roots of the equation, if they are not entirely real, must be the complex conjugates of each other. This is true for all the cases we looked at in Example 16.9.

$$x^2 + 3x + 5 = 0$$

has solutions $x \simeq -1.5 + j1.658$ and $x \simeq -1.5 - j1.658$.

$$x^2 - x + 1 = 0$$

has solutions $x \simeq 0.5 + j0.866$ and $x \simeq 0.5 - j0.866$.

$$4x^2 + 1 = 0$$

has solutions $x = j0.5$ and $x = -j0.5$.

We can show that if the coefficients a, b and c are real in the equation $ax^2 + bx + c = 0$ then the roots of the equation must either be real or complex conjugates of each other. We know from the formula that

$$ax^2 + bx + c = 0 \Leftrightarrow x = \frac{-b \pm \sqrt{b^2 - 4ac}}{2a}$$

If x has an imaginary part, then $b^2 - 4ac < 0$, so that $\sqrt{b^2 - 4ac}$ is an imaginary number. We can write this in terms of j as follows:

$$\sqrt{b^2 - 4ac} = \sqrt{-(4ac - b^2)} = \sqrt{-1}\sqrt{4ac - b^2} = j\sqrt{4ac - b^2}$$

So the solutions in the case $b^2 - 4ac < 0$ are

$$x = \frac{-b}{2a} \pm \frac{j\sqrt{4ac - b^2}}{2}$$

which can be written, using

$$p = \frac{-b}{2a}$$

and

$$q = \frac{\sqrt{4ac - b^2}}{2a}$$

as

$$x = p \pm jq \Leftrightarrow x = p - jq \vee x = p + jq$$

where p and q are real. This shows that the two solutions are complex conjugates of each other. This fact can be used to find the other root when one root is known.

Example 16.10 Given that the equation $x^2 - kx + 8 = 0$, where $k \in \mathbb{R}$, has one solution $x = 2 - j2$, then find the other solution and find the value of k.

SOLUTION We know that non-real solutions must be complex conjugates of each other, so if one solution is $x = 2 - j2$ the other one must be $x = 2 + j2$.

To find k we use the result that if an equation has exactly two solutions x_1 and x_2 then the equation must be equivalent to $(x - x_1)(x - x_2) = 0$.

We know that $x = 2 + j2$ or $x = 2 - j2$, so therefore the equation must be equivalent to

$$(x - (2 + j2))(x - (2 - j2)) = 0$$

Multiplying out the brackets gives

$$x^2 - x(2 + j2) - x(2 - j2) + (2 + j2)(2 - j2) = 0$$

$$\Leftrightarrow x^2 + x(-2 - j2 - 2 + j2) + (4 + 4) = 0$$

$$\Leftrightarrow x^2 - 4x + 8 = 0$$

Compare $x^2 - 4x + 8 = 0$ with $x^2 - kx + 8 = 0$. The coefficients of x^2 are equal in both cases, as are the constant terms, so the equations would be the same if $-k = -4 \Leftrightarrow k = 4$, giving the solution as $k = 4$.

16.5 POLAR FORM OF A COMPLEX NUMBER

From the Argand diagram in Fig. 16.8 we can see that we can express a complex number in terms of the length of the vector (the modulus) and the angle it makes with the x-axis (the argument). This is exactly the same process as that in expressing vectors in polar coordinates, as in Sec. 15.4.

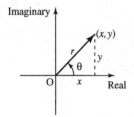

Figure 16.8 The number $x + jy$ can be expressed in polar form by the length of the line representing the number, r, and the angle it makes to the x-axis, θ; that is, $x + jy = r\underline{/\theta}$.

If $z = x + jy$, then z can be represented in polar form as $r\underline{/\theta}$, where

$$r^2 = x^2 + y^2$$

$$\tan(\theta) = \frac{x}{y}$$

Hence

$$r = \sqrt{x^2 + y^2}$$

$$\theta = \tan^{-1}(y/x) \qquad (+\pi \text{ if } x \text{ is negative})$$

We can write the complex number as

$$z = r\underline{/\theta}$$

r, the modulus of z, is also written as $|z|$.

As it is usual to give the angle between $-\pi$ and π, it may be necessary to subtract 2π from the angle given by this formula. As 2π is a complete rotation this will make no difference to the position of the complex number on the diagram.

Example 16.11 Express the following complex numbers in polar form:

(a) $3 + j2$ (b) $-2 - j5$
(c) $-4 + j2$ (d) $4 - j2$

SOLUTION To perform these conversions to polar form it is a good idea to draw a diagram of the number in order to be able to check that the angle is of the correct size (see Fig. 16.9).

(a) $3 + j2$ has modulus $r = \sqrt{3^2 + 2^2} \approx 3.61$ and the angle is given by $\tan^{-1}(2/3)$; therefore in polar form $3 + j2 \approx 3.61\underline{/0.59}$.

(b) $-2 - j5$ has modulus $r = \sqrt{(-2)^2 + (-5)^2} \approx 5.39$ and the angle is given by $\tan^{-1}(-5/-2) + \pi \approx 4.332$. As this angle is bigger than π, subtract 2π (a complete revolution)

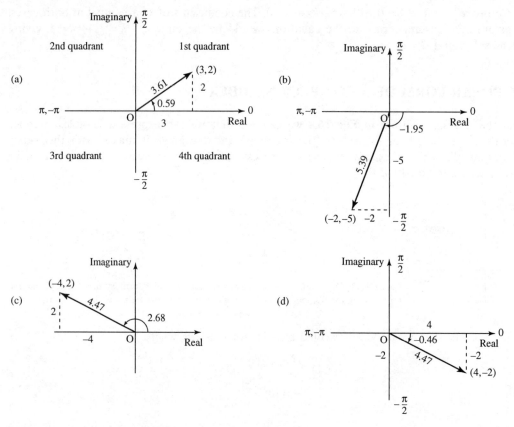

Figure 16.9 Conversion to polar form. (a) $3+j2$. (b) $-2-j5$. (c) $-4+j2$. (d) $4-j2$.

to give -1.95. Therefore in polar form $-2-j5 \simeq 5.39\underline{/-1.95}$. Note that the angle is between $-\pi$ and $-\pi/2$, meaning that the number must lie in the third quadrant, which we can see is correct from the diagram.

(c) $-4+j2$ has modulus $r=\sqrt{(-4)^2+2^2}\simeq 4.47$, and the angle is given by $\tan^{-1}(2/-4)\simeq -0.46+\pi \simeq 2.68$. Therefore, in polar form $-4+j2 \simeq 4.47\underline{/2.68}$. Note that the angle is between $\pi/2$ and π, meaning that the number must lie in the second quadrant, which we can see is correct from the diagram.

(d) $4-j2$ has modulus $r=\sqrt{4^2+(-2)^2}\simeq 4.47$ and the angle is given by $\tan^{-1}(-2/4)\simeq -0.46$. Therefore, in polar form $4-j2 \simeq 4.47\underline{/-0.46}$. Note that the angle is between $-\pi/2$ and 0, meaning that the number must lie in the fourth quadrant, which we can see is correct from the diagram.

Check the calculations by using the rectangular to polar conversion facility on your calculator.

Conversion from polar form to Cartesian (rectangular) form

If a number is given by its modulus and argument in polar form, $r\underline{/\theta}$, then we can convert back to Cartesian (rectangular) form using

$$x=r\cos(\theta) \quad \text{and} \quad y=r\sin(\theta)$$

This can be seen from Fig. 16.10, and examples are given in Fig. 16.11.

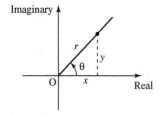

Figure 16.10 The number $r\underline{/\theta}$ can be written as $x+jy$. Using the triangle, $\cos(\theta)=x/r$, giving $x=r\cos(\theta)$. Also $\sin(\theta)=y/r$, giving $y=r\sin(\theta)$.

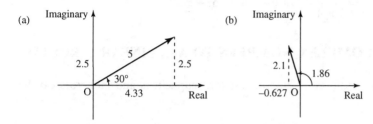

Figure 16.11 (a) $5\underline{/30°}$ in Cartesian (rectangular) form is given by $x=5\cos(30°)\simeq4.33$. $y=5\sin(30°)=2.5$, therefore $5\underline{/30°}\simeq4.33+j2.5$. (b) $2.2\underline{/1.86}$ in rectangular form is given by $x=2.2\cos(1.86)\simeq-0.627$ and $y=2.2\sin(1.86)\simeq2.1$. Therefore $2.2\underline{/1.86}\simeq-0.627+j2.1$.

Addition, subtraction, multiplication and division of complex numbers in polar form

To add and subtract two complex numbers, always express them first in rectangular form; that is, write them as $z=a+jb$. The result of the addition or subtraction can then be converted back to polar form.

To multiply two numbers in polar form, multiply the moduli and add the arguments.

To divide two numbers in polar form divide the moduli and subtract the arguments.

Example 16.12 Given $z_1=3\underline{/\pi/6}$ and $z_2=2\underline{/\pi/4}$, find z_1+z_2, z_1-z_2, z_1z_2 and z_1/z_2.

SOLUTION
To find $z_1\pm z_2$, use $r\underline{/\theta}=r(\cos(\theta)+j\sin(\theta))$:

$$z_1=3(\cos(\pi/6)+j\sin(\pi/6))\simeq2.5981+j1.5$$

$$z_2=2(\cos(\pi/4)+j\sin(\pi/4))\simeq1.4142+j1.4142$$

Then

$$z_1+z_2\simeq2.5981+j1.5+1.4142+j1.4142=4.0123+j2.9142$$

To express z_1+z_2 back in polar form, use $r=\sqrt{x^2+y^2}$ and $\theta=\tan^{-1}(y/x)(+\pi$ if x is negative):

$$r=\sqrt{(4.0123)^2+(2.9142)^2}\simeq4.959,\ \theta=\tan^{-1}(2.9142/4.0123)\simeq0.6282.$$

Hence $z_1+z_2\simeq4.959\underline{/0.6282}$.

To find z_1-z_2, we have already found (above) that $z_1=3\underline{/\pi/6}\simeq2.5981+j1.5$ and $z_2=2\underline{/\pi/4}\simeq1.4142+j1.4142$. Then

$$z_1-z_2\simeq2.5981+j1.5-(1.4142+j1.4142)=1.1839+j0.0858$$

To express z_1-z_2 back in polar form, use $r=\sqrt{x^2+y^2}$ and $\theta=\tan^{-1}(y/x)$ $(+\pi$ if x is negative). Then

$$z_1 - z_2 \simeq 1.1839 + j0.0858 \simeq 1.187\underline{/0.0723}.$$

To find $z_1 z_2$, multiply the moduli and add the arguments:

$$z_1 z_1 = 3\underline{/\pi/6} . 2\underline{/\pi/4} = 3.2\underline{/\pi/6 + \pi/4} = 6\underline{/5\pi/12}$$

To find z_1/z_2, divide the moduli and subtract the arguments:

$$\frac{z_1}{z_2} = \frac{3\underline{/\pi/6}}{2\underline{/\pi/4}} = \tfrac{3}{2}\underline{/\pi/6 - \pi/4} = 1.5\underline{/-\pi/12}$$

16.6 APPLICATIONS OF COMPLEX NUMBERS TO A.C. LINEAR CIRCUITS

An alternating current electrical circuit consisting of resistors, capacitors and inductors can be analysed using the relationship

$$V = ZI$$

where V is the voltage, Z is the impedance and I is the current and V, Z and I are all complex quantities.

Each component has a complex impedance associated with it, and for two components in series, as in Fig. 16.12(a), the resultant impedance Z_R is found by the formula

$$Z_R = Z_1 + Z_2$$

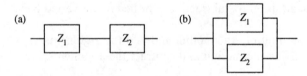

Figure 16.12 (a) Components in series. (b) Components in parallel.

For two components in parallel, as in Fig. 16.12(b), the resultant impedance is given by

$$\frac{1}{Z_R} = \frac{1}{Z_1} + \frac{1}{Z_2}$$

In the case of parallel circuit elements it may be easier to calculate the admittance, the reciprocal of the impedance $Y = 1/Z$. Then use the fact that for circuit elements in parallel

$$Y_R = Y_1 + Y_2$$

and then $V = I/Y$.

The real part of Z is called the resistance and the imaginary part is called the reactance. $Z = R + jS$ where R is the resistance in ohms and S is the reactance in ohms.

The impedances of circuit elements are as follows:

Resistor	$Z = R$	No reactive element
Capacitor	$Z = \dfrac{1}{j\omega C} = \dfrac{-j}{\omega C}$	Purely reactive
Inductor	$Z = j\omega L$	Purely reactive

where ω is the angular frequency of the source ($=2\pi f$), f is the frequency in Hz, R is the resistance (in ohms), C is the capacitance (in farads) and L is the impedance (in henries).

Example 16.13 Find the impedance of the circuit shown in Fig. 16.13(a) at 20 kHz, where $L=2$ mH, $C=100$ pF and $R=2000$ Ω.
Assuming a voltage amplitude of 300 V, calculate the current I and the relative phase.

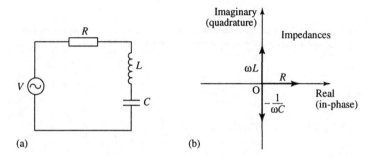

(a)   (b)

Figure 16.13 (a) The circuit for Example 16.13. (b) The impedances of the circuit elements shown on an Argand diagram. As the elements are in series, the resultant is found by taking the sum of the impedances of the components.

SOLUTION The impedances of the elements are shown in Fig. 16.13(b). As we are given that f, the frequency of the input, is 20 kHz, then

$$\omega = 2\pi f = 2\pi \times 20 \times 10^3$$

As the elements are in series we can sum the impedances, giving

$$Z = R + j\omega L - \frac{j}{\omega C} = 2000 + j2\pi \times 20 \times 10^3 \times 2 \times 10^{-3} - \frac{j}{2\pi \times 20 \times 10^3 \times 100 \times 10^{-6}}$$

giving

$$2000 + j\left(80\pi - \frac{1}{4\pi}\right) \simeq 2000 + j251.2$$

Expressing this in polar form gives $Z \simeq 2016 \underline{/7^\circ}$. Therefore from $V = ZI$ and the given $V = 300$,

$$I = \frac{300}{Z} = \frac{300}{2016\underline{/7^\circ}} \simeq 0.149\underline{/-7^\circ}$$

giving a current of magnitude of 0.149 A with a relative phase of -7°.

Example 16.14 One form of a 'tuned circuit' which can be used as a bandpass filter is given in Fig. 16.14(a). Given that $R = 300$ Ω, $L = 2$ mH and $C = 10$ pF, find the admittance of the circuit at 2 kHz. Given that the current source is of amplitude 12 A, find the voltage amplitude and its relative phase.

SOLUTION The admittances of the elements are shown in Fig. 16.14(b). As the elements are in parallel we can sum the admittances, giving

$$Y = \frac{1}{R} + j\omega C - \frac{j}{\omega L}$$

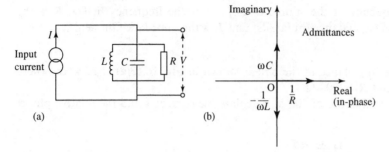

Figure 16.14 (a) The circuit for Example 16.14. (b) The admittances of the circuit elements shown on an Argand diagram. As the elements are in parallel, the resultant is found by taking the sum of the admittances of the components.

where $\omega = 2 \times 10^3 \times 2\pi$. This gives

$$Y = \frac{1}{300} + j(2 \times 10^3 \times 2\pi \times 10 \times 10^{-6}) - \frac{j}{2 \times 10^3 \times 2\pi \times 2 \times 10^{-3}}$$

$$\simeq 0.003\,333 + j\left(4\pi \times 10^{-2} - \frac{j}{8\pi}\right)$$

$$\simeq 0.003\,333 + j0.0859$$

Expressing this in polar form gives $Y = 0.086 \underline{/88°}$. Therefore from $V = ZI$ or $V = I/Y$ we have

$$V = \frac{12}{0.086 \underline{/88°}} \simeq 139.6 \underline{/-88°}$$

giving a voltage amplitude of 139.6 with a relative phase of $-88°$.

16.7 CIRCULAR MOTION

When we introduced the sine and cosine function in Chapter 11 we used the example of a rotating rod of length r. We plotted the position of the rod, its height and horizontal distance from the centre against the angle through which the rod had rotated. This defined the sine and cosine functions. We consider this problem again. This time we specify that the circular motion is at constant angular velocity ω. That is, the rate of change of the angle θ, $d\theta/dt$, is constant and equals ω. That is:

$$\frac{d\theta}{dt} = \omega \qquad \text{(where } \omega \text{ is a constant)}$$

$$\Leftrightarrow \theta = \omega t + \phi$$

where ϕ is the angle when $t = 0$. If we start with the rod horizontal, then $\phi = 0$ and we have $\theta = \omega t$. The (x, y) position of the tip of the rod of length r is given as a function of time by $x = r\cos(\omega t)$ and $y = r\sin(\omega t)$, where ω is the constant angular velocity, so the rotating vector is given by $\mathbf{r} = (r\cos(\omega t), r\sin(\omega t))$ where $\theta = \omega t$.

Considering now a ball on the end of a string with a constant angular velocity ω. Can we obtain an expression for its acceleration? The acceleration is of particular importance, because we know from Newton's second law that the force required to maintain the circuit motion can be found by using $\mathbf{F} = m\mathbf{a}$, where $\mathbf{F}$ is the force, m is the mass and $\mathbf{a}$ is the acceleration. We assume in this discussion that the effects of gravity and air resistance are negligible. The ball is being rotated in a plane which is vertical to the ground.

First, imagine lying flat on the ground in a line with the x-axis and watching the ball. It appears to oscillate back and forth and its position is given by $x = r\cos(\omega t)$. This is pictured in Fig. 16.15. Differentiating with respect to t gives the component of the velocity in the x-direction:

$$\frac{dx}{dt} = -r\omega \sin(\omega t)$$

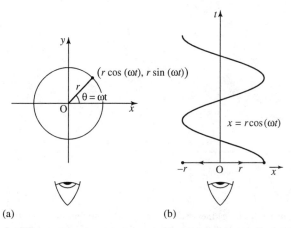

(a)   (b)

Figure 16.15 A ball on a string is moving round a circle with constant angular velocity so that $\theta = \omega t$. If we observe the ball from the position of the eye in (a) then we can only see the motion in the horizontal direction and the ball appears to oscillate back and forth as shown in (b). Also displayed is the graph of x against time, t.

The acceleration is the derivative of this velocity:

$$\frac{d}{dt}\left(\frac{dx}{dt}\right)$$

This is also written as d^2x/dt^2 (read as 'd two x by dt squared'), and it means the derivative of the derivative of x.

$$\frac{dx}{dt} = -r\omega \sin(\omega t)$$

Differentiating again gives

$$\frac{d^2x}{dt^2} = -r\omega^2 \cos(\omega t)$$

$$= -\omega^2(r\cos(\omega t))$$

and, as $x = r\cos(\omega t)$,

$$\frac{d^2x}{dt^2} = -\omega^2 x$$

This equation tells us that the horizontal acceleration is proportional to the horizontal distance from the origin in a direction towards the origin. This type of behaviour is called simple harmonic motion.

We can consider the movement in the y-direction by changing our point of view, as in Fig. 16.16. Again we can find the component of the acceleration, this time in the y-direction. We differentiate $y = r \sin(\omega t)$ to get the component of velocity in the y-direction:

$$\frac{dy}{dt} = r\omega \cos(\omega t)$$

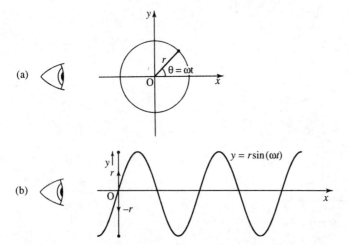

Figure 16.16 If we observe the ball from the position of the eye in (a) then we can only see the motion in the vertical direction and the ball appears to oscillate up and down, as shown in (b). Also shown is the graph of y against time, t: $y = r \sin(\omega t)$.

Differentiate again to find the acceleration:

$$\frac{d^2y}{dt^2} = -r\omega^2 \sin(\omega t)$$

and, as $y = r \sin(\omega t)$,

$$\frac{d^2y}{dt^2} = -\omega^2 y$$

Notice that this is the same equation as we had for x. We can represent the motion, both in the x-direction and in the y-direction, by using a complex number to represent the rotating vector. The real part of z represents the position in the x-direction and the imaginary part of z represents the position in the y-direction:

$$z = x + jy = r \cos(\omega t) + jr \sin(\omega t)$$

Then

$$\frac{dz}{dt} = -r\omega \sin(\omega t) + jr\omega \cos(\omega t)$$

The real part of dz/dt represents the component of velocity in the x-direction and the imaginary part represents the velocity in the y-direction. Again we can differentiate to find the acceleration:

$$\frac{d^2z}{dt^2} = -r\omega^2 \cos(\omega t) - jr\omega^2 \sin(\omega t)$$

$$= -\omega^2(r \cos(\omega t) + jr \sin(\omega t)) = -\omega^2 z$$

as $z = r \cos(\omega t) + jr \sin(\omega t)$. So we get

$$\frac{d^2z}{dt^2} = -\omega^2 z$$

This shows that the acceleration operates along the length of the vector z towards the origin, and it must be of magnitude $|-\omega^2 z| = \omega^2 r$, where r is the radius of the circle. The ball is always accelerating towards the centre of the circle. This also tells us the force that the string must provide in order to maintain the circular motion at constant angular velocity. The force towards the centre, called the centripetal force, must be $|\mathbf{F}| = m\omega^2 r$ where r is the radius of the circle and m is the mass of the ball. This has been given by Newton's second law $\mathbf{F} = m\mathbf{a}$.

We can use the equation for circular motion to show that it is possible to represent a complex number, z, in the form $z = re^{j\theta}$, where r is the modulus and θ the argument. To do this we must first establish the conditions which determine a particular solution to the equation $d^2z/dt^2 = -\omega^2 z$, where $\omega > 0$.

We know that one solution of the differential equation

$$\frac{d^2z}{dt^2} = -\omega^2 z$$

with the condition that $z = r$ when $t = 0$, is given by $z = r \cos(\omega t) + jr \sin(\omega t)$. Unfortunately, there is at least one other solution, given by the case where the string travels clockwise rather than anticlockwise, that is $z = r \cos(-\omega t) + jr \sin(-\omega t)$. However, we can pin down the solution to the anticlockwise direction of rotation by using the fact that we defined the angular velocity by $d\theta/dt = \omega$. This gives a condition on the initial velocity (at $t = 0$). From Fig. 16.17 we can see that the velocity must be positive and only have a component in the y-direction at $t = 0$. This discounts the possibility of the motion being clockwise as this would give a negative initial velocity. From $z = r \cos(\omega t) + jr \sin(\omega t)$:

$$\frac{dz}{dt} = -r\omega \sin(\omega t) + jr\omega \cos(\omega t)$$

and at $t = 0$, $dz/dt = jr\omega$. We now have enough information to say that $d^2z/dt^2 = -\omega^2 z$ and $z = r$ when $t = 0$ and $dz/dt = j\omega r$ when $t = 0$, $\Leftrightarrow z = r \cos(\omega t) + jr \sin(\omega t)$.

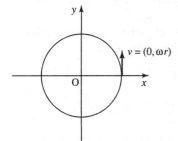

Figure 16.17 The initial velocity vector (as shown) must be in the positive y-direction if the motion is anticlockwise.

In Chapter 14 we looked at the exponential function and we found that $y = y_0 e^{kt}$ is a solution to the equation $dy/dt = ky$. This equation models the situation where the rate of change of the population is proportional to its current size: the first derivative of y is proportional to y. The equation

$$\frac{d^2z}{dt^2} = -\omega^2 z$$

is similar, only now the acceleration is related to z, i.e. the second derivative is proportional to z. As the exponential functions have the property that the derivative gives a scaled version of the original function, we must also get a scaled version of the original function if we differentiate twice. So we can true a solution of the form $z = re^{kt}$ for the equation $d^2z/dt^2 = -\omega^2 z$:

$$z = re^{kt}$$

$$\Rightarrow \frac{dz}{dt} = rke^{kt}$$

$$\Rightarrow \frac{d^2z}{dt^2} = rk^2 e^{kt}$$

Substituting into $d^2z/dt^2 = -\omega^2 z$, we get

$$rk^2 e^{kt} = -\omega^2 re^{kt}$$

Dividing both sides by re^{kt} gives

$$k^2 = -\omega^2$$

$$\Leftrightarrow k = \pm j\omega$$

This gives two possible solutions: $z = re^{j\omega t}$ when $k = j\omega$ and $z = re^{-j\omega t}$ when $k = -j\omega$. Again we can use the initial velocity to determine the solution. Using $z = re^{j\omega t}$ we get

$$\frac{dz}{dt} = jr\omega e^{j\omega t}$$

At $t = 0$ we get the velocity as $jr\omega$, which was one of the conditions we wanted to fulfil.

This shows that the two expressions $z = re^{j\omega t}$ and $z = r\cos(\omega t) + jr\sin(\omega t)$ both satisfy $d^2z/dt^2 = -\omega^2 z$ and $z = r$ when $t = 0$ and $dz/dt = j\omega r$ when $t = 0$. We have stated that these initial conditions are enough to determine the solution of the differential equation. So the only possibility is that

$$re^{j\omega t} = r\cos(\omega t) + jr\sin(\omega t)$$

This shows the equivalence of the polar form of a complex number and the exponential form. Replacing ωt by θ we get $re^{j\theta} = r\cos(\theta) + jr\sin(\theta)$, which we recognize as the polar form for a complex number: $z = r\underline{/\theta}$.

We can represent any complex number $z = x + jy$ in the form $re^{j\theta}$, where r is the modulus and θ is the argument. r and θ are found, as given before for the polar form, by

$$r = \sqrt{x^2 + y^2}$$

$$\theta = \tan^{-1}\left(\frac{y}{x}\right) \qquad (+\pi \text{ if } x \text{ is negative})$$

Conversely, to express a number given in exponential form in rectangular (Cartesian) form we can use

$$re^{j\theta} = r\cos(\theta) + jr\sin(\theta)$$

Example 16.15 Show that $z = 2e^{j3t}$ is a solution to

$$\frac{d^2z}{dt^2} = -9z$$

where $z=2$ when $t=0$ and $dz/dt=j6$ when $t=0$.

SOLUTION If $z=2e^{j3t}$, then when $t=0$, $z=2e^0=2$.

$$\frac{dz}{dt}=2(j3)e^{j3t}=j6e^{j3t}$$

when $t=0$, $dz/dt=j6e^0=j6$. Hence

$$\frac{d^2z}{dt^2}=j6(j3)e^{j3t}$$

$$\Rightarrow\frac{d^2z}{dt^2}=-18e^{j3t}$$

Substituting into $d^2z/dt^2=-9z$ gives $-18e^{j3t}=-9(2e^{j3t})\Leftrightarrow-18e^{j3t}=-18e^{j3t}$, which is true.

Example 16.16 Express $z=3+j4$ in exponential form.

SOLUTION The modulus r is given by $r=\sqrt{3^2+4^2}=\sqrt{25}=5$. The argument is $\tan^{-1}(4/3)\simeq$
0.9273. Hence $z\simeq5e^{j0.9273}$.

Example 16.17 Find the real and imaginary parts of the following

(a) $3e^{j\pi/2}$ (b) e^{-j} (c) e^{3+j2}
(d) $e^{-j(j-1)}$ (e) j^j

SOLUTION
(a) Use $re^{j\theta}=r\cos(\theta)+jr\sin(\theta)$. $3e^{j\pi/2}$ has $r=3$ and $\theta=\pi/2$, so

$$3e^{j\pi/2}=3\cos(\pi/2)+j3\sin(\pi/2)$$

$$=3.0+j(3)(1)=j3$$

$$=j3$$

The real part is 0 and the imaginary part is 3.
(b) Comparing e^{-j} with $re^{j\theta}$ gives $r=1$ and $\theta=-1$. Using $re^{j\theta}=r\cos(\theta)+jr\sin(\theta)$:

$$e^{-j}=1\cos(-1)+j\sin(-1)$$

$$\simeq0.5403-j0.8415$$

So the real part of e^{-j} is approximately 0.5403 and the imaginary part is approximately
-0.8415.
(c) Notice that e^{3+j2} is not in the form $re^{j\theta}$, because the exponent has a real part. We
therefore split the exponent into its real and imaginary parts by using the rules of powers:

$$e^{3+j2}=e^3e^{j2}$$

$e^3\simeq20.09$ is a real number, and the remaining exponent $j2$ is purely imaginary:

$$e^3e^{j2}\simeq20.09e^{j2}$$

Comparing $e^3 e^{j^2}$ with $re^{j\theta}$ gives $r = e^3$ and $\theta = 2$. Using $re^{j\theta} = r\cos(\theta) + jr\sin(\theta)$ gives

$$e^3 e^{j^2} = e^3 \cos(2) + je^3 \sin(2)$$

$$\simeq -8.359 + j18.26$$

The real part of e^{3+j^2} is approximately -8.359 and the imaginary part is approximately 18.26.

(d) For $e^{-j(j-1)}$ we need first to write the exponent in a form that allows us to split it into its real and imaginary bits. So we remove the brackets, giving

$$e^{-j(j-1)} = e^{-j^2 + j}$$

Using $j^2 = -1$ this gives

$$e^{-j(j-1)} = e^{1+j}$$

Now, using the rules of powers we can write $e^{1+j} = e^1 e^j$. e^1 is a real number and the exponent of e^j is purely imaginary. Comparing $e^1 e^j$ with $re^{j\theta}$ gives $r = e^1$ and $\theta = 1$. Using $re^{j\theta} = r\cos(\theta) + jr\sin(\theta)$ gives

$$e^1 e^{j1} = e^1 \cos(1) + je^1 \sin(1)$$

$$\simeq 1.469 + j2.287$$

The real part of $e^{-j(j-1)}$ is approximately 1.469 and the imaginary part is approximately 2.287.

(e) j^j looks a very confusing number as we have only dealt with complex powers when the base is e. We therefore begin by looking for a way of rewriting the expression so that its base is e. To do this we write the base, j, in exponential form. From the Argand diagram in Fig. 16.18, as j is represented by the point $(0,1)$ we can see that it has modulus 1 and argument $\pi/2$. Jence $j = e^{j\pi/2}$.

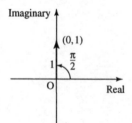

Figure 16.18 j is represented by the point $(0,1)$ in the complex plane. Therefore it has modulus 1 and argument $\pi/2$, i.e. $j = e^{j\pi/2}$.

Use this to replace the base in the expression j^j, so that

$$j^j = (e^{j\pi/2})^j = e^{(j^2 \pi/2)} = e^{-\pi/2}$$

We can now see that j^j is in fact a real number!

$$j^j = e^{-\pi/2} \simeq 0.2079$$

The real part of j^j is approximately 0.2079 and the imaginary part is 0.

16.8 THE IMPORTANCE OF BEING EXPONENTIAL

We have shown that

$$re^{j\theta} = r\cos(\theta) + jr\sin(\theta) = r\underline{/\theta}$$

Here r is the modulus of the complex number and θ is the argument. Therefore exponential form is simply another way of writing the polar form. The advantage of the exponential form is its simplicity. For circular motion, at constant regular velocity, it can represent the motion both in the real and imaginary (x and y) directions in one simple expression, $re^{j\omega t}$. The rules for multiplication and powers of complex numbers in exponential form are given by the rules of powers, as for any other number, given in Chapter 4, thus confirming the rules that we gave for polar form.

Multiplication:

$$r_1 e^{j\theta_1} r_2 e^{j\theta_2} = r_1 r_2 e^{j(\theta_1 + \theta_2)}$$

That is, we multiply the moduli and add the arguments.

Division:

$$\frac{r_1 e^{j\theta_1}}{r_2 e^{j\theta_2}} = \frac{r_1}{r_2} e^{j(\theta_1 - \theta_2)}$$

That is, we divide the moduli and subtract the arguments.

The complex conjugate of a number $re^{j\theta}$ is the number of same modulus and negative argument. That is, the complex conjugate of $re^{j\theta}$ is $re^{-j\theta}$.

Powers:

$$(re^{j\theta})^n = r^n e^{jn\theta}$$

That is, to raise a complex number to the power n, take the nth power of its modulus and multiply its argument by n.

Example 16.18 Given $z_1 = 3e^{j\pi/6}$ and $z_2 = 2e^{j\pi/4}$, find $z_1 + z_2, z_1 - z_2, z_1 z_2, z_1/z_2, z_1^* z_2^*$ and z_1^3.

SOLUTION

To find $z_1 + z_2$, use $re^{j\theta} = r(\cos(\theta) + j\sin(\theta))$:

$$z_1 = 3e^{j\pi/6} = 3(\cos(\pi/6) + j\sin(\pi/6)) \simeq 2.5981 + j1.5$$

$$z_2 = 2e^{j\pi/4} = 2(\cos(\pi/4) + j\sin(\pi/4)) \simeq 1.4142 + j1.4142$$

Then

$$z_1 + z_2 \simeq 2.5981 + j1.5 + 1.4142 + j1.4142$$

$$= 4.0123 + j2.9142$$

To express $z_1 + z_2$ back in exponential form, use $r = \sqrt{x^2 + y^2}$ and $\theta = \tan^{-1}(y/x)$ ($+\pi$ if x is negative):

$$r = \sqrt{(4.0123)^2 + (2.9142)^2} \simeq 4.959$$

and

$$\theta = \tan^{-1}\left(\frac{2.9142}{4.0123}\right) \simeq 0.6282$$

Hence $z_1 + z_2 \simeq 4.959 e^{j0.6282}$.

To find $z_1 - z_2$, we have already found (above) that $z_1 = 3e^{j\pi/6} \simeq 2.5981 + j1.5$ and $z_2 = 2e^{j\pi/4} \simeq 1.4142 + j1.4142$. Then

$$z_1 - z_2 \simeq 2.5981 + j1.5 - (1.4142 + j1.4142)$$

$$= 1.1839 + j0.0858$$

To express $z_1 - z_2$ back in polar form, use $r = \sqrt{x^2 + y^2}$ and $\theta = \tan^{-1}(y/x)(+\pi$ if x is negative):

$$z_1 - z_2 = 1.1839 + j0.0858 \simeq 1.187 e^{j0.0723}$$

To find $z_1 z_2$, multiply the moduli and add the arguments:

$$z_1 z_2 = 3e^{j\pi/6} 2e^{j\pi/4} = 3.2 e^{j(\pi/6 + \pi/4)} = 6e^{j5\pi/12}$$

To find z_1/z_2, divide the moduli and subtract the arguments:

$$\frac{z_1}{z_2} = \frac{3e^{j\pi/6}}{2e^{j\pi/4}} = \tfrac{3}{2} e^{j(\pi/6 + \pi/4)} = 1.5 e^{-j\pi/12}$$

To find $z_1^* z_2^*$, we find the complex conjugates of z_1 and z_2:

$$(z_1)^* = (3e^{j\pi/6})^* = 3e^{-j\pi/6}$$

$$(z_2)^* = (2e^{j\pi/4})^* = 2e^{-j\pi/4}$$

$$z_1^* z_2^* = 3e^{-j\pi/6} 2e^{-j\pi/4} = 3.2 e^{-j(\pi/6 + \pi/4)} = 6e^{-j5\pi/12}$$

To find z_1^3

$$z_1^3 = (3e^{j\pi/6})^3 = 3^3 e^{j\pi/6 \times 3} = 27 e^{j\pi/2}$$

Expressions for the trigonometric functions

From the exponential form we can find expressions for the cosine or sine in terms of complex numbers. We begin with a complex number z of modulus 1 and its complex conjugate:

$$e^{j\theta} = \cos(\theta) + j\sin(\theta) \tag{16.1}$$

$$e^{-j\theta} = \cos(\theta) - j\sin(\theta) \tag{16.2}$$

From these we can find the expressions for the cosine and sine in terms of the complex exponential. Adding Eqs (16.1) and (16.2) gives

$$e^{j\theta} + e^{-j\theta} = \cos(\theta) + j\sin(\theta) + \cos(\theta) - j\sin(\theta)$$

$$\Leftrightarrow e^{j\theta} + e^{-j\theta} = 2\cos(\theta)$$

Dividing both sides by 2 gives

$$\tfrac{1}{2}(e^{j\theta} + e^{-j\theta}) = \cos(\theta)$$

$$\Leftrightarrow \cos(\theta) = \tfrac{1}{2}(e^{j\theta} + e^{-j\theta})$$

Subtracting Eq. (16.2) from Eq. (16.1), we get

$$e^{j\theta} - e^{-j\theta} = \cos(\theta) + j\sin(\theta) - (\cos(\theta) - j\sin(\theta))$$

$$\Leftrightarrow e^{j\theta} - e^{-j\theta} = 2j\sin(\theta)$$

Dividing both sides by 2j gives

$$\frac{1}{2j}(e^{j\theta} - e^{-j\theta}) = \sin(\theta)$$

$$\Leftrightarrow \sin(\theta) = \frac{1}{2j}(e^{j\theta} - e^{-j\theta})$$

So we have

$$\cos(\theta) = \tfrac{1}{2}(e^{j\theta} + e^{-j\theta})$$

$$\sin(\theta) = \frac{1}{2j}(e^{j\theta} - e^{-j\theta})$$

Using $\tan(\theta) = \sin(\theta)/\cos(\theta)$ we get

$$\tan(\theta) = \frac{1}{j}\left(\frac{e^{j\theta} - e^{-j\theta}}{e^{j\theta} + e^{-j\theta}}\right)$$

Compare these with the definition of the sinh, cosh and tanh functions, given in Chapter 14:

$$\cosh(\theta) = \tfrac{1}{2}(e^{\theta} + e^{-\theta})$$

$$\sinh(\theta) = \tfrac{1}{2}(e^{\theta} - e^{-\theta})$$

$$\tanh(\theta) = \frac{(e^{\theta} - e^{-\theta})}{e^{\theta} + e^{-\theta}}$$

We see that:

$$\cos(j\theta) = \cosh(\theta)$$

$$\sin(j\theta) = j\,\sinh(\theta)$$

$$\tan(j\theta) = j\,\tanh(\theta)$$

De Moivre's theorem

Using the expression for the complex number in terms of a sine and cosine, $re^{j\theta} = r(\cos(\theta) + j\,\sin(\theta))$, and using this in $(re^{j\theta})^n = r^n e^{jn\theta}$, we get $(r(\cos(\theta) + j\,\sin(\theta))^n = r^n(\cos(n\theta) + j\,\sin(n\theta))$. This is called De Moivre's theorem and can be used to obtain multiple angle formulae.

Example 16.19 Find $\sin(3\theta)$ in terms of powers of $\sin(\theta)$ and $\cos(\theta)$.

SOLUTION We use the fact that $\sin(3\theta) = \text{Im}(\cos(3\theta) + j\,\sin(3\theta))$, where $\text{Im}(\ )$ represents 'the imaginary part of'. Hence

$$\sin(3\theta) = \text{Im}(e^{j3\theta})$$

$$= \text{Im}((\cos(\theta) + j\,\sin(\theta))^3)$$

Expanding:

$$(\cos(\theta) + j\,\sin(\theta))^3 = (\cos(\theta) + j\,\sin(\theta))(\cos(\theta) + j\,\sin(\theta))^2$$

$$= (\cos(\theta) + j\,\sin(\theta))(\cos^2(\theta) + 2j\,\cos(\theta)\sin(\theta) + j^2\,\sin^2(\theta))$$

$$= \cos^3(\theta) + j\,\sin(\theta)\cos^2(\theta) + 2j\,\cos^2(\theta)\sin(\theta)$$

$$+ j^2\,\cos(\theta)\sin^2(\theta) + 2j^2\,\cos(\theta)\sin^2(\theta) + j^3\,\sin^3(\theta)$$

$$= \cos^3(\theta) + 3j\,\sin(\theta)\cos^2(\theta) - 3\,\cos(\theta)\sin^2(\theta) - j\,\sin^3(\theta)$$

$$= \cos^3(\theta) - 3\,\cos(\theta)\sin^2(\theta) + j(3\,\sin(\theta)\cos^2(\theta) - \sin^3(\theta))$$

As $\sin(3\theta) = \text{Im}((\cos(\theta) + j\,\sin(\theta))^3)$, we take the imaginary part of the expression we have found, giving

$$\sin(3\theta) = 3\,\sin(\theta)\cos^2(\theta) - \sin^3(\theta)$$

Example 16.20 Express $\cos^3(\theta)$ in terms of cosine of multiples of θ.

SOLUTION Using $\cos(\theta) = \frac{1}{2}(e^{j\theta} + e^{-j\theta})$ and the expansion $(a+b)^3 = a^3 + 3a^2b + 3ab^2 + b^3$:

$$\cos^3(\theta) = (\tfrac{1}{2}(e^{j\theta} + e^{-j\theta}))^3$$

$$= \frac{1}{2^3}(e^{j3\theta} + 3e^{j\theta} + 3e^{-j\theta} + e^{-j3\theta})$$

$$= \frac{1}{2^2}(\tfrac{1}{2}(e^{j3\theta} + e^{-j3\theta}) + \tfrac{3}{2}(e^{j\theta} + e^{-j\theta}))$$

As $\cos(\theta) = \frac{1}{2}(e^{j\theta} + e^{-j\theta})$ and $\cos(3\theta) = \frac{1}{2}(e^{j3\theta} + e^{-j3\theta})$, we get

$$\cos^3(\theta) = \tfrac{1}{4}\cos(3\theta) + \tfrac{3}{4}\cos(\theta)$$

The exponential form can be used to solve complex equations of the form $z^n = c$, where c is a complex number. A particularly important example is the problem of finding all the solutions of $z^n = 1$, called the n roots of unity.

The n roots of unity

To solve the equation $z^n = 1$ we use the fact that 1 is a complex number with modulus 1 and argument 0, as can be seen in Fig. 16.19(a). However, we can also use an argument of 2π, 4π, 6π or any other multiple of 2π. As 2π is a complete revolution, adding 2π to the argument of any complex number does not change the position of the vector representing it and therefore does not change the value of the number.

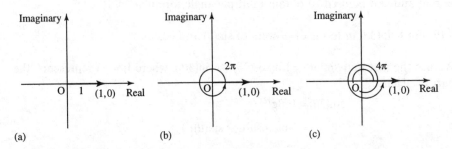

Figure 16.19 (a) 1 is the complex number e^{j0}, i.e. with a modulus of 1 and an argument of 0. (b) 1 can also be represented using an argument of 2π, i.e. $1 = e^{j2\pi}$. (c) 1 represented with an argument of 4π, i.e. $1 = e^{j4\pi}$.

The equation $z^n = 1$ can be expressed as

$$z^n = e^{j2\pi N} \qquad \text{where } N \in \mathbb{Z}$$

We can solve this equation by taking the nth root of both sides, which is the same as taking both sides to the power $1/n$:

$$(z^n)^{1/n} = e^{j2\pi N/n} \qquad \text{where } N \in \mathbb{Z}$$

We can substitute some values for N to find the various solutions, also using the fact that there should be n roots to the equation $z^n = 1$ so that we can stop after finding all n roots.

Example 16.21 Find all the solutions to $z^3 = 1$.

SOLUTION Write 1 as a complex number with argument $2\pi N$, giving the equation as

$$z^3 = e^{j2\pi N} \qquad \text{where } N \in \mathbb{Z}$$

Taking the cube root of both sides:

$$(z^3)^{1/3} = e^{j2\pi N/3} \qquad \text{where } N \in \mathbb{Z}$$

Substituting $N = 0$: $z = e^{j2\pi 0/3} = 1$
Substituting $N = 1$: $z = e^{j2\pi 1/3} = e^{j2\pi/3}$
Substituting $N = 2$: $z = e^{j2\pi 2/3} = e^{j4\pi/3}$
There is no need to use any more values of N. We use the fact that there should be three roots of a cubic equation. If we continued to substitute values for N then the values will begin to repeat. For example, substituting $N = 3$ gives $z = e^{j2\pi 3/3} = e^{j2\pi}$, which we know is the same as e^{j0} (subtracting 2π from the argument) which equals 1, a root that we have already found.

The solutions to $z^3 = 1$ are shown on an Argand diagram in Fig. 16.20. The principal root of a complex equation is the one found nearest to the position of the positive x-axis. Notice that in the case of $z^3 = 1$ the principal root is 1. Any solution can be obtained from another by rotation through $2\pi/3$. Hence another way of finding the n roots of $z^n = 1$ is to start with the principal root of $z = 1 = e^{j0}$ and add on multiples of $2\pi/n$ to the argument, in order to find the other roots.

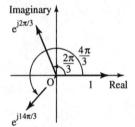

Figure 16.20 The solutions to $z^3 = 1$ are $z = 1, z = e^{j2\pi/3}, z = e^{j4\pi/3}$. Notice that one solution can be obtained from another by rotation through $2\pi/3$.

Example 16.22 Find all the roots of $z^5 = 1$.

SOLUTION One root, the principal root, is $z = 1 = e^{j0}$. The other roots can be found by rotating this around the complex plane by multiples of $2\pi/5$. Therefore we have the solutions:

$$z = 1, \ e^{j2\pi/5}, \ e^{j4\pi/5}, \ e^{j6\pi/5}, \ e^{j8\pi/5}$$

These are shown in Fig. 16.21.

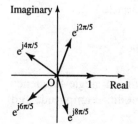

Figure 16.21 The solutions to $z^5 = 1$ are $z = 1, e^{j2\pi/5}, e^{j4\pi/5}, e^{j6\pi/5}, e^{j8\pi/5}$. Notice that one solution can be obtained from another by rotation through $2\pi/5$.

Solving some other complex equations

If we have the equation $z^n = c$ where c is any complex number, then we write the right-hand side of the equation in exponential form and use the fact that we can add a multiple of 2π to the argument without changing the value of the number. Write $c = re^{j\theta} = re^{j(\theta + 2\pi N)}$ where $N \in \mathbb{Z}$:

$$z^n = re^{j(\theta + 2\pi N)}$$

$$\Leftrightarrow z = r^{1/n}e^{\frac{j(\theta + 2\pi N)}{n}}$$

taking the nth root of both sides.

Example 16.23 Solve $z^3 = -4 + j4\sqrt{3}$.

SOLUTION Write $-4 + j4\sqrt{3}$ in exponential form, $re^{j\theta}$:

$$r = \sqrt{(-4)^2 + (4\sqrt{3})^2} = \sqrt{16 + 48} = \sqrt{64} = 8$$

$$= \tan^{-1}(-4\sqrt{3}/4) + \pi = 2\pi/3 \qquad (\text{using } \tan^{-1}(\sqrt{3}) = \pi/3)$$

So the equation becomes

$$z^3 = 8e^{j(2\pi/3 + 2\pi N)} \qquad \text{where } N \in \mathbb{Z}$$

$$\Leftrightarrow z = 8^{\frac{1}{3}}e^{\frac{j(2\pi/3 + 2\pi N)}{3}} \Leftrightarrow z = 2e^{j(2\pi/9 + 2\pi N/3)}$$

where $N \in \mathbb{Z}$.

Substituting some values for N gives

$$N = 0: \ z = 2e^{j2\pi/9}$$

$$N = 1: \ z = 2e^{j(2\pi/9 + 2\pi/3)} = 2e^{j8\pi/9}$$

$$N = 2: \ z = 2e^{j(2\pi/9 + 4\pi/3)} = 2e^{j14\pi/9}$$

The solutions are

$$z = 2e^{j2\pi/9}, \ z = 2e^{j8\pi/9}, \ z = 2e^{j14\pi/9}$$

These are shown in Fig. 16.22.

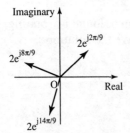

Figure 16.22 The solutions to $z^3 = -4 + j4\sqrt{3}$ are $z = 2e^{j2\pi/9}, 2e^{j8\pi/9}, 2e^{j14\pi/9}$. Notice that one solution can be obtained from another by rotation through $2\pi/3$.

16.9 SUMMARY

1. Simple systems can be represented by a complex number multiplying a single frequency input. The output then has the same frequency as the input, with a modified amplitude and a shifted phase.

2. j is the number which when multiplying a phasor has the effect of rotating the phase by $\pi/2$ (or 90°). $j \times j$ rotates the phase by π (180°) which is equivalent to multiplication by -1. Hence $j^2 = -1$ and $j = \sqrt{-1}$. Sometimes i is used instead of j to represent $\sqrt{-1}$.

3. A complex number is any number which can be represented on the complex plane. It can be written as $z = x + jy$ (x and y real), where x is the real part of z ($\text{Re}(z) = x$) and y is the imaginary part of z ($\text{Im}(z) = y$). A complex number expressed in the form $z = x + jy$ is said to be in Cartesian or rectangular form.

4. The complex conjugate of $a + jb$ is $a - jb$: $(a + jb)^* = a - jb$. The product of a number and its complex conjugate is always a real number greater than or equal to 0: $(a + jb)(a + jb)^* = (a + jb)(a - jb) = a^2 + b^2$ which is real and $\geqslant 0$.

$zz^* = |z|^2$ where z is any complex number: a number multiplied by its conjugate gives its modulus squared.

5. The operations of addition and subtraction of complex numbers are like those for vectors: simply add or subtract the real parts and then the imaginary parts. Multiply as follows, remembering $j^2 = -1$:

$$(1+j2)(-3-j3) = (1)(-3) + j2(-3) + 1(-j3) + (j2)(-j3)$$

$$= -3 - j6 - j3 - j^2 6$$

$$= -3 - j9 + 6 = 3 - j9$$

To divide, multiply the top and bottom lines by the complex conjugate of the bottom line, as follows:

$$\frac{1+j2}{-3-j3} = \frac{(1+j2)(-3+j3)}{(-3-j3)(-3+j3)} = \frac{-3-j6+j3-6}{(-3)^2+(3)^2} = \frac{-9-j3}{18} = \frac{-1}{2} - \frac{j1}{6}$$

6. All quadratic equations can be solved if $x \in \mathbb{C}$, i.e. x is a complex number. If $ax^2 + bx + c = 0$, where a, b and c are real numbers, then

$$x = \frac{-b \pm \sqrt{b^2 - 4ac}}{2a}$$

The solutions are real if $b^2 \geqslant 4ac$. If $b^2 < 4ac$, then the solutions can be written as $x = p \pm jq$, where

$$p = \frac{-b}{2a}$$

and

$$q = \frac{\sqrt{4ac - b^2}}{2a}$$

and therefore non-real roots are complex conjugates of each other.

7. Complex numbers can be written in polar form, $r \underline{/\theta}$ where r is the modulus of the number and θ is the argument. The modulus, r, is the length of the vector representing the complex number and θ is the angle made with the positive real axis.

If $z = x + jy = r \underline{/\theta}$, then

$$r = \sqrt{x^2 + y^2} \qquad \text{and} \qquad \theta = \tan^{-1}\left(\frac{y}{x}\right) \; (+\pi \text{ if } x < 0)$$

$$x = r\cos(\theta) \qquad \text{and} \qquad y = r\sin(\theta)$$

To add or subtract complex numbers expressed in polar form, first convert to rectangular form. To multiply, multiply the moduli and add the arguments, and to divide, divide the moduli and subtract the arguments:

$$r_1\underline{/\theta_1}\,r_2\underline{/\theta_2}=r_1r_2\underline{/\theta_1+\theta_2}$$

$$\frac{r_1\underline{/\theta_1}}{r_2\underline{/\theta_2}}=\frac{r_1}{r_2}\underline{/\theta_1-\theta_2}$$

8. Complex numbers are used in the analysis of alternating current (a.c.) circuits. ω is the angular frequency of the source. Resistors, capacitors and inductors have associated impedances, Z, where for a resistor $Z=R$, for a capacitor $Z=1/j\omega C$ and for an inductor $Z=j\omega L$, where R is the resistance, C is the capacitance and L is the inductance. Then the voltage and the current are related by

$$V=ZI$$

and the impedances of circuit elements obey

$$Z_R=Z_1+Z_2$$

for elements in series

$$\frac{1}{Z_R}=\frac{1}{Z_1}+\frac{1}{Z_2}$$

for elements in parallel, where Z_R is the resultant impedance. The admittance Y is the reciprocal of the impedance: $Y=1/Z$.

9. $d^2y/dt^2=-\omega^2y$ is a differential equation which defines waves as a function of time. This is an equation of motion where the acceleration is proportional to the distance from the origin. This is called simple harmonic motion. By examining the case of circular motion at constant angular velocity, where the rotating vector $z=x+jy=r\cos(\omega t)+jr\sin(\omega t)$ obeys this equation, we can show the equivalence of the polar representation of a complex wave and the exponential form:

$$re^{j\omega t}=r\cos(\omega t)+jr\sin(\omega t)=r\underline{/\omega t}$$

As $\theta=\omega t$, we have

$$re^{j\theta}=r\cos(\theta)+jr\sin(\theta)=r\underline{/\theta}$$

Here r is the modulus of the complex number and θ is the argument.

Replacing θ by $-\theta$ we also get

$$re^{-j\theta}=r\cos(\theta)-jr\sin(\theta)=r\underline{/-\theta}$$

From these, using the case where $r=1$, we can find the expressions for the cosine and sine in terms of the complex exponential:

$$\cos(\theta)=\tfrac{1}{2}(e^{j\theta}+e^{-j\theta})$$

$$\sin(\theta)=\frac{1}{2j}(e^{j\theta}-e^{-j\theta})$$

and

$$\tan(\theta)=\frac{1}{j}\left(\frac{e^{j\theta}-e^{-j\theta}}{e^{j\theta}+e^{-j\theta}}\right)$$

By comparing these with the definitions of the sinh, cosh and tanh functions we see that

$$\cos(j\theta) = \cosh(\theta)$$

$$\sin(j\theta) = j\sinh(\theta)$$

$$\tan(j\theta) = j\tanh(\theta)$$

10. The advantage of the exponential form is its simplicity. For circular motion, at constant angular velocity, it can represent the motion both in the real and imaginary (x and y) directions in one simple expression $re^{j\omega t}$. The rules for multiplication and powers of complex numbers in exponential form are given by the rules of powers, as for any other number, given in Chapter 4.

Multiplication:

$$r_1 e^{j\theta_1} r_2 e^{j\theta_2} = r_1 r_2 e^{j(\theta_1 + \theta_2)}$$

That is, we multiply the moduli and add the arguments.

Division:

$$\frac{r_1 e^{j\theta_1}}{r_2 e^{j\theta_2}} = \frac{r_1}{r_2} e^{j(\theta_1 - \theta_2)}$$

That is, we divide the moduli and subtract the arguments.

Powers:

$$(re^{j\theta})^n = r^n e^{jn\theta}$$

This last relationship can be used to show De Moivre's theorem. Using the expression for the complex number in terms of a sine and cosine, $re^{j\theta} = r(\cos(\theta) + j\sin(\theta))$, and using this in the expression above, we get

$$(r(\cos(\theta) + j\sin(\theta))^n = r^n(\cos(n\theta) + j\sin(n\theta))$$

The complex conjugate of $re^{j\theta}$ is $re^{-j\theta}$.

The derivative of a complex exponential is easy to find. As $d(e^t)/dt - e^t$, then $d(e^{j\omega t})/dt = j\omega e^{j\omega t}$

16.10 EXERCISES

16.1 Given $z_1 = 1 - 2j$, $z_2 = 3 + 3j$ and $z_3 = -1 + 4j$:

(a) Represent z_1, z_2 and z_3, on an Argand diagram.
(b) Find the following and show the results on the Argand diagram

    (i) $z_1 + z_2$ (ii) $z_3 - z_1$ (iii) z_1^*

(c) Calculate

    (i) $z_1 + z_2 + z_3$ (ii) $z_1 - z_3 + z_2$

    (iii) $z_1 z_1^*$ (iv) $\dfrac{z_1}{z_2}$ (v) $z_1 z_3$

16.2 Simplify

(a) j^8 (b) j^{11} (c) j^{28}

16.3 Find each of the following complex numbers in the form $a+jb$, where a and b are real:

(a) $(3-7j)(2+j4)$ (b) $(-1+2j)^2$ (c) $\dfrac{4-j3}{5-j}$

(d) $\dfrac{5+j3}{j(4-j9)} - \dfrac{6}{j}$

16.4 Find the real and imaginary parts of $z^2 + 1/z^2$, where

$$z = \frac{3+j}{2-j}$$

16.5 Given that x and y are real and that $2x-3+j(y-x)=x+j2$, find x and y.

16.6 Find the roots x_1 and x_2 of the following quadratic equations. In each case find the product $(x-x_1)(x-x_2)$, and show that the original equation is equivalent to $(x-x_1)(x-x_2)=0$.

(a) $x^2-3x+2=0$ (b) $-6+2x-x^2=0$
(c) $3x^2-x+1=0$ (d) $4x^2-7x-2=0$
(e) $2x^2+3=0$

16.7 The equation $x^2+bx+c=0$, where b and c are real numbers, has one complex root, $x=-1+j3$.

(a) What is the other root?
(b) Find b and c.

16.8 Convert the following to polar form:

(a) $3+j5$ (b) $-6+j3$ (c) $-4-j5$ (d) $-5-j3$

16.9 Express in rectangular (Cartesian) form:

(a) $5\underline{/225^\circ}$ (b) $4\underline{/330^\circ}$
(c) $2\underline{/2.723}$ (d) $5\underline{/-0.646}$

16.10 If x and y are real and $2x+y+j(2x-y)=15+j6$, find x and y.

16.11 If $z_1=12\underline{/3\pi/4}$ and $z_2=3\underline{/2\pi/5}$, find:

(a) z_1z_2 (b) $\dfrac{z_1}{z_2}$ (c) z_1+z_2
(d) z_2-z_1 (e) z_1^* (f) z_2^2

giving the results in polar form.

16.12 If $z=2\underline{/0.8}$, find z^4.

16.13 Find the impedance of the circuit shown in Fig. 16.23(a) at 90 kHz where $L=4$ mH, $C=20$ pF and $R=400$ kΩ. Assuming a current source of amplitude 5 A, calculate the voltage V and its relative phase.

16.14 Find the admittance of the circuit given in Fig. 16.23(b) at 20 kHz, given that $R=250$ kΩ, $L=20$ mH and $C=50$ pF. Given that the voltage source has amplitude 10 V, find the current I and its relative phase.

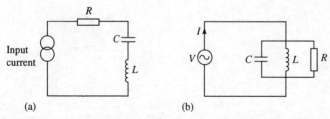

(a) (b)

Figure 16.23 (a) Circuit for Exercise 16.13. (b) Circuit for Exercise 16.14.

16.15 Feedback is applied to an amplifier such that

$$A' = \frac{A}{1 - \beta A}$$

where A', A and β are complex quantities. A is the amplifier gain, A' is the gain with feedback and β is the proportion of the output which has been fed back (see Fig. 16.24).

(a) If at 30 Hz $A = 500\underline{/180°}$ and $\beta = 0.005\underline{/160°}$, calculate A'.
(b) At a particular frequency it is desired to have $A' = 300\underline{/100°}$, where it is known that $A = 400\underline{/110°}$, Find the value of β necessary to achieve this gain modification.

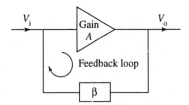

Figure 16.24 An amplifier with feedback, as in Exercise 16.15.

16.16 Write the following in exponential form:

(a) $3 + j5$ (b) $-6 + j3$ (c) $-4 - j5$
(d) $8\underline{/22°}$ (e) $3\underline{/-4.15}$ (f) $6(\cos(1.9) + j\sin(1.9))$

16.17 Express the following in polar form and in the form $a + jb$:

(a) $4e^{j2}$ (b) $e^{-j\pi/2}$ (c) $2e^{-j\pi}$ (d) $-6e^{j4}$
(e) $\frac{1}{2}e^{-j5}$ (f) $e^{j\pi/6}3e^{j3\pi/4}$ (g) $e^{j\pi/6} + 3e^{j3\pi/4}$

16.18 Find the real and imaginary parts of the following:

(a) $2e^{-j\pi}$ (b) $-3e^{j0.5}$ (c) $2.5e^{-2+j}$
(d) $5e^{j(3+j)}$ (e) $(3 - j4)^{2+j}$

16.19 Given $z_1 = 12e^{j3\pi/4}$ and $z_2 = 3e^{j2\pi/5}$, find:

(a) $z_1 z_2$ (b) z_1/z_2 (c) $z_1 + z_2$ (d) $z_2 - z_1$
(e) z_1^* (f) $\dfrac{z_1^*}{z_2}$ (g) $z_1 z_1^*$

giving the results in exponential form.

16.20 If $z = 3e^{j0.46}$ find z^3 in polar and exponential forms.

16.21 Find all the solutions of the following and show them on an Argand diagram:

(a) $z^4 = 1$ (b) $z^6 = -1$
(c) $z^5 + 32 = 0$ (d) $3z^3 + 2 = 0$

16.22 Find $\cos(3\theta)$ in powers of $\sin(\theta)$ and $\cos(\theta)$.

16.23 Express $\sin^3(\theta)$ in terms of sines of multiples of θ.

16.24 Find the fifth roots of $-2 + j3$ and represent the results on an Argand diagram.

16.25 Solve the following equations:

(a) $z^2 + 2jz - 2 = 0$ (b) $z^2 - 3jz = j$

16.26 Show that the following are solutions to the differential equation $d^2z/dt^2 = -\omega^2 z$ and find the value of ω in each case:

(a) $z = 2e^{-j4t}$ (b) $z = 4e^{j0.5t}$

SEVENTEEN

MAXIMA AND MINIMA AND SKETCHING FUNCTIONS

17.1 INTRODUCTION

Differentiation can be used to examine the behaviour of a function and to find regions where it is increasing or decreasing and where it has maximum and minimum values. For instance, we may be interested in finding the maximum height, maximum power or generating the maximum profit, or in finding ways to use the minimum amount of energy or minimum use of materials. Maximum and minimum points can also help in the process of sketching a function.

17.2 STATIONARY POINTS AND LOCAL MAXIMA AND MINIMA

Example 17.1 Throw a stone in the air and initially it will have a positive velocity as the height, s, increases; that is, $ds/dt > 0$. At some point it will start to fall back to the ground. The distance from the ground is then decreasing, and the velocity is negative, $ds/dt < 0$. In order to go from a positive velocity to a negative velocity there must be a turning point, where the stone is at its maximum height and the velocity is zero. If the stone has initial velocity of 20 m s^{-1}, how can we find the maximum height that it reaches?

In order to express the velocity of the stone we can make the assumption that air resistance is negligible and use the relationship between distance and time for motion under constant acceleration, giving

$$s = ut + \frac{at^2}{2}$$

where s is the distance travelled, u is the initial velocity, t is time and a is the acceleration. In this case, $u = 20$ m s^{-1} and $a = -g$ (acceleration due to gravity ≈ 10 m s^{-2}), so $s = 20t - 5t^2$.

At the maximum height, the rate of change of distance with time must be 0, i.e. the velocity is 0. Therefore we differentiate to find the velocity:

$$v = \frac{ds}{dt} = 20 - 10t$$

Putting $v=0$ gives

$$0=20-10t$$

$$\Leftrightarrow 10t=20 \Leftrightarrow t=2$$

We have shown that the maximum height is reached after 2 s. But what is that height? Substituting $t=2$ into the equation for s gives

$$s=20(2)-5(2)^2=20 \text{ m}$$

giving the maximum value of s as 20 m.

This example illustrates the important step in finding maximum and minimum values of a function, $y=f(x)$. That is, we differentiate and solve

$$\frac{dy}{dx}=0$$

This may give various values of x. The points where $dy/dx=0$ are called the stationary points, but having found these we still need a way of deciding whether they could be maximum or minimum values. In the example, we knew that a stone thrown into the air must reach a maximum height and then return to the ground, and so by solving $ds/dt=0$ we would find the time at the maximum. Other problems may not be so clear-cut, and then we need a method of distinguishing between different types of stationary points.

A stationary point is classified as either a local maximum, a local minimum or a point of inflexion. The plural of maximum is maxima and the plural of minimum is minima. The word 'local' is used in the description, because local maxima or local minima do not necessarily give the overall maximum or minimum values of the function. For instance, in Fig. 17.1 there is a local maximum at B, but the value of y at $x=x_1$ is actually bigger; hence the overall maximum value of the function in the range is given by y at x_1.

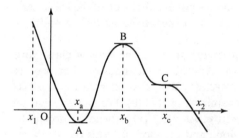

Figure 17.1 A graph of some function $y=f(x)$ plotted from $x=x_1$ to $x=x_2$. Points A, B, C in the graph are stationary points. They are points where the gradient of the tangent to the curve is zero, i.e. $dy/dx=0$.

To see how to classify stationary points, examine Fig. 17.1, where points A, B and C are all stationary points.

In order to analyse the slope of the function imagine the function as representing the cross-section of a mountain range and that we are crossing it from left to right.

At points A, B and C in Fig. 17.1 the gradient of the tangent to the curve is zero, i.e. $dy/dx=0$.

At A there is a local minimum, where the graph changes from going downhill to going uphill.

At B there is a local maximum, where the graph changes from going uphill to going downhill.

At C there is a point of inflexion, where the graph goes flat briefly before resuming its descent.

Local maxima and minima are also called turning points because the function is changing from increasing to decreasing or vice versa.

By looking at where the graph is going uphill, i.e. $dy/dx > 0$, at where it is going downhill, i.e. $dy/dx < 0$, and especially remembering to mark the points where $dy/dx = 0$, the stationary points, we can draw a very rough sketch of the derivative of any function just by looking at its graph.

This we do for our example graph in Fig. 17.2. By examining the graph of dy/dx we can see that at a local minimum point, $dy/dx = 0$ and dy/dx is negative just before the minimum and positive afterwards. For a local maximum point, $dy/dx = 0$ and dy/dx is positive just before the maximum and negative afterwards.

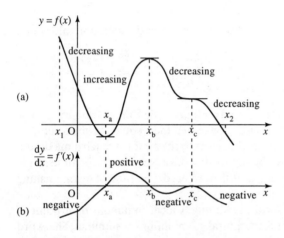

Figure 17.2 (a) The graph of Fig. 17.1 and (b) a sketch of its derivative, $dy/dx = f'(x)$. Where $y = f(x)$ has a stationary point, that is where the tangent to the curve is flat, then $dy/dx = 0$. Where $f(x)$ is increasing then the derivative is positive and where $f(x)$ is decreasing then the derivative is negative.

If at the point where $dy/dx = 0$ the derivative has the same sign on either side of the stationary point, then it must be a point of inflexion. (At point C, dy/dx is negative just before and just after $x = x_e$.)

Analysing the sign of dy/dx on either side of a stationary point is one way of classifying whether it is a maximum, minimum or point of inflexion. Another, sometimes quicker way, is to use the derivative of the derivative, the second derivative, d^2y/dx^2, also referred to as $f''(x)$.

To understand how to use the second derivative, examine Fig. 17.2(b). We can see that at x_a, $f'(x)$ is heading uphill; hence its slope, $f''(x)$, is positive. At x_b, $f'(x)$ is heading downhill; hence its slope, $f''(x)$, is negative. At x_c $f'(x)$ has zero slope and therefore $f''(x)$ is 0.

We can now summarize the steps involved in finding and classifying the stationary points of a function $y = f(x)$ and in finding the overall maximum or minimum value of a function.

Step 1
Find the values of x at the stationary points. First, find dy/dx and then solve for x such that $dy/dx = 0$.

Step 2
To classify the stationary points there is a choice of method, although Method 2 does not always give a conclusive result.

Method 1
For each of the values of x found in Step 1, find out whether the derivative is positive or

negative just before the stationary point and just after the stationary point dy/dx.

Before	→	0	→	after	
+	→	0	→	−	gives a maximum point
−	→	0	→	+	gives a minimum point
+	→	0	→	+	or
−	→	0	→	−	gives a point of inflexion

These results are summarized in Fig. 17.3.

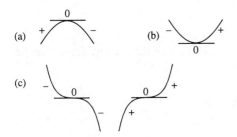

(a) (b)

(c)

Figure 17.3 Distinguishing stationary points: (a) a maximum point; (b) minimum point; (c) two points of inflexion. The sign of the slope of the curve, given by dy/dx, are marked on either side of the stationary point.

Method 2
Find the second derivative, d^2y/dx^2, and substitute in turn each of the values of x found in Step 1 (the values of x at the stationary points).
    If d^2y/dx^2 is negative, this indicates that there is a maximum point.
    If d^2y/dx^2 is positive, this indicates that there is a minimum point.
    However, if d^2y/d$x^2 = 0$, then the test is inconclusive and we must revert to Method 1.

Step 3
Find the values of y at the maximum and minimum points to give the coordinates of the turning points.

Step 4
The overall maximum or minimum values of the function can be found in the following way, as long the function is continuous over the values of x of interest. Substitute the boundary values of x into the function to find the corresponding values for y. Compare the values of y found in Step 3 to these boundary values to find the overall maximum and minimum.

Example 17.2 Find and classify the stationary points of $y = x^3 - 9x^2 + 24x + 3$ and find the overall maximum and minimum value of the function in the range $x = 0$ to $x = 5$.

SOLUTION
Step 1
First we must solve dy/d$x = 0$:

$$y = x^3 - 9x^2 + 24x + 3$$

$$\Rightarrow \frac{dy}{dx} = 3x^2 - 18x + 24$$

So we put

$$3x^2 - 18x + 24 = 0$$

$$\Leftrightarrow x^2 - 6x + 8 = 0 \qquad \text{(dividing by 3)}$$

$$\Leftrightarrow (x - 2)(x - 4) = 0$$

(factorizing to find the roots, although we could also use the formula for solving a quadratic equation if no factorization can easily be found)

$$\Leftrightarrow x - 2 = 0 \text{ or } x - 4 = 0$$

$$\Leftrightarrow x = 2 \text{ or } x = 4$$

Therefore the stationary points occur when $x = 2$ or $x = 4$.

Step 2
To classify these we use Method 2 as outlined above and look at the second derivative, i.e. we differentiate dy/dx:

$$\frac{dy}{dx} = 3x^2 - 18x + 24$$

$$\Rightarrow \frac{d^2y}{dx^2} = 6x - 18$$

At $x = 2$, $d^2y/dx^2 = 12 - 18 = -6$, which is negative, showing that at $x = 2$, $f''(x)$ is negative, and therefore we have a local maximum.

At $x = 4$, $d^2y/dx^2 = 24 - 18 = 6$, which is positive, showing that at $x = 4$, $f''(x)$ is positive and we therefore have a local minimum.

Step 3
We still need to know the function value, the value of y at these stationary points. To find the value of y we substitute into the original expression.

At $x = 2$ we get

$$y = (2)^3 - 9(2)^2 + 24(2) + 3 = 8 - 36 + 48 + 3 = 23$$

Therefore the local maximum occurs at the point (2,23).

At $x = 4$ we get

$$y = (4)^3 - 9(4)^2 + 24(4) + 3 = 64 - 144 + 96 + 3 = 19$$

Hence the local minimum occurs at the point (4,19).

Step 4
To find the overall maximum value and minimum value then substitute the boundary values of x. These are given as $x = 0$ and $x = 5$.

$$\text{At } x = 0: y = (0)^3 - 9(0)^2 + 24(0) + 3 = 0 - 0 + 0 + 3 = 3$$

$$\text{At } x = 5: y = (5)^3 - 9(5)^2 + 24(5) + 3 = 125 - 225 + 120 + 3 = 23$$

So the boundary points are (0,3) and (5,23).

Comparing the numbers 3 and 23 with the values of the function at the maximum and minimum in Step 3 i.e. 23 and 19, we can see that the overall maximum value occurs at $x = 5$ and $x = 2$, where $y = 23$. The overall minimum value occurs at $x = 0$, where $y = 3$. These

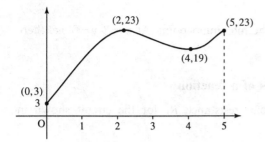

Figure 17.4 Sketch of $y=x^3-9x^2+24x+3$.

findings are confirmed by the sketch of the function, which we now have sufficient information to make, as in Fig. 17.4.

Example 17.3 Find and classify the stationary points of $y=-(2-x)^4$.

SOLUTION
Step 1

$$y=-(2-x)^4$$

$$\Rightarrow \frac{dy}{dx}=4(2-x)^3$$

Stationary points occur where $dy/dx=0$:

$$4(2-x)^3=0$$

$$\Leftrightarrow x=2$$

Step 2
To classify this stationary point, we differentiate again:

$$d^2y/dx^2=-12(2-x)^2$$

At $x=2$ we get

$$d^2y/dx^2=-12(2-2)^2=0.$$

So the second derivative is zero.

We cannot use the second derivative test to classify the stationary point because a zero value is inconclusive, so we go back to the first derivative and examine its sign at values of x just less than $x=2$ and just greater than $x=2$. This can be done with the help of a table. Choose any values of x less than $x=2$ and greater than $x=2$; here we choose $x=1$ and $x=3$. Be careful if the function is discontinuous at any point not to cross the discontinuity.

x	1	2	3
$dy/dx=4(2-x)^3$	4	0	-4

At $x=1$, $dy/dx=4(2-1)^3=4$ (positive).
At $x=3$, $dy/dx=4(2-3)^3=-4$ (negative).
Therefore near the point $x=2$ the derivative goes from positive to zero to negative. Therefore the graph of the function goes from travelling uphill to travelling downhill, showing that we have a maximum value.

Step 3
Finally, we find the value of the function at the maximum point. At $x=2$, $y=0$, i.e. there is a maximum at $(2,0)$.

Applications of maximum and minimum values of a function

Example 17.4 The power delivered to the load resistance R_L for the circuit shown in Fig. 17.5 is defined by

$$P = \frac{25R_L}{(2000 + R_L)^2}$$

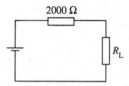

2000 Ω

R_L

Figure 17.5 Circuit for Example 17.4.

Show that the maximum power delivered to the load occurs for $R_L = 2000$.

SOLUTION For a maximum value, $dP/dR_L = 0$.

$$P = \frac{25R_L}{(2000 + R_L)^2}$$

Using the quotient rule to find the derivative we get

$$\frac{dP}{dR_L} = \frac{25(2000 + R_L)^2 - 25R_L(2)(2000 + R_L)}{(2000 + R_L)^4}$$

$$\Leftrightarrow \quad \frac{dP}{dR_L} = \frac{(2000 + R_L)(50\,000 + 25R_L - 50R_L)}{(2000 + R_L)^4}$$

As R_L is positive, $2000 + R_L$ is non-zero, so we can cancel the common factor of $2000 + R_L$. Also, simplifying the top line gives

$$\frac{dP}{dR_L} = \frac{50\,000 - 25R_L}{(2000 + R_L)^3}$$

Setting $dP/dR_L = 0$ gives

$$\frac{50\,000 - 25R_L}{(2000 + R_L)^3} = 0$$

Multiplying by $(2000 + R_L)^3$, then

$$50\,000 - 25R_L = 0 \Leftrightarrow R_L = 2000$$

We have shown that there is a stationary value of the function P when $R_L = 2000$, but now we need to check that it is in fact a maximum value. To do this, we substitute values above and below $R_L = 2000$ (say, $R_L = 1000$ and $R_L = 3000$) into dP/dR_L:

$$\frac{dP}{dR_L} = \frac{50\,000 - 25R_L}{(2000 + R_L)^3}$$

When $R_L = 1000$ the top line is positive and the bottom line is positive, so the sign of dP/dR_L is $+/+$, which is positive. When $R_L = 3000$ the top line is negative and the bottom line is positive so the sign of dP/dR_L is $-/+$, which is negative.

So the derivative of the power with respect to the load resistance goes from positive to 0 to negative when $R_L = 2000$, indicating a maximum point.

We have shown that the maximum power to the load occurs when $R_L = 2000$.

Example 17.5 A rectangular field is to be surrounded by a fence of length 400 m. What are the dimensions of the field such that it has maximum area?

SOLUTION The field is pictured in Fig. 17.6. Call the length of the sides a metres and b metres. Then the area is given by $A = ab$.

Figure 17.6 A rectangular field.

The perimeter is given by $P = 2a + 2b$, and as the length of the fence is 400 m we get

$$400 = 2a + 2b$$

$$\Leftrightarrow 200 = a + b$$

We wish to find the maximum area, and to do this we need to be able to differentiate A in terms of one of the variables, a or b. We use the information given about the length of the perimeter to express a in terms of b:

$$a = b - 200$$

and substitute this into the expression for the area, giving

$$A = b(200 - b) = 200b - b^2$$

To find the maximum value for A differentiate with respect to b, giving

$$\frac{dA}{db} = 200 - 2b$$

and solve $dA/db = 0$:

$$\Rightarrow 200 - 2b = 0 \Leftrightarrow 2b = 200$$

$$\Leftrightarrow b = 100$$

Check that this is indeed a local maximum value by differentiating a second time: $d^2A/db^2 = -2$, which is negative.

Now we find the length of the other side of the field:

$$a = 200 - b = 100$$

Therefore the field with maximum area is a square of side 100. The area of the field is $100 \times 100 = 10\,000$ m^2.

To check that this agrees with the original condition that $2a + 2b = 400$, substitute $a = 100$ and $b = 100$, giving $200 + 200 = 400$, which is correct.

17.3 GRAPH SKETCHING BY ANALYSING THE FUNCTION'S BEHAVIOUR

To sketch the graph of any function $y = f(x)$ we first analyse the main features of the function's behaviour.

Look at the graph in Fig. 17.7 and list the 'important features' of the graph. List these in a way that would enable someone else to sketch the graph from your description.

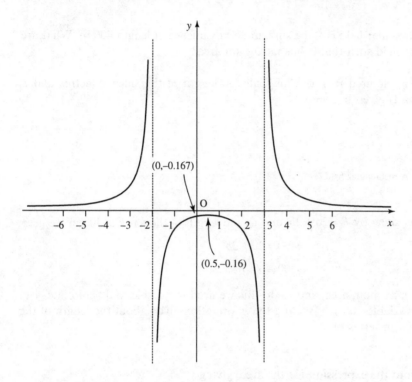

Figure 17.7

Here are some of the important features. The list is not exhaustive, but should be enough to enable someone to reproduce the graph.

1. The graph is positive for $x < -2$ and $x > 3$ and negative for x between -2 and $+3$. There are no values of x for which y is zero and when $x = 0$, $y = -0.167$.
2. The graph is discontinuous at $x = -2$ and $x = 3$. y keeps getting larger as x approaches -2 from the left. A more precise way of expressing this is to say that as x tends to -2, with x less than -2, y tends to infinity. The symbol for infinity is ∞, and the symbol for 'tends to' is $\rightarrow$. x tends to -2, with x less than -2, can be expressed more briefly as $x \rightarrow 2^-$. This gives:

$$\text{as } x \rightarrow -2^-, \ y \rightarrow \infty$$

Similarly:

$$\text{as } x \rightarrow -2^+, \ y \rightarrow -\infty$$

For the discontinuity at $x = 3$ we have

$$\text{as } x \rightarrow 3^-, \ y \rightarrow -\infty$$

$$\text{as } x \rightarrow 3^+, \ y \rightarrow \infty$$

3. There is a local maximum at $x = 0.5$, where $y = -0.16$.

4. For large values of x, y gets nearer to 0. This can be expressed as

$$\text{as } x \to \infty, \ y \to 0^+$$

Similarly:

$$\text{as } x \to -\infty, \ y \to 0^+$$

This list of important features indicates the steps that should be taken in order to sketch a graph.

Step 1
Find the value of y when the graph cross the y-axis; that is, when $x = 0$.
   If possible, find where the graph crosses the x-axis; that is, the values of x where $y = 0$.

Step 2
Find any discontinuities in the function, i.e. are there values of x where there is no value of y? Infinite discontinuities (a 'divide by zero') will lead to vertical asymptotes. These are vertical lines which the function approaches but does not meet.
   We must decide whether the function is positive or negative on either side of the asymptote.

Step 3
Find the coordinates of the maxima and minima.

Step 4
Find the behaviour of the function as x tends to plus and minus infinity.

Step 5
Mark the features found in Steps 1 to 4 on the graph and join them, where appropriate, to give the sketch of the graph.

Example 17.6 Sketch the curve whose equation is

$$y = \frac{(2x+1)}{(x+1)(x-5)}$$

Solution
Step 1
When $x = 0$:

$$y = \frac{(2.0+1)}{(0+1)(0-5)} = \frac{1}{-5} = -0.2$$

When $y = 0$:

$$\frac{(2x+1)}{(x+1)(x-5)} = 0 \text{ then } 2x + a = 0 \Leftrightarrow x = -0.5$$

Step 2
If the bottom line of the function were 0 this would lead to a 'divide by zero' which is undefined. This happens when $x + 1 = 0$, i.e. $x = -1$, and when $x - 5 = 0$, i.e. $x = 5$.
   These are infinite discontinuities: y will tend to plus or minus infinity as x approaches these values. The lines at $x = -1$ and $x = 5$ are the asymptotes.

To find the sign of y on either side of the asymptotes, substitute values of x on either side of them (avoiding including any values where $y=0$). This can be done using a table, as in Table 17.1.

Table 17.1 Finding the sign on either side of the asymptotes

x	-2	-1	-0.75	-0.5	4	5	6
$y = \dfrac{(2x+1)}{(x+1)(x-5)}$	-0.4286	Not defined	0.348	0	-1.8	Not defined	0.93
Sign of y	$-$	Not defined	$+$	0	$-$	Not defined	$+$

Using Table 17.1 we can conclude that as y is negative to the left of the asymptote at $x=-1$ and positive to the right, then

$$\text{as } x \to -1^- \qquad y \to -\infty$$

$$\text{as } x \to -1^+ \qquad y \to +\infty$$

Similarly, as y is negative to the left of the asymptote at $x=5$ and positive to the right of it, then

$$\text{as } x \to 5^- \qquad y \to -\infty$$

$$\text{as } x \to 5^+ \qquad y \to +\infty$$

Step 3
To find the turning points, look for points where $dy/dx=0$

$$y = \frac{(2x+1)}{(x+1)(x-5)}$$

To differentiate, first multiply out the brackets on the bottom line of the expression and then use the formula for finding the derivative of a quotient:

$$y = \frac{(2x+1)}{(x+1)(x-5)} \Leftrightarrow y = \frac{(2x+1)}{x^2-4x-5}$$

This gives

$$\frac{dy}{dx} = \frac{2(x^2-4x-5)-(2x+1)(2x-4)}{(x^2-4x-5)^2}$$

$$\frac{dy}{dx} = \frac{2x^2-8x-10-4x^2+6x+4}{(x^2-4x-5)^2}$$

$$= \frac{-2x^2-2x-6}{(x^2-4x-5)^2}$$

We now solve $dy/dx=0$:

$$\frac{-2x^2-2x-6}{(x^2-4x-5)^2} = 0 \Rightarrow -2x^2-2x-6=0$$

$$\Rightarrow x^2+x+3=0$$

Using the formula to solve the quadratic equation gives

$$x = \frac{-1 \pm \sqrt{1-12}}{2} = \frac{-1 \pm \sqrt{-11}}{2}$$

Because of the square root of a negative number in this expression we can see that there are no real solutions, so there are no turning points.

Step 4
When x is large in magnitude then the highest powers of x on the top and bottom lines of the function expression will dominate. In this case, ignore all the other terms. Consider

$$y = \frac{2x+1}{(x+1)(x-5)} = \frac{2x+1}{x^2-4x-5}$$

For x large in magnitude:

$$y \sim \frac{2x}{x^2} = \frac{2}{x}$$

which tends to 0 and is positive as $x \to \infty$ and tends to 0 and is negative as $x \to -\infty$. We can say that as $x \to \infty$, $y \to 0^+$, and as $x \to -\infty$, $y \to 0^-$.

Step 5
Sketch the graph. This is done in two stages, shown in Fig. 17.8.

Example 17.7 A spring of modulus of elasticity k has a mass m suspended from it and is subjected to a oscillating force, $F_0 \cos(\omega t)$, where $\omega > 0$. The motion of the mass is damped by the use of a dashpot of damping constant c. This is displayed in Fig. 17.9. After some time the spring force is found to have oscillations of angular frequency ω and of magnitude

$$F = \frac{kF_0}{\sqrt{\omega^2 c^2 + (\omega^2 m - k)^2}}$$

Taking $m = 1$ and $k = 1$, sketch the graph of F/F_0 against ω for

(a) $c = 2$
(b) $c = \frac{1}{2}$
(c) $c = \frac{1}{4}$
(d) $c = 0$

SOLUTION We are given that

$$F = \frac{kF_0}{\sqrt{\omega^2 c^2 + (\omega^2 m - k)^2}}$$

Therefore

$$\frac{F}{F_0} = \frac{k}{\sqrt{\omega^2 c^2 + (\omega^2 m - k)^2}}$$

(a)

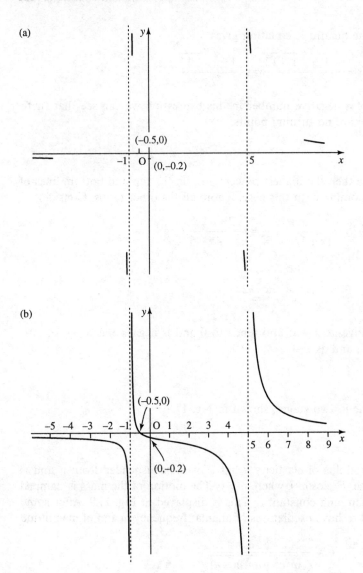

(b)

Figure 17.8 Sketching the graph of $y=(2x+1)/((x+1)(x-5))$. (a) First, mark the important points, as found in Example 17.6. (b) Join up these points, where relevant, to give the sketch of the function.

Figure 17.9 A spring subjected to damped, forced motion.

Substituting $m=1$ and $k=1$ gives

$$\frac{F}{F_0}=\frac{1}{\sqrt{\omega^2 c^2+(\omega^2-1)^2}}$$

Call this function H.

One method to solve this problem would be to consider cases (a), (b), (c) and (d) separately.

However, it is quicker to leave c as an unknown constant and substitute in for the particular cases later.

Step 1

$$H = \frac{1}{\sqrt{\omega^2 c^2 + (\omega^2 - 1)^2}}$$

When $\omega = 0$, $H = 1$. The denominator is a positive square root, which means that H is always positive where it is defined.

Step 2
Consider any points where H is not defined.

$$H = \frac{1}{\sqrt{\omega^2 c^2 + (\omega^2 - 1)^2}}$$

As the term inside the square root in the expression for H is a sum of squares, it is always ≥ 0. This means that H is always defined, except where the denominator is 0.
   H is not defined when

$$\sqrt{\omega^2 c^2 + (\omega^2 - 1)^2} = 0$$
$$\Rightarrow \omega^2 c^2 + (\omega^2 - 1)^2 = 0$$
$$\omega^2 c^2 + \omega^4 - 2\omega^2 + 1 = 0$$
$$\omega^4 + \omega^2(c^2 - 2) + 1 = 0$$

By substituting the values of c of interest we see that in case (a), when $c = 2$, (b) when $c = \frac{1}{2}$ and case (c) when $c = \frac{1}{4}$ there are no real solutions of this equation for ω. This means that there are no points where H is undefined. However, for the case (d), where $c = 0$, we get the equation

$$\omega^4 - 2\omega^2 + 1 = 0$$
$$\Leftrightarrow (\omega^2 - 1)^2 = 0$$
$$\omega^2 - 1 = 0 \Leftrightarrow \omega^2 = 1 \Leftrightarrow \omega = 1 \text{ or } \omega = -1$$

As the frequency of the forcing function is positive, we just have the one value where H is discontinuous, at $\omega = 1$. For case (d), where $c = 0$, H is undefined when $\omega = 1$. This is an infinite 'divide by zero' discontinuity. We have already noted that H is always positive where it is defined, and so we know that as $\omega \to 1^- H \to +\infty$ and as $\omega \to 1^+ H \to +\infty$.

Step 3
To find the stationary points solve $dH/d\omega = 0$.

$$H = \frac{1}{\sqrt{\omega^2 c^2 + (\omega^2 - 1)^2}}$$

is easier to differentiate if we use the rules of powers to give

$$H = (\omega^2 c^2 + (\omega^2 - 1)^2)^{-1/2}$$

Using the function of a function rule we get

$$\frac{dH}{d\omega} = -\tfrac{1}{2}(\omega^2 c^2 + (\omega^2 - 1)^2)^{-3/2}(2\omega c^2 + 2(\omega^2 - 1)(2\omega))$$

$$= -\frac{2\omega c^2 + 2(\omega^2 - 1)(2\omega)}{2(\omega^2 c^2 + (\omega^2 - 1)^2)^{3/2}} = -\frac{2\omega c^2 + 4\omega^3 - 4\omega}{2(\omega^2 c^2 + (\omega^2 - 1)^2)^{3/2}}$$

Dividing the top and bottom lines by 2 and rearranging the terms on the top line gives

$$\frac{dH}{d\omega} = -\frac{2\omega^3 + \omega(c^2 - 2)}{(\omega^2 c^2 + (\omega^2 - 1)^2)^{3/2}}$$

Setting $dH/d\omega = 0$ gives

$$-\frac{2\omega^3 + \omega(c^2 - 2)}{(\omega^2 c^2 + (\omega^2 - 1)^2)^{3/2}} = 0$$

Multiplying both sides by -1 times the denominator of the left-hand side gives

$$\Rightarrow 2\omega^3 + \omega(c^2 - 2) = 0$$

$$\Leftrightarrow \omega(2\omega^2 + c^2 - 2) = 0$$

$$\Leftrightarrow \omega = 0 \vee 2\omega^2 + c^2 - 2 = 0$$

$$\Leftrightarrow \omega = 0 \vee 2\omega^2 = 2 - c^2$$

$$\Leftrightarrow \omega = 0 \vee \omega^2 = \frac{2 - c^2}{2}$$

$$\Leftrightarrow \omega = 0 \vee \omega = \pm\sqrt{\frac{2 - c^2}{2}}$$

As we assume that the frequency is positive (or zero) there are two possibilities: $\omega = 0$ and $\omega = \sqrt{(2 - c^2)/2}$. The second case does not give a real solution if $c = 2$.

We wait until specific values of c are substituted in order to analyse the type of stationary points and the value of H at these points. Also note that this analysis is not valid for the case $c = 0$, as this completely changes the nature of the function H.

Step 4
When ω is large in magnitude

$$H \sim \frac{1}{\sqrt{\omega^4}}$$

and this tends to 0 for large ω. Therefore as $\omega \to \infty$, $H \to 0^+$.

Step 5
Now we can sketch the graph using the information found and after substituting the various values of c.

Case (a): $c = 2$
Substitute $c = 2$, we have that

$$H = \frac{1}{\sqrt{4\omega^2 + (\omega^2 - 1)^2}}$$

The graph passes through $(0,1)$ (Step 1), and the function is defined for all $\omega \geqslant 0$ (Step 2). The stationary point is at $\omega = 0$. In this case

$$\frac{\mathrm{d}H}{\mathrm{d}\omega} = -\frac{2\omega^3 + 2\omega}{(4\omega^2 + (\omega^2 - 1)^2)^{3/2}} = \frac{-2\omega(\omega^2 + 1)}{(4\omega^2 + (\omega^2 - 1)^2)^{3/2}}$$

This is positive for $\omega < 0$ and negative for $\omega > 0$ and therefore there is a maximum value at $\omega = 0$.

Case (b): $c = \frac{1}{2}$
Here

$$H = \frac{1}{\sqrt{\omega^2/4 + (\omega^2 - 1)^2}}$$

The graph passes through $(0,1)$ (Step 1), and there are no discontinuities (Step 2). The stationary points are at $\omega = 0$ and $\omega = \sqrt{(2 - c^2)/2} = \sqrt{7/8} \approx 0.935$ (Step 3).

$$\frac{\mathrm{d}H}{\mathrm{d}\omega} = \frac{-\omega(2\omega^2 - 7/4)}{(\omega^2/4 + (\omega^2 - 1)^2)^{3/2}}$$

$\mathrm{d}H/\mathrm{d}\omega < 0$ for $\omega = 0$ and $\mathrm{d}H/\mathrm{d}\omega > 0$ for $\omega > 0$; therefore there is a minimum value at $\omega = 0$.

$\mathrm{d}H/\mathrm{d}\omega > 0$ for ω just less than $\sqrt{7/8}$.
$\mathrm{d}H/\mathrm{d}\omega < 0$ for ω just greater than $\sqrt{7/8}$.
Therefore there is a maximum value of H at $\omega = \sqrt{7/8} \approx 0.935$. At the maximum

$$H = \frac{1}{\sqrt{7/32 + 1/64}} = \frac{1}{\sqrt{15/64}} \simeq 2.06$$

Case (c): $c = \frac{1}{4}$

$$H = \frac{1}{\sqrt{\omega^2/16 + (\omega^2 - 1)^2}}$$

The graph passes through $(0,1)$ (Step 1), and there are no discontinuities (Step 2). The stationary points are at $\omega = 0$ and $\omega = \sqrt{(2 - c^2)/2} = \sqrt{31/32} \approx 0.98$. There is a minimum value at $\omega = 0$ and a maximum at $\omega = \sqrt{31/32}$, where

$$H = \frac{1}{\sqrt{31/512 + 1/1024}} \simeq 4.03$$

Case (d): $c = 0$

$$H = \frac{1}{\sqrt{(\omega^2 - 1)^2}} = \begin{cases} \dfrac{1}{1 - \omega^2} & \text{for } \omega < 1 \\[2ex] \dfrac{1}{\omega^2 - 1} & \text{for } \omega > 1 \end{cases}$$

The graph passes through $(0,1)$ (Step 1), there is an infinite discontinuity at $\omega = 1$, and for

$$\omega \to 1^-, \ H \to +\infty$$

$$\omega \to 1^+, \ H \to +\infty$$

The stationary points have to be analysed separately as the general case always assumed $c > 0$. Differentiating H we get

$$\frac{dH}{d\omega} = \frac{2\omega}{(\omega^2 - 1)^2} \qquad (\omega < 1)$$

$$\frac{dH}{d\omega} = \frac{-2\omega}{(\omega^2 - 1)^2} \qquad (\omega > 1)$$

which have a zero value at $\omega = 0$. $dH/d\omega$ is negative for ω just less than 0 and positive for ω just greater than 0. Therefore there is a minimum point at $\omega = 0$.

We can now sketch the graphs as in Fig. 17.10.

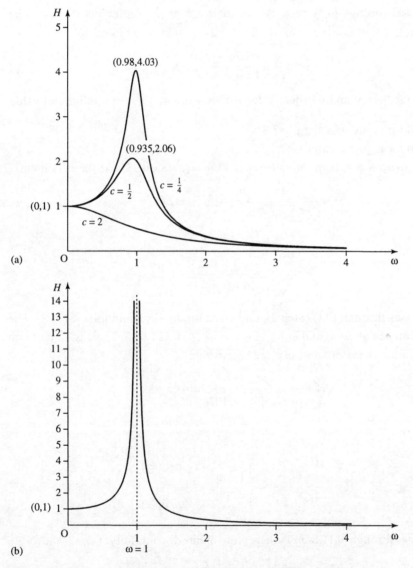

(a)

(b)

Figure 17.10 (a) The graph of $H = 1/\sqrt{\omega^2 c^2 + (\omega^2 - 1)^2}$, as in Example 17.7, for $c = 2$, $c = \frac{1}{2}$ and $c = \frac{1}{4}$. (b) The graph of H for $c = 0$. $H = 1/(1 - \omega^2)$ for $\omega < 1$ and $H = 1/(\omega^2 - 1)$ for $\omega > 1$.

17.4 SUMMARY

1. To find and classify stationary values of a function $y = f(x)$ then:

Step 1: find dy/dx and solve for x such that $dy/dx = 0$
Step 2: Classify the stationary points either

Method 1
By examining the sign of dy/dx near the point, in which case $+ \to 0 \to -$ indicates a local maximum point, $- \to 0 \to +$ indicates a local minimum point, and $+ \to 0 \to +$ or $- \to 0 \to -$ indicate a point of inflexion; or

Method 2
By finding d^2y/dx^2 at the point.

If $d^2y/dx^2 < 0$ then there is a local maximum point.
If $d^2y/dx^2 > 0$ then there is a local minimum point. However, if $d^2y/dx^2 = 0$ then this test is inconclusive and Method 1 must be used instead.
   Substitute the values of x at the stationary points to find the relevant values of y. Local maxima and minima are also called turning points.
2. If the function, defined for a range of values of x, is continuous, then the overall maximum and minimum values can be found by finding the values of y at the local maxima and minima and the values of y at the boundary points. The maximum of all of these is the global maximum and the minimum of all of these is the global minimum value.
3. There are many practical problems that involve the need to find the maximum or minimum value of a function.
4. The stationary values are used when sketching a graph. We also look for:

   (a) Values where the graph crosses the axes.
   (b) Points of discontinuity and the behaviour near discontinuities.
   (c) Behaviour as x tends to $\pm \infty$.

17.5 EXERCISES

17.1 Find and classify the stationary points of the following functions:

(a) $y = x^2 - 5x + 2$ (b) $y = -3x^2 + 4x$ (c) $y = 3x^3 - x$
(d) $x = 2t + \dfrac{200}{t}$ (e) $w = z^4 + 4z^3 - 8z^2 + 2$

17.2 Find the overall maximum and minimum values of

$$\frac{x}{2x^2 + 1}$$

in the range $x \in [-1, 1]$.

17.3 Sketch the graphs of the following functions:

(a) $y = \dfrac{(x-3)(x+5)}{(x+2)}$ (b) $y = x + \dfrac{1}{x}$

(c) $y = x^3 - 3x - 1$ (d) $y = \dfrac{(x-1)(x+4)}{(x-2)(x-3)}$

17.4 Sketch the graph of $y = 2\sin(x) - \sin(2x)$ for x between -2π and 2π.

17.5 Find the overall maximum and minimum values of $y = x^3(x-1)$ in the range $x \in [0,2]$,

17.6 An open box of variable height and width is to have a length of 3 m. It should not use more than a total of 20 m^2 of surface area. Find the height and width which give the maximum volume.

17.7 An *LRC* series circuit has an impedance of magnitude

$$Z = \sqrt{R^2 + \left[\omega L - \frac{1}{\omega C}\right]^2}$$

where R is the resistance, L the inductance, C the capacitance and ω is the angular frequency of the voltage source. Sketch Z against ω for the case where $R = 200\ \Omega$, $C = 0.03\ \mu F$, $L = 2$ mH.

17.8 A crank is used to drive a piston as in Fig. 17.11. The angular velocity of the crank shaft is the rate of change of the angle θ, $\omega = d\theta/dt$. The piston moves horizontally with velocity v_p and acceleration a_p. The crankpin performs circular motion with a velocity of v_c and centripetal acceleration of $\omega^2 r$. The acceleration a_p of the piston varies with θ and is related by

$$a_p = \omega^2 r \left(\cos(\theta) + \frac{r\cos(2\theta)}{l} \right)$$

where r is the length of the crank and l is the length of the connecting rod. Substituting $r = 150$ mm and $l = 375$ mm, find the maximum and minimum values of the acceleration a_p.

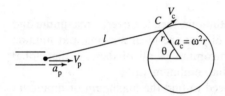

Figure 17.11 A crank is used to drive a piston (Exercise 17.8).

17.9 A water wheel is constructed with symmetrical curved vanes of angle of curvature θ. Assuming that friction can be taken as negligible, the efficiency, η, i.e. the ratio of output power to input power, is calculated as

$$\eta = \frac{2(V-v)(1+\cos(\theta))v}{V^2}$$

where V is the velocity of the jet of water as it strikes the vane, v is the velocity of the vane in the direction of the jet and θ is constant. Find the ratio v/V which gives maximum efficiency and find the maximum efficiency.

17.10 Power is transmitted by a fluid of density ρ moving with positive velocity V along a pipeline of constant cross-sectional area A. Assuming that the loss of power is mainly attributable to friction, and that the friction coefficient f can be taken to be a constant, then the power transmitted is given by P, where $P = \rho g A(hV - cV^3)$, g is acceleration due to gravity and h is the head (the energy per unit weight). $c = 4fl/2gd$ where l is the length of the pipe and d is the diameter of the pipe. Assuming h is a constant find the value of V which gives a maximum value for P, and given that the input power is $P_i = \rho g A V h$, find the maximum efficiency.

EIGHTEEN

SEQUENCES AND SERIES

18.1 INTRODUCTION

Sequences have two main applications: serving as a digital representation of a signal after analogue to digital (A/D) conversion or as a method of solving a numerical problem by getting a sequence of answers, each one being closer to the exact, correct solution.

The advance of digital communications has resulted from the increased accuracy in reproduction of the stored or transmitted signal in digital form. For instance, the reproduction of the stored signal by a compact disc player is far superior to the analogue reproduction of the old vinyl records. It is also very convenient to be able to apply filters and other processing techniques in digital form using computers or dedicated microprocessors.

Sequences are often defined in the form of a recurrence relation, special sorts of which are also called difference equations. Recurrence relations can be found which will solve certain problems numerically, or they may be derived by modelling the physical processes in a digital system.

The sum of a sequence of terms is called a series. An important example of a series is the Taylor series, which can be used to approximate a function. Later in the book, we shall look at other examples of series, such as z transforms and Fourier series.

Many problems involving sequences and series are solved using a computer. However, it is useful to be able to solve a few simple cases without the aid of a computer, as this can often help check a result for some special cases. Some examples of special sequences are the arithmetic progression and the geometric progression.

18.2 SEQUENCES AND SERIES DEFINITIONS

A sequence is a collection of objects (not necessarily all different) arranged in a definite order. Some examples of sequences are:

1. The numbers 1 to 10, i.e.

$$1, 2, 3, 4, 5, 6, 7, 8, 9, 10$$

2. Red, Red and Amber, Green, Amber, Red
3. 3, 6, 9, 12, 15, 18

When a sequence follows some 'obvious' rule then three dots (...) are used to indicate 'and so on', e.g. list 1 above may be rewritten as

$$1, 2, 3, \ldots, 10$$

The examples so far have all been finite sequences. Infinite sequences may use dots at the end, meaning carry on indefinitely in the same fashion, e.g.

$$2, 4, 6, 8, 10, \ldots$$

$$1, 2, 4, 8, 16, \ldots$$

$$4, 9, 16, 25, 36, \ldots$$

$$1, 1, 2, 3, 5, 8, \ldots$$

and the (...) indicates that there is no end to the sequence of values and that they carry on in the same fashion. The elements of a sequence can be represented using letters, e.g.

$$a_1, a_2, a_3, a_4, \ldots, a_n, \ldots$$

The first term is called a_1, the second a_2 etc. (Sometimes it is more convenient to say that a sequence begins with a zeroth term, a_0.)

If a rule exists by which any term in the sequence can be found, then this may be expressed by the 'general term' of the sequence, usually called a_n or a_r. This rule may be expressed in the form of a recurrence relation, giving a_{n+1} in terms of a_n, a_{n-1},.... In this case it may be quite difficult to find the explicit function definition, that is to solve the recurrence relation. We look at solving recurrence relations in Chapter 20.

A sequence is a function of natural numbers, or integers. The function expression is given by the general term.

Example 18.1 Find the general term of the sequence of numbers from 1 to 10.

SOLUTION 1, 2, 3, 4, 5, 6, 7, 8, 9, 10 has the general term $a_n = n$, where $n = 1$ to 10. We can also write this in 'standard' function notation as

$$a(n) = n \qquad \text{where } n = 1 \text{ to } 10$$

Check
To check that the correct general term or function expression has been found, reproduce a few of the members of the sequence by substituting values for n in the general term and checking that the sequence found is the same as the given values. Wherever n occurs in the function expression or general term replace it by a value. $a_n = n$ for $n = 1$ gives $a_1 = 1$ and for $n = 2$ gives $a_2 = 2$ etc.

Example 18.2 Find the general term of the sequence

$$1, 4, 9, 16, 25, 36, \ldots$$

and also define the sequence in terms of a recurrence relation.

SOLUTION Notice that each term in the sequence is a complete square. The first term is 1^2, the second term 2^2 etc. We therefore speculate that the general term is

$$a_n = n^2 \qquad \text{where } n = 1 \text{ to } \infty$$

In function notation this is

$$a(n) = n^2$$

To define the sequence in terms of a recurrence relation means that we must find a way of getting to the $(n+1)$th term if we know the nth term. There is no prescribed way of doing this: we merely have to try out a few ideas of how to see a pattern in the sequence. In this case we can best see the pattern with the aid of a diagram where we represent the 'square numbers' using a square, as in Fig. 18.1. Here we can see that to get from 2^2 to 3^2 we need to add a row of 2 dots and a column of 3 dots. To get from 3^2 to 4^2 we need to add a row of 3 dots and a column of 4 dots. In general. to get from n^2 to $(n+1)^2$ we need to add a row of n dots and a column of $n+1$ dots.

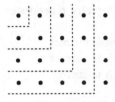

Figure 18.1 The 'square numbers'.

As n^2 is a_n and $(n+1)^2$ is a_{n+1}, the rule can be expressed as

$$a_{n+1} = a_n + n + n + 1$$
$$\Leftrightarrow a_{n+1} = a_n + 2n + 1$$

However, we need also to give the starting value in order to define the sequence using a recurrence relation, so we can say that

$$a_{n+1} = a_n + 2n + 1 \qquad \text{where } a_1 = 1$$

Check
To check, we substitute a few values into both the explicit definition and the recurrence relation to see if we correctly reproduce the terms in the sequence.

Substituting $n=1$, $n=2$, $n=3$, $n=4$ and $n=5$ into $a_n = n^2$ gives 1, 4, 9, 16, 25, correctly reproducing the first five terms of the sequence.

Substituting $n=1$, $n=2$, $n=3$ and $n=4$ into $a_{n+1} = a_n + 2n + 1$, where $a_1 = 1$ gives the following.

$n=1$: $a_2 = a_1 + 2 + 1$, as $a_1 = 1$, this gives $a_2 = 1 + 2 + 1 = 4$, which is correct.

$n=2$: $a_3 = a_2 + 4 + 1$, as $a_2 = 4$, this gives $a_3 = 4 + 4 + 1 = 9$, which is correct.

$n=3$: $a_4 = a_3 + 6 + 1$, as $a_3 = 9$, this gives $a_4 = 9 + 6 + 1 = 16$, which is correct.

$n=4$: $a_5 = a_4 + 8 + 1$, as $a_4 = 16$, this gives $a_5 = 16 + 8 + 1 = 25$, which is correct.

Example 18.3 Find a recurrence relation to define the Fibonacci sequence:

$$1, 1, 2, 3, 5, 8, 13, 21, \ldots$$

SOLUTION After some trial and error attempts to spot the rule, we should be able to see that the way to get the next number is to add up the last two numbers, so the next term after 13, 21, is $13 + 21 = 44$ and the next is $21 + 44 = 65$, etc.

The recurrence relation is therefore

$$a_{n+1} = a_n + a_{n-1}$$

and because we have two previous values of the sequence used in the recurrence relation then we also need to give two initial values. So we define $a_1 = 1$ and $a_2 = 1$.

The recurrence relation definition of the Fibonacci sequence is

$$a_{n+1} = a_n + a_{n-1} \qquad \text{where } a_1 = 1 \text{ and } a_2 = 1$$

Check

Substitute a few values for n into the recurrence relation to see if it correctly reproduces the given values of the sequence.

$n = 2$: $a_3 = a_2 + a_1$, where $a_1 = 1$ and $a_2 = 1$, gives $a_3 = 1 + 1 = 2$, which is correct.
$n = 3$: $a_4 = a_3 + a_2$, where $a_2 = 1$ and $a_3 = 2$, gives $a_4 = 2 + 1 = 3$, which is correct.
$n = 4$: $a_5 = a_4 + a_3$, where $a_4 = 3$ and $a_3 = 2$, gives $a_5 = 3 + 2 = 5$, which is correct.

Digital representation of signals

Suppose that we would like to give a digital representation of the sine wave of angular frequency 3:

$$f(t) = \sin(3t)$$

Then we might choose $a(n) = \sin(3n)$ to give the sequence of values. Substituting some values for n gives the sequence (to 2 s.f.):

$$n = 0: \ a(0) = \sin(0) = 0$$

$$n = 1: \ a(1) = \sin(3) = 0.14$$

$$n = 2: \ a(2) = \sin(6) = -0.27$$

$$n = 3: \ a(3) = \sin(9) = 0.41$$

This sequence is shown on a graph in Fig. 18.2. This graph does not look like the sine wave it is supposed to represent. This is due to 'undersampling'. $a(n) = \sin(3n)$ is the function $f(t) = \sin(3t)$ sampled at a sample rate of 1, which is very inadequate to see a good representation. Digital signals are usually expressed in terms of the sampling interval T so that a suitable sample rate can be chosen. For the function $f(t) = \sin(3t)$ this gives the sequence $a(n) = \sin(3Tn)$ where $n = 0, 1, 2, 3, \ldots$.

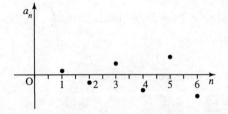

Figure 18.2 The sequence given by $a(n) = \sin(3n)$.

The original variable, usually time, t, can be given by $t = Tn$. Choosing $T = 0.1$, for instance, gives

$$a(n) = \sin(3 \times 0.1n) = \sin(0.3n) \qquad \text{where } t = nT = 0.1n$$

Substituting a few values for n gives

$$n = 0: \ a(0) = 0, \qquad t = 0$$
$$n = 1: \ a(1) = 0.3, \quad t = 0.1$$
$$n = 2: \ a(2) = 0.56, \ t = 0.2$$
$$n = 3: \ a(3) = 0.78, \ t = 0.3$$
$$n = 4: \ a(4) = 0.93, \ t = 0.4$$
$$n = 5: \ a(5) = 1, \qquad t = 0.5$$
$$n = 6: \ a(6) = 0.97, \ t = 0.6$$

The values are plotted against t in Fig. 18.3. We can see that this picture is a reasonable representation of the function. The digital representation of $f(t) = \sin(3t)$ is therefore $f(nT) = \sin(3nT)$, where n is an integer.

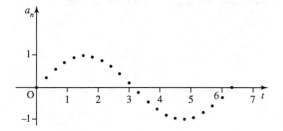

Figure 18.3 The function $f(t) = \sin(3t)$ sampled at an interval of $T = 0.1$, giving the sequence $a(n) = \sin(0.3n)$.

The problem of undersampling which we saw in Fig 18.2 leads to a phenomenon called aliasing. Instead of looking like $\sin(3t)$, Fig. 18.2 looks like a sine wave of much lower frequency. This same phenomenon is the one that makes car wheels, pictured on television, apparently rotate backwards and at the wrong frequency. The television picture is scanned 30 times per second, whereas the wheel on the car is probably revolving in excess of 30 times per second. The sample rate is insufficient to give a good representation of the movement of the wheel. The sampling theorem states that the sampling interval must be less than $T = 1/(2f)$ seconds in order to be able to represent a frequency of f hertz.

Example 18.4 Represent the function $y = 4\cos(10\pi t)$ as a sequence using a sampling interval of $T = 0.01$. What is the maximum sampling interval that could be used to represent this signal?

SOLUTION The digital representation of $y = 4\cos(10\pi t)$ is given by $y(0.01n) = 4\cos(0.1\pi n)$. Substituting some values for n gives (to 3 s.f.):

n	0	1	2	3	4	5	6	7	8	9	10
y	1	0.951	0.809	0.588	0.309	0	−0.309	−0.588	−0.809	−0.951	−1

The maximum sampling interval that could be used is $1/(2f)$, where f, the frequency in this case is 5, giving $T = 1/(2 \times 5) = 0.1$.

Example 18.5 A triangular wave of period 2 is given by the function

$$f(t)=t \qquad 0 \leqslant t < 1$$
$$f(t)=2-t \qquad 1 \leqslant t < 2$$

Draw a graph of the function and give a sequence of values for $t \geqslant 0$ at a sampling interval of 0.1.

SOLUTION To draw the continuous function use the definition $y=t$ between $t=0$ and 1 and draw the function $y=2-t$ in the region where t lies between 1 and 2. The function is of period 2, so that section of the graph is repeated between $t=2$ to 4, $t=4$ to 6 etc.

The sequence of values found by using a sampling interval of 0.1 is given by substituting $t = Tn = 0.1n$ into the function definition, giving

$$a(n) = 0.1n \qquad 0 < 0.1n < 1 \qquad \text{(for } n \text{ between 0 and 10)}$$
$$a(n) = 2 - 0.1n \qquad 1 < 0.1n < 2 \qquad \text{(for } n \text{ between 10 and 20)}$$

The sequence then repeates periodically.
This gives the sequence:

$$0, 0.1, 0.2, 0.3, 0.4, 0.5, 0.6, 0.7, 0.8, 0.9, 1, 0.9, 0.8, 0.7, 0.6, 0.5, 0.4, 0.3, 0.2, 0.1, 0, 0.1, 0.2, \ldots$$

The continuous function is plotted in Fig. 18.4(a) and the digital function in 18.4(b).

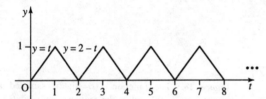

(a)

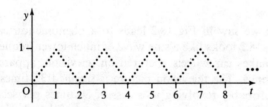

(b)

Figure 18.4 (a) A triangular wave of period 2 given by $f(t)=t, 0 < t \leqslant 1, f(t)=2-t, 1 < t < 2$, and (b) the function sampled at a sampling interval of 0.1.

Series

A series is the sum of a sequence of numbers or of functions. If the series contains a finite number of terms then it is a finite series; otherwise it is infinite. For example

$$1+2+3+4+5+6+7+8+10$$

is a finite series, while

$$1+1/2+1/4+1/8+1/16+ \ldots +1/2^n + \ldots$$

is an infinite series.

To represent series we may use the sigma notation introduced earlier.

$$\sum_{n=0}^{n=10} \frac{1}{2^n}$$

means 'sum all the terms $1/2^n$ for n from 0 to 10'.

Example 18.6 Express the following series in sigma notation:

$$-1+4-9+16+\ldots+256$$

SOLUTION To write in sigma notation we need first to express the general term in the sequence. We notice here that the pattern is that each term is a complete square with every other term multiplied by -1. The general term is therefore

$$(-1)^n n^2$$

The $(-1)^n$ part of this will just cause the sign of the term to be negative or positive depending on whether n is odd or even. We can now write

$$-1+4-9+16+\ldots+256 = \sum_{n=1}^{n=16} (-1)^n n^2$$

The limits of the summation are found by considering the value of n to use for the first term and the value of n to use for the last term. Check that the expression is correct by substituting a few values of n, which should recover terms in the original series.

We now look at two commonly encountered types of sequences and series, the arithmetic and geometric progressions.

18.3 ARITHMETIC PROGRESSIONS

An arithmetic progression (AP) is a sequence where each term is found by adding a fixed amount on to the previous term. This fixed amount is called the common difference. Some examples of arithmetic progressions are:

1. $-1, 3, 7, 11, 15, 19, 23, 27, \ldots$

Notice that successive terms can be found by adding 4 to the previous term

$$-1+4=3$$

$$3+4=7$$

$$7+4=11$$

$$\vdots$$

showing that the common difference is 4.

2. $25, 15, 5, -5, -15, \ldots$

Notice that successive terms can be found by adding -10 to the previous term:

$$25 + -10 = 15$$
$$15 + -10 = 5$$
$$5 + -10 = -5$$
$$-5 + -10 = -15$$

showing that the common difference is -10.

It is not difficult to obtain the recurrence relation for the arithmetic progression. If we call the common difference d then the $(n+1)$th term can be found from the previous term, the nth, by adding on d; that is, $a_{n+1} = a_n + d$.

This can also be expressed using the difference operator, Δ (Δ is the Greek capital letter delta) so that $\Delta a_n = d$, where $\Delta a_n = a_{n+1} - a_n$.

If the first term is a and the common difference is d, then the sequence is

$$a, a+d, a+2d, a+3d, a+4d, a+5d, a+6d, \ldots$$

The second term is $a + d$, the fourth is $a + 3d$, the seventh term is $a + 6d$, and the general term is

$$a_n = a + (n-1)d$$

Example 18.7 The seventh term of an AP is 11 and the 16th term is 29. Find the common difference, the first term of the sequence and the nth term.

SOLUTION If the first term of the sequence is a and the common difference is d, then the seventh term is given by $a_n = a + (n-1)d$, with $n = 7$ so

$$a + 6d = 11 \tag{18.1}$$

Similarly, the 16th term is $a + 15d$ and as we are given that this is 29 we have

$$a + 15d = 29 \tag{18.2}$$

Eq. (18.1) $-$ Eq. (18.2) gives $9d = 18 \Leftrightarrow d = 2$, and substituting this into Eq. (18.1) gives

$$a + 6 \times 2 = 11 \Leftrightarrow a = 11 - 12 \Leftrightarrow a = -1$$

i.e. the first term is -1 and the common difference is 2. Hence the nth term is $a + (n-1)d = -1 + (n-1)2 = -1 + 2n - 2$, giving $a_n = 2n - 3$.

Check
To check that the general term is correct for this sequence, substitute $n = 7$, giving $a_7 = 2(7) - 3 = 14 - 3 = 11$; substitute $n = 16$, giving $a_{16} = 2(16) - 3 = 29$, which are the values given in the problem.

The sum of n terms of an arithmetic progression (AP)

There are some simple formulae which can be used to find the sum of the first n terms of an AP. These can be found by writing out all the terms of a general AP from the first term to the last term, l, and then adding on the same series again, but this time reversing it. We will begin by finding the sum of 20 terms of an AP with first term 1 and common difference 3, giving the general term as $1 + (n-1) \times 3$ and the last term as $1 + 19 \times 3$:

$$S_{20} = 1 + (1+3) + (1+2 \times 3) + \ldots + (1+18 \times 3) + (1+19 \times 3)$$

Reversing:

$$S_{20} = (1+19 \times 3) + (1+18 \times 3) + (1+17 \times 3) + \ldots + (1+1 \times 3) + 1$$

adding

$$2S_{20} = (2+19 \times 3) + (2+19 \times 3) + (2+19 \times 3) + \ldots + (2+19 \times 3) + (2+19 \times 3)$$

Notice that each term in the last line is the same, and is equal to the sum of the first term (1) and the last term $(1+19 \times 3)$, giving $2+19 \times 3$. As there are 20 terms, we have

$$2S_{20} = 20 \times (2+19 \times 3)$$

$$S_{20} = 10 \times (2+19 \times 3) = 590$$

It would obviously be simpler to be able to use a formula to calculate this rather than having to repeat this process for every AP. Therefore we go through the same process for an AP of first term a and common difference d, with last term l.

The sum of the first n terms of an AP is given by

$$S_n = a + (a+d) + (a+2d) + \ldots + (a+(n-2)d) + (a+(n-1)d)$$

$$S_n = (a+(n-1)d) + (a+(n-2)d) + (a+(n-3)d) + \ldots + (a+d) + a$$

$$2S_n = (2a+(n-1)d) + (2a+(n-1)d) + (2a+(n-1)d) + \ldots + (2a+(n-1)d) + (2a+(n-1)d)$$

Using the fact that there are n terms we have

$$2S_n = n(2a+(n-1)d)$$

$$S_n = \frac{n(2a+(n-1)d)}{2}$$

giving the first of two formulae that can be used to find the sum of n terms of an AP.

An alternative formula is found by noticing that the sum is given by the number of terms multiplied by the average term. The average term is half the sum of the first term and the last term:

$$\text{average term} = \tfrac{1}{2}(a+l)$$

giving the sum of n terms as

$$S_n = \frac{n}{2}(a+l)$$

This is the second of the two formulae which may be used to find the sum of the first n terms of an AP.

Example 18.8 Find the sum of an AP whose first term is 3 and which has 12 terms ending with -15.

SOLUTION Using the formula

$$S_n = \frac{n}{2}(a+l)$$

then substituting $n = 12$, $a = 3$ and $l = -15$:

$$S_n = \frac{12}{2}(3-15) = 6(-12) = -72$$

Example 18.9 Find

$$\sum_{r=1}^{r=9}\left(1-\frac{r}{4}\right)$$

SOLUTION Write out the series by substituting values for r:

$$\sum_{r=1}^{r=9}\left(1-\frac{r}{4}\right) = \left(1-\frac{1}{4}\right) + \left(1-\frac{2}{4}\right) + \left(1-\frac{3}{4}\right) + \left(1-\frac{4}{4}\right) + \left(1-\frac{5}{4}\right) + \left(1-\frac{6}{4}\right) + \dots + \left(1-\frac{9}{4}\right)$$

This is the sum of an arithmetic progression with nine terms, where

$$a = \tfrac{3}{4} \quad \text{and} \quad d = -\tfrac{1}{4}$$

Using

$$S_n = \frac{n}{2}(2a + (n-1)d)$$

gives

$$S_9 = \frac{9}{2}\left(\frac{6}{4} + (9-1)\left(-\frac{1}{4}\right)\right) = \frac{9}{2}\left(\frac{6}{4} - \frac{8}{4}\right) = -\frac{9}{4}$$

18.4 GEOMETRIC PROGRESSIONS

A geometric progression (GP) is a sequence where each term is found by multiplying the previous term by a fixed number. This fixed number is called the common ratio, r. We have already come across examples of geometric progressions in Chapter 14, where we looked at exponential growth. There we had the example of £1 deposited in a bank with a real rate of growth of 3 per cent, so we get the sequence 1, 1.03, 1.06, 1.09, 1.13, 1.16, 1.19,... (expressed to the nearest penny), where each year the amount in the bank is multiplied by 1.03.

Some more examples of GPs are

1. 16, 8, 4, 2, 1, 0.5, 0.25, 0.125,...

Notice that successive terms can be found by multiplying the previous term by $\tfrac{1}{2}$:

$$16 \times \tfrac{1}{2} = 8$$

$$8 \times \tfrac{1}{2} = 4$$

$$4 \times \tfrac{1}{2} = 2$$

$$\vdots$$

showing that the common ratio is $\tfrac{1}{2}$.

2. 1, 3, 9, 27, 81,...

Notice that successive terms can be found by multiplying the previous term by 3:

$$1 \times 3 = 3$$

$$3 \times 3 = 9$$

$$9 \times 3 = 27$$

$$\vdots$$

showing that the common ratio is 3.

3. $-1, 2, -4, 8, -16,...$

Notice that successive terms can be found by multiplying the previous term by -2:

$$(-1) \times (-2) = 2$$

$$2 \times (-2) = -4$$

$$(-4) \times (-2) = 8$$

$$\vdots$$

showing that the common ratio is -2.

It is not difficult to obtain the recurrence relation for the geometric progression. If we call the common ratio r, then the $(n+1)$th term can be found from the previous term, the nth, by multiplying by r, that is $a_{n+1} = ra_n$. This can also be expressed using the difference operator, Δ, so that $\Delta a_n = (r-1)a_n$, where $\Delta a_n = a_{n+1} - a_n$.

If a GP has first term a and common ratio r, the sequence is

$$a, ar, ar^2, ar^3, ar^4,..., ar^{n-1},...$$

The second term is ar, the fourth term is ar^3, the seventh term is ar^6, and the general term is given by

$$a_n = ar^{n-1}$$

Example 18.10 Find the general term of the geometric progression

$$16, 8, 4, 2, 1, 0.5, 0.25, 0.125,...$$

SOLUTION This GP has first term 16. The common ratio is found by taking the ratio of any two successive terms. Take the ratio of the first two terms (second term divided by the first term), giving

$$r = \frac{8}{16} = \frac{1}{2}$$

The general term is then given by $ar^{n-1} = 16(\frac{1}{2})^{n-1}$.

Example 18.11 A geometric progression has third term 12 and fifth term 48. Find the first term and the common ratio.

SOLUTION Call the first term a and the common ratio r. We know that the nth term is given by

$$a_n = ar^{n-1}$$

The fact that the third term is 12 gives the equation

$$ar^2 = 12 \tag{18.3}$$

and the fact that the fifth term is 48 gives the equation

$$ar^4 = 48 \tag{18.4}$$

Dividing Eq. (18.3) by Eq. (18.4) gives

$$r^2 = 4$$

This means that there are two possible values for the common ratio: either 2 or -2.
To find the first term, substitute for r into Eq. (18.3), giving

$$ar^2 = 12 \quad \text{and} \quad r = \pm 2$$

$$\Rightarrow 4a = 12$$

$$\Leftrightarrow a = 3$$

So the first term is 3.

The sum of a geometric progression

Consider the sum of the first six terms of the GP with first term 2 and common ratio 4.

To try to find the sum we first write out the original series and then multiply the whole series by the common ratio, as this will reproduce the same terms in the series, only shifted up one place. We are then able to subtract the two expressions:

$$S_6 = 2 + 8 + 32 + 128 + 512 + 2048$$

$$4S_6 = 8 + 32 + 128 + 512 + 2048 + 9192$$

$$S_6(1 - 4) = 2 - 9192$$

So

$$S_6 = \frac{2 - 9192}{1 - 4}$$

As $9192 = 2 \times 4^6$, this gives the sum of the first six terms as

$$S_6 = \frac{2(1 - 4^6)}{1 - 4}$$

Applying this process to a general GP gives a formula for the sum of the first n terms.

Consider the sum, S_n, of the first n terms of a GP whose first term is a and whose common ratio is r. Multiply this by r and subtract.

$$S_n = a + ar + ar^2 + \ldots + ar^{n-2} + ar^{n-1}$$

$$rS_n = ar + ar^2 + ar^3 + \ldots + ar^{n-1} + ar^n$$

This gives

$$S_n(1 - r) = a - ar^n = a(1 - r^n)$$

$$\Leftrightarrow S_n = \frac{a(1 - r^n)}{1 - r}$$

If $r > 1$ it may be more convenient to write

$$S_n = \frac{a(r^n - 1)}{r - 1}$$

Example 18.12 The second term of a geometric progression is 2 and the fifth term is 0.031 25. Find the first term, the common ratio and the sum of the first 10 terms.

SOLUTION Calling the first term a and the common ratio r, then the nth term is ar^{n-1}. The second term is 2, giving the equation

$$ar = 2 \qquad (18.5)$$

The fifth term is 0.031 25 giving the equation

$$ar^4 = 0.031\,25 \qquad (18.6)$$

Dividing Eq. (18.6) by Eq. (18.5) gives

$$\frac{ar^4}{ar} = \frac{0.031\,25}{2}$$

$$\Leftrightarrow r^3 = 0.015\,625$$

$$\Leftrightarrow r = (0.015\,625)^{1/3}$$

$$\Leftrightarrow r = 0.25$$

Substituting this value for r into Eq. (18.5) gives

$$a(0.25) = 2 \Leftrightarrow a = \frac{2}{0.25}$$

$$\Leftrightarrow a = 8$$

Therefore the first term is 8 and the common ratio is 0.25.

The sum of the first n terms is given by

$$S_n = \frac{a(1 - r^n)}{1 - r}$$

Substituting $a = 8$, $r = 0.25$ and $n = 10$ gives

$$S_{10} = \frac{10(1 - 0.25^{10})}{1 - 0.25} = 13.33 \text{ to 4 s.f.}$$

Example 18.13 The general term of a series is given by

$$a_n = \frac{2^{n+1}}{3^n}$$

Show that the terms of the series form a geometric progression and find the sum of the first n terms.

SOLUTION To show that this is a geometric progression we must show that consecutive terms have a common ratio. Take two terms: the mth term and the $(m+1)$th term. The mth term is

$$\frac{2^{m+1}}{3^m} = a_m$$

and the $(m+1)$th term is found by substituting $n = m + 1$ into the expression for the general term, $a_n = 2^{n+1}/3^n$, which gives

$$a_{m+1} = \frac{2^{(m+1)+1}}{3^{(m+1)}} = \frac{2^{m+2}}{3^{m+1}}$$

We can now spot that

$$a_{m+1} = \tfrac{2}{3} a_m$$

meaning that the $(m+1)$th term is found by multiplying the mth term by $\tfrac{2}{3}$.

Alternatively, we could divide the $(m+1)$th term by the mth term:

$$\frac{a_{m+1}}{a_m} = \frac{2^{m+2}/3^{m+1}}{2^{m+1}/3^m} = \frac{2^{m+2}}{3^{m+1}} \times \frac{3^m}{2^{m+1}} = \frac{2}{3}$$

giving the common ratio as $\tfrac{2}{3}$.

To find the sum of n terms we need to know the first term. To find this, substitute $n = 1$ into $2^{n+1}/3^n$, giving $2^2/3 = \tfrac{4}{3}$. Thus the sum of n terms is given by

$$\frac{\tfrac{4}{3}(1 - (\tfrac{2}{3})^n)}{1 - \tfrac{2}{3}} = \frac{\tfrac{4}{3}(1 - (\tfrac{2}{3})^n)}{\tfrac{1}{3}} = 4(1 - (\tfrac{2}{3})^n)$$

The sum to infinity of a geometric progression

Consider the last example concerning the series whose general term is $2^{n+1}/3^n$. This can be written as

$$\frac{4}{3} + \frac{4}{3}\left(\frac{2}{3}\right) + \frac{4}{3}\left(\frac{2}{3}\right)^2 + \frac{4}{3}\left(\frac{2}{3}\right)^3 + \frac{4}{3}\left(\frac{2}{3}\right)^4 + \ldots + \frac{2^{n+1}}{3^n} + \ldots$$

and the sum of n terms is

$$4(1 - (\tfrac{2}{3})^n)$$

We can write out this sum for various values of n to seven significant figures, as in Table 18.1. After 40 terms the sum has become 4 to seven significant figures, and however many more terms are considered the sum is found to be 4 to seven significant figures. This shows that the limit of the sum is 4 to 7 s.f. The limit of the sum of n terms as n tends to infinity is called the sum to infinity.

Table 18.1 The sum of the first n terms of the GP expressed to seven significant figures for various values of n

n	5	10	15	20	25	30	35	40	45	50
$4(1 - (\tfrac{2}{3})^n)$	3.473 25	3.930 634	3.990 865	3.999 842	3.999 842	3.999 979	3.999 997	4	4	4

We can see that the limit is exactly 4 in this case by looking at what happens to $4(1 - (\tfrac{2}{3})^n)$ as n tends to infinity.

$$4(1 - (\tfrac{2}{3})^n) = 4 - 4(\tfrac{2}{3})^n$$

The second term becomes smaller and smaller as n gets bigger and bigger, and we can see that $(\tfrac{2}{3})^n \to 0$ as $n \to \infty$. Therefore

$$S_\infty = \lim_{n \to \infty} S_n = \lim_{n \to \infty} (4 - 4(\tfrac{2}{3})^n) = 4$$

This approach can also be applied to the general GP, where

$$S_n = \frac{a(1-r^n)}{1-r} = \frac{a}{1-r} - \frac{ar^n}{1-r}$$

Then if $|r| < 1$ we have

$$\lim_{n \to \infty} r^n = 0$$

which gives

$$\lim_{n \to \infty} (S_n) = \lim_{n \to \infty} \left(\frac{a}{1-r} \right) - \lim_{n \to \infty} \left(\frac{ar^n}{1-r} \right) = \frac{a}{1-r}$$

We can write

$$S_\infty = \frac{a}{1-r} \qquad |r| < 1$$

Example 18.14 Find the sum to infinity of a GP with first term -10 and common ratio 0.1.

SOLUTION The formula for the sum to infinity gives

$$S_\infty = \frac{a}{1-r}$$

Substituting $a = -10$ and $r = 0.1$ gives

$$S_\infty = \frac{-10}{1-0.1} = \frac{-10}{0.9} = -11\tfrac{1}{9}$$

Example 18.15 Express the recurring decimal $0.0\dot{2} = (0.022\,222\,2\ldots)$ as a fraction.

SOLUTION

$$0.0\dot{2} = 0.02 + 0.002 + 0.0002 + \ldots$$

This is the sum to infinity of a GP with first term 0.02 and common ratio 0.1. The sum to infinity is therefore

$$S_\infty = \frac{0.02}{1-0.1} = \frac{0.02}{0.9} = \frac{2}{90}$$

giving $0.0\dot{2} = \tfrac{2}{90}$.

Example 18.16 Find the sum to infinity of

$$1 + z - z^2 + z^3 - z^4 + \ldots$$

where $0 < z < 1$.

SOLUTION The first term of this series is 1 and the common ratio is $-z$, giving the sum to infinity as

$$S_\infty = \frac{1}{1-(-z)} = \frac{1}{1+z}$$

Note that in the case of a geometrical progression with common ratio $|r| \geqslant 1$, the series sum will not tend to a finite limit. For instance, the sum of the GP

$$2+4+8+16+32+ \ldots +2^n$$

gets much larger each time a new term is added. We say that the sum of this series tends to infinity.

18.5 PASCAL'S TRIANGLE AND THE BINOMIAL SERIES

Expressions like $(3+2y)^5$ are called binomial expressions. Expanding these expressions can be very tedious, as we need to multiply out $(3+2y)(3+2y)(3+2y)(3+2y)(3+2y)$ term by term. To speed up this process we analyse the coefficients of the terms in the expansion and find that they make a triangular pattern, called Pascal's triangle.

Pascal's triangle

Consider a simpler case of $(1+x)^5$. To work this out we would first start with $(1+x)$ and multiply by $(1+x)$ to get $(1+x)^2$. We then multiply $(1+x)^2$ by $(1+x)$ to get $(1+x)^3$ etc. Carrying on this process gives the following

$$
\begin{aligned}
&1+x &\quad (1)\\
&1+2x+x^2 &\quad (2)\\
&1+3x+3x^2+x^3 &\quad (3)\\
&1+4x+6x^2+4x^3+x^4 &\quad (4)\\
&1+5x+10x^2+10x^3+5x^4+x^5 &\quad (5)
\end{aligned}
$$

If we write out the coefficients of each line of this in a triangular form we get

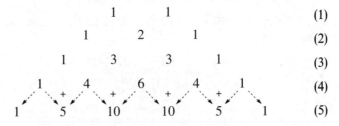

Notice that each line can be found from the line above by adding pairs of numbers, where the outer numbers are always 1. For instance, in line 5 the first number is 1, and the others are found by adding the two numbers above, $5=1+4$, $10=4+6$, $10=6+4$, $5=4+1$. The last number is 1.

Example 18.17 Expand $(1+x)^7$ in powers of x.

SOLUTION Write out the first seven lines of Pascal's triangle in order to find the coefficients in the expansion, giving

			1		1				(1)
		1		2		1			(2)
	1		3		3		1		(3)
1		4		6		4		1	(4)

$$\begin{array}{ccccccccc} & & & 1 & & 1 & & & & (1)\\ & & 1 & & 2 & & 1 & & & (2)\\ & 1 & & 3 & & 3 & & 1 & & (3)\\ 1 & & 4 & & 6 & & 4 & & 1 & (4)\\ \end{array}$$

Row (5): 1   5   10   10   5   1   (5)
Row (6): 1   6   15   20   15   6   1   (6)
Row (7): 1   7   21   35   35   21   7   1   (7)

Now write out the expansion with the powers of x, giving

$$(1+x)^7 = 1 + 7x + 21x^2 + 35x^3 + 35x^4 + 21x^5 + 7x^6 + x^7$$

As we can now easily find expansions of the form $(1+x)^n$, we now move on to the more difficult problem of expressions such as $(1+2y)^5$. We can find this by substituting $x = 2y$ into the expression for $(1+x)^5$:

$$(1+x)^5 = 1 + 5x + 10x^2 + 10x^3 + 5x^4 + x^5$$

Substitute $x = 2y$:

$$(1+2y)^5 = 1 + 5(2y) + 10(2y)^2 + 10(2y)^3 + 5(2y)^4 + (2y)^5$$

Remembering to take the powers of 2 as well as y this gives

$$(1+2y)^5 = 1 + 10y + 40y^2 + 80y^3 + 80y^4 + 32y^5$$

Finally, we look at $(3+2y)^5$, and to do this we need to be able to expand expressions like $(a+b)^5$. To find $(a+b)^5$, start by dividing inside the bracket by a, giving

$$(a+b)^5 = a^5\left(1 + \frac{b}{a}\right)^5$$

substitute $x = b/a$ and use the expansion for $(1+x)^5$:

$$a^5\left(1 + \frac{b}{a}\right)^5 = a^5\left(1 + 5\left(\frac{b}{a}\right) + 10\left(\frac{b}{a}\right)^2 + 10\left(\frac{b}{a}\right)^3 + 5\left(\frac{b}{a}\right)^4 + \left(\frac{b}{a}\right)^5\right)$$

Multiplying out gives

$$(a+b)^5 = a^5 + 5a^4b + 10a^3b^2 + 10a^2b^3 + 5ab^4 + b^5$$

Notice the pattern on the powers of a and b. The powers of a are decreasing term by term as the powers of b are increasing. Always the sum of the powers of a and b is 5.

We can now expand $(3+2x)^5$ by using

$$(a+b)^5 = a^5 + 5a^4b + 10a^3b^2 + 10a^2b^3 + 5ab^4 + b^5$$

and substituting $a = 3$ and $b = 2x$, giving

$$(3+2x)^5 = 3^5 + 5 \times 3^4(2x) + 10 \times 3^3(2x)^2 + 10 \times 3^2(2x)^3 + 5 \times 3(2x)^4 + (2x)^5$$

$$= 243 + 810x + 1080x^2 + 720x^3 + 240x^4 + 32x^5$$

Example 18.18 Expand $(2x-y)^6$.

SOLUTION Find the sixth row of Pascal's triangle:

$$
\begin{array}{ccccccccccccc}
& & & & & & 1 & & 1 & & & & \\
& & & & & 1 & & 2 & & 1 & & & \\
& & & & 1 & & 3 & & 3 & & 1 & & \\
& & & 1 & & 4 & & 6 & & 4 & & 1 & \\
& & 1 & & 5 & & 10 & & 10 & & 5 & & 1 \\
& 1 & & 6 & & 15 & & 20 & & 15 & & 6 & & 1
\end{array}
$$

This gives the expansion of $(a+b)^6$ as

$$a^6 + 6a^5b + 15a^4b^2 + 20a^3b^3 + 15a^2b^4 + 6ab^5 + b^6$$

Now substitute $a = 2x$ and $b = -y$

$$(2x-y)^6 = (2x)^6 + 6(2x)^5(-y) + 15(2x)^4(-y)^2 + 20(2x)^3(-y)^3$$
$$+ 15(2x)^2(-y)^4 + 6(2x)(-y)^5 + (-y)^6$$
$$(2x-y)^6 = 64x^6 - 192x^5y + 240x^4y^2 - 160x^3y^3 + 60x^2y^4 - 12xy^5 + y^6$$

Example 18.19 Expand

$$\left(x+\frac{1}{x}\right)^3$$

SOLUTION Find the third row of Pascal's triangle:

$$
\begin{array}{ccccccc}
& & & 1 & & 1 & \\
& & 1 & & 2 & & 1 \\
& 1 & & 3 & & 3 & & 1
\end{array}
$$

This gives the expansion of $(a+b)^3$ as

$$(a+b)^3 = a^3 + 3a^2b + 3ab^2 + b^3$$

Now substitute $a = x$ and $b = 1/x$ to give

$$\left(x+\frac{1}{x}\right)^3 = x^3 + 3x^2\left(\frac{1}{x}\right) + 3x\left(\frac{1}{x}\right)^2 + \left(\frac{1}{x}\right)^3$$

$$= x^3 + 3x + \frac{3}{x} + \frac{1}{x^3}$$

Example 18.20 Expand

$$(e^x - e^{-x})^4$$

SOLUTION Find the fourth row of Pascal's triangle:

$$
\begin{array}{ccccccc}
 & & & 1 & & 1 & \\
 & & 1 & & 2 & & 1 \\
 & 1 & & 3 & & 3 & & 1 \\
1 & & 4 & & 6 & & 4 & & 1
\end{array}
$$

This gives the expansion of $(a+b)^4$ as

$$(a+b)^4 = a^4 + 4a^3b + 6a^2b^2 + 4ab^3 + b^4$$

Now substitute $a = e^x$ and $b = -e^{-x}$ to give

$$(e^x - e^{-x})^4 = (e^x)^4 + 4(e^x)^3(-e^{-x}) + 6(e^x)^2(-e^{-x})^2 + 4(e^x)(-e^{-x})^3 + (-e^{-x})^4$$

$$= e^{4x} - 4e^{2x} + 6 - 4e^{-2x} + e^{-4x}$$

The binomial theorem

The binomial theorem gives a way of writing the terms which we have found for the binomial expansion without having to write out all the lines of Pascal's triangle to find the coefficients. The rth coefficient in the binomial expansion of $(1+x)^n$ is expressed by ^nC_r or

$$\binom{n}{r} = \frac{n!}{(n-r)!r!}$$

where '!' is the factorial sign. The factorial function is defined by

$$n! = n(n-1)(n-2)\ldots 1$$

For example

$$3! = 3 \times 2 \times 1 = 6$$

$$6! = 6 \times 5 \times 4 \times 3 \times 2 \times 1 = 720$$

The binomial expansion then gives

$$(1+x)^n = 1 + \binom{n}{1}x + \binom{n}{2}x^2 + \binom{n}{3}x^3 + \binom{n}{4}x^4 + \ldots + \binom{n}{r}x^r + \ldots + x^n$$

and

$$(a+b)^n = a^n + \binom{n}{1}a^{n-1}b + \binom{n}{2}a^{n-2}b^2 + \binom{n}{3}a^{n-3}b^3 + \ldots + \binom{n}{r}a^{n-r}b^r + \ldots + b^n$$

This can also be written as

$$(1+x)^n = 1 + nx + \frac{n(n-1)}{2!}x^2 + \frac{n(n-1)(n-2)}{3!}x^3 + \ldots + x^n$$

and the expansion for $(a+b)^n$ becomes

$$(a+b)^n = a^n + na^{n-1}b + \frac{n(n-1)}{2!}a^{n-2}b^2 + \frac{n(n-1)(n-2)}{3!}a^{n-3}b^3 + \ldots + b^n$$

Example 18.21 Expand $(1+x)^4$.

SOLUTION Using the binomial expansion

$$(1+x)^n = 1 + nx + \frac{n(n-1)}{2!}x^2 + \frac{n(n-1)(n-2)}{3!}x^3 + \ldots + x^n$$

substituting $n=4$ gives

$$(1+x)^4 = 1 + 4x + \frac{4(3)}{2!}x^2 + \frac{4(3)(2)}{3!}x^3 + x^4 = 1 + 4x + 6x^2 + 4x^3 + x^4$$

Example 18.22 Expand

$$\left(2 - \frac{1}{x}\right)^5$$

SOLUTION Using the binomial expansion of $(a+b)^n$

$$(a+b)^n = a^n + na^{n-1}b + \frac{n(n-1)}{2!}a^{n-2}b^2 + \frac{n(n-1)(n-2)}{3!}a^{n-3}b^3 + \ldots + b^n$$

this gives the expansion of $(a+b)^5$ as

$$(a+b)^5 = a^5 + 5a^4b + \frac{5(4)}{2!}a^3b^2 + \frac{5(4)(3)}{3!}a^2b^3 + \frac{5(4)(3)(2)}{4!}ab^4 + b^5$$

$$= a^5 + 5a^4b + 10a^3b^2 + 10a^2b^3 + 5ab^4 + b^5$$

Substitute $a=2$ and $b=-1/x$ to give

$$\left(2 - \frac{1}{x}\right)^5 = 2^5 + 5(2)^4\left(-\frac{1}{x}\right) + 10(2^3)\left(-\frac{1}{x}\right)^2 + 10(2)^2\left(-\frac{1}{x}\right)^3 + 5(2)\left(-\frac{1}{x}\right)^4 + \left(-\frac{1}{x}\right)^5$$

$$= 32 - \frac{80}{x} + \frac{80}{x^2} - \frac{40}{x^3} + \frac{10}{x^4} - \frac{1}{x^5}$$

Example 18.23 Find to 4 s.f., without using a calculator, $(2.95)^4$.

SOLUTION Write $2.95 = 3 - 0.05$. Thus we need to find $(3-0.05)^4$. Using the expansion for $(a+b)^4$:

$$(a+b)^4 = a^4 + 4a^3b + \frac{4(3)}{2!}a^2b^2 + \frac{(4)(3)(2)}{3!}ab^3 + b^4$$

Substitute $a=3$, $b=-0.05$:

$$(3-0.05)^4 = 3^4 + 4(3^3)(-0.05) + 6(3)^2(-0.05)^2 + 4(3)(-0.05)^3 + (-0.05)^4$$

$$= 81 - 5.4 + 0.135 - 0.0015 + 0.000\,006\,25$$

$$= 75.73 \text{ to 4 s.f.}$$

18.6 POWER SERIES

A power series is of the form

$$a_0 + a_1 x + a_2 x^2 + a_3 x^3 + a_4 x^4 + \ldots + a_n x^n + \ldots$$

Many functions can be approximated by a power series. To find a series we use repeated differentiation.

Suppose we wanted to find a power series for $\sin(x)$. We could write

$$\sin(x) = a_0 + a_1 x + a_2 x^2 + a_3 x^3 + a_4 x^4 + a_5 x^5 + \ldots + a_n x^n + \ldots \tag{18.7}$$

For $x = 0$ we know that $\sin(0) = 0$. Hence substituting $x = 0$ in Eq. (18.7) we find

$$0 = a_0$$

To find a_1 we differentiate both sides of Eq. (18.7), giving

$$\cos(x) = a_1 + 2a_2 x + 3a_3 x^2 + 4a_4 x^3 + 5a_5 x^4 + \ldots + na_n x^{n-1} + \ldots \tag{18.8}$$

Substitute $x = 0$, and as $\cos(0) = 1$ this gives

$$1 = a_1$$

Differentiating Eq. (18.8) we get

$$-\sin(x) = 2a_2 + 3 \times 2a_3 x^1 + 4 \times 3a_4 x^2 + 5 \times 4a_5 x^3 + \ldots + n(n-1)a_n x^{n-2} + \ldots \tag{18.9}$$

Therefore at $x = 0$

$$0 = 2a_2 \Leftrightarrow a_2 = 0$$

Differentiating Eq. (18.9) we get

$$-\cos(x) = 3 \times 2 \times 1 a_3 + 4 \times 3 \times 2a_4 x + \ldots + n(n-1)(n-2)a_n x^{n-3} + \ldots \tag{18.10}$$

Substituting $x = 0$ gives

$$-1 = 3! a_3 \Leftrightarrow a_3 = -1/3!$$

A pattern is emerging. We can write

$$\sin(x) = x - \frac{x^3}{3!} + \frac{x^5}{5!} - \frac{x^7}{7!} + \ldots$$

This is a power series for $\sin(x)$ which we have found by expanding around $x = 0$.

When we expand around $x = 0$, we find a special case of the Taylor series expansion, called a Maclaurin series.

Maclaurin series: definition

If a function $f(x)$ is defined for values of x around $x = 0$, within some radius R, that is for $-R < x < R$ (or $|x| < R$), and if all its derivatives are defined, then:

$$f(x) = f(0) + f'(0)x + \frac{f''(0)x^2}{2!} + \frac{f'''(0)x^3}{3!} + \ldots + \frac{f^{(n)}(0)x^n}{n!} + \ldots \tag{18.11}$$

Notice that this is a power series with the coefficient sequence

$$a_n = \frac{f^{(n)}(0)}{n!}$$

where $f^{(n)}(0)$ is found by finding the nth derivative of $f(x)$ with respect to x and then substituting $x=0$.

Example 18.24 Find the Maclaurin series for $f(x)=e^x$ and give the range of values of x for which the series is valid.

SOLUTION Find all order derivatives of $f(x)$ and substitute $x=0$:

$$f(x)=e^x,\ f'(x)=e^x,\ f''(x)=e^x,\ f'''(x)=e^x,\ f^{(iv)}(x)=e^x$$

$$f(0)=e^0=1,\ f'(0)=1,\ f''(0)=1,\ f'''(0)=1,\ f^{(iv)}(0)=1$$

Therefore, substituting into Eq. (18.11) the Maclaurin series is

$$e^x=1+x+\frac{x^2}{2!}+\frac{x^3}{3!}+\frac{x^4}{4!}+\dots+\frac{x^n}{n!}+\dots$$

As e^x exists for all values of x we can use this series for all values of x.

Example 18.25 Find a power series for $\sinh(x)$ and give the values of x for which it is valid.

SOLUTION Find all order derivatives of $\sinh(x)$ and substitute $x=0$ in each one:

$$f(x)=\sinh(x),\ f'(x)=\cosh(x),\ f''(x)=\sinh(x),\ f'''(x)=\cosh(x),\ \text{etc.}$$

$$f(0)=\sinh(0)=0,\ f'(0)=1,\ f''(0)=0,\ f'''(0)=1,\ f^{(iv)}(0)=0$$

Then, substituting into Eq. (18.11)

$$\sinh(x)=x+\frac{x^3}{3!}+\frac{x^5}{5!}+\dots$$

As $\sinh(x)$ is defined for all values of x, the series is valid for all values of x.

Example 18.26 Expand

$$f(x)=\frac{1}{1+x}$$

in powers of x.

SOLUTION Find all order derivatives of $f(x)$ and substitute $x=0$:

$$f(x)=\frac{1}{1+x}=(1+x)^{-1},\ f'(x)=-1(1+x)^{-2},\ f''(x)=(-1)(-2)(1+x)^{-3}$$

$$f(0)=1,\ f'(0)=-1,\ f''(0)=2!,\ f'''(0)=-3!$$

Then, substituting into Eq. (18.11)

$$\frac{1}{1+x}=1-x+\frac{2!}{2!}x^2-\frac{3!}{3!}x^3+\dots$$

$$=1-x+x^2-x^3+\dots+(-1)x^n+\dots$$

As $1/(1+x)$ is not defined at $x=-1$ we can only use this series for $|x|<1$.

The binomial theorem revisited

The binomial theorem as stated in the previous section was only given for n as a whole positive number. We can now find the binomial expansion for $(1+x)^n$ for all values of n using the Maclaurin series.

$$f(x)=(1+x)^n$$

Then

$$f'(x)=n(1+x)^{n-1}$$
$$f''(x)=n(n-1)(1+x)^{n-2}$$
$$f'''(x)=n(n-1)(n-2)(1+x)^{n-3}$$

Substituting $x=0$ we get

$$f(0)=1,\ f'(0)=n,\ f''(0)=n(n-1),\ f'''(0)=n(n-1)(n-2)$$

Therefore, using Eq. (18.11) for the Maclaurin series, we find

$$(1+x)^n=1+nx+\frac{n(n-1)}{2!}x^2+\frac{n(n-1)(n-2)}{3!}x^3+\dots \tag{18.12}$$

Notice that n can take fractional or negative values, but if n is negative $|x|<1$ (as found in Example 18.26), for many fractional values of n we also need to keep to the restriction $|x|<1$.

Example 18.27 Expand $(1+x)^{1/2}$ in powers of x.

SOLUTION Using the binomial expansion

$$(1+x)^n=1+nx+\frac{n(n-1)}{2!}x^2+\frac{n(n-1)(n-2)}{3!}x^3+\dots$$

substituting $n=\frac{1}{2}$ gives

$$(1+x)^{1/2}=1+\frac{x}{2}+\frac{(\frac{1}{2})(-\frac{1}{2})}{2!}x^2+\frac{(\frac{1}{2})(-\frac{1}{2})(-\frac{3}{2})}{3!}x^3+\frac{(\frac{1}{2})(-\frac{1}{2})(-\frac{3}{2})(-\frac{5}{2})}{4!}x^4+\dots$$

$$=1+\frac{x}{2}-\frac{1}{8}x^2+\frac{x^3}{16}-\frac{5}{128}x^4+\dots$$

Notice that $(1+x)^{1/2}=\sqrt{1+x}$ is not defined for $x<-1$ so the series is only valid for $|x|<1$.

Series to represent products and quotients

Example 18.28 Find the Maclaurin series up to the term in x^3 for the function

$$f(x)=\frac{(1+x)^{1/2}}{1-x}$$

SOLUTION As this function would be difficult to differentiate three times (to use the Maclaurin series directly) we use

$$f(x)=(1+x)^{1/2}(1-x)^{-1}$$

and then find series for the two terms in the product and multiply them together.

$$(1+x)^{1/2} = 1 + \frac{x}{2} - \frac{1}{8}x^2 + \frac{x^3}{16} - \frac{5}{128}x^4 + \ldots$$

$$(1-x)^{-1} = 1 + x + \frac{(-1)(-2)}{2!}(-x)^2 + \frac{(-1)(-2)(-3)}{3!}(-x)^3 + \ldots$$

$$= 1 + x + x^2 + x^3 + \ldots$$

Then

$$(1+x)^{1/2}(1-x)^{-1} = \left(1 + \frac{x}{2} - \frac{1}{8}x^2 + \frac{x^3}{16} - \frac{5}{128}x^4 + \ldots\right)(1 + x + x^2 + x^3 + \ldots)$$

Up to the term in x^3:

$$(1+x)^{1/2}(1-x)^{-1} = 1 + x + x^2 + x^3 + \frac{x}{2} + \frac{x^2}{2} + \frac{x^3}{2} - \frac{x^2}{8} - \frac{x^3}{8} + \frac{x^3}{16} + \ldots$$

$$= 1 + \left(1 + \frac{1}{2}\right)x + \left(1 + \frac{1}{2} - \frac{1}{8}\right)x^2 + \left(1 + \frac{1}{2} - \frac{1}{8} + \frac{1}{16}\right)x^3 + \ldots$$

$$= 1 + \frac{3}{2}x + \frac{11}{8}x^2 + \frac{23}{16}x^3 + \ldots$$

Approximation

Power series can be used for approximations.

Example 18.29 Use a series expansion to find $\sqrt{1.06}$ correct to five significant figures.

SOLUTION We need to write $\sqrt{1.06}$ in a way that we could use the binomial expansion and so we use $\sqrt{1.06} = \sqrt{1+0.06} = (1+0.06)^{1/2}$ When doing this, it is important that the second term, in this case 0.06, should be a small number so that its higher powers will tend towards zero.

We can now use the binomial expansion, taking terms up to x^3, as we estimate that terms beyond that will be very small. Using Eq. (18.12), we find

$$(1+x)^{1/2} = 1 + \frac{x}{2} - \frac{1}{8}x^2 + \frac{x^3}{16} - \ldots$$

Now substitute $x = 0.06$, giving

$$\sqrt{1.06} \simeq 1 + \frac{0.06}{2} - \frac{(0.06)^2}{8} + \frac{(0.06)^3}{16}$$

$$= 1 + 0.03 - 0.000\,45 + 0.000\,013\,5 = 1.029\,563\,5$$

$$\Rightarrow \sqrt{1.06} = 1.0296 \text{ to 5 s.f.}$$

Example 18.30 Find $\sin(0.1)$ correct to five decimal places by using a power series expansion.

SOLUTION At the beginning of this section we found that the power series expansion for

$\sin(x)$ was as follows:

$$\sin(x) = x - \frac{x^3}{3!} + \frac{x^5}{5!} - \ldots$$

We substitute $x = 0.1$ to find $\sin(0.1)$ and continue until the next term is small compared with $0.000\,005$, which means that it would not effect the result when calculated to five decimal places:

$$\sin(0.1) = 0.1 - \frac{(0.1)^3}{3!} + \frac{(0.1)^5}{5!} - \ldots$$

$$= 0.1 - 0.000\,16\dot{6} + 0.000\,000\,83\dot{3} - \ldots$$

$$= 0.099\,83 \text{ to 5 d.p.}$$

Taylor series

The Maclaurin series is just a special case of the Taylor series. A Taylor series is a series expansion of a function not necessarily taken around $x = 0$.

If a function $f(x)$ is defined for values of x around $x = a$, within some radius R, that is for $a - R < x < a + R$ (or $|x - a| < R$) and if all its derivatives are defined, then:

$$f(x) = f(a) + f'(a)(x - a) + \frac{f''(a)}{2!}(x - a)^2 + \frac{f'''(a)}{3!}(x - a)^3 + \ldots + \frac{f^{(n)}(a)}{n!}(x - a)^n \ldots \quad (18.13)$$

Substituting $x = a + h$ gives, where h is usually considered to be a small value,

$$f(a + h) = f(a) + f'(a)h + \frac{f''(a)}{2!}h^2 + \frac{f'''(a)}{3!}h^3 + \ldots + \frac{f^{(n)}(a)}{n!}h^n \ldots \quad (18.14)$$

Example 18.31 Given $\sin(45°) = 1/\sqrt{2}$ and $\cos(45°) = 1/\sqrt{2}$, approximate $\sin 44°$ by using a power series expansion.

SOLUTION

$$\sin(44°) = \sin(45° - 1°)$$

Remember that the sine function is defined as a function of radians, so we must convert the angles to radians in order to use the Taylor series: $45° = \pi/4$ and $1° = \pi/180$. Expand using the Taylor series for $\sin(a + h)$, where $a = \pi/4$ and using Eq. (18.14):

$$f(a + h) = f(a) + f'(a)h + \frac{f''(a)}{2!}h^2 + \frac{f'''(a)}{3!}h^3 + \ldots$$

$$f(x) = \sin(x), \ f'(x) = \cos(x), \ f''(x) = -\sin(x), \ f'''(x) = \cos(x)$$

$$f(\pi/4) = \frac{1}{\sqrt{2}}, \ f'(\pi/4) = \frac{1}{\sqrt{2}}, \ f''(\pi/4) = -\frac{1}{\sqrt{2}}, \ f'''(\pi/4) = -\frac{1}{\sqrt{2}}$$

So

$$\sin(\pi/4 + h) = \frac{1}{\sqrt{2}} + \frac{1}{\sqrt{2}}h - \frac{1}{\sqrt{2}}\frac{h^2}{2!} - \frac{1}{\sqrt{2}}\frac{h^3}{3!} + \ldots$$

Substituting $h = -\pi/180$ radians we get

$$\sin(44°) = \frac{1}{\sqrt{2}}\left(1 - \frac{\pi}{180} - \frac{1}{2}\left(\frac{\pi}{180}\right)^2 + \frac{1}{6}\left(\frac{\pi}{180}\right)^3 + \ldots\right)$$

$$= \frac{1}{\sqrt{2}}(1 - 0.017\,45 - 0.000\,152\,3 + 0.000\,000\,9 + \ldots) = 0.694\,66 \text{ to 5 d.p.}$$

L'Hôpital's rule

When sketching the graphs of functions in Chapter 17 we looked at sketching graphs where the function is undefined for some values of x. The function $f(x) = 1/x$, for instance is not defined when $x = 0$ and tends to $-\infty$ as $x \to 0^-$ and tends to $+\infty$ as $x \to 0^+$. Not all functions that have undefined points tend to $\pm\infty$ near the point where they are undefined. For example, consider the function $f(x) = \sin(x)/x$. The graph of this function is shown in Fig. 18.5. The function is not defined for $x = 0$, which we can see by substituting $x = 0$ into the function expression. This gives a zero in the denominator and hence an attempt to divide by 0, which is undefined. However, we can see from the graph that the function, rather than tending to plus or minus infinity as $x \to 0$, just tends to 1. This is very useful, because we are able to 'patch' the function by giving it a value at $x = 0$ and the new function is defined for all values of x.

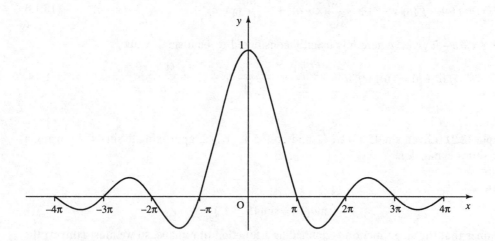

Figure 18.5 The graph of the function $f(x) = \sin(x)/x$.

We can define a new function. This particular function is quite famous, and is called the sinc function:

$$\mathrm{sinc}(x) = \begin{cases} \dfrac{\sin(x)}{x} & \text{where } x \neq 0 \\ 1 & \text{where } x = 0 \end{cases}$$

The points where functions may tend to a finite limit can be identified by looking out for points which lead to 0/0. These are called indeterminate points, indicating that they are a special type of undefined point.

We can examine what happens to the function near the indeterminate point by using the power series expansions of the denominator and numerator. Substituting the series for $\sin(x)$ into the expression $\sin(x)/x$ gives

$$\frac{\sin(x)}{x} = \frac{1}{x}\left(x - \frac{x^3}{3!} + \frac{x^5}{5!} - \frac{x^7}{7!} - \cdots\right)$$

and therefore

$$\lim_{x \to 0} \frac{\sin(x)}{x} = \lim_{x \to 0} \frac{1}{x}\left(x - \frac{x^3}{3!} + \frac{x^5}{5!} - \cdots\right) = \lim_{x \to 0}\left(1 - \frac{x^2}{3!} + \frac{x^4}{4!} - \cdots\right)$$

The last expression is defined at $x = 0$, so we can substitute $x = 0$ in order to find the limit. This gives the value 1. L'Hôpital's rule is a quick way of finding this limit without needing to write out the series specifically.

L'Hôpital's rule states that if a function $f(x) = g(x)/h(x)$ is indeterminate at $x = a$, then

$$\lim_{x \to a} \frac{g(x)}{h(x)} = \lim_{x \to a} \frac{g'(x)}{h'(x)}$$

If $g'(a)/h'(a)$ is defined we can then use

$$\lim_{x \to a} \frac{g'(x)}{h'(x)} = \frac{g'(a)}{h'(a)}$$

and if $g'(a)/h'(a)$ is indeterminate we can use the rule again.

We can show this to be true by using expression (18.13) for the Taylor series expansion about a:

$$\frac{g(x)}{h(x)} = \frac{g(a) + (x-a)g'(a) + ((x-a)/2!)g''(a) + \cdots}{h(a) + (x-a)h'(a) + ((x-a)/2!)h''(a) + \cdots}$$

and given that $g(a) = 0$ and $h(a) = 0$

$$\lim_{x \to a} \frac{g(x)}{h(x)} = \lim_{x \to a} \frac{(x-a)g'(a) + ((x-a)^2/2!)g''(a) + \cdots}{(x-a)h'(a) + ((x-a)^2/2!)h''(a) + \cdots}$$

$$= \lim_{x \to a} \frac{g'(a) + ((x-a)/2!)g''(a) + \cdots}{h'(a) + ((x-a)/2!)h''(a) + \cdots} = \frac{g'(a)}{h'(a)} \qquad \text{(if } h'(a) \neq 0\text{)}$$

Example 18.32 Find

$$\lim_{x \to 1} \frac{x^3 - 2x^3 + 4x - 3}{4x^2 - 5x + 1}$$

SOLUTION Substituting $x = 1$ into

$$\frac{x^3 - 2x^2 + 4x - 3}{4x^2 - 5x + 1}$$

gives 0/0, which is indeterminate. Using L'Hôpital's rule, we differentiate the top and bottom lines:

$$\lim_{x \to 1} \frac{x^3 - 2x^2 + 4x - 3}{4x^2 - 5x + 1} = \lim_{x \to 1} \frac{3x^2 - 4x + 4}{8x - 5}$$

We find that the new expression is defined at $x=1$, so

$$\lim_{x \to 1} \frac{3x^2 - 4x + 4}{8x - 5} = \frac{3 - 4 + 4}{8 - 5} = \frac{3}{3} = 1$$

Example 18.33 Find

$$\lim_{x \to 0} \frac{\cos(x) - 1}{x^2}$$

SOLUTION Substituting $x=0$ into $(\cos(x) - 1)/x^2$ gives $0/0$, which is indeterminate. Therefore, using L'Hôpital's rule, we differentiate the top and bottom lines:

$$\lim_{x \to 0} \frac{\cos(x) - 1}{x^2} = \lim_{x \to 0} \frac{-\sin(x)}{2x}$$

We find that the new expression is also indeterminate at $x=0$, so we use L'Hôpital's rule again:

$$\lim_{x \to 0} \frac{-\sin(x)}{2x} = \lim_{x \to 0} \frac{-\cos(x)}{2}$$

The last expression is defined at $x=0$, so we can substitute $x=0$ to give

$$\lim_{x \to 0} \frac{-\cos(x)}{2} = \frac{-1}{2}$$

Hence

$$\lim_{x \to 0} \frac{\cos(x) - 1}{x^2} = -\tfrac{1}{2}$$

18.7 LIMITS AND CONVERGENCE

We have already briefly mentioned ideas of limits in various contexts. We will now look at the idea in more detail. When we looked at the sum to infinity of a geometric progression in Table 18.1 we looked at $S_n = 4(1 - (\frac{2}{3})^n)$ as n was made larger and discovered that $S_{40} = 4$ to seven significant figures, and that all values of $n > 40$ also gave $S_n = 4$ to seven significant figures. This can give us an idea of how to find out if a sequence tends to a limit:

1. Choose a number of significant figures.
2. Write all terms in the sequence to that number of significant figures.
3. The sequence tends to a limit if the terms in the sequence, when expressed to the agreed number of significant figures, become constant, i.e. does not change after some value of n.

Theoretically this procedure must work for all possible numbers of significant figures. As my calculator only displays eight significant figures I cannot go through this process for more than 7 s.f. A computer using double precision arithmetic could perform the calculations to far more (usually up to 18 significant figures).

Consider the series

$$1 + z + z^2 + z^3 + \dots$$

which is a geometric progression with first term 1 and common ratio z. The sum to n terms gives

$$\frac{1-z^n}{1-z}$$

For $z = \frac{1}{2}$ this gives the series

$$1 + \frac{1}{2} + \left(\frac{1}{2}\right)^2 + \left(\frac{1}{2}\right)^3 + \ldots$$

and the sum of n terms gives

$$S_n = \frac{1 - \left(\frac{1}{2}\right)^n}{1 - \frac{1}{2}} = 2\left(1 - \left(\frac{1}{2}\right)^n\right)$$

We can write S_n as a sequence of values to 3 s.f., 5 s.f. and 7 s.f., as is done in Table 18.2. From these results we can see that the limit appears to be 2. The limit is reached to 3 s.f. for $n = 9$, to 5 s.f. for $n = 13$ and to 7 s.f. for $n = 21$.

Table 18.2 The values in the sequence $S_n = 2(1 - \left(\frac{1}{2}\right)^n)$ expressed to 3, 5 and 7 s.f.

n	S_n (3 s.f.)	S_n (5 s.f.)	S_n (7 s.f.)
1	1	1	1
2	1.5	1.5	1.5
3	1.75	1.75	1.75
4	1.88	1.875	1.875
5	1.94	1.9375	1.9375
6	1.97	1.9688	1.96875
7	1.98	1.9844	1.984375
8	1.99	1.9922	1.992188
9	2	1.9961	1.996094
10	2	1.998	1.998047
11	2	1.999	1.999023
12	2	1.9995	1.999512
13	2	2	1.999756
14	2	2	1.999878
15	2	2	1.99994
16	2	2	1.99997
17	2	2	1.999985
18	2	2	1.999992
19	2	2	1.999996
20	2	2	1.999998
21	2	2	2
22	2	2	2

The more terms taken in a sequence which converges, the nearer we will get to the limit. However, we can only get as near as the number of significant figures, usually limited by the calculator, permits.

When using a numerical method to solve a problem we use these ideas about convergence.

18.8 NEWTON–RAPHSON METHOD FOR SOLVING EQUATIONS

Although we already know how to solve linear equations and quadratic equations, other equations may need to be solved by using a numerical method. One such method is the Newton–Raphson method. The method consists of an algorithm which can be expressed as follows:

Step 1: Take an equation and write it in the form $f(x)=0$.
Step 2: Take a guess at a solution.
Step 3: Calculate a new value for x using

$$x \leftarrow x - \frac{f(x)}{f'(x)}$$

Step 4: Repeat Step 3 until some convergence criterion has been satisfied or until it is decided that the method has failed to find a solution.

Here the '$\leftarrow$' symbol has been used to represent the 'assignment operator'.

$$x \leftarrow x - \frac{f(x)}{f'(x)}$$

means replace x by a value found by taking the old value of x and calculating $x-(f(x)/f'(x))$. We shall return to the problems in Steps 1 and 4 later; first we shall look at a simple example of using the Newton–Raphson method.

Example 18.34 Use the Newton–Raphson method to find $\sqrt{5}$ correct to seven significant figures.

SOLUTION $\sqrt{5}$ is one solution to the equation $x^2=5$.
  Step 1: Write the equation in the form $f(x)=0$:

$$x^2=5 \Leftrightarrow x^2-5=0$$

  Step 2: Take a guess at the solution. We know that $\sqrt{5}$ is slightly bigger than $\sqrt{4}$ so take a first guess as $x=2$.
  Steps 3 and 4: Calculate

$$x \leftarrow x - \frac{f(x)}{f'(x)}$$

until some convergence criterion is satisfied. As $f(x)=x^2-5$, then

$$f'(x)=2x$$

$$x \leftarrow x - \frac{f(x)}{f'(x)}$$

gives

$$x \leftarrow x - \frac{x^2-5}{2x}$$

which can be written over a common denominator, giving

$$x \leftarrow \frac{x^2+5}{2x}$$

SEQUENCES AND SERIES **359**

which is the Newton–Raphson formula for solving $x^2 - 5 = 0$.

Start with $x = 2$	$x \leftarrow \dfrac{4+5}{4}$	$x = 2.25$
Substitute $x = 2.25$	$x \leftarrow \dfrac{(2.25)^2 + 5}{2(2.25)}$	$x = 2.36\,111\,1$
Substitute $x = 2.236\,111\,1$	$x \leftarrow \dfrac{(2.236\,111\,1)^2 + 5}{2(2.236\,111\,1)}$	$x = 2.236\,068$
Substitute $x = 2.236\,068$	$x \leftarrow \dfrac{(2.236\,068)^2 + 5}{2(2.236\,068)}$	$x = 2.236\,068$

We notice in the last iteration that there has been no change in the value of x, so we assume that the algorithm has converged, giving $\sqrt{5} = 2.236\,068$ to 7 s.f.

The sequence of values we have found is: 2, 2.25, 2.236 111 1, 2.236 068, 2.236 068, and we need to stop at this point because the value of x has not changed in the last iteration.

Example 18.35 Find a solution to the equation $x^3 - 3x^2 + 2x + 1 = 0$.

SOLUTION

Step 1: The equation is already expressed in the correct form.

Step 2: We need to find a first guess for the solution. To do this we could sketch the graph to see roughly where it crosses the x-axis, or we could try substituting a few values into the function $f(x) = x^3 - 3x^2 + 2x + 1$ and look for a change of sign, which we have done in Table 18.3. As the function is continuous, the function must pass through zero in order to change from positive to negative, or vice versa. There is a change of sign between $x = -0.5$ and $x = 0$, so we take as a first guess a point half-way between these two values, giving $x = -0.25$.

Table 18.3 The function $f(x) = x^3 - 3x^2 + 2x + 1$ evaluated for a few values of x

x	-1	-0.5	0	0.5	1	1.5	2
$f(x) = x^3 - 3x^2 + 2x + 1$	-7	-0.875	1	1.375	1	0.625	1

Step 3: Using the Newton–Raphson formula

$$x \leftarrow x - \frac{f(x)}{f'(x)}$$

and substituting $f(x) = x^3 - 3x^2 + 2x + 1$ gives

$$x \leftarrow x - \frac{x^3 - 3x^2 + 2x + 1}{3x^2 - 6x + 2}$$

and simplifying gives

$$x \leftarrow \frac{2x^3 - 3x^2 - 1}{3x^2 - 6x + 2}$$

Starting by substituting $x = -0.25$ gives

$$x \leftarrow \frac{2(-0.25)^3 - 3(0.25) - 1}{3(-0.25)^2 - 6(-0.25) + 2} = \frac{-1.218\,75}{3.6875} = -0.330\,508\,4$$

Now substitute $x = -0.330\,584$, giving

$$x \leftarrow \frac{2(-0.330\,584)^3 - 3(-0.330\,584) - 1}{3(-0.330\,584)^2 - 6(-0.330\,584) + 2} = \frac{-1.399\,914\,1}{4.310\,757\,8} = -0.324\,748\,9$$

Substitute $x = -0.324\,748\,9$, giving

$$x \leftarrow \frac{2(-0.324\,789)^3 - 3(-0.324\,748\,9) - 1}{3(-0.324\,789)^2 - 6(-0.324\,748\,9) + 2} = \frac{-1.384\,882\,9}{4.264\,879\,4} = -0.324\,717\,9$$

Substitute $x = -0.324\,717\,9$ giving

$$x \leftarrow \frac{2(-0.324\,717\,9)^3 - 3(-0.324\,717\,9) - 1}{3(-0.324\,717\,9)^2 - 6(-0.324\,717\,9) + 2} = \frac{-1.384\,802\,9}{4.264\,633} = -0.324\,717\,9$$

As the last two numbers are the same to the degree of accuracy we have used there is no point in continuing. We have obtained the sequence of values

$$-0.25, \ -0.330\,508\,4, \ -0.324\,748\,9, \ -0.324\,717\,9, \ -0.324\,717\,9$$

Finally, we can check that we have found a good approximation to a solution of the equation by substituting $x = -0.324\,717\,9$ into the function $f(x) = x^3 - 3x^2 + 2x + 1$, which gives 2.441×10^{-7}. As this value is very close to 0 this confirms that we have found a reasonable approximation to a solution of the equation $f(x) = 0$.

The convergence criterion

In Examples 18.34 and 18.35 we decided to stop the calculation when the last two values found were equal: we had found the limit of the recurrence relation to seven significant figures. In a computer algorithm we could test whether the last two calculated values of x differ by a very small amount.

Example 18.36 A convergent sequence is defined by a recurrence relation. The calculation should stop when the limit has been found to an accuracy of at least three decimal places. Give a condition that could be used in this case.

SOLUTION Assuming that two consecutive terms are x_{n-1} and x_n, then the absolute difference between them is given by $|x_n - x_{n-1}|$. To test whether this is small enough to accept x_n as the limit to three decimal places, we remember that in Chapter 5 we saw that a number expressed to 3 d.p. could have an absolute error of just less than 0.0005. So the condition we can use to stop the algorithm could be $|x_n - x_{n-1}| < 0.0005$. If this condition is satisfied we could then assume that x_n is the limit to 3 d.p. To be on the safe side, however, it is better to perform the calculation one final time and check that it is also true that $|x_{n+1} - x_n| < 0.0005$, and then use x_{n+1} as the value of the limit, which should be accurate to at least 3 d.p.

Example 18.37 A convergent sequence is defined by a recurrence relation. The calculation should stop when the limit has been found to an accuracy of at least four significant figures. Give a condition that could be used in this case.

SOLUTION Assuming that two consecutive terms are x_{n-1} and x_n, then the absolute difference between them is given by $|x_n - x_{n-1}|$. To test whether this is small enough to accept x_n as the limit to four significant figures use the fact that a number expressed to 4 s.f. figures can have an absolute relative error of just less than 0.00005. As we do not know that value of the limit we must approximate it by the last value calculated in the sequence, so the absolute relative error is approximately

$$\frac{|x_n - x_{n-1}|}{|x_n|}$$

so an appropriate condition would be

$$\frac{|x_n - x_{n-1}|}{|x_n|} < 0.00005$$

or

$$|x_n - x_{n-1}| < 0.00005 |x_n|$$

As in the previous example it would be preferable to test that this condition holds on at least two successive iterations. Hence we could also check that

$$|x_{n+1} - x_n| < 0.00005 |x_{n+1}|$$

and take x_{n+1} as the limit of the sequence correct to four significant figures.

Divergence

A divergent sequence is one that does not tend to a finite limit. Some examples of divergent sequences are:

1. 1, 0, 1, 0, 1, 0, 1, 0,... which is an oscillating sequence.
2. 1, 2, 4, 8, 16,... which tends to plus infinity.
3. −1, −3, −5,... which tends to minus infinity.

Recurrence relations that are used for some numerical methods may not always converge, particularly if the initial value is chosen inappropriately. To check for this eventuality it is usual to stop the algorithm after some finite number of steps, maybe 100 or 1000 iterations, depending on the problem. If no convergence has been found after that number of iterations, then it is considered that the sequence is failing to converge.

18.9 SUMMARY

1. A sequence is a collection of objects arranged in a definite order. The elements of a sequence can be represented by $a_1, a_2, a_3, \ldots, a_n, \ldots$.
2. If a rule exists by which any term in the sequence can be found then this may be used to express the general term of the sequence, usually represented by a_n or $a(n)$. This rule can also be expressed in the form of a recurrence relation, where a_{n+1} is expressed in terms of $a_n, a_{n-1}, a_{n-2}, \ldots$.
3. During analogue to digital (A/D) conversion a signal is sampled and can then be represented by a sequence of numbers. $f(t)$ can be represented by $a(n) = f(nT)$, where T is the sampling interval and $t = nT$. The sampling theorem states that the sampling interval must be less than $T = 1/2f$ seconds in order to be able to represent a frequency of f Hz.

4. A series is the sum of a sequence. To represent a series we may use sigma notation, using the capital Greek letter sigma, Σ, to indicate the summation process, e.g.

$$\sum_{n=0}^{n=10} \frac{1}{2^n}$$

which means 'sum all the terms $1/2^n$ for n from 0 to 10'.

5. An arithmetic progression (AP) is a sequence where each term is found by adding a fixed amount, called the common difference, to the previous term. If the first term is a and the common difference is d, then the general term is $a_n = a + (n-1)d$ and the sum of the first n terms is given by

$$S_n = \frac{n}{2}(2a + (n-1)d)$$

or

$$S_n = \frac{n}{2}(a+l)$$

where l is the last term in the sequence and n is the number of terms.

6. A geometric progression (GP) is a sequence where each term is found by multiplying the previous term by a fixed amount, called the common ratio. If the first term is a and the common ratio is r, then the general term is $a_n = ar^{n-1}$, and the sum of the first n terms is given by

$$S_n = \frac{a(1-r^n)}{1-r}$$

The sum to infinity of a geometric progression can be found if $|r| < 1$, and is given by

$$S_\infty = \frac{a}{1-r}$$

7. The binomial expansion gives

$$(a+b)^n = a^n + na^{n-1}b + \frac{n(n-1)}{2!}a^{n-2}b^2 + \frac{n(n-1)(n-2)}{3!}a^{n-3}b^3 + \ldots$$

where n can be a whole number or a fraction.

8. The Maclaurin series is a series expansion of a function about $x=0$.
   If a function $f(x)$ is defined for values of x around $x=0$, within some radius R, that is for $-R < x < R$ (or $|x| < R$), and if all its derivatives are defined, then:

$$f(x) = f(0) + f'(0)x + \frac{f''(0)}{2!}x^2 + \frac{f'''(0)}{3!}x^3 + \ldots + \frac{f^{(n)}(0)}{n!}x^n + \ldots$$

This gives a power series with the coefficient sequence:

$$a_n = \frac{f^{(n)}(0)}{n!}$$

where $f^{(n)}(0)$ is found by finding the nth derivative of $f(x)$ with respect to x and then substituting $x=0$.

9. The Maclaurin series is just a special case of the Taylor series. A Taylor series is a series expansion of a function not necessarily taken around $x=0$.
   If a function $f(x)$ is defined for values of x around $x=a$, within some radius R, that is

for $a-R<x<a+R$ (or $|x-a|<R$), and if all its derivatives are defined, then:

$$f(x)=f(a)+f'(a)(x-a)+\frac{f''(a)}{2!}(x-a)^2+\frac{f'''(a)}{3!}(x-a)^3+\ldots+\frac{f^{(n)}(a)}{n!}(x-a)^n$$

or, substituting $x=a+h$, where h is usually considered to be a small value, we get

$$f(a+h)=f(a)+f'(a)h+\frac{f''(a)}{2!}h^2+\frac{f'''(a)}{3!}h^3+\ldots+\frac{f^{(n)}(a)}{n!}h^n+\ldots$$

10. L'Hôpital's rule is a way of finding the limit of a function at a point where its value is indeterminate (giving 0/0 at that point). The rule states that if a function $f(x)=g(x)/h(x)$ is indeterminate at $x=a$, then:

$$\lim_{x\to a}\frac{g(x)}{h(x)}=\lim_{x\to a}\frac{g'(x)}{h'(x)}$$

If $g'(a)/h'(a)$ is defined we can then use

$$\lim_{x\to a}\frac{g'(x)}{h'(x)}=\frac{g'(a)}{h'(a)}$$

and if $g'(a)/h'(a)$ is indeterminate we can use the rule again.
11. To test if a sequence tends to a limit, adopt the following procedure:

(a) Choose a number of significant figures.
(b) Write all the terms in the sequence to that number of significant figures.
(c) The sequence tends to a limit if the terms in the sequence, when expressed to the agreed number of significant figures, become constant, i.e. do not change after some value of n.

This procedure must theoretically work for any chosen number of significant figures.
12. The algorithm for solving an equation using the Newton–Raphson method can be described as:

Step 1: Take an equation and write it in the form $f(x)=0$.
Step 2: Take a guess at a solution.
Step 3: Calculate a new value for x using

$$x\leftarrow x-\frac{f(x)}{f'(x)}$$

Step 4: Repeat Step 3 until some convergence criterion has been satisfied or until it is decided that the method has failed to find a solution.
13. Convergence criteria can either be based on testing the size of the absolute error or the relative absolute error. To find the limit of a convergent sequence defined by a recurrence relation, correct to three decimal places, then we can test for $|x_n-x_{n-1}|<0.0005$, and to be correct to three significant figures we could test for $|x_n-x_{n-1}|<0.0005|x_n|$.

It is also necessary to put a limit on the number of iterations of some algorithm to check for the eventuality that the sequence fails to converge (is divergent).

18.10 EXERCISES

18.1 Find the next three terms in the following sequences. In each case express the rule for the sequence as a recurrence relation:

(a) $-3, 1, 5, 9, 13, 17,\ldots$
(b) $8, 4, 2, 1, 0.5,\ldots$

(c) $18, 15, 12, 9, 6, 3, \ldots$
(d) $6, -6, 6, -6, \ldots$
(e) $10, 8, 6, 4, \ldots$
(f) $1, 2, 4, 7, 11, 16, 22, \ldots$
(g) $1, 3, 6, 10, 15, \ldots$

18.2 Given the following definitions of sequences, write out the first five terms:

(a) $a_n = 3n - 1$
(b) $x_n = \dfrac{720}{n}$
(c) $b_n = 1 - n^2$
(d) $a_{n+1} = a_n + 2, \ a_1 = 6$
(e) $a_{n+1} = 3a_n, \ a_1 = 2$
(f) $a_{n+1} = -2a_n, \ a_1 = -1$
(g) $b_{n+1} = 2b_n - b_{n-1}, \ b_1 = \frac{1}{2}, \ b_2 = 1$
(h) $\Delta y_n = 3, \ y_0 = 2$
(i) $\Delta y_n = 2y_n, \ y_0 = 1$

18.3 Express the following using sigma notation:

(a) $1 + x + x^2 + x^3 + \ldots + x^{10}$
(b) $-2 + 4 - 8 + 16 - \ldots + 256$
(c) $1 + 8 + 27 + 64 + 125 + 216$
(d) $-\dfrac{1}{3} + \dfrac{1}{9} - \dfrac{1}{27} + \ldots + \dfrac{1}{6561}$
(e) $\dfrac{1}{4} + \dfrac{1}{9} + \dfrac{1}{16} + \dfrac{1}{25} + \dfrac{1}{36} + \ldots + \dfrac{1}{100}$
(f) $-4 - 1 - \dfrac{1}{4} - \dfrac{1}{16} - \ldots - \dfrac{1}{4096}$

18.4 Sketch the following functions and give the first 10 terms of their sequence representation ($t \geqslant 0$) at the sampling interval T given:

(a) $f(t) = \sin(2t), \ T = 0.1$
(b) $f(t) = \cos(30t), \ T = 0.01$
(c) $f(t)$ is the periodic function of period 16 defined for $0 \leqslant t \leqslant 16$ by

$$f(t) = \begin{cases} 2t & 0 \leqslant t \leqslant 4 \\ 16 - 2t & 4 < t \leqslant 12 \\ 2t - 32 & 12 < t < 16 \end{cases}$$

with sampling interval $T = 1$.
(d) The square wave of period 2 given for $0 \leqslant t < 2$ by

$$f(t) = \begin{cases} 1 & 0 \leqslant t < 1 \\ -1 & 1 \leqslant t < 2 \end{cases}$$

with a sampling interval of $T = 0.25$.

18.5 The following are arithmetic progressions. Find the fifth, tenth and general term of the sequence in each case:

(a) $6, 10, 14, \ldots$
(b) $3, 2.5, 2, \ldots$
(c) $-7, -1, 5, \ldots$

18.6 a_n is an arithmetic progression. Given the terms indicated, find the general term and find the sum of the first 20 terms in each case:

(a) $a_5 = 6, \ a_{10} = 26$
(b) $a_7 = -2, \ a_{16} = 2.5$
(c) $a_6 = 10, \ a_{12} = -8$

18.7 The sum of the first 10 terms of an arithmetic progression is 50 and the first term is 2. Find the common difference and the general term, and list the first six terms of the sequence.

18.8 How many terms are required in the arithmetic series $2+4+6+8+ \ldots$ to make a sum of 1056?

18.9 The following are geometric progressions. Find the fourth, eighth and general term in each case:

(a) $1, 2, 4, \ldots$

(b) $\dfrac{1}{3}, \dfrac{1}{12}, \dfrac{1}{48}, \ldots$

(c) $-9, 3, -1, \ldots$

(d) $15, 18.75, 23.4375, \ldots$

18.10 a_n is a geometric progression. Given the terms indicated, find the general term and find the sum of the first eight terms in each case:

(a) $a_3 = 8, a_6 = 1000$

(b) $a_6 = 54, a_9 = -486$

(c) $a_2 = -32, a_7 = 1$

18.11 How many terms are required in the geometric series $8+4+2+ \ldots$ to make a sum of 15.9375?

18.12 A loan of £40 000 is repaid by annual instalments of £5000, except in the final year when the outstanding debt (if less than £5000) is repaid. Interest is charged at 10 per cent per year, calculated at the end of each year on the outstanding amount of the debt. The first repayment is one year after the loan was taken out. Calculate the number of years required to repay the loan.

18.13 Evaluate the following

(a) $\displaystyle\sum_{n=1}^{n=4} 2^n$ (b) $\displaystyle\sum_{r=0}^{r=8} \dfrac{1}{2^r}$ (c) $\displaystyle\sum_{j=1}^{j=10} (-1)^j \left(\dfrac{1}{3}\right)^{j-2}$

18.14 Find the sum of the first n terms of the following:

(a) $1+z+z^2+z^3+ \ldots$

(b) $1-y^2+y^4- \ldots$

(c) $2x+\dfrac{4}{x}+\dfrac{8}{x^2}- \ldots$

18.15 State whether the following series are convergent, and if they are find the sum to infinity:

(a) $2+1+\frac{1}{2}+\frac{1}{4}+ \ldots$

(b) $3+0-3-6- \ldots$

(c) $27-9+3-1+ \ldots$

(d) $0.3+0.03+0.003+ \ldots$

18.16 Find the following recurring decimals as fractions:

(a) $0.\dot{4}$ (b) $0.1\dot{6}$ (c) $0.0\dot{2}$

18.17 Expand the following expressions:

(a) $\left(1+\dfrac{x}{2}\right)^3$ (b) $(1-x)^4$

(c) $(x-1)^3$ (d) $(1-2y)^4$

(e) $(1+x)^8$ (f) $(2x+1)^3$

(g) $(2a+b)^3$ (h) $\left(x+\dfrac{1}{x}\right)^7$

(i) $(a-2b)^4$

18.18 Find the following using the expansion indicated:

(a) $(1.1)^3$ using $(1+0.1)^3$

(b) $(0.9)^4$ using $(1-0.1)^4$

(c) $(2.01)^3$ using $(2+0.01)^3$

18.19 Give the first four terms in the binomial expansions of the following:

(a) $(1+2x)^5$ (b) $(1-3x)^8$

(c) $(2+z)^6$ (d) $\left(1+\dfrac{x}{2}\right)^{16}$

(e) $(1-x)^6$ (f) $(\tfrac{1}{2}-2x)^5$

18.20 Use $\sin(5\theta)=\mathrm{Im}(e^{j5\theta})=\mathrm{Im}((\cos(\theta)+j\sin(\theta))^5)$ to find $\sin(5\theta)$ in terms of powers of $\cos(\theta)$ and $\sin(\theta)$.

18.21 Find the real and imaginary parts of the following:

(a) $(1-j)^6$ (b) $(1+j2)^4$ (c) $(3+j)^5$

18.22 Use a binomial expansion to find the following correct to four decimal places:

(a) $(0.99)^8$ (b) $(1.01)^7$ (c) $(2.05)^6$

18.23 Find the first four non-zero terms in a power series expansion of the following functions and state for what values of x they are valid in each case:

(a) $\cos(x)$ (b) $\cosh(x)$ (c) $\ln(1+x)$

(d) $(1+x)^{1.5}$ (e) $\dfrac{1}{(1+x)^2}$

18.24 Find the first four non-zero terms in a power series expansion for the following functions:

(a) $\cos^2(x)$ (b) $\tan^{-1}(x)$

(c) $e^x\sin(x)$ (d) $\dfrac{(1-x)^{1.5}}{1+x}$

18.25 Using a series expansion find the following correct to four significant figures:

(a) $\sqrt{1.05}$ (b) $\tan^{-1}(0.1)$

(c) $\sin(0.03)$ (d) $\dfrac{1}{\sqrt{1.06}}$

18.26 Using a series expansion and the given value of the function at $x=a$, evaluate the following correct to four significant figures:

(a) $\cos\left(\dfrac{7\pi}{16}\right)$ using $\cos\left(\dfrac{\pi}{2}\right)=0$

(b) $\sqrt{4.02}$ using $\sqrt{4}=2$

18.27 Find the following limits:

(a) $\displaystyle\lim_{x\to2}\dfrac{x^2-x-2}{4x^3-4x^2-7x-2}$

(b) $\displaystyle\lim_{x\to0}\dfrac{x^3}{x-\sin(x)}$

(c) $\displaystyle\lim_{x\to-3}\dfrac{x^2+6x+9}{4x^2+11x-3}$

(d) $\displaystyle\lim_{x\to\pi/2}\dfrac{\pi/2-x}{\cos(x)}$

(e) $\displaystyle\lim_{x\to0}\dfrac{\tan(x)}{x}$

(f) $\displaystyle\lim_{x\to0}\dfrac{\sin(x-2)}{x^2-4x+4}$

18.28 Use the Newton–Raphson method to find a solution to the following equations correct to six significant figures.

(a) $x^3 - 2x = 1$ (b) $x^4 = 5$ (c) $\cos(x) = 2x$

18.29 Suggest a test for convergence that could be used in a computer program so that the limit of a sequence, defined by some recurrence relation, could be assumed to be correct to

(a) six decimal places
(b) six significant figures.

PART
THREE

SYSTEMS

SYSTEMS OF LINEAR EQUATIONS,
MATRICES AND DETERMINANTS

19.1 INTRODUCTION

The widespread use of computers to solve engineering problems means that it is important to be able to represent problems in a form suitable for solution by a computer. Matrices are used to represent systems of linear equations; transformations used in computer graphs or for robotic control; road, electrical and communication networks; and stresses and strains in materials. A matrix is a rectangular array of numbers of dimension $m \times n$, where m is the number of rows and n is the number of columns in the matrix. Matrices are also useful because they enable us to consider an array of many numbers as a single object, denote it by a single symbol and perform calculations with these symbols in a very compact form. Matrix algebra is therefore an important form of mathematical shorthand. In this chapter we look at applications of matrices, arithmetic operations on matrices and some common numerical methods. We shall also look at the problem of solving systems of linear equations. The methods of solving linear systems of equations are well understood and we only need to be able to solve simple cases of such problems 'by hand'. However, it is important to be able to express a problem in matrix form and also to appreciate situations where no solution exists or where more than one solution exists. This allows us to analyse the problems of ill-conditioning of systems of equations, which can lead to instability in the solution and the problem of over- or underdeterminancy, where either we have too much information, leading to possibly contradictory conditions, or we have not got enough to produce a single set of solutions for the unknowns.

We shall also look at the eigenvalue problem. The technique of finding eigenvalues will become particularly important when applied to systems of differential equations, which we meet in Chapter 20.

19.2 MATRICES

A matrix is a rectangular array of numbers. It may also be used as a simple store of information, as in the following example.

Every weekday, a household orders pints of milk, loaves of bread and yoghurts from a milk lorry. The orders for the week can be displayed as follows:

	Milk	Bread	Yoghurt
Monday	3	2	4
Tuesday	4	1	0
Wednesday	2	2	4
Thursday	5	1	0
Friday	1	1	4

This information forms a matrix.

Transformations in a plane can be represented by using matrices. For example, a reflection about the x-axis can be represented by the matrix

$$\begin{pmatrix} 1 & 0 \\ 0 & -1 \end{pmatrix}$$

and rotation through the angle θ by

$$\begin{pmatrix} \cos(\theta) & -\sin(\theta) \\ \sin(\theta) & \cos(\theta) \end{pmatrix}$$

We shall return to these examples later. Also in this chapter we shall see that linear equations can be written in matrix form.

Notation

A matrix is represented by a capital letter $\mathbf{A}$ (bold) or by $[a_{ij}]$ where a_{ij} represents a typical element in the ith row and jth column of the matrix. We represent a general matrix in the following form:

$$\begin{array}{c} \text{Column number} \\ \begin{array}{ccccc} 1 & 2 & 3 & \cdots & n \end{array} \\ \text{Row number} \begin{pmatrix} \begin{array}{c} 1 \\ 2 \\ 3 \\ \vdots \\ m \end{array} & \begin{array}{ccccc} a_{11} & a_{12} & a_{13} & \cdots & a_{1n} \\ a_{21} & a_{22} & a_{23} & \cdots & a_{2n} \\ a_{31} & a_{32} & a_{33} & \cdots & a_{3n} \\ \vdots & \vdots & \vdots & & \vdots \\ a_{m1} & a_{m2} & a_{m3} & \cdots & a_{mn} \end{array} \end{pmatrix} \end{array}$$

In order to refer to the element which is in the third row and the second column, we can say a_{32}.

The matrix

$$\begin{pmatrix} 3 & 2 \\ 6 & 1 \\ 8 & 2 \end{pmatrix}$$

is a 3×2 matrix (read as 3 by 2) as it has three rows and two columns.

The sum and difference of matrices

The sum and difference of matrices is found by adding or subtracting corresponding elements of the matrix. Only matrices of exactly the same dimensions can be added or subtracted.

Example 19.1

$$A = (2 \quad 1) \qquad B = (6 \quad 2) \qquad C = \begin{pmatrix} 3 & 7 \\ 8 & 3 \end{pmatrix}$$

$$D = \begin{pmatrix} 1 & 2 \\ 2 & 1 \end{pmatrix} \qquad E = \begin{pmatrix} 8 & 2 & 1 \\ 6 & 1 & 3 \end{pmatrix} \qquad F = \begin{pmatrix} 2 & 6 & 3 \\ 12 & -2 & -6 \end{pmatrix}$$

Find where possible

(a) $A + B$ (b) $C + D$ (c) $E - F$ (d) $A + D$

SOLUTION
(a) $A + B = (2 \quad 1) + (6 \quad 2) = (8 \quad 3)$

(b) $C + D = \begin{pmatrix} 3 & 7 \\ 8 & 3 \end{pmatrix} + \begin{pmatrix} 1 & 2 \\ 2 & 1 \end{pmatrix} = \begin{pmatrix} 4 & 9 \\ 10 & 4 \end{pmatrix}$

(c) $E - F = \begin{pmatrix} 8 & 2 & 1 \\ 6 & 1 & 3 \end{pmatrix} - \begin{pmatrix} 2 & 6 & 3 \\ 12 & -2 & -6 \end{pmatrix} = \begin{pmatrix} 6 & -4 & -2 \\ -6 & 3 & 9 \end{pmatrix}$

(d) $A + D$ cannot be found because the two matrices are of different dimensions.

Multiplication of a matrix by a scalar

To multiply a matrix by a scalar, every element is multiplied by the scalar.

Example 19.2 If

$$A = \begin{pmatrix} 2 & 5 \\ 6 & 1 \end{pmatrix}$$

find $2A$ and $\frac{1}{3}A$.

SOLUTION

$$2A = 2\begin{pmatrix} 2 & 5 \\ 6 & 1 \end{pmatrix} = \begin{pmatrix} 4 & 10 \\ 12 & 2 \end{pmatrix}$$

$$\tfrac{1}{3}A = \tfrac{1}{3}\begin{pmatrix} 2 & 5 \\ 6 & 1 \end{pmatrix} = \begin{pmatrix} \frac{2}{3} & \frac{5}{3} \\ 2 & \frac{1}{3} \end{pmatrix}$$

Multiplication of two matrices

To multiply two matrices, every row is multiplied by every column. For instance, if $C = AB$, to find the element in the second row and the third column of the product, C, we take the second row of A and the third column of B and multiply them together, like taking the scalar product of two vectors. Multiplication is only possible if the number of columns in A is the same as the number of rows in B. For instance if A is 2×3 it can only multiply matrices that are $3 \times n$, where

n could be any dimension. The result of a 2×3 matrix multiplying a 3×4 matrix is a 2×4 matrix. Notice the pattern:

$$(2 \times 3) \text{ multiplying } (3 \times 4) \text{ gives } 2 \times 4$$

Must be equal

Example 19.3

$$A = \begin{pmatrix} 1 & -1 \\ 3 & 1 \end{pmatrix} \qquad B = \begin{pmatrix} 6 & 0 & -1 \\ 2 & 2 & 3 \end{pmatrix}$$

Find if possible **AB** and **BA**.

SOLUTION

$$AB = \begin{pmatrix} 1 & -1 \\ 3 & 1 \end{pmatrix}\begin{pmatrix} 6 & 0 & -1 \\ 2 & 2 & 3 \end{pmatrix} = \begin{pmatrix} 1 \times 6 + (-1) \times 2 & 1 \times 0 + (-1) \times 2 & 1 \times (-1) + (-1) \times (3) \\ 3 \times 6 + 1 \times 2 & 3 \times 0 + 1 \times 2 & 3 \times (-1) + 1 \times 3 \end{pmatrix}$$

$$= \begin{pmatrix} 4 & -2 & -4 \\ 20 & 2 & 0 \end{pmatrix}$$

BA cannot be found because the number of columns in **B** is not equal to the number of rows in **A**.

We can justify the practical reasons for this method of matrix multiplication as in the following two examples. In the first one we return to our household shopping example.

Example 19.4 Every weekday a household orders pints of milk, loaves of bread and yoghurts from a milk lorry. The orders for the week are as follows:

	Milk	Bread	Yoghurt
Monday	3	2	4
Tuesday	4	1	0
Wednesday	2	2	4
Thursday	5	1	0
Friday	1	1	4

Next week the dairy is introducing a special offer and reducing its prices. The prices for this week and next week are as follows:

	This week	Next week
Milk	0.34	0.32
Bread	0.60	0.50
Yoghurt	0.33	0.30

Calculate the cost each day for this week and next week.

SOLUTION The cost each day is made up of the number of pints of milk times the cost of a pint plus the number of loaves of bread times the cost of a loaf plus the number of cartons of yoghurt times the cost of the yoghurt. In other words we can find the cost each day by performing matrix multiplication:

$$
\begin{pmatrix} 3 & 2 & 4 \\ 4 & 1 & 0 \\ 2 & 2 & 4 \\ 5 & 1 & 0 \\ 1 & 1 & 4 \end{pmatrix}
\begin{pmatrix} 0.34 & 0.32 \\ 0.60 & 0.50 \\ 0.33 & 0.30 \end{pmatrix}
$$

$$
= \begin{pmatrix}
3 \times (0.34) + 2 \times (0.60) + 4 \times (0.33) & 3 \times (0.32) + 2 \times (0.50) + 4 \times (0.30) \\
4 \times (0.34) + 1 \times (0.60) + 0 \times (0.33) & 4 \times (0.32) + 1 \times (0.50) + 0 \times (0.30) \\
2 \times (0.34) + 2 \times (0.60) + 4 \times (0.33) & 2 \times (0.32) + 2 \times (0.50) + 4 \times (0.30) \\
5 \times (0.34) + 1 \times (0.60) + 0 \times (0.33) & 5 \times (0.32) + 1 \times (0.50) + 0 \times (0.30) \\
1 \times (0.34) + 1 \times (0.60) + 4 \times (0.33) & 1 \times (0.32) + 1 \times (0.50) + 4 \times (0.30)
\end{pmatrix}
$$

$$
= \begin{pmatrix}
3.54 & 3.16 \\
1.96 & 1.78 \\
3.20 & 2.84 \\
2.30 & 2.10 \\
2.26 & 2.02
\end{pmatrix}
$$

The rows now represent the days of the week and the columns represent this week and next week. Hence, for instance, the cost for Thursday of next week is given by the element $a_{42} = 2.10$.

Example 19.5 Figure 19.1 represents a communication network where the vertices a, b, f, g represent offices and vertices c, d, e represent switching centres. The numbers marked along the edges represent the number of connections between any two vertices. Calculate the number of routes from a, b to f, g.

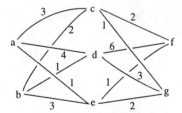

Figure 19.1 A representation of a communication network.

SOLUTION The number of routes from a to f can be calculated by taking the number via c plus the number via d plus the number via e. In each case this is given by multiplying the number of connections along the edges connecting a to c, c to f etc., giving:

$$\text{Number of routes from a to } f = 3 \times 2 + 4 \times 6 + 1 \times 1$$

We can see that we can get the number of routes by matrix multiplication.

The network from ab to cde is represented by:

$$
\begin{array}{c}
 \quad c \quad d \quad e \\
\begin{array}{c} a \\ b \end{array}
\begin{pmatrix} 3 & 4 & 1 \\ 2 & 1 & 3 \end{pmatrix}
\end{array}
$$

and from cde to fg by

$$\begin{array}{cc} & \begin{array}{cc} f & g \end{array} \\ \begin{array}{c} c \\ d \\ e \end{array} & \begin{pmatrix} 2 & 1 \\ 6 & 3 \\ 1 & 2 \end{pmatrix} \end{array}$$

So the total number of routes is given by

$$\begin{pmatrix} 3 & 4 & 1 \\ 2 & 1 & 3 \end{pmatrix} \begin{pmatrix} 2 & 1 \\ 6 & 3 \\ 1 & 2 \end{pmatrix} = \begin{pmatrix} 3\times2+4\times6+1\times1 & 3\times1+4\times3+1\times2 \\ 2\times2+1\times6+3\times1 & 2\times1+3\times1+3\times2 \end{pmatrix}$$

$$= \begin{array}{c} \\ a \\ b \end{array} \begin{array}{cc} \begin{array}{cc} f & g \end{array} \\ \begin{pmatrix} 31 & 17 \\ 13 & 11 \end{pmatrix} \end{array}$$

Hence by interpreting the rows and columns of the resulting matrix we can see that there are 31 routes from a to f, 17 from a to g, 13 from b to f and 11 from b to g.

The unit matrix

The unit matrix is a square matrix which leaves any matrix, **A**, unchanged under multiplication. If **A** is square then

$$\mathbf{AI} = \mathbf{IA} = \mathbf{A}$$

The unit matrix has 1s on its leading diagonal and 0s elsewhere. In two dimensions:

$$\mathbf{I} = \begin{pmatrix} 1 & 0 \\ 0 & 1 \end{pmatrix}$$

In three dimensions:

$$\mathbf{I} = \begin{pmatrix} 1 & 0 & 0 \\ 0 & 1 & 0 \\ 0 & 0 & 1 \end{pmatrix}$$

Example 19.6

$$\mathbf{A} = \begin{pmatrix} 2 & -1 \\ 0 & 1 \end{pmatrix} \qquad \mathbf{B} = \begin{pmatrix} 3 \\ 2 \end{pmatrix}$$

Show that **AI** = **IA** = **A** and **IB** = **B**.

SOLUTION

$$\mathbf{AI} = \begin{pmatrix} 2 & -1 \\ 0 & 1 \end{pmatrix} = \begin{pmatrix} 1 & 0 \\ 0 & 1 \end{pmatrix}\begin{pmatrix} 2\times1+(-1)\times0 & 2\times0+(-1)\times1 \\ 0\times1+1\times0 & 0\times0+1\times1 \end{pmatrix} = \begin{pmatrix} 2 & -1 \\ 0 & 1 \end{pmatrix} = \mathbf{A}$$

$$\mathbf{IA} = \begin{pmatrix} 1 & 0 \\ 0 & 1 \end{pmatrix} = \begin{pmatrix} 2 & -1 \\ 0 & 1 \end{pmatrix}\begin{pmatrix} 1\times2+0\times0 & 1\times(-1)+0\times1 \\ 0\times2+1\times0 & 0\times(-1)+1\times1 \end{pmatrix} = \begin{pmatrix} 2 & -1 \\ 0 & 1 \end{pmatrix} = \mathbf{A}$$

$$\mathbf{IB} = \begin{pmatrix} 1 & 0 \\ 0 & 1 \end{pmatrix}\begin{pmatrix} 3 \\ 2 \end{pmatrix} = \begin{pmatrix} 1\times3+0\times2 \\ 0\times3+1\times2 \end{pmatrix} = \begin{pmatrix} 3 \\ 2 \end{pmatrix} = \mathbf{B}$$

The transpose of a matrix

The transpose of a matrix is obtained by interchanging the rows and the columns. The transpose of a matrix $\mathbf{A}$ is represented by $\mathbf{A}^{\mathrm{T}}$.

Example 19.7 Given

$$\mathbf{A} = \begin{pmatrix} 2 & -1 \\ 6 & 3 \end{pmatrix} \qquad \mathbf{B} = \begin{pmatrix} 2 & 1 & 8 \\ -1 & 0 & 1 \end{pmatrix}$$

find $\mathbf{A}^{\mathrm{T}}$ and $\mathbf{B}^{\mathrm{T}}$.

SOLUTION The first row of $\mathbf{A}$ is $(2 \quad -1)$; therefore this is the first column of $\mathbf{A}^{\mathrm{T}}$. The second row of $\mathbf{A}$ is $(6 \quad 3)$; therefore this is the second column of $\mathbf{A}^{\mathrm{T}}$. This gives $\mathbf{A}^{\mathrm{T}}$ as follows:

$$\mathbf{A}^{\mathrm{T}} = \begin{pmatrix} 2 & 6 \\ -1 & 3 \end{pmatrix}$$

Similarly

$$\mathbf{B}^{\mathrm{T}} = \begin{pmatrix} 2 & -1 \\ 1 & 0 \\ 8 & 1 \end{pmatrix}$$

Some special types of matrices

A **square matrix** has the same number of rows as columns.

$\begin{pmatrix} 2 & -1 \\ 6 & 3 \end{pmatrix}$ is a square matrix of dimension 2

$\begin{pmatrix} 8 & 6 & 2 \\ -3 & 1 & 0 \\ 3 & 2 & 1 \end{pmatrix}$ is a square matrix of dimension 3

A square matrix has a leading diagonal, which comprises the terms lying along the diagonal from the top left-hand corner to the bottom right-hand corner. These terms have the same row number as they have column number.

$$\begin{pmatrix} 8 & 6 & 2 \\ -3 & 1 & 0 \\ 3 & 2 & 1 \end{pmatrix}$$

the leading diagonal

A **diagonal matrix** is a square matrix which has zero elements everywhere except, possibly, on its leading diagonal, e.g.

$$\begin{pmatrix} 4 & 0 & 0 \\ 0 & -2 & 0 \\ 0 & 0 & 3 \end{pmatrix}$$

An **upper triangular matrix** is a square matrix which has zeros below the leading diagonal, e.g.

$$\begin{pmatrix} 1 & 1 & 2 \\ 0 & 6 & 6 \\ 0 & 0 & 8 \end{pmatrix}$$

A **lower triangular matrix** has zeros above the leading diagonal, e.g.

$$\begin{pmatrix} 1 & 0 & 0 \\ 3 & -1 & 0 \\ 6 & 8 & 2 \end{pmatrix}$$

A **symmetric matrix** is such that $A^T = A$, i.e. the elements are symmetric about the leading diagonal; e.g.

$$A = \begin{pmatrix} 1 & 6 & -3 \\ 6 & 0 & -2 \\ -3 & -2 & 8 \end{pmatrix} \qquad B = \begin{pmatrix} 1 & 6 \\ 6 & 1 \end{pmatrix}$$

are symmetric matrices. If you take the transpose of one of these matrices it results in the original matrix.

A **skew-symmetric matrix** is such that $A^T = -A$.

Example 19.8 Show that

$$A = \begin{pmatrix} 0 & 6 \\ -6 & 0 \end{pmatrix}$$

is skew-symmetric.

SOLUTION

$$A^T = \begin{pmatrix} 0 & -6 \\ 6 & 0 \end{pmatrix}$$

Multiplying A by -1 we get

$$-A = \begin{pmatrix} 0 & -6 \\ 6 & 0 \end{pmatrix}$$

We can see that $A^T = -A$, and hence we have shown that A is skew-symmetric.

Hermitian matrix

A Hermitian matrix is such that $A^{*T} = A$ (where $*$ represents the complex conjugate).

Example 19.9 Show that

$$A = \begin{pmatrix} 3 & 7+j2 \\ 7-j2 & -2 \end{pmatrix} \qquad \text{and} \qquad B = \begin{pmatrix} 2 & 3e^{-j2} \\ 3e^{j2} & 1 \end{pmatrix}$$

are Hermitian.

SOLUTION Taking the complex conjugates of each of the terms in **A** and **B** gives

$$\mathbf{A}^* = \begin{pmatrix} 3 & 7-j2 \\ 7+j2 & -2 \end{pmatrix}$$

$$\mathbf{B}^* = \begin{pmatrix} 2 & 3e^{+j2} \\ 3e^{-2j} & 1 \end{pmatrix}$$

Now taking the transpose gives

$$\mathbf{A}^{*T} = \begin{pmatrix} 3 & 7+j2 \\ 7-j2 & -2 \end{pmatrix}$$

$$\mathbf{B}^{*T} = \begin{pmatrix} 2 & 3e^{-j2} \\ 3e^{2j} & 1 \end{pmatrix}$$

So we can see that

$$\mathbf{A}^{*T} = \mathbf{A}$$

$$\mathbf{B}^{*T} = \mathbf{B}$$

Showing that they are Hermitian.

In the rest of this chapter we shall assume that our matrices are real.
A **column vector** is a matrix with only one column, e.g.

$$\mathbf{v} = \begin{pmatrix} 1 \\ 2 \\ 3 \end{pmatrix}$$

A **row vector** is a matrix with only one row, e.g.

$$\mathbf{v} = (1 \quad 2 \quad 3)$$

The inverse of a matrix

The inverse of a matrix **A** is a matrix $\mathbf{A}^{-1}$ such that $\mathbf{A}\mathbf{A}^{-1} = \mathbf{A}^{-1}\mathbf{A} = \mathbf{I}$ (the unit matrix).

Example 19.10 Show that

$$\begin{pmatrix} \frac{1}{3} & \frac{1}{3} \\ \frac{1}{3} & -\frac{2}{3} \end{pmatrix}$$

is the inverse of

$$\begin{pmatrix} 2 & 1 \\ 1 & -1 \end{pmatrix}$$

SOLUTION Multiply:

$$\begin{pmatrix} \frac{1}{3} & \frac{1}{3} \\ \frac{1}{3} & -\frac{2}{3} \end{pmatrix}\begin{pmatrix} 2 & 1 \\ 1 & -1 \end{pmatrix} = \begin{pmatrix} (\frac{1}{3})(2)+(\frac{1}{3})(1) & (\frac{1}{3})(1)+(\frac{1}{3})(-1) \\ (\frac{1}{3})(2)+(-\frac{2}{3})(1) & (\frac{1}{3})(1)+(-\frac{2}{3})(-1) \end{pmatrix}$$

$$= \begin{pmatrix} 1 & 0 \\ 0 & 1 \end{pmatrix}$$

Also

$$\begin{pmatrix} 2 & 1 \\ 1 & -1 \end{pmatrix}\begin{pmatrix} \frac{1}{3} & \frac{1}{3} \\ \frac{1}{3} & -\frac{2}{3} \end{pmatrix} = \begin{pmatrix} 2(\frac{1}{3})+1(\frac{1}{3}) & 2(\frac{1}{3})+1(-\frac{2}{3}) \\ 1(\frac{1}{3})+(-1)\frac{1}{3} & 1(\frac{1}{3})+(-1)(-\frac{2}{3}) \end{pmatrix} = \begin{pmatrix} 1 & 0 \\ 0 & 1 \end{pmatrix}$$

Not all matrices have inverses, and only square matrices can possibly have inverses. A matrix does not have an inverse if its determinant is 0.

The determinant of

$$\begin{pmatrix} a & b \\ c & d \end{pmatrix}$$

is given by

$$\begin{vmatrix} a & b \\ c & d \end{vmatrix} = ad - cb$$

If the determinant of a matrix is 0 then it has no inverse and the matrix is said to be singular. If the determinant is non-zero then the inverse exists. The inverse of the 2×2 matrix

$$\begin{pmatrix} a & b \\ c & d \end{pmatrix}$$

is

$$\frac{1}{(ad-cb)}\begin{pmatrix} d & -b \\ -c & a \end{pmatrix}$$

That is, to find the inverse of a 2×2 matrix, we swap the leading diagonal terms, negate the off-diagonal elements and divide the resulting matrix by the determinant.

Example 19.11 Find the determinants of the following matrices and state if the matrix has an inverse or is singular. Find the inverse in the cases where it exists and check that $\mathbf{AA}^{-1} = \mathbf{A}^{-1}\mathbf{A} = \mathbf{I}$.

(a) $\begin{pmatrix} -1 & 3 \\ 2 & 1 \end{pmatrix}$ (b) $\begin{pmatrix} 6 & -2 \\ -3 & 1 \end{pmatrix}$ (c) $\begin{pmatrix} \dfrac{1}{\sqrt{2}} & \dfrac{-1}{\sqrt{2}} \\ \dfrac{1}{\sqrt{2}} & \dfrac{1}{\sqrt{2}} \end{pmatrix}$

SOLUTION

(a) $\begin{vmatrix} -1 & 3 \\ 2 & 1 \end{vmatrix} = (-1) \times 1 - 2 \times 3 = -7$

As the determinant is not zero, the matrix

$$\begin{pmatrix} -1 & 3 \\ 2 & 1 \end{pmatrix}$$

has an inverse, given by

$$\frac{1}{-7}\begin{pmatrix} 1 & -3 \\ -2 & -1 \end{pmatrix} = \frac{1}{7}\begin{pmatrix} -1 & 3 \\ 2 & 1 \end{pmatrix}$$

Check that $\mathbf{AA}^{-1}=\mathbf{I}$:

$$\begin{pmatrix} -1 & 3 \\ 2 & 1 \end{pmatrix}\frac{1}{7}\begin{pmatrix} -1 & 3 \\ 2 & 1 \end{pmatrix}=\frac{1}{7}\begin{pmatrix} (-1)(-1)+(3)(2) & (-1)3+3(1) \\ 2(-1)+1(2) & 2(3)+(1)(1) \end{pmatrix}=\begin{pmatrix} 1 & 0 \\ 0 & 1 \end{pmatrix}$$

and that $\mathbf{A}^{-1}\mathbf{A}=\mathbf{I}$:

$$\frac{1}{7}\begin{pmatrix} -1 & 3 \\ 2 & 1 \end{pmatrix}\begin{pmatrix} -1 & 3 \\ 2 & 1 \end{pmatrix}=\frac{1}{7}\begin{pmatrix} (-1)(-1)+3(2) & (-1)(3)+3(1) \\ 2(-1)+1(2) & (2)(3)+(1)(1) \end{pmatrix}=\begin{pmatrix} 1 & 0 \\ 0 & 1 \end{pmatrix}$$

(b) $\begin{vmatrix} 6 & -2 \\ -3 & 1 \end{vmatrix}=6\times1-(-3)(-2)=0$

As the determinant is zero, the matrix

$$\begin{pmatrix} 6 & -2 \\ -3 & 1 \end{pmatrix}$$

has no inverse. It is singular.

(c) $\begin{vmatrix} \dfrac{1}{\sqrt{2}} & \dfrac{-1}{\sqrt{2}} \\ \dfrac{1}{\sqrt{2}} & \dfrac{1}{\sqrt{2}} \end{vmatrix}=\left(\dfrac{1}{\sqrt{2}}\right)\left(\dfrac{1}{\sqrt{2}}\right)-\left(\dfrac{-1}{\sqrt{2}}\right)\left(\dfrac{1}{\sqrt{2}}\right)=1$

Therefore the matrix is invertible. Its inverse is given by swapping the leading diagonal terms, negating the off-diagonal terms, and then dividing by the determinant. This gives

$$\begin{pmatrix} \dfrac{1}{\sqrt{2}} & \dfrac{1}{\sqrt{2}} \\ \dfrac{-1}{\sqrt{2}} & \dfrac{1}{\sqrt{2}} \end{pmatrix}$$

Check that $\mathbf{AA}^{-1}=\mathbf{I}$:

$$\mathbf{AA}^{-1}=\begin{pmatrix} \dfrac{1}{\sqrt{2}} & \dfrac{-1}{\sqrt{2}} \\ \dfrac{1}{\sqrt{2}} & \dfrac{1}{\sqrt{2}} \end{pmatrix}\begin{pmatrix} \dfrac{1}{\sqrt{2}} & \dfrac{1}{\sqrt{2}} \\ \dfrac{-1}{\sqrt{2}} & \dfrac{1}{\sqrt{2}} \end{pmatrix}=\begin{pmatrix} \dfrac{1}{\sqrt{2}\sqrt{2}}+\dfrac{-1}{\sqrt{2}\sqrt{2}} & \dfrac{1}{\sqrt{2}\sqrt{2}}+\left(\dfrac{-1}{\sqrt{2}}\right)\left(\dfrac{1}{\sqrt{2}}\right) \\ \dfrac{1}{\sqrt{2}\sqrt{2}}+\dfrac{1}{\sqrt{2}}\left(\dfrac{-1}{\sqrt{2}}\right) & \dfrac{1}{\sqrt{2}\sqrt{2}}+\left(\dfrac{1}{\sqrt{2}}\right)\left(\dfrac{1}{\sqrt{2}}\right) \end{pmatrix}$$

$$=\begin{pmatrix} 1 & 0 \\ 0 & 1 \end{pmatrix}$$

Similarly $\mathbf{AA}^{-1}=\mathbf{I}$.

Solving matrix equations

To solve matrix equations we use the same ideas about equivalent equations that we have used before. As in ordinary equations we can 'do the same things to both sides' in order to find equivalent equations. It is important to remember that division by a matrix has not been defined. In order to 'undo' matrix multiplication we have to multiply by an inverse matrix, where it exists, and we need to specify whether we are pre-multiplying or post-multiplying. This is necessary because matrices do not obey the commutative law ($\mathbf{AB} \neq \mathbf{BA}$). If we pre- or post-multiply both sides of an equation by a matrix we must also be able to justify that the dimensions of the expressions are such that the multiplication is possible. Also, if we add or subtract a matrix from both sides of the equation, it must have exactly the same dimensions as the current matrix expression.

Example 19.12 Given that $\mathbf{A}$, $\mathbf{B}$ and $\mathbf{C}$ are matrices and $\mathbf{AB} = \mathbf{C}$, where $\mathbf{A}$ and $\mathbf{B}$ are non-singular, find an expression for $\mathbf{B}$ and an expression for $\mathbf{A}$.

SOLUTION In this case we are told that $\mathbf{A}$ and $\mathbf{B}$ are invertible, so they must be square, and therefore $\mathbf{C}$ must also be square and of the same dimension.

To find $\mathbf{B}$ we wish to 'get rid' of the $\mathbf{A}$ term on the left-hand side. We pre-multiply both sides of the equation by $\mathbf{A}^{-1}$:

$$\mathbf{AB} = \mathbf{C}$$

and given that $\mathbf{A}$ is invertible:

$$\Leftrightarrow \mathbf{A}^{-1}\mathbf{AB} = \mathbf{A}^{-1}\mathbf{C}$$

Now using $\mathbf{A}^{-1}\mathbf{A} = \mathbf{I}$, the unit matrix, we have

$$\mathbf{IB} = \mathbf{A}^{-1}\mathbf{C}$$

As the unit matrix multiplied by any matrix leaves it unchanged, we have

$$\Leftrightarrow \mathbf{B} = \mathbf{A}^{-1}\mathbf{C}$$

Similarly, now post-multiplying by $\mathbf{B}^{-1}$:

$$\mathbf{AB} = \mathbf{C}$$

and given that $\mathbf{B}$ is invertible:

$$\Leftrightarrow \mathbf{ABB}^{-1} = \mathbf{CB}^{-1}$$

Now using $\mathbf{BB}^{-1} = \mathbf{I}$, the unit matrix, we have

$$\mathbf{AI} = \mathbf{CB}^{-1}$$

As the unit matrix multiplied by any matrix leaves it unchanged, we have

$$\Leftrightarrow \mathbf{A} = \mathbf{CB}^{-1}$$

Remember that it is always important to specify whether you are pre-multiplying or post-multiplying when solving matrix equations. A term like $\mathbf{B}^{-1}\mathbf{AB}$ cannot be simplified because we cannot swap the order, as we would do with numbers.

19.3 TRANSFORMATIONS

On a computer graphics screen an object is represented by a set of coordinates, either with reference to the screen origin or with reference to the origin of some window created by the graphical user interface (GUI). We may wish to move the object around inside its window. We shall consider in this section only two-dimensional objects, as dealing with three-dimensional objects would add the complication of needing to represent a perspective view. Ideas about transformations are also important when considering the movement of a robotic arm.

There are three ways of moving an object without affecting its overall size or shape: rotation, reflection and translation. We could also stretch it or compress it in some direction: the operation of scaling.

We shall look at how to perform these operations using matrices and vectors. We can check that the operations performed are those that we expected by looking at the effect on some simple shape. In most of these examples we look at the effect on a unit square at the origin, defined by points A (0,0), B (1,0), C (1,1) and D (0,1). The outcome of the transformation is called the image, which we shall represent by the points A′, B′, C′ and D′. The transformation, T, is a function whose domain and codomain are the plane (which is referred to as $\mathbb{R}^2$). The term 'mapping' is also used in this context. It has exactly the same meaning as function, but is more often used when referring to geometrical problems.

Rotation

To perform a rotation through an angle θ about the origin we multiply the position vector of a point $\begin{pmatrix} x \\ y \end{pmatrix}$ by the matrix

$$\begin{pmatrix} \cos(\theta) & -\sin(\theta) \\ \sin(\theta) & \cos(\theta) \end{pmatrix}$$

Example 19.13 Find and draw the image of the unit square with vertices A (0,0), B (1,0), C (1,1) and D (0,1) after rotation through 30° about the origin.

SOLUTION Rotation through 30° about the origin is found by multiplying the position vectors of the points by

$$\begin{pmatrix} \cos(30°) & -\sin(30°) \\ \sin(30°) & \cos(30°) \end{pmatrix} \simeq \begin{pmatrix} 0.866 & -0.5 \\ 0.5 & 0.866 \end{pmatrix}$$

To find the image of the unit square we multiply the position vectors of the vertices by this matrix

$$\begin{pmatrix} 0.866 & -0.5 \\ 0.5 & 0.866 \end{pmatrix}\begin{pmatrix} 0 \\ 0 \end{pmatrix} = \begin{pmatrix} 0 \\ 0 \end{pmatrix}$$

$$\begin{pmatrix} 0.866 & -0.5 \\ 0.5 & 0.866 \end{pmatrix}\begin{pmatrix} 1 \\ 0 \end{pmatrix} = \begin{pmatrix} 0.866 \\ 0.5 \end{pmatrix}$$

$$\begin{pmatrix} 0.866 & -0.5 \\ 0.5 & 0.866 \end{pmatrix}\begin{pmatrix} 1 \\ 1 \end{pmatrix} = \begin{pmatrix} 0.366 \\ 1.366 \end{pmatrix}$$

$$\begin{pmatrix} 0.866 & -0.5 \\ 0.5 & 0.866 \end{pmatrix}\begin{pmatrix} 0 \\ 1 \end{pmatrix} = \begin{pmatrix} -0.5 \\ 0.866 \end{pmatrix}$$

This transformation is shown in Fig. 19.2.

Sometimes it is useful to be able to rotate the axes rather than the object. For instance, the object may be held by a robotic arm and we want the arm to rotate but keep the orientation of the object the same. This is pictured for the tea-drinking robot in Fig. 19.3. In this case, if we rotate the axes Ox, Oy, by θ the position of the object remains the same, but even so has new coordinates relative to the transformed axes OX, OY. If the axes rotate through 30° then

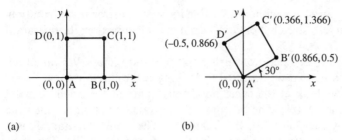

(a)   (b)

Figure 19.2 (a) The unit square with vertices A (0,0), B (1,0), C (1,1) and D (0,1). (b) The same unit square after rotation by 30°.

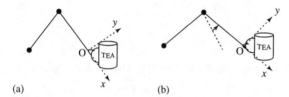

(a)   (b)

Figure 19.3 In order not to spill the tea, the axes, defined with reference to the lower arm, rotate, but the orientation of the cup must stay the same.

the object moves relative to the axes by $-30°$. So to rotate the axes by θ we multiply the position vectors of the points $\binom{x}{y}$ by the matrix

$$\begin{pmatrix} \cos(-\theta) & -\sin(-\theta) \\ \sin(-\theta) & \cos(-\theta) \end{pmatrix} = \begin{pmatrix} \cos(\theta) & \sin(\theta) \\ -\sin(\theta) & \cos(\theta) \end{pmatrix}$$

Example 19.14 A unit square has vertices A (0,0), B (1,0), C (1,1) and D (0,1) relative to axes Ox, Oy. The axes are rotated through 30° to OX, OY, without moving the square. Find the coordinates of the vertices relative to the new axes OX, OY.

SOLUTION The effect of rotating the axes through 30° is found by multiplying the position vectors of the points by

$$\begin{pmatrix} \cos(30°) & \sin(30°) \\ \sin(-30°) & \cos(30°) \end{pmatrix} \simeq \begin{pmatrix} 0.866 & 0.5 \\ -0.5 & 0.866 \end{pmatrix}$$

To find the coordinates of the unit square relative to the new axes, we multiply the position vectors of the vertices by this matrix:

$$\begin{pmatrix} 0.866 & 0.5 \\ -0.5 & 0.866 \end{pmatrix}\begin{pmatrix} 0 \\ 0 \end{pmatrix} = \begin{pmatrix} 0 \\ 0 \end{pmatrix}$$

$$\begin{pmatrix} 0.866 & 0.5 \\ -0.5 & 0.866 \end{pmatrix}\begin{pmatrix} 1 \\ 0 \end{pmatrix} = \begin{pmatrix} 0.866 \\ -0.5 \end{pmatrix}$$

$$\begin{pmatrix} 0.866 & 0.5 \\ -0.5 & 0.866 \end{pmatrix}\begin{pmatrix} 1 \\ 1 \end{pmatrix} = \begin{pmatrix} 1.366 \\ 0.366 \end{pmatrix}$$

$$\begin{pmatrix} 0.866 & 0.5 \\ -0.5 & 0.866 \end{pmatrix}\begin{pmatrix} 0 \\ 1 \end{pmatrix} = \begin{pmatrix} 0.5 \\ 0.866 \end{pmatrix}$$

This is shown in Fig. 19.4.

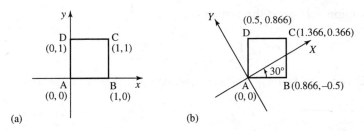

(a)   (b)

Figure 19.4 (a) The unit square with vertices A (0,0), B (1,0), C (1,1) and D (0,1) relative to axes Ox, Oy. (b) The same unit square shown relative to axes OX, OY found by rotating Ox, Oy through 30°.

Reflection

To perform a reflection in the x-axis we multiply the position vectors of the points $\begin{pmatrix} x \\ y \end{pmatrix}$ by the matrix

$$\begin{pmatrix} 1 & 0 \\ 0 & -1 \end{pmatrix}$$

This has the effect of keeping the x-coordinate the same while changing the sign of the y-coordinate, hence turning the object upside down.

To perform a reflection in the y-axis we multiply the position vectors of the points $\begin{pmatrix} x \\ y \end{pmatrix}$ by the matrix

$$\begin{pmatrix} -1 & 0 \\ 0 & 1 \end{pmatrix}$$

which keeps the y-value constant while changing the sign of the x-coordinate. The effect on the unit square is shown in Fig. 19.5.

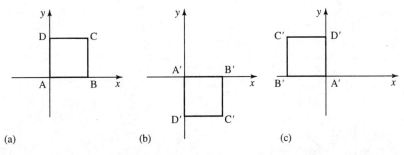

(a)   (b)   (c)

Figure 19.5 (a) The unit square with vertices A (0,0), B (1,0), C (1,1) and D (0,1). (b) The same unit square after reflection in the x-axis. (c) After reflection in the y-axis.

Translation

Translation in the plane cannot be represented by multiplying by a 2×2 matrix. To perform a translation we add the vector representing the translation to the original position vectors of the points.

Example 19.15 Find and draw the image of the unit square with vertices A (0,0), B (1,0), C (1,1), D (0,1) after translation through $\begin{pmatrix} 3 \\ 4 \end{pmatrix}$.

SOLUTION Add $\binom{3}{4}$ to the position vectors of the vertices, i.e.

$$\mathbf{v} + \binom{3}{4}$$

giving A′ as (3,4), B′ (4,4), C′ (4,5) and D′ (3,5). This transformation is shown in Fig. 19.6.

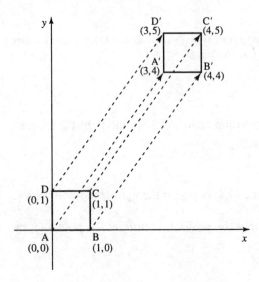

Figure 19.6 The unit square with vertices A (0,0), B (1,0), C (1,1) and D (0,1). The same unit square after translation through (3,4) becomes A′ (3,4), B′ (4,4), C′ (4,5) and D′ (3,5).

It is again often useful to consider what happens if the object stays where it is and the axes are translated. If the axes are translated through $\binom{3}{4}$ then the object appears to move relative to the axes by $-\binom{3}{4}$. Therefore we subtract $\binom{3}{4}$ from the coordinates defining it. This is shown in Fig. 19.7.

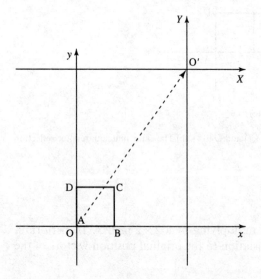

Figure 19.7 The unit square with vertices, relative to Ox, Oy, A (0,0), B (1,0), C (1,1) and D (0,1). The unit square has coordinates $(-3,-4), (-2,-4), (-2,-3), (-3,-3)$ relative to the axes OX, OY, which have been translated through (3,4).

Scaling

To scale in the x-direction we multiply the position vectors of the points $\binom{x}{y}$ by a matrix

$$\begin{pmatrix} S_x & 0 \\ 0 & 1 \end{pmatrix}$$

where S_x is the scale factor. Under this transformation, vectors that have no x-component will be unaffected.

To scale in the y-direction we multiply the position vectors of the points $\binom{x}{y}$ by a matrix

$$\begin{pmatrix} 1 & 0 \\ 0 & S_y \end{pmatrix}$$

where S_y is the scale factor. Under this transformation vectors that have no y-component will be unaffected.

The effect on the unit square of scaling by 2 in the x-direction is shown in Fig. 19.8(b) and the affect of scaling by 3 in the y-direction is shown in Fig. 19.8(c).

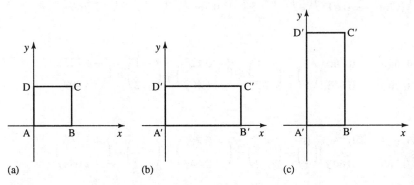

(a)   (b)   (c)

Figure 19.8 (a) The unit square with vertices A (0,0), B (1,0), C (1,1) and D (0,1). (b) The same unit square after scaling in the x-direction by a factor of 2. (c) The unit square after scaling in the x-direction by a factor of 3.

Combined transformations

Example 19.16 Find the coordinates of the vertices of the unit square after

(a) rotation about the origin through 50° followed by a translation of $(-1,2)$.
(b) Translation of $(-1,2)$ followed by rotation about the origin through 50°.

SOLUTION
(a) We can write this combined transformation as

$$\mathbf{p}' = \mathbf{R}\mathbf{p} + \mathbf{t}$$

where $\mathbf{p}'$ is the position vector of the image point, $\mathbf{p}$ is the position vector of the original point, $\mathbf{R}$ is the matrix representing the rotation and $\mathbf{t}$ is the vector representing the translation. In this case

$$\mathbf{R} = \begin{pmatrix} \cos(50°) & -\sin(50°) \\ \sin(50°) & \cos(50°) \end{pmatrix} \simeq \begin{pmatrix} 0.643 & -0.766 \\ 0.766 & 0.643 \end{pmatrix}$$

and

$$\mathbf{t}=\begin{pmatrix} -1 \\ 2 \end{pmatrix}, \ \mathbf{p}'=\begin{pmatrix} x' \\ y' \end{pmatrix}, \ \text{and} \ \mathbf{p}=\begin{pmatrix} x \\ y \end{pmatrix}$$

So we have

$$\begin{pmatrix} x' \\ y' \end{pmatrix}=\begin{pmatrix} 0.643 & -0.766 \\ 0.766 & 0.643 \end{pmatrix}\begin{pmatrix} x \\ y \end{pmatrix}+\begin{pmatrix} -1 \\ 2 \end{pmatrix}$$

For the coordinates of A' substitute $x=0$ and $y=0$, giving

$$\begin{pmatrix} x' \\ y' \end{pmatrix}=\begin{pmatrix} 0.643 & -0.766 \\ 0.766 & 0.643 \end{pmatrix}\begin{pmatrix} 0 \\ 0 \end{pmatrix}+\begin{pmatrix} -1 \\ 2 \end{pmatrix}=\begin{pmatrix} 0 \\ 0 \end{pmatrix}+\begin{pmatrix} -1 \\ 2 \end{pmatrix}=\begin{pmatrix} -1 \\ 2 \end{pmatrix}$$

For B':

$$\begin{pmatrix} x' \\ y' \end{pmatrix}=\begin{pmatrix} 0.643 & -0.766 \\ 0.766 & 0.643 \end{pmatrix}\begin{pmatrix} 1 \\ 0 \end{pmatrix}+\begin{pmatrix} -1 \\ 2 \end{pmatrix}=\begin{pmatrix} 0.643 \\ 0.766 \end{pmatrix}+\begin{pmatrix} -1 \\ 2 \end{pmatrix}=\begin{pmatrix} -0.357 \\ 2.766 \end{pmatrix}$$

For C':

$$\begin{pmatrix} x' \\ y' \end{pmatrix}=\begin{pmatrix} 0.643 & -0.766 \\ 0.766 & 0.643 \end{pmatrix}\begin{pmatrix} 1 \\ 1 \end{pmatrix}+\begin{pmatrix} -1 \\ 2 \end{pmatrix}=\begin{pmatrix} -0.123 \\ 1.409 \end{pmatrix}+\begin{pmatrix} -1 \\ 2 \end{pmatrix}=\begin{pmatrix} -1.123 \\ 3.409 \end{pmatrix}$$

For D':

$$\begin{pmatrix} x' \\ y' \end{pmatrix}=\begin{pmatrix} 0.643 & -0.766 \\ 0.766 & 0.643 \end{pmatrix}\begin{pmatrix} 0 \\ 1 \end{pmatrix}+\begin{pmatrix} -1 \\ 2 \end{pmatrix}=\begin{pmatrix} -0.766 \\ 0.643 \end{pmatrix}+\begin{pmatrix} -1 \\ 2 \end{pmatrix}=\begin{pmatrix} -1.766 \\ 2.643 \end{pmatrix}$$

The image of the unit square is pictured in Fig. 19.9(b).

(b) We can write this combined transformation as

$$\mathbf{p}''=\mathbf{R}(\mathbf{p}+\mathbf{t})$$

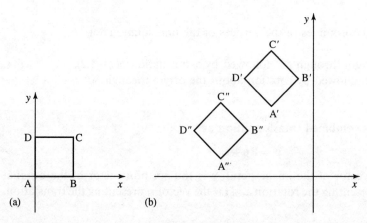

Figure 19.9 (a) The unit square with vertices A (0,0), B (1,0), C (1,1) and D (0,1). (b) The same unit square after rotation through 50° about the origin and translation through (−1,2), and the unit square after translation through (−1,2) and then rotation of 50° about the origin.

where $\mathbf{p}''$ is the position vector of the image point, $\mathbf{p}$ is the position vector of the original point, $\mathbf{R}$ is the matrix representing the rotation and $\mathbf{t}$ is the vector representing the translation. We have put the brackets in to indicate that the translation is performed first. As before:

$$\mathbf{R}=\begin{pmatrix} \cos(50°) & -\sin(50°) \\ \sin(50°) & \cos(50°) \end{pmatrix} \simeq \begin{pmatrix} 0.643 & -0.766 \\ 0.766 & 0.643 \end{pmatrix}$$

and

$$\mathbf{t}=\begin{pmatrix} -1 \\ 2 \end{pmatrix}, \mathbf{p}''=\begin{pmatrix} x'' \\ y'' \end{pmatrix}, \text{ and } \mathbf{p}=\begin{pmatrix} x \\ y \end{pmatrix}$$

So we have

$$\begin{pmatrix} x'' \\ y'' \end{pmatrix}=\begin{pmatrix} 0.643 & -0.766 \\ 0.766 & 0.643 \end{pmatrix}\left(\begin{pmatrix} x \\ y \end{pmatrix}+\begin{pmatrix} -1 \\ 2 \end{pmatrix}\right)$$

which is the same as

$$\begin{pmatrix} x'' \\ y'' \end{pmatrix}=\begin{pmatrix} 0.643 & -0.766 \\ 0.766 & 0.643 \end{pmatrix}\begin{pmatrix} x-1 \\ y+2 \end{pmatrix}$$

For the coordinates of A″, substitute $x=0$ and $y=0$, giving

$$\begin{pmatrix} x'' \\ y'' \end{pmatrix}=\begin{pmatrix} 0.643 & -0.766 \\ 0.766 & 0.643 \end{pmatrix}\begin{pmatrix} 0-1 \\ 0+2 \end{pmatrix}=\begin{pmatrix} 0.643 & -0.766 \\ 0.766 & 0.643 \end{pmatrix}\begin{pmatrix} -1 \\ 2 \end{pmatrix}=\begin{pmatrix} -2.175 \\ 0.52 \end{pmatrix}$$

For B″:

$$\begin{pmatrix} x'' \\ y'' \end{pmatrix}=\begin{pmatrix} 0.643 & -0.766 \\ 0.766 & 0.643 \end{pmatrix}\begin{pmatrix} 1-1 \\ 0+2 \end{pmatrix}=\begin{pmatrix} 0.643 & -0.766 \\ 0.766 & 0.643 \end{pmatrix}\begin{pmatrix} 0 \\ 2 \end{pmatrix}=\begin{pmatrix} -1.532 \\ 1.286 \end{pmatrix}$$

For C″:

$$\begin{pmatrix} x'' \\ y'' \end{pmatrix}=\begin{pmatrix} 0.643 & -0.766 \\ 0.766 & 0.643 \end{pmatrix}\begin{pmatrix} 1-1 \\ 1+2 \end{pmatrix}=\begin{pmatrix} 0.643 & -0.766 \\ 0.766 & 0.643 \end{pmatrix}\begin{pmatrix} 0 \\ 3 \end{pmatrix}=\begin{pmatrix} -2.298 \\ 1.929 \end{pmatrix}$$

For D″:

$$\begin{pmatrix} x'' \\ y'' \end{pmatrix}=\begin{pmatrix} 0.643 & -0.766 \\ 0.766 & 0.643 \end{pmatrix}\begin{pmatrix} 0-1 \\ 1+2 \end{pmatrix}=\begin{pmatrix} 0.643 & -0.766 \\ 0.766 & 0.643 \end{pmatrix}\begin{pmatrix} -1 \\ 3 \end{pmatrix}=\begin{pmatrix} -2.941 \\ 1.163 \end{pmatrix}$$

The image of the unit square is pictured in Fig. 19.9(b).
Note that the order of the transformations is important.

Sometimes we might need to use a trick of temporarily moving the axes in order to perform certain transformations. Suppose we want to scale by 2 along the line $x=y$. We can rotate the axes temporarily so that the new X-axis lies along the line that was previously $x=y$, then perform X scaling, and then rotate back again, so that the axes are back in their original position. This is done in the next example.

Example 19.17 Find a matrix that performs scaling by a factor of 2 along the direction $x = y$ and draw the image of the unit square defined by the vertices

$$A\left(-\frac{1}{2}, -\frac{1}{2}\right), B\left(\frac{1}{2}, -\frac{1}{2}\right), C\left(\frac{1}{2}, \frac{1}{2}\right), D\left(-\frac{1}{2}, \frac{1}{2}\right)$$

SOLUTION First we rotate the axes by 45°, so that the OX axis will lie along the line that was previous $x = y$. This is pictured in Fig. 19.10. The matrix that transforms the coordinates so that they are relative to the new axes at an angle of 45° is given by:

$$\begin{pmatrix} \cos(45°) & \sin(45°) \\ -\sin(45°) & \cos(45°) \end{pmatrix}$$

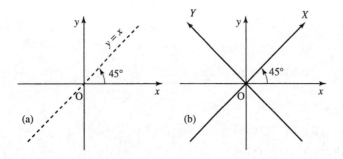

Figure 19.10 (a) The line $x = y$ is at 45° to the Ox axis. If we rotate the axes by 45°, the new OX axis will lie in this direction. This is shown in (b).

A scaling of 2 in the X-direction is then performed by multiplying by

$$\begin{pmatrix} 2 & 0 \\ 0 & 1 \end{pmatrix}$$

We then need to rotate the axes back to their original position, i.e. rotate the axes by −45°. This is done by multiplying by

$$\begin{pmatrix} \cos(-45°) & \sin(-45°) \\ -\sin(-45°) & \cos(-45°) \end{pmatrix} = \begin{pmatrix} \cos(45°) & -\sin(45°) \\ \sin(45°) & \cos(45°) \end{pmatrix}$$

Putting the three transformation matrices together we get

$$\begin{pmatrix} \cos(45°) & -\sin(45°) \\ \sin(45°) & \cos(45°) \end{pmatrix}\begin{pmatrix} 2 & 0 \\ 0 & 1 \end{pmatrix}\begin{pmatrix} \cos(45°) & \sin(45°) \\ -\sin(45°) & \cos(45°) \end{pmatrix}$$

which gives the matrix which represents a scaling along the line $x = y$. Using

$\cos(45°)=1/\sqrt{2}=\sin(45°)$ we get

$$\begin{pmatrix} \dfrac{1}{\sqrt{2}} & \dfrac{-1}{\sqrt{2}} \\ \dfrac{1}{\sqrt{2}} & \dfrac{1}{\sqrt{2}} \end{pmatrix} \begin{pmatrix} 2 & 0 \\ 0 & 1 \end{pmatrix} \begin{pmatrix} \dfrac{1}{\sqrt{2}} & \dfrac{1}{\sqrt{2}} \\ \dfrac{-1}{\sqrt{2}} & \dfrac{1}{\sqrt{2}} \end{pmatrix}$$

Taking out the two factors of $1/\sqrt{2}$ gives

$$\frac{1}{2}\begin{pmatrix} 1 & -1 \\ 1 & 1 \end{pmatrix}\begin{pmatrix} 2 & 0 \\ 0 & 1 \end{pmatrix}\begin{pmatrix} 1 & 1 \\ -1 & 1 \end{pmatrix}$$

Multiplying the second two matrices gives

$$\frac{1}{2}\begin{pmatrix} 1 & -1 \\ 1 & 1 \end{pmatrix}\begin{pmatrix} 2 & 2 \\ -1 & 1 \end{pmatrix}$$

and multiplying out the remaining two matrices gives

$$\frac{1}{2}\begin{pmatrix} 3 & 1 \\ 1 & 3 \end{pmatrix} = \begin{pmatrix} \frac{3}{2} & \frac{1}{2} \\ \frac{1}{2} & \frac{3}{2} \end{pmatrix}$$

We can now multiply the position vectors representing the vertices of the square

$$\begin{pmatrix} \frac{3}{2} & \frac{1}{2} \\ \frac{1}{2} & \frac{3}{2} \end{pmatrix}\begin{pmatrix} -\frac{1}{2} \\ -\frac{1}{2} \end{pmatrix} = \begin{pmatrix} -1 \\ -1 \end{pmatrix}$$

$$\begin{pmatrix} \frac{3}{2} & \frac{1}{2} \\ \frac{1}{2} & \frac{3}{2} \end{pmatrix}\begin{pmatrix} \frac{1}{2} \\ -\frac{1}{2} \end{pmatrix} = \begin{pmatrix} \frac{1}{2} \\ -\frac{1}{2} \end{pmatrix}$$

$$\begin{pmatrix} \frac{3}{2} & \frac{1}{2} \\ \frac{1}{2} & \frac{3}{2} \end{pmatrix}\begin{pmatrix} \frac{1}{2} \\ \frac{1}{2} \end{pmatrix} = \begin{pmatrix} 1 \\ 1 \end{pmatrix}$$

$$\begin{pmatrix} \frac{3}{2} & \frac{1}{2} \\ \frac{1}{2} & \frac{3}{2} \end{pmatrix}\begin{pmatrix} -\frac{1}{2} \\ \frac{1}{2} \end{pmatrix} = \begin{pmatrix} -\frac{1}{2} \\ \frac{1}{2} \end{pmatrix}$$

The transformed figure is shown in Fig. 19.11. We can see that it has been stretched along the $x=y$ direction, but has not been scaled along the other diagonal. The image is no longer a square but a rhombus.

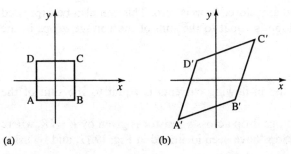

(a)   (b)

Figure 19.11 (a) The unit square with vertices A $(-\frac{1}{2},-\frac{1}{2})$, B $(\frac{1}{2},-\frac{1}{2})$, C $(\frac{1}{2},\frac{1}{2})$, D$(-\frac{1}{2},\frac{1}{2})$. (b) The image after scaling by 2 along the line $y=x$.

Example 19.18 Find a transformation that will rotate any point **p** about (1,1) through an angle of 90°.

SOLUTION To rotate about a point not at the origin we translate the origin temporarily, rotate, and then translate the origin back again.

Rotation through 90° is performed by multiplying by

$$\begin{pmatrix} \cos(90°) & -\sin(90°) \\ \sin(90°) & \cos(90°) \end{pmatrix} = \begin{pmatrix} 0 & -1 \\ 1 & 0 \end{pmatrix}$$

The combined transformation on a point **p** can be represented by

$$\mathbf{p}' = \begin{pmatrix} 0 & -1 \\ 1 & 0 \end{pmatrix}\left(\mathbf{p} - \begin{pmatrix} 1 \\ 1 \end{pmatrix}\right) + \begin{pmatrix} 1 \\ 1 \end{pmatrix}$$

19.4 SYSTEMS OF EQUATIONS

Example 19.19 Using Ohm's law and Kirchhoff's laws for the electrical network in Fig. 19.12, show that

$$I_1 - I_2 - I_3 = 0$$
$$3I_2 - 2I_3 = 0$$
$$7I_1 \qquad + 2I_3 = 0$$

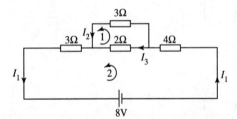

Figure 19.12 The electrical network for Example 19.19.

SOLUTION Kirchhoff's laws for an electrical network are as follows.

Kirchhoff's voltage law (KVL)
The sum of all the voltage drops around any closed loop is zero. This can also be expressed as: the voltage impressed on a closed loop is equal to the sum of the voltage drops in the rest of the loop.

Kirchhoff's current law (KCL)
At any point of a circuit, the sum of the in-flowing currents is equal to the sum of the out-flowing currents.

By Ohm's law we know that the voltage drop across a resistor is given by $V = IR$, where R is the resistance of the resistor. Two loops have been identified in Fig. 19.12, and by using KVL and Ohm's law in loop 1 we get

$$3I_2 - 2I_3 = 0$$

Now looking at loop 2 we get

$$3I_1 - 8 + 4I_1 + 2I_3 = 0$$
$$\Leftrightarrow 7I_1 + 2I_3 = 8$$

Finally, we use the current law at one of the nodes to give

$$I_1 = I_2 + I_3$$
$$\Leftrightarrow I_1 - I_2 - I_3 = 0$$

We can list all the equations we have found:

$$I_1 - I_2 \ - I_3 = 0$$
$$3I_2 - 2I_3 = 0$$
$$7I_1 \quad + 2I_3 = 8$$

and the problem is now to find a solution which satisfies all of these equations simultaneously.

This is called a system of equations. In many electrical networks there will be far more than three unknown currents. In such situations it is impractical to solve the equations without the use of a computer. However, we can discover a number of important principles and problems involved in solving systems of linear equations by looking at some simple cases. The first problem we have is that it is possible to get more than these three equations from the network given in Fig. 19.12.

Using KVL in the outer loop would give

$$7I_1 + 3I_2 = 8$$

and using KCL at the other node gives

$$I_2 + I_3 = I_1$$
$$\Leftrightarrow -I_1 + I_2 + I_3 = 0$$

We therefore have five equations and only three unknowns.

Luckily it is possible to show that these equations are a consistent set, that is, it is possible to find a solution. We shall return to solve for I_1, I_2 and I_3 later. First, we shall examine all the possibilities when we have only two unknown quantities.

Systems of equations in two unknowns

The equation

$$ax + by = c$$

where a, b and c are constants, is a linear equation in two unknowns (or variables) x and y.

Because there are two unknowns we need two axes to represent it, and therefore the graph can be drawn in a plane.

Because the graph only involves terms in x, y and the constant term, and no other powers of either x or y, we know that the graph of the equation is a straight line, as we met in Chapter 8. Examples of graphs of linear equations in two unknowns are given in Fig. 19.13.

We call a solution to the equation a pair of values for x and y which satisfy the equation; i.e. when they are substituted they give a true expression. A solution to the equation $x + y = 1$ is $x = 0.5$, $y = 0.5$, because if we substitute these values we obtain a true expression:

$$0.5 + 0.5 = 1$$

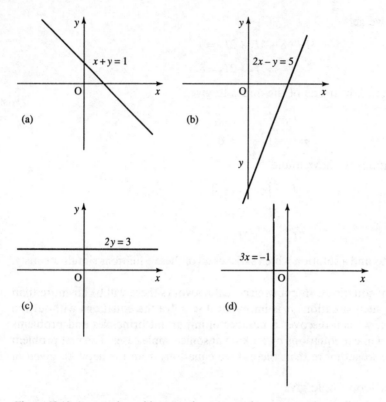

Figure 19.13 An equation with two unknowns can be represented as a line in a plane: (a) $x+y=1$; (b) $2x-y=5$; (c) $2y=3$; (d) $3x=-1$.

However, there are many other solutions to $x+y=1$, for instance $x=2$, $y=-1$ or $x=2.5$, $y=-1.5$ etc. We say that the equation is indeterminate because there are any number of solutions to the equation $x+y=1$. In fact, any point on the line $x+y=1$ is a solution to the equation. We can express the solutions in terms of x or y (e.g. $x=1-y$), and therefore the solutions are $(1-y,y)$, where y can be any number. Alternatively, $y=1-x$ gives solutions $(x,1-x)$, where x can be any number.

A system of two linear equations with two unknowns

We want to find values for x and y which solve both $a_1x+b_1y=c_1$ and $a_2x+b_2y=c_2$ simultaneously. The problem could be expressed as

$$(a_1x+b_1y=c_1) \land (a_2x+b_2y=c_2)$$

When we talk of systems of equations it is understood that we want all of the equations to hold simultaneously, so they are usually just listed as

$$a_1x+b_1y=c_1$$

$$a_2x+b_2y=c_2$$

Each equation can be represented geometrically by a straight line. For example, the system

$$3x+4y=7$$

$$x+2y=2$$

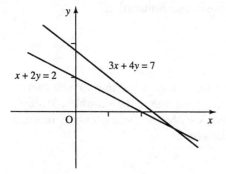

Figure 19.14 A system of two equations in two unknowns: $3x+4y=7$ and $x+2y=2$.

can be represented by the pair of straight lines in Fig. 19.14. We can find the point where the two straight lines cross by using substitution, as follows.

Example 19.20 Solve the following system of equations using substitution:

$$3x+4y=7$$

$$x+2y=2$$

SOLUTION We begin by numbering the equations in order to identify them

$$3x+4y=7 \qquad (19.1)$$

$$x+2y=2 \qquad (19.2)$$

From Eq. (19.2) we can express x in terms of y, giving

$$x+2y=2 \Leftrightarrow x=2-2y \text{ (subtracting } 2y \text{ from both sides)}$$

Now substitute $x=2-2y$ into Eq. (19.1), giving

$$3(2-2y)+4y=7$$

$$\Leftrightarrow 6-6y+4y=7$$

$$\Leftrightarrow 6-2y=7$$

$$\Leftrightarrow -2y=1 \qquad \text{(subtracting 6 from both sides)}$$

$$\Leftrightarrow y=-\tfrac{1}{2} \qquad \text{(dividing both sides by } -2)$$

$$y=-0.5 \qquad \Leftrightarrow$$

Now we can use $x=2-2y$ to find x by substituting $y=-0.5$, giving

$$x=2-2(-0.5)$$

$$\Leftrightarrow x=2+1$$

$$\Leftrightarrow x=3$$

The solution is given by $x=3$ and $y=-0.5$, which can be represented by the pair of values for (x,y) of $(3,-0.5)$.

An alternative method of solution is to use elimination.

Example 19.21 Solve the following system of equations using elimination:

$$3x + 4y = 7$$

$$x + 2y = 2$$

SOLUTION To solve the system of equations we look for a way of adding or subtracting multiples of one equation from the other in order to eliminate one of the variables. Multiply the second equation by 3 and leave the first the same. We choose these numbers in order to get the coefficients of x in both equations to be the same.

$$3x + 4y = 7$$

$$3x + 6y = 6$$

Subtract the equations to give

$$-2y = 1 \Leftrightarrow y = -\tfrac{1}{2}$$

Substitute this into the first equation, giving

$$3x + 4(-\tfrac{1}{2}) = 7$$

$$\Leftrightarrow 3x - 2 = 7$$

$$\Leftrightarrow 3x = 9$$

$$\Leftrightarrow x = 3$$

The solution is given by $x = 3$ and $y = -0.5$, which can be represented by the pair of values for (x,y) of $(3, -0.5)$.

The point $(3, -0.5)$ is the point on the graph where the two lines cross. This is the only point which lies both on the first graph and on the second graph. It is the only point which satisfies both equations simultaneously. Hence we say there is a unique solution to the system of equations. The system of equations is said to be determined because there is a single solution. The system of equations is also said to be consistent because it is possible to find a solution.

We could have the system of equations:

$$x + 2y = 1$$

$$2x + 4y = 2$$

If we plot these lines we find that they are coincident, i.e. one line lies on top of the other, as in Fig. 19.15. In this case, the second equation, $2x + 4y = 2$, can be obtained by multiplying the first equation by 2. We say that the equations are dependent. Two equations are dependent if one can be obtained from the other by multiplying by a constant or by adding a constant to both sides. As any point that lies on $x + 2y = 1$ also lies on $2x + 4y = 2$, there is no unique solution to the system of equations.

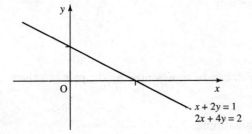

$x + 2y = 1$
$2x + 4y = 2$

Figure 19.15 The graphs of equations $x + 2y = 1$ and $2x + 4y = 2$ are coincident. The graph of one lies on top of the other.

We say that the system of equations is indeterminate as there exist any number of solutions to the system, i.e. the system reduces to only one equation. However, the system of equations is said to be consistent, because at least one solution exists.

Example solutions are:

$$x=2, \ y=0.5 \quad \text{or} \ (2,-0.5)$$
$$x=3, \ y=-1 \quad \text{or} \ (3,-1)$$
$$x=4, \ y=-1.5 \ \text{or} \ (4,-1.5)$$

The solution set can be written as $(x,(1-x)/2)$, where x can take any value, or as $(1-2y,y)$, where y can take any value.

If we try to use elimination to solve the equations

$$x+2y=1$$
$$2x+4y=2$$

we find that one equation reduces to $0=0$, i.e. a condition that is always true.

Example 19.22 Solve, using elimination:

$$x+2y=1$$
$$2x+4y=2$$

SOLUTION

$$x+2y=1$$
$$2x+4y=2$$

Multiply the first equation by 2, giving

$$2x+4y=2$$
$$2x+4y=2$$

Subtract, and this gives

$$0=0$$

which is always true, thus indicating that the system of equations is indeterminate. The solutions are therefore any points lying on the line $x+2y=1$.

The third possibility for a system of equations is one that has no solutions at all. Such a system is as follows:

$$x+2y=1$$
$$2x+4y=5$$

If we plot these equations we find that they are parallel, as in Fig. 19.16.

From the geometrical interpretation it is therefore clear that no solution exists to this system of equations, as no point on the line $x+2y=1$ lies on the line $2x+4y=5$. We say that the system of equations is inconsistent because no solutions exist.

If we used elimination to attempt to solve an inconsistent system of equations like these then we would find that we would get an impossible condition, such as

$$0=3$$

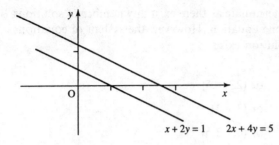

Figure 19.16 The graphs of $x+2y=1$ and $2x+4y=5$ are parallel. There are no points in common between the two graphs.

which is false. This situation indicates that there are no solutions and the equations are inconsistent.

Example 19.23 Solve, using elimination:

$$x+2y=1$$
$$2x+4y=5$$

SOLUTION Multiply the first equation by 2, giving

$$2x+4y=2$$
$$2x+4y=5$$

Subtracting the equations we get

$$0=-3$$

This condition is false. This indicates that the system of equations is inconsistent and there are no solutions.

We can express systems of equations in matrix form as

$$\mathbf{Av}=\mathbf{b}$$

where $\mathbf{A}$ is the matrix of coefficients $\mathbf{v}=\binom{x}{y}$ and $\mathbf{b}$ is the vector of constants on the right-hand side of the equations. For the three cases we have looked at, in Examples 19.21, 19.22 and 19.23, we get the following.

Case 1

$$3x+4y=7$$
$$x+2y=2$$

which can be represented in matrix form as

$$\begin{pmatrix} 3 & 4 \\ 1 & 2 \end{pmatrix}\begin{pmatrix} x \\ y \end{pmatrix}=\begin{pmatrix} 7 \\ 2 \end{pmatrix}$$

Case 2

$$x+2y=1$$
$$2x+4y=2$$

which can be represented in matrix form as

$$\begin{pmatrix} 1 & 2 \\ 2 & 4 \end{pmatrix}\begin{pmatrix} x \\ y \end{pmatrix}=\begin{pmatrix} 1 \\ 2 \end{pmatrix}$$

Case 3

$$x + 2y = 1$$

$$2x + 4y = 5$$

which can be represented in matrix form as

$$\begin{pmatrix} 1 & 2 \\ 2 & 4 \end{pmatrix} \begin{pmatrix} x \\ y \end{pmatrix} = \begin{pmatrix} 1 \\ 5 \end{pmatrix}$$

We can look at the determinants of the coefficient matrices in order to help us analyse the system.

For Case 1 (the system with a unique solution) we find

$$\begin{vmatrix} 3 & 4 \\ 1 & 2 \end{vmatrix} = 3.2 - 1.4 = 6 - 4 = 2$$

The fact that the determinant of the matrix of coefficients is non-zero shows that the system of equations has a unique solution.

For Case 2 (the system with many solutions) we find

$$\begin{vmatrix} 1 & 2 \\ 2 & 4 \end{vmatrix} = 1.4 - 2.2 = 0$$

If we replace any column in this determinant by the constant terms $\left(\frac{1}{2}\right)$ we get the determinants

$$\begin{vmatrix} 1 & 1 \\ 2 & 2 \end{vmatrix} = 1.2 - 2.1 = 0 \quad \text{and} \quad \begin{vmatrix} 1 & 2 \\ 2 & 4 \end{vmatrix} = 0$$

This can be shown to hold in general. If all the determinants formed in this way are 0 then we have indeterminacy in the solutions; that is, there will be many solutions to the system.

For Case 3 (the system with no solutions) we find

$$\begin{vmatrix} 1 & 2 \\ 2 & 4 \end{vmatrix} = 1.4 - 2.2 = 0$$

If we replace any column in this determinant by the constant terms $\left(\frac{1}{5}\right)$ we get the determinants

$$\begin{vmatrix} 1 & 1 \\ 2 & 5 \end{vmatrix} = 1.5 - 2.1 = 3 \quad \text{and} \quad \begin{vmatrix} 1 & 2 \\ 5 & 4 \end{vmatrix} = 1.4 - 2.5 = -6$$

This can be shown to hold in general. If the determinant of the matrix of coefficients is zero but any one of the determinants formed using the vector of constant terms is non-zero then this shows that the system is inconsistent, and there are no solutions.

We can summarize the results of this section as follows. For a system of equations (assuming we have as many equations as unknowns) there are three possibilities.

Case 1: A determined system has a unique solution which can be found by using elimination. Geometrically, the solution is a single point which (in the case of a system in two unknowns) represents the intersection of the two lines. The determinant of the coefficients is non-zero. The system is both consistent and determined.

Case 2: An undetermined system has many solutions. If elimination is used to solve the system it will result in a condition like $0 = 0$, which is always true. Geometrically the solutions lie (for a system in two unknowns) anywhere along a line. The determinant of the coefficients is zero, as are any determinants found by replacing a column in the matrix of coefficients by the vector of constant terms. The system is indeterminate but consistent (as there are solutions).

Case 3: An inconsistent system has no solutions. If elimination is used to solve the system it will result in a condition like $0 = 3$, which is always false. Geometrically, for a system in two unknowns, the system is represented by parallel lines which have no points in common, and hence no solutions. The determinant of the coefficients is zero, but at least one of the determinants found by replacing a column in the matrix of coefficients by the vector of constant terms is non-zero. The system is inconsistent.

For Case 1 the solution of the system can be found by using the inverse of the matrix of coefficients.

We can represent the system by

$$\mathbf{Av} = \mathbf{b}$$

As the determinant of $\mathbf{A}$ is non-zero we know that $\mathbf{A}$ has an inverse $\mathbf{A}^{-1}$. We pre-multiply both sides of the matrix equation by $\mathbf{A}^{-1}$, giving

$$\mathbf{A}^{-1}\mathbf{Av} = \mathbf{A}^{-1}\mathbf{b}$$

As $\mathbf{A}^{-1}\mathbf{A} = \mathbf{I}$, the unit matrix, and $\mathbf{Iv} = \mathbf{v}$, we get

$$\mathbf{v} = \mathbf{A}^{-1}\mathbf{b}$$

We have shown that

$$\mathbf{Av} = \mathbf{b} \wedge |\mathbf{A}| \neq 0 \Leftrightarrow \mathbf{v} = \mathbf{A}^{-1}\mathbf{b}$$

Example 19.24 Solve

$$3x + 4y = 7$$
$$x + 2y = 2$$

by finding the inverse of the matrix of coefficients.

SOLUTION The system can be expressed as

$$\begin{pmatrix} 3 & 4 \\ 1 & 2 \end{pmatrix} \begin{pmatrix} x \\ y \end{pmatrix} = \begin{pmatrix} 7 \\ 2 \end{pmatrix}$$

As we know that if $\mathbf{Av} = \mathbf{b}$ and $\mathbf{A}$ is invertible then

$$\mathbf{v} = \mathbf{A}^{-1}\mathbf{b}$$

and in this case we have

$$\mathbf{A} = \begin{pmatrix} 3 & 4 \\ 1 & 2 \end{pmatrix}$$

and

$$\mathbf{b} = \begin{pmatrix} 7 \\ 2 \end{pmatrix}$$

Then to find the solution we find the inverse of

$$\mathbf{A} = \begin{pmatrix} 3 & 4 \\ 1 & 2 \end{pmatrix}$$

We know that the inverse of

$$\begin{pmatrix} a & b \\ c & d \end{pmatrix}$$

is

$$\frac{1}{(ad-bc)}\begin{pmatrix} d & -b \\ -c & a \end{pmatrix}$$

This gives the inverse of

$$\begin{pmatrix} 3 & 4 \\ 1 & 2 \end{pmatrix}$$

as

$$\frac{1}{(3.2-4.1)}\begin{pmatrix} 2 & -4 \\ -1 & 3 \end{pmatrix} = \tfrac{1}{2}\begin{pmatrix} 2 & -4 \\ -1 & 3 \end{pmatrix}$$

Using $\mathbf{x} = \mathbf{A}^{-1}\mathbf{b}$ gives

$$\begin{pmatrix} x \\ y \end{pmatrix} = \tfrac{1}{2}\begin{pmatrix} 2 & -4 \\ -1 & 3 \end{pmatrix}\begin{pmatrix} 7 \\ 2 \end{pmatrix} = \tfrac{1}{2}\begin{pmatrix} 6 \\ -1 \end{pmatrix} = \begin{pmatrix} 3 \\ -0.5 \end{pmatrix}$$

So the solution of this system of equations is $x = 3$ and $y = -0.5$.

For a 2×2 system this method of solving a system of equations is quite straightforward. However, for larger systems a solution by finding the inverse involves nearly twice as many operations as that by elimination of variables, and therefore should not be used as a method of solving equations.

Equations with three unknowns

For three unknowns we need three axes to represent the equations. Each equation is represented by a plane. For example, Fig. 19.17 shows the plane which represents the equation $x + y + z = 1$. Two planes, if they intersect, will intersect along a line and if a third independent equation is given then the three planes will intersect at a point. More than three unknowns cannot be represented geometrically.

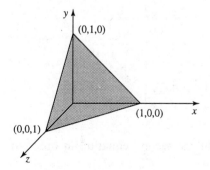

Figure 19.17 The plane given by the equation $x + y + z = 1$.

However many unknowns there are in a system of equations the three types of systems which we identified as Cases 1, 2 and 3 remain, as do the methods to be used to distinguish between determined, indeterminate and inconsistent systems.

We shall later look at finding the determinant and inverse of larger matrices, but first we look at a systematic way of doing elimination which is suitable for computer solution of a system of equations.

19.5 GAUSSIAN ELIMINATION

Gaussian elimination is a structured process for the elimination of variables in one of the equations. It is easy to generalize to larger systems of equations and it is relatively numerically stable, making it suitable for use with a computer.

Gaussian elimination is performed in stages. At Stage 1 we concentrate on the first column, the coefficients of x. The idea is to make the coefficient of the first equation 1 and eliminate the variable from the other equation(s) by using multiples of the first equation. Equation 1 is therefore the pivotal equation for Stage 1.

Example 19.25 Solve

$$3x + 4y = 7$$
$$5x - 8y = 8$$

using Gaussian elimination.

SOLUTION We can either write out the equation each time we perform a step or we can abbreviate the solution by expressing the equations in shorthand as an augmented matrix:

$$\begin{pmatrix} 3 & 4 & 7 \\ 5 & -8 & 8 \end{pmatrix}$$

We shall present both notations at the same time, and refer to the elements in the augmented matrix by

$$\begin{pmatrix} a_{11} & a_{12} & b_1 \\ a_{21} & a_{22} & b_2 \end{pmatrix}$$

For the first step the first equation is the pivotal equation:

$$3x + 4y = 7 \qquad \begin{pmatrix} 3 & 4 & 7 \\ 5 & -8 & 8 \end{pmatrix}$$
$$5x - 8y = 8$$

Stage 1

   Step 1: Divide the first equation by a_{11}:

$$x + \tfrac{4}{3}y = \tfrac{7}{3} \qquad \begin{pmatrix} 1 & \tfrac{4}{3} & \tfrac{7}{3} \\ 5 & -8 & 8 \end{pmatrix}$$
$$5x - 8y = 8$$

   Step 2: Take 5 times the first equation away from the second equation in order to eliminate the term in x in the second equation:

$$x + \tfrac{4}{3}y = \tfrac{7}{3}$$
$$-8y - (\tfrac{4}{3}y \times 5) = 8 - \tfrac{7}{3} \times 5$$

which is the same as

$$x + \tfrac{4}{3}y = \tfrac{7}{3} \qquad \begin{pmatrix} 1 & \tfrac{4}{3} & \tfrac{7}{3} \\ 0 & -\tfrac{44}{3} & -\tfrac{11}{3} \end{pmatrix}$$
$$-\tfrac{44}{3}y = -\tfrac{11}{3}$$

Stage 2

Divide the second equation through by the coefficient of $y(a_{22})$,

$$x + \tfrac{4}{3}y = \tfrac{7}{3} \qquad \begin{pmatrix} 1 & \tfrac{4}{3} & \tfrac{7}{3} \\ 0 & 1 & \tfrac{1}{4} \end{pmatrix}$$
$$y = \tfrac{1}{4}$$

We now already have the solution for y and we can obtain the solution for x by substitution into the first equation. This is called 'back-substitution'.

$$x + \tfrac{4}{3}(\tfrac{1}{4}) = \tfrac{7}{3}$$
$$\Leftrightarrow x = \tfrac{7}{3} - \tfrac{1}{3}$$
$$\Leftrightarrow x = 2$$

Therefore the solution is $(2, 0.25)$.

We see that the point of the exercise is to write the system of equations so that the final equation contains only one variable and the last but one equation has up to two variables etc. The matrix of coefficients should be in upper triangular form:

$$\begin{pmatrix} 1 & 4/3 \\ 0 & 1 \end{pmatrix}$$

The augmented matrix is then like

$$\begin{pmatrix} 1 & 4/3 & \vdots & 7/3 \\ 0 & 1 & \vdots & 1/4 \end{pmatrix}$$

which is said to be in echelon form. Once this form has been achieved then back-substitution can be performed to find the value of the variables.

Example 19.26 Solve the system of equations

$$2x + y - 2z = -1$$
$$2x - 3y + 2z = 9$$
$$-x + y - z = -3.5$$

SOLUTION

$$\begin{array}{c} 2x + y - 2z = -1 \\ 2x - 3y + 2z = 9 \\ -x + y - z = -3.5 \end{array} \qquad \begin{pmatrix} 2 & 1 & -2 & -1 \\ 2 & -3 & 2 & 9 \\ -1 & 1 & -1 & -3.5 \end{pmatrix}$$

Stage 1

Stage 1 concerns the first column. We use the first row (the pivotal row) to eliminate the elements below a_{11}.

 Step 1: Divide the first equation by a_{11}.

$$\begin{array}{c} x + 0.5y - z = -0.5 \\ 2x - 3y + 2z = 9 \\ -x + y - z = -3.5 \end{array} \qquad \begin{pmatrix} 1 & 0.5 & -1 & -0.5 \\ 2 & -3 & 2 & 9 \\ -1 & 1 & -1 & -3.5 \end{pmatrix}$$

Step 2: Eliminate x from equations 2 and 3 by taking away multiples of equation 1. To do this we take Row $2-2\times$(Row 1) and Row $3-(-1)\times$(Row 1):

$$x+0.5y-z=-0.5 \qquad \begin{pmatrix} 1 & 0.5 & -1 & -0.5 \\ 0 & -4 & 4 & 10 \\ 0 & 1.5 & -2 & -4 \end{pmatrix}$$
$$-4y+4z=10$$
$$1.5y-2z=-4$$

The calculations can be done 'in the margin' and were:
Row $2-2\times$ (Row 1):

$$2x-3y+2z=9$$
$$\underline{-2\times(x+0.5y \quad -z=-0.5)}$$
$$-4y+4z=10$$

Row $3-(-1)\times$ (Row 1):

$$-x \quad +y-z=-3.5$$
$$\underline{-(-1)\times(x+0.5y-z=-0.5)}$$
$$1.5y-2z=-4$$

Stage 2
Stage 2 concerns the second column. We use the second row to eliminate the elements below a_{22}. Here row 2 is the pivotal row.
    Step 1: Divide equation 2 by the coefficient a_{22}:

$$x+0.5y-z=-0.5 \qquad \begin{pmatrix} 1 & 0.5 & -1 & -0.5 \\ 0 & 1 & -1 & -2.5 \\ 0 & 1.5 & -2 & -4 \end{pmatrix}$$
$$y-z=-2.5$$
$$1.5y-2z=-4$$

    Step 2: Eliminate y from equation 3 by taking away multiples of equation 2:

$$x+0.5y-z=-0.5 \qquad \begin{pmatrix} 1 & 0.5 & -1 & -0.5 \\ 0 & 1 & -1 & -2.5 \\ 0 & 0 & -0.5 & -0.25 \end{pmatrix}$$
$$y-z=-2.5$$
$$-0.5z=-0.25$$

Here the calculation was Row $3-1.5\times$(Row 2) and the calculation was as follows:

$$1.5y-2z=-4$$
$$\underline{-1.5(y \quad -z=-2.5)}$$
$$-0.5z=-0.25$$

Stage 3
Divide equation 3 by the coefficient of z:

$$x+0.5y-z=-0.5 \qquad \begin{pmatrix} 1 & 0.5 & -1 & -0.5 \\ 0 & 1 & -1 & -2.5 \\ 0 & 0 & 1 & 0.5 \end{pmatrix}$$
$$y-z=-2.5$$
$$z=0.5$$

Back-substitution

We have now finished the elimination stage and we can easily solve the equations using back-substitution.

From equation 3, $z=0.5$.

Find y from equation 2:

$$y=-2.5+z \Leftrightarrow y=-2.5+0.5$$

$$\Leftrightarrow y=-2$$

Substitute into equation 1 to find x:

$$x+0.5(-2)-0.5=-0.5$$

$$\Leftrightarrow x-1.5=-0.5$$

$$\Leftrightarrow x=1$$

So the solution of the system of equations is $(1,-2,0.5)$.

Check

To check, substitute $x=1$, $y=-2$ and $z=0.5$ into the original equations:

$$2x+y-2z=-1$$

$$2x-3y+2z=9$$

$$-x+y-z=-3.5$$

giving

$$2(1)+(-2)-2(0.5)=-1 \qquad \text{which is true}$$

$$2(1)-3(-2)+2(0.5)=9 \qquad \text{which is true}$$

$$-(1)+(-2)-0.5=-3.5 \qquad \text{which is true}$$

Now we can solve the system of equations for the electrical network, which was the introductory example of Section 19.4.

Example 19.27 Solve, using Gaussian elimination, the system of equations

$$I_1-I_2-I_3=0$$

$$3I_2-2I_3=0$$

$$7I_1 \quad +2I_3=8$$

SOLUTION We shall only show the augmented matrix in this example, so we begin with

$$\begin{pmatrix} 1 & -1 & -1 & 0 \\ 0 & 3 & -2 & 0 \\ 7 & 0 & 2 & 8 \end{pmatrix}$$

Stage 1

Stage 1 concerns the first column. We use the first row to eliminate the elements below a_{11}.

    Step 1: Divide the first equation by a_{11}. As this is already 1 we do not need to divide by it.

    Step 2: Eliminate elements in the first column below a_{11} by taking away multiples of row 1 from rows 2 and 3. Row 2 already has no entry in the first column, so we leave it alone. We take (Row 3)−(7)×(Row 1):

$$\begin{pmatrix} 1 & -1 & -1 & 0 \\ 0 & 3 & -2 & 0 \\ 0 & 7 & 9 & 8 \end{pmatrix}$$

The calculation performed here was (Row 3)−7×(Row 1):

$$\begin{array}{cccc} 7 & 0 & 2 & 8 \\ -7(1 & -1 & -1 & 0) \\ \hline 0 & 7 & 9 & 8 \end{array}$$

Stage 2
Stage 2 concerns the second column. We use the second row to eliminate the elements below a_{22}.

Step 1: Divide equation 2 by the coefficient a_{22}:

$$\begin{pmatrix} 1 & -1 & -1 & 0 \\ 0 & 1 & -\frac{2}{3} & 0 \\ 0 & 7 & 9 & 8 \end{pmatrix}$$

Step 2: Eliminate the element in the second column below a_{22} by taking away multiples of Row 2 from Row 3.

$$\begin{pmatrix} 1 & -1 & -1 & 0 \\ 0 & 1 & -\frac{2}{3} & 0 \\ 0 & 0 & \frac{41}{3} & 8 \end{pmatrix}$$

Here the calculation was (Row 3)−7×(Row 2), and the calculation was as follows:

$$\begin{array}{cccc} 0 & 7 & 9 & 8 \\ -(7)(0 & 1 & -\frac{2}{3} & 0) \\ \hline 0 & 0 & \frac{41}{3} & 8 \end{array}$$

Stage 3
Divide equation 3 by the coefficient of z:

$$\begin{pmatrix} 1 & -1 & -1 & 0 \\ 0 & 1 & -\frac{2}{3} & 0 \\ 0 & 0 & 1 & \frac{24}{41} \end{pmatrix}$$

Back-substitution
We have now finished the elimination stage and we can easily solve the equations using back-substitution.

From equation 3, $I_3 = 24/41$.

Find I_2 from equation 2:

$$I_2 - \tfrac{2}{3} I_3 = 0$$

$$I_2 - \tfrac{2}{3} \times \tfrac{24}{41} = 0 \Leftrightarrow I_2 = \tfrac{16}{41}$$

Substitute into equation 1 to find I_1:

$$I_1 - \tfrac{16}{41} - \tfrac{24}{41} = 0$$

$$\Leftrightarrow I_1 = \tfrac{40}{41}$$

So the solution of the system of equations is $(\tfrac{40}{41}, \tfrac{16}{41}, \tfrac{24}{41})$.

Check

To check, substitute $I_1 = \tfrac{40}{41}$, $I_2 = \tfrac{16}{41}$ and $I_3 = \tfrac{24}{41}$ into the original equations:

$$I_1 - I_2 - I_3 = 0$$

$$3I_2 - 2I_3 = 0$$

$$7I_1 + 2I_3 = 8$$

giving

$$\tfrac{40}{41} - \tfrac{16}{41} - \tfrac{24}{41} = 0 \qquad \text{which is true}$$

$$3(\tfrac{16}{41}) - 2(\tfrac{24}{41}) = 0 \qquad \text{which is true}$$

$$7(\tfrac{40}{41}) + 2(\tfrac{24}{41}) = 8 \qquad \text{which is true}$$

Indeterminacy and inconsistency

When we analysed systems of equations in Sec. 19.4 we saw that one of the equations reducing to $0 = 0$ indicates that we have an indeterminate system; that is, there will be many solutions. If, in the process of performing Gaussian elimination, we find a row of zeros then we know that we have an indeterminate system. We can use the remaining equations to eliminate as many of the unknowns as possible, giving a solution which will still involve one or more of the variables. This will give a whole line, or possibly (in three dimensions) a plane of solutions.

If we come across a row that is zero everywhere in the matrix of coefficients, but which has a non-zero constant term we have found an equation $0 = c$, which is false. This is an inconsistent system and has no solutions.

Order of the equations

At the beginning of each stage in Gaussian elimination the order of the rows may be swapped (other than those already used in previous stages as the pivotal equation). We will have to swap the equations if the next 'pivotal' equation has a zero coefficient for the next variable to be eliminated. It is usual, although we have not illustrated this point, always to consider swapping the order of the equations in order to choose the equation with the largest absolute value of the coefficient in the term to be used for eliminating as the pivotal equation. This is called partial pivoting. This procedure is an attempt to avoid problems with equations that may become ill-conditioned in the course of performing the elimination. The equations are ill-conditioned when a small change in the coefficients of the equations causes a large change in the values of the solutions, and in such equations rounding errors can become large and cause significant inaccuracies in the solutions. We have not performed partial pivoting in these examples as they are only presented to give an idea of the method. It is assumed that for any real-life problem a computer algorithm will be used to solve the system of equations, and that such an algorithm will incorporate partial pivoting.

19.6 THE INVERSE AND DETERMINANT OF A 3 × 3 MATRIX

Finding the inverse by elimination

To find the inverse using elimination we write the matrix we need to invert on the left and the unit matrix on the right. We perform operations on both matrices at the same time. The method, called Gauss–Jordan elimination, begins in the same way as Gaussian elimination. When we have upper triangular form for the matrix we metaphorically turn the problem upside down and eliminate the upper triangle also.

Example 19.28 Find the inverse of

$$\begin{pmatrix} 4 & 0 & -4 \\ 3 & 4 & 2 \\ -1 & -1 & 1 \end{pmatrix}$$

SOLUTION We start by writing the matrix along with the unit matrix

$$\begin{pmatrix} 4 & 0 & -4 & 1 & 0 & 0 \\ 3 & 4 & 2 & 0 & 1 & 0 \\ -1 & -1 & 1 & 0 & 0 & 1 \end{pmatrix}$$

Stage 1

Step 1: Divide the first row by a_{11}:

$$\begin{pmatrix} 1 & 0 & -1 & 0.25 & 0 & 0 \\ 3 & 4 & 2 & 0 & 1 & 0 \\ -1 & -1 & 1 & 0 & 0 & 1 \end{pmatrix}$$

Step 2: Eliminate the first column below a_{11} by subtracting multiples of the first row from the second and third rows:

$$\begin{pmatrix} 1 & 0 & -1 & 0.25 & 0 & 0 \\ 0 & 4 & 5 & -0.75 & 1 & 0 \\ 0 & -1 & 0 & 0.25 & 0 & 1 \end{pmatrix}$$

The calculations were as follows:

Row 2 − 3 × Row 1

$$
\begin{array}{cccccc}
3 & 4 & 2 & 0 & 1 & 0 \\
-3(1 & 0 & -1 & 0.25 & 0 & 0) \\
\hline
0 & 4 & 5 & -0.75 & 1 & 0
\end{array}
$$

Row 3 − (−1) × Row 1

$$
\begin{array}{cccccc}
-1 & -1 & 1 & 0 & 0 & 1 \\
-(-1)(1 & 0 & -1 & 0.25 & 0 & 0 \\
\hline
0 & -1 & 0 & 0.25 & 0 & 1
\end{array}
$$

Stage 2
    Step 1: Divide the second row by a_{22} (4):

$$\begin{pmatrix} 1 & 0 & -1 & 0.25 & 0 & 0 \\ 0 & 1 & 1.25 & -0.1875 & 0.25 & 0 \\ 0 & -1 & 0 & 0.25 & 0 & 1 \end{pmatrix}$$

    Step 2: Eliminate the elements in the second column below a_{22} by subtracting multiples of the second row from the third row:

$$\begin{pmatrix} 1 & 0 & -1 & 0.25 & 0 & 0 \\ 0 & 1 & 1.25 & -0.1875 & 0.25 & 0 \\ 0 & 0 & 1.25 & 0.0625 & 0.25 & 1 \end{pmatrix}$$

The calculations were as follows:

Row 3 − (− 1) + Row 2

$$
\begin{array}{rrrrrr}
0 & -1 & 0 & 0.25 & 0 & 1 \\
-(-1)(0 & 1 & 1.25 & -0.1875 & 0.25 & 0) \\
\hline
0 & 0 & 1.25 & 0.0625 & 0.25 & 1
\end{array}
$$

Stage 3
    Step 1: Divide the third row by a_{33} (1.25):

$$\begin{pmatrix} 1 & 0 & -1 & 0.25 & 0 & 0 \\ 0 & 1 & 1.25 & -0.1875 & 0.25 & 0 \\ 0 & 0 & 1 & 0.05 & 0.2 & 0.8 \end{pmatrix}$$

    Step 2: Turn the problem metaphorically upside down and use the third row to eliminate elements in the third column above a_{33} by subtracting multiples of the third row from the first row and the second row:

$$\begin{pmatrix} 1 & 0 & 0 & 0.3 & 0.2 & 0.8 \\ 0 & 1 & 0 & -0.25 & 0 & -1 \\ 0 & 0 & 1 & 0.05 & 0.2 & 0.8 \end{pmatrix}$$

The calculations were as follows:

Row 1 − (− 1) × row 3

$$
\begin{array}{rrrrrr}
1 & 0 & -1 & 0.25 & 0 & 0 \\
-(-1)(0 & 0 & 1 & 0.05 & 0.2 & 0.8) \\
\hline
1 & 0 & 0 & 0.3 & 0.2 & 0.8
\end{array}
$$

Row 2 − 1.25 × row 3

$$
\begin{array}{rrrrrr}
0 & 1 & 1.25 & -0.1875 & 0.25 & 0 \\
-1.25(0 & 0 & 1 & 0.05 & 0.2 & 0.8) \\
\hline
0 & 1 & 0 & -0.25 & 0 & -1
\end{array}
$$

The matrix on the right-hand side is now the inverse of the original matrix. The inverse is

$$\begin{pmatrix} 0.3 & 0.2 & 0.8 \\ -0.25 & 0 & -1 \\ 0.05 & 0.2 & 0.8 \end{pmatrix}$$

Check

Multiply the original matrix by its inverse:

$$\begin{pmatrix} 4 & 0 & -4 \\ 3 & 4 & 2 \\ -1 & -1 & 1 \end{pmatrix}\begin{pmatrix} 0.3 & 0.2 & 0.8 \\ -0.25 & 0 & -1 \\ 0.05 & 0.2 & 0.8 \end{pmatrix}$$

$$=\begin{pmatrix} 4(0.3)+0(-0.25)-4(0.05) & 4(0.2)+(0)(0)-4(0.2) & 4(0.8)+0(-1)-4(0.8) \\ 3(0.3)+4(-0.25)+2(0.05) & 3(0.2)+4(0)+2(0.2) & 3(0.8)+4(-1)+2(0.8) \\ -1(0.3)-1(-0.25)+1(0.05) & -1(0.2)-1(0)+1(0.2) & -1(0.8)-1(-1)+1(0.8) \end{pmatrix}$$

$$=\begin{pmatrix} 1 & 0 & 0 \\ 0 & 1 & 0 \\ 0 & 0 & 1 \end{pmatrix}$$

Therefore we have correctly found the inverse of the matrix.

The determinant of a 3×3 matrix

The definition of the (2×2) determinant has been given as

$$\begin{vmatrix} a_1 & b_1 \\ a_2 & b_2 \end{vmatrix} = a_1b_2 - a_2b_1$$

Each of the terms on the right-hand side of this definition is of the form a_ib_j where i and j are different choices of the numbers 1 and 2. We can define higher order determinants by using ideas of permutations. We notice that the term a_1b_2 above has a positive sign because the indices 1 and 2 appear in order, whereas the term a_2b_1 has a negative sign because the indices 2,1 are reversed.

To define

$$\begin{vmatrix} a_1 & b_1 & c_1 \\ a_2 & b_2 & c_2 \\ a_3 & b_3 & c_3 \end{vmatrix}$$

we write down all terms of the form $a_ib_jc_k$ and give each term a $+$ sign or a $-$ sign depending on whether the permutation ijk is even or odd. A permutation of 123 is even if it can be achieved by an even number of swaps of the numbers, beginning with the order 123. If it can only be obtained by an odd number of swaps then the permutation is odd. For example 2,3,1 is even because we an reach it by first swapping 1 and 2, giving 2,1,3, and then swapping 1 and 3. Alternatively, we could have interchanged 2 and 3, giving 1,3,2, 1 and 3, giving 3,1,2, 2 and 1, giving 3,2,1, and 3 and 2, giving 2,3,1. Whatever way we use to get to the order 2,3,1 involves an even number of steps. Similarly, we say that a permutation of 1,2,3 is odd if it involves an odd number of adjacent interchanges.

This definition gives the determinant of a 3×3 array as

$$\begin{vmatrix} a_1 & b_1 & c_1 \\ a_2 & b_2 & c_2 \\ a_3 & b_3 & c_3 \end{vmatrix} = a_1b_2c_3 - a_1b_3c_2 - a_2b_1c_3 + a_2b_3c_1 + a_3b_1c_2 - a_3b_2c_1$$

This expression may be written in such a way that it involves 2×2 determinants as follows:

$$a_1b_2c_3 - a_1b_3c_2 - a_2b_1c_3 + a_3b_1c_2 + a_2b_3c_1 - a_3b_2c_1$$

$$= a_1(b_2c_3 - b_3c_2) - b_1(a_2c_3 - a_3c_1) + c_1(a_2b_3 - a_3b_2)$$

$$= a_1 \begin{vmatrix} b_2 & c_3 \\ b_3 & c_3 \end{vmatrix} - b_1 \begin{vmatrix} a_2 & c_2 \\ a_3 & c_3 \end{vmatrix} + c_1 \begin{vmatrix} a_2 & b_2 \\ a_3 & b_3 \end{vmatrix}$$

The 2×2 determinants that appear in this expression are called minors. This formula for the determinant is called the expansion by the first row, because the numbers a_1, b_1, c_1 which multiply the minors are from the first row of the matrix.

Note that the minor multiplying a_1 is the (2×2) determinant obtained from the original array by crossing out the row and column in which a_1 appears, as follows:

$$\begin{matrix} \cancel{a_1} & b_1 & c_1 \\ a_2 & b_2 & c_2 \\ a_3 & b_3 & c_3 \end{matrix}$$

gives the minor of a_1 as

$$\begin{vmatrix} b_2 & c_2 \\ b_3 & c_3 \end{vmatrix}$$

Similarly, the number multiplying b_1 is the determinant found by crossing out the row and the column in which b_1 appears.

We could also find the determinant by expanding about the first column:

$$a_1b_2c_3 - a_1b_3c_2 - a_2b_1c_3 + a_2b_3c_1 + a_3b_1c_2 - a_3b_2c_1$$

$$= a_1(b_2c_3 - b_3c_2) - a_2(b_1c_3 - b_3c_1) + a_3(b_1c_2 - b_2c_1)$$

$$= a_1 \begin{vmatrix} b_2 & c_2 \\ b_3 & c_3 \end{vmatrix} - a_2 \begin{vmatrix} b_1 & c_1 \\ b_3 & c_3 \end{vmatrix} + a_3 \begin{vmatrix} b_1 & c_1 \\ b_2 & c_2 \end{vmatrix}$$

Again we see that the minor of a_2, for instance, can be found by crossing out the row and column in which a_2 appears in the original array.

To find the sign multiplying each term in the expansion for the determinant we can remember the following pattern:

$$\begin{matrix} + & - & + \\ - & + & - \\ + & - & + \end{matrix}$$

To find the determinant, we can expand about any row and column, multiplying each term a_{ij} by its respective minor and find the sign either by using the pattern above or, equivalently, using the expression $(-1)^{i+j}$.

Example 19.29 Find the following determinant:

$$\begin{vmatrix} -1 & 2 & 3 \\ 6 & -1 & 2 \\ 4 & 0 & -1 \end{vmatrix}$$

SOLUTION Expanding about the first row:

$$\begin{vmatrix} -1 & 2 & 3 \\ 6 & -1 & 2 \\ 4 & 0 & -1 \end{vmatrix} = -1\begin{vmatrix} -1 & 2 \\ 0 & -1 \end{vmatrix} -2\begin{vmatrix} 6 & 2 \\ 4 & -1 \end{vmatrix} +3\begin{vmatrix} 6 & -1 \\ 4 & 0 \end{vmatrix}$$

$$= -1(1-0)-2(-6-8)+3(0+4)$$

$$= -1+28+12=39$$

Alternatively, expanding about the first column we get

$$-1\begin{vmatrix} -1 & 2 \\ 0 & -1 \end{vmatrix} -6\begin{vmatrix} 2 & 3 \\ 0 & -1 \end{vmatrix} +4\begin{vmatrix} 2 & 3 \\ -1 & 2 \end{vmatrix} = -1(1-0)-6(-2-0)+4(4-(-3))$$

$$= -1+12+28=39$$

The inverse of a matrix using (adjoint(A))/|A|

We have already seen how to find the inverse of a matrix by using elimination. It is also possible to find the inverse by the following procedure:

1. Find the matrix of minors.
2. Multiply the minor for row i and column j by $(-1)^{i+j}$. This is then called the matrix of cofactors.
3. Take the transpose of the matrix of cofactors to find the adjoint matrix.
4. Divide by the determinant of the original matrix.

This procedure should rarely need to be used and only usually if we have a matrix which involves some unknown variables or expresses some formula and we would like to find the inverse formula. It would never be used as a numerical procedure, as it is both numerically unstable and also uses a very large number of operations (of the order of n^3 operations, where n is the dimension of the matrix, whereas elimination is only of the order of n^2).

Example 19.30 Find the inverse of

$$\begin{pmatrix} 4 & 0 & -4 \\ 3 & 4 & 2 \\ -1 & -1 & 1 \end{pmatrix}$$

using

$$\mathbf{A}^{-1} = \frac{1}{|\mathbf{A}|}\text{Adjoint}(\mathbf{A})$$

SOLUTION Find the matrix of minors for each term in the matrix. The minor for the *i*th row and *j*th column is found by crossing out that row and column and finding the determinant of the remaining elements.

This gives the matrix of minors as

$$\begin{pmatrix} \begin{vmatrix} 4 & 2 \\ -1 & 1 \end{vmatrix} & \begin{vmatrix} 3 & 2 \\ -1 & 1 \end{vmatrix} & \begin{vmatrix} 3 & 4 \\ -1 & -1 \end{vmatrix} \\ \begin{vmatrix} 0 & -4 \\ -1 & 1 \end{vmatrix} & \begin{vmatrix} 4 & -4 \\ -1 & 1 \end{vmatrix} & \begin{vmatrix} 4 & 0 \\ -1 & -1 \end{vmatrix} \\ \begin{vmatrix} 0 & -4 \\ 4 & 2 \end{vmatrix} & \begin{vmatrix} 4 & -4 \\ 3 & 2 \end{vmatrix} & \begin{vmatrix} 4 & 0 \\ 3 & 4 \end{vmatrix} \end{pmatrix} = \begin{pmatrix} 6 & 5 & 1 \\ -4 & 0 & -4 \\ 16 & 20 & 16 \end{pmatrix}$$

To find the matrix of cofactors we multiply by the pattern

$$\begin{matrix} + & - & + \\ - & + & - \\ + & - & + \end{matrix}$$

giving

$$\begin{pmatrix} 6 & -5 & 1 \\ 4 & 0 & 4 \\ 16 & -20 & 16 \end{pmatrix}$$

To find the adjoint we take the transpose of the above, giving

$$\begin{pmatrix} 6 & 4 & 16 \\ -5 & 0 & -20 \\ 1 & 4 & 16 \end{pmatrix}$$

Now we find the determinant. Expanding about the first row this gives

$$4(4-(-2))-0(3(1)-2(-1))-4(3(-1)-(-1)(4))=20$$

Finally we divide the adjoint by the determinant to find the inverse, giving

$$\frac{1}{20}\begin{pmatrix} 6 & 4 & -16 \\ -5 & 0 & -20 \\ 1 & 4 & 16 \end{pmatrix} = \begin{pmatrix} 0.3 & 0.2 & 0.8 \\ -0.25 & 0 & -1 \\ 0.05 & 0.2 & 0.8 \end{pmatrix}$$

Check

To check that the calculation is correct we multiply the original matrix by the inverse. If we get the unit matrix as the result we can conclude that we have indeed found the inverse:

$$\begin{pmatrix} 4 & 0 & -4 \\ 3 & 4 & 2 \\ -1 & -1 & 1 \end{pmatrix}\begin{pmatrix} 0.3 & 0.2 & 0.8 \\ -0.25 & 0 & -1 \\ 0.05 & 0.2 & 0.8 \end{pmatrix} = \begin{pmatrix} 1 & 0 & 0 \\ 0 & 1 & 0 \\ 0 & 0 & 1 \end{pmatrix}$$

which is correct.

19.7 EIGENVALUES AND EIGENVECTORS

In Example 19.17 we looked at the problem of scaling along the line $x = y$ and we saw that the matrix

$$\begin{pmatrix} \frac{3}{2} & \frac{1}{2} \\ \frac{1}{2} & \frac{3}{2} \end{pmatrix}$$

represents a scaling along the line $y = x$, and it leaves points along the line $y = -x$ unchanged. This means that any vector in the direction $(1,1)$ will simply be multiplied by 2 and any vector in the direction $(-1,1)$ will remain unchanged after multiplication by this matrix. Other vectors will undergo a mixed effect.

Suppose that we know that a matrix **A** represents a scaling but without knowing the direction of the scaling or by how much it scales. Is there any way we can find that direction and the scaling constant?

The problem then is to find a vector **v** which is simply scaled by some currently unknown amount λ when multiplied by **A**. λ and **v** must be such that

$$\mathbf{Av} = \lambda\mathbf{v}$$

If we manage to find values of λ and **v** we call these the eigenvalues and eigenvectors of the matrix **A**. We will solve this for

$$\begin{pmatrix} \frac{3}{2} & \frac{1}{2} \\ \frac{1}{2} & \frac{3}{2} \end{pmatrix}$$

as we know the result that we expect to get.

Example 19.31 Find λ and **v** such that

$$\mathbf{Av} = \lambda\mathbf{v}$$

where

$$\mathbf{A} = \begin{pmatrix} \frac{3}{2} & \frac{1}{2} \\ \frac{1}{2} & \frac{3}{2} \end{pmatrix}$$

SOLUTION Subtract $\lambda\mathbf{v}$ from both sides of the equation

$$\mathbf{Av} = \lambda\mathbf{v}$$

$$\Leftrightarrow \mathbf{Av} - \lambda\mathbf{v} = \mathbf{0}$$

$$\begin{pmatrix} \frac{3}{2} & \frac{1}{2} \\ \frac{1}{2} & \frac{3}{2} \end{pmatrix}\mathbf{v} - \lambda\mathbf{v} = \mathbf{0}$$

We put in the unit matrix as $\mathbf{v} = \mathbf{Iv}$ and we want to combine terms:

$$\begin{pmatrix} \frac{3}{2} & \frac{1}{2} \\ \frac{1}{2} & \frac{3}{2} \end{pmatrix}\mathbf{v} - \lambda\begin{pmatrix} 1 & 0 \\ 0 & 1 \end{pmatrix}\mathbf{v} = \begin{pmatrix} 0 \\ 0 \end{pmatrix}$$

$$\Leftrightarrow \begin{pmatrix} \frac{3}{2}-\lambda & \frac{1}{2} \\ \frac{1}{2} & \frac{3}{2}-\lambda \end{pmatrix}\mathbf{v} = \begin{pmatrix} 0 \\ 0 \end{pmatrix}$$

Now substitute

$$\mathbf{v} = \begin{pmatrix} x \\ y \end{pmatrix}$$

giving

$$(\tfrac{3}{2} - \lambda)x + \tfrac{1}{2}y = 0$$

$$\tfrac{1}{2}x + (\tfrac{3}{2} - \lambda)y = 0$$

Unfortunately the solution to this gives $x=0$ and $y=0$, which is not very enlightening (it is called the trivial solution). However, we started by saying that we wanted to find the direction in which this matrix scaled any vector. That is, we want to find a whole line of solutions. We can use a result that we found from solving systems of equations. The equations may have a whole line of solutions if the determinant of the coefficients is 0.

Hence we need to find λ such that

$$\begin{vmatrix} \tfrac{3}{2} - \lambda & \tfrac{1}{2} \\ \tfrac{1}{2} & \tfrac{3}{2} - \lambda \end{vmatrix} = 0$$

Expanding the determinant gives

$$(\tfrac{3}{2} - \lambda)(\tfrac{3}{2} - \lambda) - \tfrac{1}{4} = 0$$

$$\tfrac{9}{4} - 3\lambda + \lambda^2 - \tfrac{1}{4} = 0 \Leftrightarrow \lambda^2 - 3\lambda + 2 = 0$$

This factorizes to

$$(\lambda - 2)(\lambda - 1) = 0$$

$$\Leftrightarrow \lambda = 2 \quad \text{or} \quad \lambda = 1$$

Hence the eigenvalues of the matrix $\mathbf{A}$ are 1 and 2. To find the vectors which go with each of these eigenvalues we substitute into the equations

$$(\tfrac{3}{2} - \lambda)x + \tfrac{1}{2}y = 0$$

$$\tfrac{1}{2}x + (\tfrac{3}{2} - \lambda)y = 0$$

For $\lambda = 2$:

$$-\tfrac{1}{2}x + \tfrac{1}{2}y = 0$$

$$\tfrac{1}{2}x - \tfrac{1}{2}y = 0$$

We notice that these equations are dependent, which we would have expected, as by setting the determinant $=0$ we were looking for an undetermined system.

We have

$$-\tfrac{1}{2}x + \tfrac{1}{2}y = 0 \Leftrightarrow x = y$$

This means that any vector (x,y), where $y=x$, will be scaled by 2 if multiplied by the matrix $\mathbf{A}$. The eigenvector can be given as $(1,1)$, as it is only necessary to indicate the direction.

The other eigenvalue, $\lambda = 1$, gives

$$\tfrac{1}{2}x + \tfrac{1}{2}y = 0$$

$$\tfrac{1}{2}x + \tfrac{1}{2}y = 0$$

Again the equations are dependent and we have $x + y = 0$. The eigenvector is any vector (x,y) where $x = -y$, so this gives the direction $(-1,1)$.

As not all matrices represent scaling it is not always possible to find real eigenvalues. In particular, a rotation matrix has no real eigenvalues for $\theta \neq 0$. Another point to note is that in this example the eigenvectors were at right angles to each other. This is only true for symmetric matrices (which we had in this case).

The method can be summarized as follows. To find the eigenvalues and eigenvectors of **A**:

1. Solve $|\mathbf{A} - \lambda \mathbf{I}| = 0$ to find the eigenvalues. This is called the characteristic equation.
2. For each value of λ found, substitute into $(\mathbf{A} - \lambda \mathbf{I})\mathbf{v} = \mathbf{0}$ and find **v**. This will be an undetermined system, so we will find at least a whole line of solutions. Choose any vector lying in the direction of the line.

Example 19.32 Find the eigenvectors and eigenvalues of

$$\begin{pmatrix} 1 & 3 \\ 2 & -4 \end{pmatrix}$$

SOLUTION Solve $|\mathbf{A} - \lambda \mathbf{I}| = 0$, which gives

$$\begin{vmatrix} 1 - \lambda & 3 \\ 2 & -4 - \lambda \end{vmatrix} = 0$$

$$\Leftrightarrow (1 - \lambda)(-4 - \lambda) - 6 = 0$$

$$\Leftrightarrow \lambda^2 + 3\lambda - 10 = 0$$

$$\Leftrightarrow (\lambda + 5)(\lambda - 2) = 0 \Leftrightarrow \lambda = -5 \text{ or } \lambda = 2$$

For each value of λ solve

$$\begin{pmatrix} 1 - \lambda & 3 \\ 2 & -4 - \lambda \end{pmatrix} \begin{pmatrix} x \\ y \end{pmatrix} = \begin{pmatrix} 0 \\ 0 \end{pmatrix}$$

For $\lambda = -5$ this gives

$$\begin{pmatrix} 6 & 3 \\ 2 & 1 \end{pmatrix} \begin{pmatrix} x \\ y \end{pmatrix} = \begin{pmatrix} 0 \\ 0 \end{pmatrix} \Rightarrow \begin{matrix} 6x + 3y = 0 \\ 2x + y = 0 \end{matrix}$$

We see that these equations are dependent. Solving the first one

$$6x + 3y = 0$$

$$\Leftrightarrow 2x + y = 0$$

$$\Leftrightarrow y = -2x$$

The vector is therefore (x, y) where $y = -2x$, giving $(x, -2x)$. We only need the direction of the vector, so choose $(1, -2)$ by substituting $x = 1$.

For $\lambda = 2$ we get

$$\begin{pmatrix} -1 & 3 \\ 2 & -6 \end{pmatrix} \begin{pmatrix} x \\ y \end{pmatrix} = \begin{pmatrix} 0 \\ 0 \end{pmatrix} \Rightarrow \begin{matrix} -x + 3y = 0 \\ 2x - 6y = 0 \end{matrix}$$

Solving

$$-x + 3y = 0$$

$$\Leftrightarrow x = 3y$$

Hence we have (x,y) where $x=3y$ giving $(3y,y)$. Substitute $y=1$, giving the vector as $(3,1)$.

We have shown that the matrix

$$\begin{pmatrix} 1 & 3 \\ 2 & -4 \end{pmatrix}$$

has eigenvalue -5 with eigenvector $(1,-2)$ and eigenvalue 2 with eigenvector $(3,1)$.

19.8 LEAST SQUARES DATA FITTING

Matrix methods can be employed to the problem of finding the 'best fit' line through a set of data. In Chapter 8 we performed a fit by eye to a set of data points. We drew a scatter diagram of the data and if the data appeared, more or less, to fit on a line then we would draw the line by hand and then use any two points lying on the line to find the equation of the line. We do not expect experimental data to be exact, and that is the reason that the data points do not lie exactly on a line. We shall now look at the method of least squares, which can be used to compute the 'best fit' line. The method is called 'least squares' because it minimizes the squared error between the data points and the equation of the line found. The method is justified in the following example.

We start with two sets of data which we suspect are related linearly.

Example 19.33 A student was late for a particularly interesting engineering maths lecture and was therefore walking briskly towards the lecture hall in a straight line at approximately constant speed b.

The student's position x (metres) at time t (seconds) is given by:

t	0	5	10	15	20	25
x	100	111	119	132	140	151

We can plot these points on a scatter diagram, as in Fig. 19.18. We can see that they lie on an approximate straight line. We need to decide which is the 'best' straight line to draw.

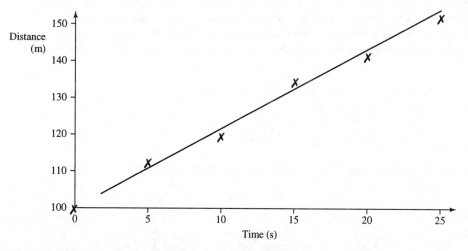

Figure 19.18 A scatter diagram of the data for Example 19.33.

SOLUTION One way is to fit the straight line $y=a+bx$ to a set of data points (x_i,y_i) so that the sum of the squares of the vertical distances of the points from the straight line drawn is minimum. This is illustrated in Fig. 19.19.

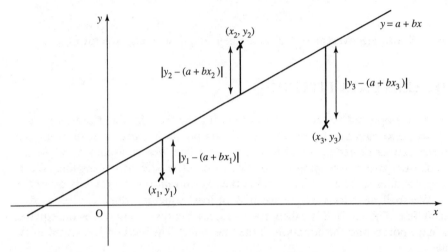

Figure 19.19 The vertical distances between the data points and the line represent the error for each point. The method of least squares minimizes the sum of the square of these errors.

In order to fit the line $y=a+bx$ we can vary the values of a and b until it satisfies our condition for minimum squared error. If the data points are (x_i,y_i), then the sum of the squares of the errors is given by

$$E= \sum_i (y_i-a-bx_i)^2$$

We can vary a and b to make this a minimum. E is a function of two variables a and b. We know how to find the minima or maxima with respect to one variable, which we looked at in Chapter 17. To minimize with respect to two variables we begin by differentiating with respect to each variable in turn, keeping the other variable constant. This is called partial differentiation, which we have not dealt with in this book. The ideas used in finding stationary values are the same as those for one variable, although the method of distinguishing between types of stationary value is slightly more involved because the function represents a two-dimensional surface drawn in three dimensions rather than a simple curve. A partial derivative is indicated by using a curly, d, ∂, for the derivative and $\partial E/\partial a$ is read as 'partial dE by da'

$$\frac{\partial E}{\partial a}=-2\sum_i (y_i-a-bx_i)$$

$$\frac{\partial E}{\partial b}=-2\sum_i x_i(t_i-a-bx_i)$$

For a minimum value we must have $\partial E/\partial a=0$ and $\partial E/\partial b=0$, giving

$$-2\sum_i y_i-a-bx_i=0 \Leftrightarrow \sum_i y_i- \sum_i a-b\sum_i x_i=0$$

$$-2\sum_i x_i(y_i-a-bx_i)=0 \Leftrightarrow \sum_i x_iy_i-a\sum_i x_i-b\sum_i x_i^2 =0$$

Finally we get the normal equations:

$$an + b\sum_{i} x_i = \sum_{i} y_i$$

$$a\sum_{i} x_i + b\sum_{i} x_i^2 = \sum_{i} x_i y_i$$

where n is the number of data points.

We have not attempted to justify that this is actually a minimum point (we have only shown it to give a stationary point). This is beyond the scope of this book. We can now illustrate the method for finding the values of a and b which minimize the sum of the squared errors.

At the start of this example the dependent variable was x and the independent variable was t. We wish to find the values of a and b so that $x = a + bt$ gives a least squares fit to the data. The number of data points is 6, so we have as the normal equations:

$$6a + b\sum_{i=1}^{6} t_i = \sum_{i=1}^{6} x_i$$

$$a\sum_{i=1}^{6} t_i + b\sum_{i=1}^{6} t_i^2 = \sum_{i=1}^{6} t_i x_i$$

Make a table from the data, as in Table 19.1. Then the normal equations become

$$6a + 75b = 753$$

$$75a + 1375b = 10\,300$$

Solving:

$$450a + 5625b = 56\,475$$

$$450a + 8250b = 61\,800$$

$$2625b = 5325$$

$$\Rightarrow b \simeq 2.03$$

$$6a + 75(2.03) = 754 \Leftrightarrow a \simeq 100.14$$

We have $a = 100.14$, $b = 2.03$. Hence the line of best fit is $x = 100.14 + 2.03t$.

Table 19.1 A table made from the data of Example 19.32

t	x	t^2	tx
0	100	0	0
5	111	25	555
10	119	100	1 190
15	132	225	1 980
20	151	625	3 775
75	753	1375	10 300

Curve fitting

The same method for fitting a straight line can be generalized to fit any polynomial. For example, it could appear that our data would be better fitted to a parabola.

$$y = b_0 + b_1 x + b_2 x^2$$

The normal equations in this case are

$$b_0 n + b_1 \sum_i x_i + b_2 \sum_i x_i^2 = \sum_i y_i$$

$$b_0 \sum_i x_i + b_1 \sum_i x_i^2 + b_2 \sum_i x_i^3 = \sum_i x_i y_i$$

$$b_0 \sum_i x_i^2 + b_1 \sum_i x_i^3 + b_2 \sum_i x_i^4 = \sum_i x_i^2 y_i$$

We can solve these system of equations using Gaussian elimination.

Example 19.34 Find the best fit parabola by the method of least squares for (0,3), (1,1), (2,0), (4,1), (6,4).

SOLUTION The table is given in Table 19.2 and the normal equations are:

$$5b_0 + 13b_1 + 57b_2 = 9$$

$$13b_0 + 57b_1 + 289b_2 = 29$$

$$57b_0 + 289b_1 + 1569b_2 = 161$$

Table 19.2 Table made from the data of Example 19.34

x	y	x^2	x^3	x^4	xy	$x^2 y$
0	3	0	0	0	0	0
1	1	1	1	1	1	1
2	0	4	8	16	0	0
4	1	16	64	256	4	16
6	4	36	216	1296	24	144
13	9	57	289	1569	29	161

Solving these using Gaussian elimination gives

$$\begin{pmatrix} 5 & 13 & 57 & 9 \\ 13 & 57 & 289 & 29 \\ 57 & 289 & 1569 & 161 \end{pmatrix}$$

Stage 1

$$\begin{pmatrix} 1 & 2.6 & 11.4 & 1.8 \\ 13 & 57 & 289 & 29 \\ 57 & 289 & 1569 & 161 \end{pmatrix}$$

$$\begin{pmatrix} 1 & 2.6 & 11.4 & 1.8 \\ 0 & 23.2 & 140.8 & 5.6 \\ 0 & 140.8 & 919.2 & 58.4 \end{pmatrix}$$

Stage 2

$$\begin{pmatrix} 1 & 2.6 & 11.4 & 1.8 \\ 0 & 1 & 6.069 & 0.241 \\ 0 & 140.8 & 919.2 & 58.4 \end{pmatrix}$$

$$\begin{pmatrix} 1 & 2.6 & 11.4 & 1.8 \\ 0 & 1 & 6.069 & 0.241 \\ 0 & 0 & 64.685 & 24.467 \end{pmatrix}$$

Stage 3

$$\begin{pmatrix} 1 & 2.6 & 11.4 & 1.8 \\ 0 & 1 & 6.069 & 0.241 \\ 0 & 0 & 1 & 0.378 \end{pmatrix}$$

Back-substitution

$$b_0 + 2.6b_1 + 11.4b_2 = 1.8$$

$$b_1 + 6.069b_2 = 0.241$$

$$b_2 = 0.378$$

gives $b_2 = 0.378$, $b_1 = -2.054$ and $b_0 = 2.831$. Hence the best fit parabola is $y = 2.831 - 2.054x + 0.378x^2$.

19.9 SUMMARY

1. Matrices are used to represent information in a way suitable for use by a computer. They can represent, among other things, systems of linear equations, transformations and networks. A matrix is a rectangular array of numbers of dimension $m \times n$, where m is the number of rows and n is the number of columns.
2. To add or subtract matrices, add or subtract each corresponding element. The matrices must be of exactly the same dimension. To multiply two matrices, $\mathbf{C} = \mathbf{AB}$, the number of columns in matrix $\mathbf{A}$ must equal the number of rows in matrix $\mathbf{B}$. The ijth element of $\mathbf{C}$ is found by multiplying the ith row of $\mathbf{A}$ by the jth column of $\mathbf{B}$.
3. The unit matrix, $\mathbf{I}$, leaves any matrix unchanged under multiplication.

$$\mathbf{I} = \begin{pmatrix} 1 & 0 \\ 0 & 1 \end{pmatrix} \text{ (2 dimensions)}$$

$$\mathbf{I} = \begin{pmatrix} 1 & 0 & 0 \\ 0 & 1 & 0 \\ 0 & 0 & 1 \end{pmatrix} \text{ (3 dimensions)}$$

$\mathbf{IA} = \mathbf{AI} = \mathbf{A}$, where $\mathbf{A}$ is any matrix.
4. The inverse of a matrix is represented by $\mathbf{A}^{-1}$ and can be found for square, non-singular matrices. A matrix is singular if its determinant is 0.

5. The 2×2 determinant is defined by

$$\begin{vmatrix} a & b \\ c & d \end{vmatrix} = ad - bc$$

6. The inverse of a 2×2 non-singular matrix

$$\begin{pmatrix} a & b \\ c & d \end{pmatrix}$$

is

$$\frac{1}{(ad-bc)} \begin{pmatrix} d & -b \\ -c & a \end{pmatrix}$$

7. Transformations of the plane ($\mathbb{R}^2$) can be defined using vectors and matrices as in Section 19.3.
8. Systems of linear equations may be determined (a single solution), indeterminate (many solutions) or inconsistent (no solutions).
9. Systems of linear equations can be solved using Gaussian elimination.
10. The inverse of a matrix, if it exists, can be found using Gauss–Jordan elimination.
11. The determinant of a 3×3 matrix may be found by expanding about any row or column, where

$$\begin{vmatrix} a_1 & b_1 & c_1 \\ a_2 & b_2 & c_2 \\ a_3 & b_3 & c_3 \end{vmatrix} = a_1 \begin{vmatrix} b_2 & c_3 \\ b_3 & c_3 \end{vmatrix} - b_1 \begin{vmatrix} a_2 & c_2 \\ a_3 & c_3 \end{vmatrix} + c_1 \begin{vmatrix} a_2 & b_2 \\ a_3 & b_3 \end{vmatrix}$$

gives the expansion about the first row.
12. The inverse of a non-singular matrix can be found using

$$\mathbf{A}^{-1} = \frac{1}{|\mathbf{A}|} \text{Adjoint}(\mathbf{A})$$

13. The eigenvalues and eigenvectors of a matrix $\mathbf{A}$ are the values of λ and $\mathbf{v}$ such that $\mathbf{Av} = \lambda \mathbf{v}$.
14. The method of least squares is used to fit a line or a curve through experimental data in such a way that the sum of the square of the errors is a minimum.

19.10 EXERCISES

19.1 $\mathbf{A} = \begin{pmatrix} 2 & 3 \\ 0 & -4 \\ 0 & 2 \end{pmatrix}$ $\mathbf{B} = \begin{pmatrix} 2 & -1 & 0 \\ 1 & 6 & 1 \\ 0 & 3 & 4 \end{pmatrix}$ $\mathbf{C} = (4 \quad 0 \quad -1)$

$\mathbf{D} = \begin{pmatrix} 0 & 3 \\ -1 & 4 \end{pmatrix}$ $\mathbf{E} = \begin{pmatrix} 3 & 2 \\ -1 & -\frac{2}{3} \end{pmatrix}$

Find the following, where possible

(a) $\mathbf{A}^T$
(b) $\mathbf{A}^T\mathbf{B}$
(c) $\mathbf{AB}$
(d) $\mathbf{BA}$
(e) $\mathbf{C}^T$
(f) $\mathbf{AC}$
(g) $\mathbf{CA}$
(h) $\mathbf{CB}$
(i) $3\mathbf{CA}$
(j) $\mathbf{D}+\mathbf{E}$
(k) $3\mathbf{D}-\frac{1}{2}\mathbf{E}$
(l) $\frac{1}{2}\mathbf{B}+\mathbf{A}$
(m) $\mathbf{B}^2$
(n) $\mathbf{E}^3$
(o) $\mathbf{AC}^2$
(p) $\mathbf{A}^T\mathbf{BC}^T$

19.2 Use

$$A = \begin{pmatrix} a_{11} & a_{12} \\ a_{21} & a_{22} \end{pmatrix}$$

$$B = \begin{pmatrix} b_{11} & b_{12} \\ b_{21} & b_{22} \end{pmatrix}$$

$$C = \begin{pmatrix} c_{11} & c_{12} \\ c_{21} & c_{22} \end{pmatrix}$$

to justify the associative law for 2×2 matrices:

$$A(BC) = (AB)C$$

19.3 Find the real square matrix which is both symmetric and skew symmetric.

19.4 A network, as in Fig. 19.20(a), (b) or (c) can be used to represent an electrical network, a system of one way streets, or a communication system. An incidence matrix can be defined for a network in the following way (the lines are called arcs and the dots are called vertices:

$$a_{ij} = 1 \qquad \text{if the arc } j \text{ is leaving the vertex } i$$

$$a_{ij} = -1 \qquad \text{if the arc } j \text{ is entering the vertex } i$$

$$a_{ij} = 0 \qquad \text{if the arc } j \text{ does not touch the vertex } i$$

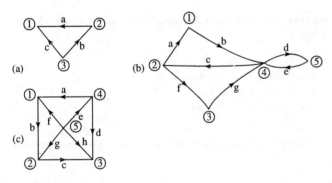

Figure 19.20 Networks for Exercise 19.4.

For instance, Fig. 19.20(a) has incidence matrix

$$
\begin{array}{c}
\phantom{\text{Vertex}} \quad \text{Arc} \\
\phantom{\text{Vertex}} \quad
\begin{array}{ccc}
a & b & c
\end{array} \\
\text{Vertex}
\begin{array}{c}
1 \\
2 \\
3
\end{array}
\begin{pmatrix}
-1 & 0 & -1 \\
1 & -1 & 0 \\
0 & 1 & 1
\end{pmatrix}
\end{array}
$$

(a) Find incidence matrices for the networks in Figs 19.20(b) and (c).
(b) Draw networks that have the following incidence matrices:

(i)

$$
\begin{array}{c}
\phantom{\text{Vertex}} \quad \text{Arc} \\
\phantom{\text{Vertex}} \quad
\begin{array}{ccc}
a & b & c
\end{array} \\
\text{Vertex}
\begin{array}{c}
1 \\
2 \\
3 \\
4
\end{array}
\begin{pmatrix}
-1 & 1 & 1 \\
0 & -1 & 0 \\
1 & 0 & -1 \\
0 & 0 & 0
\end{pmatrix}
\end{array}
$$

(ii)

Arc

$$\text{Vertex } \begin{array}{c} \\ 1 \\ 2 \\ 3 \end{array} \begin{array}{ccccc} a & b & c & d & e \\ \left(\begin{array}{ccccc} -1 & 0 & -1 & 1 & 0 \\ 1 & -1 & 0 & -1 & 1 \\ 0 & 1 & 1 & 0 & -1 \end{array} \right) \end{array}$$

19.5 Represent the following transformations using matrices and vectors. In each case apply the transformations to A (0,0), B (1,0), C (1,1) and D (0,1) to find their images A', B', C' and D' and draw your result.

(a) Rotation about the origin through 120°.
(b) Translation by $(-4,1)$.
(c) Reflection in the x-axis.
(d) Scaling in the y-direction by 5.
(e) Rotation through 120° about the origin followed by a translation by $(-2,3)$.
(f) Translation by $(-1,3)$ followed by rotation through 120° about the origin.
(g) Rotation through 120° about the origin followed by scaling by 5 in the y-direction.
(h) Reflection in the line $y=x$.
(i) Scaling along a line at a 30° angle to the x-axis, by a factor of 4.
(j) Find inverse transformations for those given in (a), (b), (c), (e), (g) and (i) and check in each case that the inverse transformation returns A', B', C' and D' to A, B, C and D.

19.6 Sketch the following systems of equations and solve them using Gaussian elimination. In each case, state whether the system is determined, indeterminate or inconsistent:

(a) $\begin{aligned} 4x + y &= 3 \\ -2x - y &= -3 \end{aligned}$ (b) $\begin{aligned} x + y &= 3 \\ x - y &= 7 \end{aligned}$ (c) $\begin{aligned} 3x + 2y &= 1 \\ -2x - \tfrac{4}{3}y &= 0 \end{aligned}$ (d) $\begin{aligned} -5x - y &= 5 \\ 10x + 2y &= -10 \end{aligned}$

(e) $\begin{aligned} 3x + 2y &= -17 \\ 10x + y &= 0 \end{aligned}$ (f) $\begin{aligned} x + 6y &= 4 \\ 3x + 18y &= 10 \end{aligned}$ (g) $\begin{aligned} -3x + 2y &= 1 \\ 1.5x - y &= 0.5 \end{aligned}$ (h) $\begin{aligned} -7x + 8y &= -10 \\ -2x + y &= -8 \end{aligned}$

19.7 Solve the following using Gaussian elimination. In each case state whether the system is determined, indeterminate or inconsistent:

(a) $\begin{aligned} 2x - 3y - z &= 4 \\ 5x + 5y + 2z &= -23 \\ x \qquad + z &= -1 \end{aligned}$ (b) $\begin{aligned} 4x - y - 2z &= 40 \\ 3x + y + 9z &= 5 \\ x - y + z &= -55 \end{aligned}$

(c) $\begin{aligned} x + y - 3z &= 8 \\ -10x + 6y &= -14 \\ 12x - 4y - 6z &= 30 \end{aligned}$ (d) $\begin{aligned} x - y + z &= 3 \\ x \qquad + z &= 3 \\ -4x - y - 4z &= -10 \end{aligned}$

19.8 Find inverses of the following matrices A, if they exist, and check that $A^{-1}A = I$:

(a) $\begin{pmatrix} 1 & -2 \\ 0 & 1 \end{pmatrix}$ (b) $\begin{pmatrix} 2 & 3 \\ -1 & 6 \end{pmatrix}$ (c) $\begin{pmatrix} 2 & -1 \\ 10 & -5 \end{pmatrix}$

(d) $\begin{pmatrix} 1 & -1 & 0 \\ -2 & 4 & 2 \\ -3 & 1 & 2 \end{pmatrix}$ (e) $\begin{pmatrix} 0 & 5 & 1 \\ -1 & 5 & 3 \\ 2 & 0 & 2 \end{pmatrix}$ (f) $\begin{pmatrix} 1 & -1 & 2 \\ 6 & 1 & 3 \\ -5 & -2 & -1 \end{pmatrix}$

19.9 Find the following determinants:

(a) $\begin{vmatrix} 1 & 6 \\ 2 & 3 \end{vmatrix}$ (b) $\begin{vmatrix} 5 & 1 \\ -6 & 2 \end{vmatrix}$

(c) $\begin{vmatrix} 3 & -2 & 1 \\ 0 & 1 & 0 \\ 3 & 2 & 1 \end{vmatrix}$ (d) $\begin{vmatrix} 6 & -3 & -2 \\ 1 & -1 & 8 \\ 0 & -1 & 0 \end{vmatrix}$

19.10 The vector product of two three-dimensional vectors can be defined using a determinant as follows:

$$(a_1, a_2, a_3) \times (b_1, b_2, b_3) = \begin{vmatrix} \mathbf{i} & \mathbf{j} & \mathbf{k} \\ a_1 & a_2 & a_3 \\ b_1 & b_2 & b_3 \end{vmatrix}$$

where $\mathbf{i}$, $\mathbf{j}$ and $\mathbf{k}$ are the unit vectors in the x, y, z directions. Use this definition to find the following:

(a) $(1,3,6) \times (-1,2,2)$
(b) $(\frac{1}{2}, \frac{1}{2}, -1) \times (0,0,3)$

19.11 The scalar triple product of three vectors $\mathbf{a}$, $\mathbf{b}$, $\mathbf{c}$, given by $\mathbf{a} \cdot (\mathbf{b} \times \mathbf{c})$ can be found using a determinant as follows:

$$(a_1, a_2, a_3) \cdot ((b_1, b_2, b_3) \times (c_1, c_2, c_3)) = \begin{vmatrix} a_1 & a_2 & a_3 \\ b_1 & b_2 & b_3 \\ c_1 & c_2 & c_3 \end{vmatrix}$$

The absolute value of this can be interpreted as the volume of the parallelepiped which has $\mathbf{a}, \mathbf{b}, \mathbf{c}$ as its adjacent edges.
   (a) Find the volume of the parallelepiped with adjacent edges given by the vectors:

(i) $(1,0,-3)$, $(0,1,1)$ and $(3,0,1)$
(ii) $(1,1,-2)$, $(3,2,-1)$ and $(2,1,1)$

   (b) Explain why in case (a)(ii) you could conclude that the three vectors lie in the same plane.

19.12 In a homogeneous, isotropic and linearly elastic material it is found that the strains on a section of the material, represented ε_x, ε_y, ε_z for the x-, y-, z-directions, respectively, can be related to the stresses, σ_x, σ_y, σ_z, by the following matrix equation:

$$\begin{pmatrix} \varepsilon_x \\ \varepsilon_y \\ \varepsilon_z \end{pmatrix} = \frac{1}{E} \begin{pmatrix} 1 & -v & -v \\ -v & 1 & -v \\ -v & -v & 1 \end{pmatrix} \begin{pmatrix} \sigma_x \\ \sigma_y \\ \sigma_z \end{pmatrix}$$

where E is the modulus of elasticity (also called Young's modulus) and v is Poisson's ratio, which relates the lateral and axial strains.
   Find σ_x, σ_y and σ_z in terms of ε_x, ε_y and ε_z, expressing the relationship in matrix form.

19.13 Find eigenvalues and eigenvectors of the following:

(a) $\begin{pmatrix} 6 & -1 \\ 0 & 2 \end{pmatrix}$ (b) $\begin{pmatrix} 1 & 5 \\ 1 & -3 \end{pmatrix}$ (c) $\begin{pmatrix} 3 & 2 \\ 1 & 2 \end{pmatrix}$

19.14 A simple circuit comprises a variable voltage V_S, a diode and a resistor, as shown in Fig. 19.21. As V_S is varied, values of V_R and I are recorded. The values are tabulated in Table 19.3. Using the method of least squares, determine Z and V_D in the equation $V_R = ZI + V_D$.

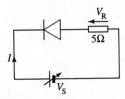

Figure 19.21 A simple circuit with a variable voltage V_S (Exercise 19.14).

Table 19.3 Voltage (V_R) against current (I) for Exercise 19.14

V_R (V)	I (A)
2	0.18
4	0.58
6	0.98
10	1.81

Table 19.4 Loads against spring length for Exercise 19.15

Load (kg)	Length (cm)
0	10
0.5	10.8
1	11.5
2	14
3	15.5
4	17.5

Table 19.5 Power dissipation of an n–p–n silicon transistor recorded against temperature

Temperature (°C)	Power dissipation (W)
25	10
60	7.9
100	5.7
120	4.8
140	3.5

19.15 In an attempt to measure the stiffness of a spring, the length of the spring under different loads was measured and the data tabulated as in Table 19.4.

(a) Use the data to find an equation that could be used to find the length of the spring given the weight.

(b) From your equation, estimate, if possible, the length of the spring when:

(i) the load is 2.5 kg
(ii) the load is 5 kg.

19.16 The power dissipation of an n–p–n silicon transistor is thought to vary linearly with temperature. The data given in Table 19.5 was recorded experimentally. Plot these points on a scatter diagram and use the method of least squares to obtain a and b in the equation relating the power (P) to the temperature (T); $P = a + bT$.

19.17 Fit parabolas to the following sets of data:

(a) $(-1, 0)$, $(0, -1)$, $(1, 4)$, $(2, 14)$, $(3, 32)$
(b) $(-1, 5.5)$, $(0, 1.5)$, $(0.5, 0.5)$, $(2, 5)$

ORDINARY DIFFERENTIAL EQUATIONS
AND DIFFERENCE EQUATIONS

20.1 INTRODUCTION

In order to model most physical situations we need to simplify their description. The idea of a system is central to this simplification. We identify the important elements that interact within the system. The only external interactions are then characterized as system inputs and the outputs are the quantities related to the system behaviour which we are interested in measuring. For instance, in examining the suspension system of a car we consider the system input to be forces exerted on the wheels of the car. The important elements in the system are the mass of the car and the springs and dampers used to connect the body of the car to the suspension links. We are interested in measuring the amount of vertical movement of the car and the velocity of that movement in response, mainly, to a bumpy road. We can examine the behaviour of the system for various external forces – building up a picture of how the system behaves in normal driving conditions. We do not need to concern ourselves with an exact model of the surface of the major motorways. In this way we have managed to separate the problem of modelling the system from modelling the environment in which it is operating. A system with a single input and output is pictured in Fig. 20.1.

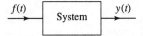

Figure 20.1 A system has an input $f(t)$, which produces an output response $y(t)$.

Dynamic physical processes involve variables which are interdependent and constantly changing. When modelling the system we shall obtain relationships between variables, many of which will be related through derivatives. Even in a simple system with a mass, spring and damper there are many variables which can be identified, e.g. the extension of the spring, its velocity and acceleration, the potential energy stored in the spring and the force exerted by it, and the force exerted by the damper. A choice of variables that completely describe the system is an important task in systems analysis. These variables are then called the state variables. They can be related to each other and the system inputs through a system of first-order differential equations, which are found by using scientific laws and engineering principles to model the system. The system output, which is some quantity that we wish to compare with the response of a real system, can then be calculated from the values of the state variables.

A differential equation is one that involves some derivative of the dependent variable, e.g. $dy/dt = ky$ is a differential equation. We have already looked at a number of examples of differential equations. The simplest case, when the solution can be found by integration, was looked at in Chapter 13; then we found that exponential functions are solutions to differential equations of the form $dy/dt = ky$ in Chapter 14 and in Chapter 16 we looked at the problem of circular motion and found that exponential functions are also solutions to equations of the form $d^2y/dt^2 = -\omega^2 y$. In simple equations we look for a solution which is a number. When solving a differential equation our idea of a solution is to find an explicit function for y in terms of t.

We shall only be concerned with single input, single output (SISO) systems, in which case there are two ways of proceeding in order to solve the system of differential equations. The system may be reduced to a single differential equation and then solved, or the system may be solved directly. These two methods correspond, in systems theory, to the transfer function description of the system or a state variable description.

The types of differential equation we shall look at are all linear; that is, they involve no powers of y, the dependent variable or powers of the derivatives or products of any of these. Any real system will, of course, have some non-linear behaviour. However, we usually assume that either the non-linearities present are not very important or that it is possible to analyse a system as locally linear. Another important assumption used is that of time invariance: that is, operating the system now or in an hour's time with exactly the same input and initial conditions will yield the same result.

The method of solution we use for finding a particular solution of differential equations in this chapter is called the method of unknown coefficients. An alternative method, using Laplace transforms, is more widely applied, and we shall look at that in the next chapter.

Differential equations are used to analyse systems that are subject to continuous or piecewise continuous inputs. It may be necessary, or preferable, to develop a discrete model of the system. A discrete model may be used if the exact nature of the system is unknown and we attempt to predict its nature by recording its response to various inputs. The recording of the system inputs and response will generally be stored in digital form and the data processed in digital form. Another major application of digital systems is for digital filter design. Here we wish to design digital filters with certain desirable characteristics. We can find the characteristics of digital systems by solving difference equations. We give a method of solution of simple difference equations. More frequently, z-transforms are used, and these are met in the next chapter.

20.2 MODELLING SIMPLE SYSTEMS

Damped forced motion of a spring

A spring of length l has a mass m attached to it and a dashpot damping the motion. It is subject to a force $f(t)$ forcing the motion (see Fig. 20.2). The input to this system is the forcing function $f(t)$ and the output is the displacement of the spring from its original length, x. In order to model this system we make a number of assumptions about its behaviour.

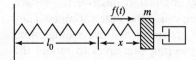

Figure 20.2 Damped forced motion of a spring.

1. We assume Newton's second law, $F_T = ma$, where $a = m\,d^2x/dt^2$ and F_T is the total force operating on the mass. m is the mass on the spring.
2. The spring does not become distorted, i.e. it is perfectly elastic. This means, from Hooke's law, that $F_S = -kx$, where x is the displacement from the original spring length l_0 and k is the spring constant.
3. The damping force due to the dashpot is $F_D = -r\,dx/dt$.

To relate these quantities together we use the fact that the total force on the mass is made up of the spring force, the damping force and the external force $F_T = F_D + F_S + f(t)$.

If we wish to combine these equations to express them using a single differential equation, we simply substitute for F_T in terms of x and the derivatives of x, giving

$$m\frac{d^2x}{dt^2} = -r\frac{dx}{dt} - kx + f(t)$$

$$\Leftrightarrow m\frac{d^2x}{dt^2} + r\frac{dx}{dt} + kx = f(t)$$

Should we wish to solve the system as a system of first-order differential equations, then we choose some state variables, usually the displacement and the velocity. Call these x_1 and x_2, and as x_1 is the displacement, $x_1 = x$, and as x_2 is the velocity this is the derivative of the displacement, giving $x_2 = dx_1/dt$.

As the acceleration is the derivative of the velocity, we can write the acceleration as the derivative of x_2:

$$a = \frac{d^2x_1}{dt^2} = \frac{d}{dt}\frac{dx_1}{dt} = \frac{dx_2}{dt}$$

Now $F_T = ma$ can be written in terms of x_2 as $F_T = m\,dx_2/dt$, and the force due to the dashpot becomes $F_D = -rx_2$. So $F_T = F_D + F_S + f(t)$ gives

$$m\frac{dx_2}{dt} = -rx_2 - kx_1 + f(t)$$

$$\Leftrightarrow \frac{dx_2}{dt} = -\frac{k}{m}x_1 - \frac{r}{m}x_2 + \frac{1}{m}f(t)$$

Hence the system of equations that represents the system is

$$\frac{dx_1}{dt} = x_2$$

$$\frac{dx_2}{dt} = -\frac{k}{m}x_1 - \frac{r}{m}x_2 + \frac{1}{m}f(t)$$

The first equation expresses the fact that x_2 is the velocity and the second equation was obtained by considering the total force on the mass and using Newton's second law.

This system of equations may be expressed in matrix form as

$$\frac{d}{dt}\begin{pmatrix} x_1 \\ x_2 \end{pmatrix} = \begin{pmatrix} 0 & 1 \\ -k/m & -r/m \end{pmatrix}\begin{pmatrix} x_1 \\ x_2 \end{pmatrix} + \begin{pmatrix} 0 \\ 1/m \end{pmatrix}f(t)$$

Writing

$$\mathbf{x} = \begin{pmatrix} x_1 \\ x_2 \end{pmatrix}$$

$$\mathbf{A} = \begin{pmatrix} 0 & 1 \\ -k/m & -r/m \end{pmatrix}$$

and

$$\mathbf{B} = \begin{pmatrix} 0 \\ 1/m \end{pmatrix}$$

we can give the system of equations as a matrix differential equation:

$$\frac{d\mathbf{x}}{dt} = \mathbf{Ax} + \mathbf{B}f$$

The system input is f and the output $y = x_1$.

To write the derivative of a variable when it is clear that it is with respect to its independent variable we may write x' (read as 'x dashed') or x'' (read as 'x double dashed'). In addition, when the independent variable is time we often write $\dot{x}$ (read as 'x dot') and for the second derivative $\ddot{x}$ (read as 'x double dot'). Using this notation we have

$$\dot{\mathbf{x}} = \mathbf{Ax} + \mathbf{B}f$$

as the matrix differential equation.

Modelling electrical circuits

We can model an *LRC* circuit as shown in Fig. 20.3 using the following assumptions:

1. Kirchhoff's voltage law (KVL): the sum of all the voltage drops around any closed loop is zero.
2. The voltage drop v_R across a resistor is proportional to the current i, i.e.

$$v_R = Ri \qquad \text{(Ohm's law)}$$

where the constant of proportionality R is called the resistance of the resistor.
3. The voltage drop across a capacitor is proportional to the electric charge q on the capacitor:

$$v_C = \frac{1}{C}q$$

where C is the capacitance and is measured in farads, and the charge q is measured in coulombs.
4. The voltage drop across an inductor is proportional to the rate of change of the current i:

$$v_L = L\frac{di}{dt}$$

where L is the inductance, measured in henries.

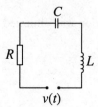

Figure 20.3 An *LRC* circuit.

Then from Kirchhoff's voltage law:

$$L\frac{di}{dt} + Ri + \frac{q}{C} = v(t)$$

Since

$$i(t) = \frac{dq}{dt}$$

then also

$$\frac{di}{dt} = \frac{d}{dt}\left(\frac{dq}{dt}\right) = \frac{d^2q}{dt^2}$$

So we can write this differential equation in terms of q:

$$L\frac{d^2q}{dt^2} + R\frac{dq}{dt} + \frac{q}{C} = v(t)$$

Here the system input is v and the output is the charge on the capacitor q.

To obtain a differential equation in terms of i we can differentiate the whole of this equation and use $i = dq/dt$, giving

$$L\frac{d^2i}{dt^2} + R\frac{di}{dt} + \frac{i}{C} = \frac{dv}{dt}$$

giving the input as dv/dt and the output is i. Notice that L, R and $1/C$ are constants.

Should we wish to solve the system as a system of first-order differential equations, then we choose some state variables. The general procedure is to choose capacitor voltages and inductor currents. Hence we choose

$$x_1 = i$$

and

$$x_2 = v_C = \frac{q}{C}$$

We can obtain one equation relating x_1 and x_2 by using

$$\frac{dq}{dt} = i \qquad \frac{1}{C}\frac{dq}{dt} = \frac{i}{C}$$

$$\frac{dx_2}{dt} = \frac{x_1}{C}$$

Now we use KVL to give the other equation:

$$L\frac{di}{dt} + Ri + \frac{q}{C} = v(t)$$

Substituting $i = x_1$ and $x_2 = q/C$ gives

$$\frac{dx_1}{dt} = -\frac{R}{L}x_1 - \frac{1}{L}x_2 + \frac{1}{L}v(t)$$

So the system of equations that gives the state variables is

$$\frac{dx_1}{dt} = -\frac{R}{L}x_1 - \frac{1}{L}x_2 + \frac{1}{L}v(t)$$

$$\frac{dx_2}{dt} = \frac{1}{C}x_1$$

The first equation expresses Kirchhoff's voltage law and the second equation expresses the relationship between the current and the voltage across the capacitor.

This system of equations may be expressed in matrix form as

$$\frac{d}{dt}\begin{pmatrix} x_1 \\ x_2 \end{pmatrix} = \begin{pmatrix} -R/L & -1/L \\ 1/C & 0 \end{pmatrix}\begin{pmatrix} x_1 \\ x_2 \end{pmatrix} + \begin{pmatrix} 1/L \\ 0 \end{pmatrix}v$$

Writing

$$\mathbf{x} = \begin{pmatrix} x_1 \\ x_2 \end{pmatrix}$$

$$\mathbf{A} = \begin{pmatrix} -R/L & -1/L \\ 1/C & 0 \end{pmatrix}$$

and

$$\mathbf{B} = \begin{pmatrix} 1/L \\ 0 \end{pmatrix}$$

we can give the system of equations as a matrix differential equation:

$$\frac{d\mathbf{x}}{dt} = \mathbf{A}\mathbf{x} + \mathbf{B}v$$

A rotational mechanical system

A rotor of moment of inertia J is supported by a shaft of torsional stiffness k and the motion is damped by a rotational damper of torque c per unit angular velocity (see Fig. 20.4). T is the external torque applied to the rotor.

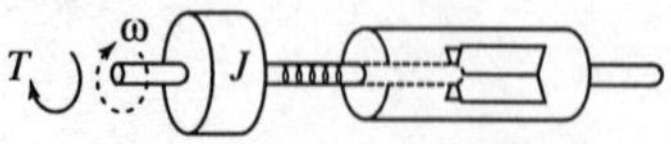

Figure 20.4 A damped rotational system.

The torque due to the rotational spring is proportional to the angle through which it has been twisted, giving $T_S = -k\theta$. The damper provides a torque of $-c\omega$ and the total rotational torque of the system is given by j dω/dt, where ω is the angular velocity and J is the moment of inertia of the rotating body.

$$J\frac{d\omega}{dt} = -c\omega - k\theta + T$$

$$J\frac{d\omega}{dt} + c\omega + k\theta = T$$

Using $\omega = d\theta/dt$ we can write this equation in terms of the angle of rotation as

$$J\frac{d^2\theta}{dt^2} + c\frac{d\theta}{dt} + k\theta = T$$

Alternatively, we can choose state variables as $x_1 = \theta$ and $x_2 = \omega = dx_1/dt$. This gives the system of equations

$$\frac{d}{dt}\begin{pmatrix} x_1 \\ x_2 \end{pmatrix} = \begin{pmatrix} 0 & 1 \\ -k/J & -c/J \end{pmatrix}\begin{pmatrix} x_1 \\ x_2 \end{pmatrix} + \begin{pmatrix} 0 \\ 1/J \end{pmatrix}T$$

Initial conditions and boundary values

To solve these equations we also need knowledge of the state of the system at some moment or moments in time. This is usually given in terms of the initial displacement and the initial velocity, i.e. x and dx/dt at $t=0$. This is then called an initial value problem. If two values of the displacement are given at different times or if the velocity and the displacement are given at different times, then the problem is called a boundary value problem. Boundary value problems are more difficult to specify correctly to ensure that they determine a solution.

20.3 ORDINARY DIFFERENTIAL EQUATIONS

An ordinary differential equation is an equation which involves derivatives of y (the dependent variable) as well as functions of y and t.

$$\frac{dy}{dt} = \cos(t)$$

$$\frac{d^2y}{dt^2} + 9t = 0$$

$$t^2\frac{d^3y}{dt^3}\frac{dy}{dt} + e^t = t$$

are all ordinary differential equations.

The other type of differential equation, which we do not consider in this book, is the partial differential equation, which is used to describe a dependence on two independent variables and involves partial derivatives like $\partial y/\partial t$ (read as 'partial dy by dt').

The order of differential equations

The order of the differential equation is the order of the highest derivative of y (the dependent variable) in the equation.

$$R\frac{dq}{dt} + \frac{q}{C} = 3 \qquad \text{is first-order in } q$$

$$\frac{d\theta}{dt} = \sin(\theta) \qquad \text{is first-order in } \theta$$

$$x'' + 4t^2 = 0 \qquad \text{is second-order in } x$$

$$\frac{d^3u}{dt^3} - \frac{du}{dt} + u = 4t^4 \qquad \text{is third-order in } u$$

The solution of a differential equation

If a function $y = g(t)$ is a solution to a differential equation in y, then if, in the differential equation, we replace y by $g(t)$ and all the derivatives of y by the corresponding derivatives of $g(t)$ then the resulting predicate should be an identity for t. That is, it should be true for all values of t where it is defined.

Example 20.1 Show that $x = t^3$ is a solution to the differential equation $dx/dt = 3t^2$.

SOLUTION Replacing x by t^3 in the differential equation gives

$$\frac{d}{dt}(t^3) = 3t^2$$

$$\Leftrightarrow 3t^2 = 3t^2$$

As this is true for all values of t, we have shown that $x = t^3$ is a solution to $dx/dt = 3t^2$.

Example 20.2 Show that $y = t^2 - 3t + 3.5$ is a solution to $y'' + 3y' + 2y = 2t^2$.

SOLUTION Set $y = t^2 - 3t + 3.5$. Then differentiating gives

$$y' = 2t - 3$$

and differentiating again gives

$$y'' = 2$$

We substitute these expressions for y and the derivatives of y into $y'' + 3y' + 2y = 2t^2$, so we get

$$2 + 3(2t - 3) + 2(t^2 - 3t + 3.5) = 2t^2$$

$$\Leftrightarrow 2 + 6t - 9 + 2t^2 - 6t + 7 = 2t^2$$

$$\Leftrightarrow 2t^2 = 2t^2$$

which is true for all values of t, showing that $y = t^2 - 3t + 3.5$ is a solution to $y'' + 3y' + 2y = 2t^2$.

A differential equation has many solutions. For instance, the equation $dx/dt = 3t^2$ has solutions $x = t^3$, $x = t^3 + 4$, $x = t^3 - 5$, etc. These are called particular solutions. A general solution is one which contains some arbitrary constants and encompasses all possible solutions to the differential equation. This means that by choosing values of these arbitrary constants we are able to find any of the particular solutions of the equation. For instance, $x = t^3 + C$ is a general solution to $dx/dt = 3t^2$.

Linear differential equations

A linear differential equation can be recognized by its form. It is linear if the coefficients of y (the dependent variable), and all order derivatives of y, are functions of t, or constant terms, only.

$$\frac{dy}{dt} = 4t$$

$$\frac{d^2y}{dt^2} = 6t$$

$$t\frac{dy}{dt} = 6$$

$$ay'' + by' + cy = f(t)$$

$$3\frac{d^2y}{dt^2} + t^2\frac{dy}{dt} + 6y = t^5$$

are all linear.

$$\left(\frac{dy}{dt}\right)^2 + 3 = 12t + y\frac{dy}{dt} - t^2$$

$$\theta'' + k\sin(\theta) = 0$$

$$u'' - (1 - u^2)u' + u = 0$$

are all non-linear.

Linearity and superposition of solutions

The property of linearity is very important. We have applied the term *linear* to equations whose graphs are straight *lines*. However, in an algebraic sense the term does not mean that everything represents a straight line! The property of linearity is defined for an operator, which we shall call O. A linear operator is one for which operating on the sum of two terms is the same as operating on the terms and then taking the sum. This can be expressed as

$$O(f_1 + f_2) = O(f_1) + O(f_2)$$

The other condition is that

$$O(af) = aO(f)$$

The two conditions can be combined to say that a linear operation on a linear combination of inputs produces a linear combination of the outputs. A linear combination of f_1 and f_2 is $af_1 + bf_2$, where a and b are constants, and therefore if an operator O is linear then

$$O(af_1 + bf_2) = aO(f_1) + bO(f_2)$$

Examples of linear operators are:

Matrices

$$A(x_1 + x_2) = Ax_1 + Ax_2$$

and

$$A(ax) = a(Ax)$$

where a is a number and A is a matrix, and x_1, x_2 and x are vectors.

The differential operator

$$\frac{d}{dt}(3t + t^2) = 3\frac{d}{dt}(t) + \frac{d}{dt}(t^2)$$

The integral operator

$$\int (3t + t^2)\,dt = \int 3t\,dt + \int t^2\,dt$$

An example of a non-linear operator is the sine function, as $\sin(A + B) \neq \sin(A) + \sin(B)$.

For a differential equation the fact that it is linear leads to the fact that if a solution to the differential equation with input function $f_1(t)$ is y_1 and a solution to the differential equation with input $f_2(t)$ is y_2, then one solution to the equation with an input function $af_1 + bf_2$ is $ay_2 + by_2$, where a and b are constants. This result is central to the method of solving differential equations used in this chapter and also to the Laplace transform method of the next chapter.

Example 20.3 A differential equation

$$\frac{d^2x}{dt^2} + 9x = f(t)$$

is found to have a particular solution $x = t/9$ when $f(t) = t$ and a particular solution $\frac{1}{5}\sin(2t)$ when $f(t) = \sin(2t)$.

Suggest a particular solution when $f(t) = t + 5\sin(2t)$ and check that your hypothesis is correct.

SOLUTION As the differential equation is linear, if we sum the input functions we should be able to sum the solutions. Hence a solution to

$$\frac{d^2x}{dt^2} + 9x = t + 5\sin(2t)$$

should be given by

$$x = \frac{t}{9} + 5(\tfrac{1}{5}\sin(2t)) = \frac{t}{9} + \sin(2t)$$

Check

$$x = \frac{t}{9} + \sin(2t)$$

$$\frac{dx}{dt} = \tfrac{1}{9} + 2\cos(2t)$$

$$\frac{d^2x}{dt^2} = -4\sin(2t)$$

Then substituting these into

$$\frac{d^2x}{dt^2} + 9x = t + 5\sin(2t)$$

gives

$$-4\sin(2t) + 9\left(\frac{t}{9} + \sin(2t)\right) = t + 5\sin(2t)$$

$$-4\sin(2t) + t + 9\sin(2t) = t + 5\sin(2t)$$

which is true for all values of t, showing that

$$x = \frac{t}{9} + \sin(2t)$$

is a solution to

$$\frac{d^2x}{dt^2} + 9x = t + 5\sin(2t)$$

Linear equations with constant coefficients

A linear equation is said to have constant coefficients if the coefficients multiplying y and the derivatives of y (the dependent variable) are all constants, i.e. do not involve functions of t.

$$3\frac{d^2y}{dt^2}+2\frac{dy}{dt}+4y=f(t)$$

has constant coefficients

$$t^2\frac{d^2y}{dt^2}+3t\frac{dy}{dt}+4y=f(t)$$

does not have constant coefficients, as t^2 and $3t$ are functions of time.

Time invariance

A linear differential equation with constant coefficients displays time invariance. If we use the same input and initial conditions for a system now or at some later time then the result relative to its respective starting time will be identical. Another way of expressing this is that if the input is time-shifted then so is the output. This idea is represented in Fig. 20.5.

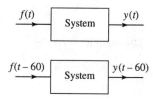

Figure 20.5 If a system is time-invariant, then a time-shifted input yields a time-shifted output.

Example 20.4 For the differential equation

$$\frac{d^2y}{dt^2}+4y=\sin(3t)$$

show that $y=\sin(2t)-\frac{1}{5}\sin(3t)$ is a solution and find a solution for the equation with the same input function delayed by 1 s, i.e. find a solution to

$$\frac{d^2y}{dt^2}+4y=\sin(3(t-1))$$

SOLUTION First we check that

$$y=\sin(2t)-\tfrac{1}{5}\sin(3t)$$

is a solution to the differential equation. To do this, we must find the first and second derivatives:

$$\frac{dy}{dt}=2\cos(2t)-\tfrac{3}{5}\cos(3t)$$

$$\frac{d^2y}{dt^2}=-4\sin(2t)+\tfrac{9}{5}\sin(3t)$$

Substitute into

$$\frac{d^2y}{dt^2}+4y=\sin(3t)$$

giving

$$-4\sin(2t) + \tfrac{9}{5}\sin(3t) + 4(\sin(2t) - \tfrac{1}{5}\sin(3t)) = \sin(3t)$$

$$\Leftrightarrow \sin(3t) = \sin(3t)$$

which is true for all t. Hence $y = \sin(2t) - \tfrac{1}{5}\sin(3t)$ is a solution to $d^2y/dt^2 + 4y = \sin(3t)$.

To find a solution to $d^2y/dt^2 + 4y = \sin(3(t-1))$, we use the property of time invariance, which means that a solution should be given by a time-shifted version of the solution to the first equation, i.e.

$$y = \sin(2(t-1)) - \tfrac{1}{5}\sin(3(t-1))$$

then

$$\frac{dy}{dt} = 2\cos(2(t-1)) - \tfrac{3}{5}\cos(3(t-1))$$

$$\frac{d^2y}{dt^2} = -4\sin(2(t-1)) + \tfrac{9}{5}\sin(3(t-1))$$

Substitute into

$$\frac{d^2y}{dt^2} + 4y = \sin(3(t-1))$$

giving

$$-4\sin(2(t-1)) + \tfrac{9}{5}\sin(3(t-1)) + 4(\sin(2(t-1)) - \tfrac{1}{5}\sin(3(t-1))) = \sin(3(t-1))$$

$$\Leftrightarrow \sin(3(t-1)) = \sin(3(t-1))$$

which is true for all t.

Example 20.5 For the differential equation

$$t\frac{dy}{dt} + y = 6t^2$$

(a) Show that $y = 2t^2$ is a solution.
(b) Show that the equation

$$t\frac{dy}{dt} + y = f(t)$$

cannot represent a time-invariant system.

SOLUTION
(a) To show that $y = 2t^2$ is a solution to

$$t\frac{dy}{dt} + y = 6t^2$$

we need to find dy/dt:

$$y = 2t^2$$

$$\Rightarrow \frac{dy}{dt} = 4t$$

Substituting into

$$t\frac{dy}{dt}+y=6t^2$$

gives

$$t(4t)+2t^2=6t^2$$
$$\Leftrightarrow 6t^2=6t^2$$

which is true for all t.

(b) To show that this cannot represent a time-invariant system, we take an equation with a time-shifted input, for instance shifted by 2 s, giving

$$t\frac{dy}{dt}+y=6(t-2)^2$$

as this equation.

If it were to be time-invariant, then a solution to this equation would be a time-shifted solution of the solution to the equation in part (a), i.e. $y=2(t-2)^2$. To show that the equation does not represent a time-invariant system, we just need to show that $y=2(t-2)^2$ is not a solution.

$$y=2(t-2)^2$$
$$\Rightarrow \frac{dy}{dt}=4(t-2)$$

Substitute into

$$t\frac{dy}{dt}+y=6(t-2)^2$$

giving

$$t(4(t-2))+2(t-2)^2=6(t-2)^2$$
$$\Leftrightarrow 4t(t-2)+2(t-2)^2-6(t-2)^2=0$$
$$\Leftrightarrow (t-2)(4t+2(t-2)-6(t-2))=0$$
$$\Leftrightarrow (t-2)(4t+2t-4-6t+12)=0$$
$$\Leftrightarrow 8(t-2)=0$$

which is not true for all values of t, showing that $y=2(t-2)^2$ is not a solution to $t\,dy/dt+y=6(t-2)^2$, and therefore we have shown that $t\,dy/dt+y=f(t)$ does not represent a time-invariant system.

We have seen that linear differential equations with constant coefficients represent linear time invariant (LTI) systems.

20.4 SOLVING FIRST-ORDER LTI SYSTEMS

To solve $ay'+by=f(t)$, we begin with by ignoring the forcing function, setting it equal to 0.

This is then called the homogeneous equation:

$$ay' + by = 0$$

The general solution to this equation is called the complementary function.

We then find any particular solution, and the sum of the complementary function and the particular solution gives us the general solution to the differential equation. In summing the two solutions in this way we are making use of the linearity property of the differential equation.

It may seem strange that a system with no input, as represented by the homogeneous equation may have a non-zero output! This is due to the fact that the initial conditions may be non-zero. For instance, a pendulum released from rest and performing small oscillations approximately obeys the differential equation $d^2y/dt^2 + \omega^2 y = 0$, where y is the distance the tip of the pendulum has moved from its rest position and ω is its angular velocity. The pendulum will only begin to move if either it is given a non-zero initial velocity or it is lifted slightly and then released. This means that either the initial position or the initial velocity or both must be non-zero for y to be non-zero.

The method we shall use for solving the complementary function relies on the fact that we know that $y' = ky$ has the solution $y = Ce^{kt}$. That is, the derivative of an exponential function is a scaled version of the original function. This was discussed in Chapter 14. We can also show that any order derivative of an exponential function is a scaled version of the original function. Consider $dy/dt = ky$. Then differentiating this equation gives

$$\frac{d^2y}{dt^2} = k\frac{dy}{dt}$$

Substituting $dy/dt = ky$ we get

$$\frac{d^2y}{dt^2} = k^2 y$$

and hence the second derivative is a scaled version of the original exponential function with scaling factor k^2. Differentiating again we get

$$\frac{d^3y}{dt^3} = k^2\frac{dy}{dt}$$

and substituting $dy/dt = ky$ we get

$$\frac{d^2y}{dt^3} = k^3 y$$

and hence the third derivative is a scaled version of the original function with scaling factor k^3. We use this property of exponential functions to solve the homogeneous equation by trying a solution of the form $y = Ce^{\lambda t}$. We then find value(s) for λ which are appropriate for the equation under consideration.

The equations we found in Sec. 20.2 were all second-order equations or systems of equations involving more than one state variable. However, in some circumstances, we get simple first-order equations. For instance if there is no inductor present in an LRC circuit we can put $L = 0$ to give

$$R\frac{dq}{dt} + \frac{q}{C} = v(t)$$

which is first-order in q.

The method of solution can be characterized by four steps. To find the solution to

$ay' + by = f(t)$:

Step 1: Find the complementary function
The solution of the homogeneous equation

$$ay' + by = 0$$

is found by assuming a solution of the form $y = Ae^{\lambda t}$. Hence $y' = A\lambda e^{\lambda t}$ and substituting into $ay' + by = 0$ we get

$$aA\lambda e^{\lambda t} + bAe^{\lambda t} = 0$$

We are not interested in the trivial solution $A = 0$, and $e^{\lambda t}$ is never zero, so we can divide through by $Ae^{\lambda t}$, giving

$$a\lambda + b = 0$$

which is called the auxiliary, or characteristic equation.

The auxiliary equation has the solution $\lambda = -b/a$. Hence the complementary function is

$$y = Ae^{-(b/a)t}$$

Step 2: Find a particular solution
The system output, after the effects of any initial conditions have died out, would be expected to mimic the system input. To find a particular solution we try a trial solution with some undetermined coefficients. Some good guesses to use for the trial solution, depending on the type of input function $f(t)$, are given in Table 20.1.

Table 20.1 A table of trial solutions to be used to find a particular solution. The form of the trial solution is suggested by the form of the forcing function $f(t)$. The coefficients $c_0, c_1, \ldots$ and c, d are to be determined

Input function	Trial solution
Polynomial of order n: $f(t) = a_0 + a_1 t + a_2 t^2 + \ldots + a_n t^n$	$y = c_0 + c_1 t + c_2 t^2 + \ldots + c_n t^n$ (Use all terms up to n for the trial solution)
Exponential function: $f(t) = ae^{\alpha t}$	$y = ce^{\alpha t}$
Sine or cosine function: $f(t) = a\cos(\omega t)$ or $b\sin(\omega t)$ or $f(t) = a\cos(\omega t) + b\sin(\omega t)$	$y = c\cos(\omega t) + d\sin(\omega t)$

The coefficients in the trial solution are determined by substituting into $ay' + by = f(t)$. These trial solutions will not always work. Laplace transforms may be used to find a particular solution in such cases (see Chapter 21).

Step 3
The general solution to the equation is given by the sum of the complementary function plus the particular solution.

Step 4
The final step is to use the initial condition to find the appropriate value of A, the arbitrary constant. This solution is then the particular solution which solves the differential equation with the given initial condition.

Example 20.6 The RC circuit in Fig. 20.6 is initially relaxed and is closed at time $t = 0$, hence applying a d.c. voltage of v_0. The charge of the capacitor obeys the differential equation

$$R\frac{dq}{dt} + \frac{q}{C} = v_0$$

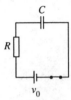

Figure 20.6 An RC circuit with an applied d.c. voltage.

Solve for q and find an expression for the voltage across the capacitor at a time t after the circuit was closed.

SOLUTION
Step 1
To find the complementary function we solve the homogeneous equation

$$R\frac{dq}{dt} + \frac{q}{C} = 0$$

Substitute $q = Ae^{\lambda t}$ and $dq/dt = A\lambda e^{\lambda t}$, giving

$$RA\lambda e^{\lambda t} + A\frac{e^{\lambda t}}{C} = 0$$

Divide by $Ae^{\lambda t}$:

$$R\lambda + \frac{1}{C} = 0 \Leftrightarrow \lambda = -\frac{1}{RC}$$

The complementary function is

$$q = Ae^{-(1/RC)t}$$

Step 2
Try a particular solution of the form $q = a$, where a is a constant. Then $q' = 0$, and substituting into

$$R\frac{dq}{dt} + \frac{q}{C} = v_0$$

we get

$$\frac{a}{C} = v_0$$

$$\Leftrightarrow a = v_0 C$$

Step 3
The general solution is therefore the sum of the complementary function (found in Step 1) and a particular solution (found from Step 2). So the general solution for q is given by

$$q = Ae^{-t/RC} + Cv_0$$

Step 4
Finally we use the initial conditions to find the arbitrary constant A. We are told that initially the circuit was relaxed, meaning that all voltages and currents are zero. In this case, this means that when $t=0$, $q=0$. Substituting this condition into

$$q = Ae^{-t/RC} + Cv_0$$

we get

$$0 = A + Cv_0$$

$$\Leftrightarrow A = -Cv_0$$

Therefore the solution is

$$q = -Cv_0 e^{-t/RC} + Cv_0$$

$$\Leftrightarrow q = Cv_0(1 - e^{-t/RC})$$

This gives the voltage across the capacitor $v_c = q/C$ as $v_c = v_0(1 - e^{-t/RC})$, where v_0 is the applied d.c. voltage.

Check
To check that $q = Cv_0(1 - e^{-t/RC})$ is in fact a solution to

$$R\frac{dq}{dt} + \frac{q}{C} = v_0$$

we carry out the following steps:

$$q = Cv_0(1 - e^{-t/RC})$$

$$\Rightarrow \frac{dq}{dt} = \frac{Cv_0}{RC}e^{-t/RC} \Leftrightarrow \frac{dq}{dt} = \frac{v_0}{R}e^{-t/RC}$$

Substituting into

$$R\frac{dq}{dt} + \frac{q}{C} = v_0$$

gives

$$\frac{Rv_0}{R}e^{-t/RC} + \frac{Cv_0}{C}(1 - e^{-t/RC}) = v_0$$

$$\Leftrightarrow v_0 = v_0$$

which is an identity for t, showing that $q = Cv_0(1 - e^{-t/RC})$ is a solution to the differential equation.

We also check that the solution satisfies the given initial condition by substituting $q=0$ and $t=0$ into $q = Cv_0(1 - e^{-t/RC})$. We find $0 = Cv_0(1-1) \Leftrightarrow 0 = 0$, which is true.

Example 20.7 Solve the differential equation $y' - 3y = 6 \cos(2t)$, given $y(0) = 2$.

SOLUTION
Step 1
To find the complementary function we solve the homogeneous equation

$$y' - 3y = 0$$

Substitute $y = Ae^{\lambda t}$ so that $dy/dt = A\lambda e^{\lambda t}$, giving

$$A\lambda e^{\lambda t} - 3Ae^{\lambda t} = 0$$

Therefore we get the auxiliary equation

$$\lambda - 3 = 0 \Leftrightarrow \lambda = 3$$

The complementary function is $y = Ae^{3t}$.

Step 2
Try a particular solution (chosen from Table 20.1) of the form

$$y = c \cos(2t) + d \sin(2t)$$

Then

$$y' = -2c \sin(2t) + 2d \cos(2t)$$

Substituting into $y' - 3y = 6 \cos(2t)$ we get

$$-2c \sin(2t) + 2d \cos(2t) - 3(c \cos(2t) + d \sin(2t)) = 6 \cos(2t)$$
$$\Leftrightarrow (-2c - 3d) \sin(2t) + (2d - 3c) \cos(2t) = 6 \cos(2t)$$

We want this to be true for all values of t, so we can equate the coefficients of the $\sin(2t)$ terms on both sides of the equation and also the $\cos(2t)$ terms.

Equating the coefficients of $\sin(2t)$ gives

$$-2c - 3d = 0$$

Equating the coefficients $\cos(2t)$ gives

$$2d - 3c = 6$$

Therefore

$$-2c - 3d = 0$$
$$-3c + 2d = 6$$

Solving these equations simultaneously gives

$$c = -\tfrac{18}{13} \qquad d = \tfrac{12}{13}$$

Hence a particular solution is

$$y = -\tfrac{18}{13} \cos(2t) + \tfrac{12}{13} \sin(2t)$$

Step 3
The general solution is given by the sum of the complementary function and a particular solution, giving

$$y = Ae^{3t} - \tfrac{18}{13} \cos(2t) + \tfrac{12}{13} \sin(2t)$$

Step 4

Use the initial condition to find A. When $t=0$, $y=2$. Substituting this into

$$y = Ae^{3t} - \tfrac{18}{13}\cos(2t) + \tfrac{12}{13}\sin(2t)$$

gives

$$2 = A - \tfrac{18}{13}\cos(0) + \tfrac{12}{13}\sin(0)$$

$$\Leftrightarrow 2 = A - \tfrac{18}{13} \Leftrightarrow A = \tfrac{44}{13}$$

Therefore the solution is

$$y = \tfrac{44}{13}e^{3t} - \tfrac{18}{13}\cos(2t) + \tfrac{12}{13}\sin(2t)$$

Check

Substitute

$$y = \tfrac{44}{13}e^{3t} - \tfrac{18}{13}\cos(2t) + \tfrac{12}{13}\sin(2t)$$

and

$$y' = \tfrac{132}{13}e^{3t} + \tfrac{36}{13}\sin(2t) + \tfrac{24}{13}\cos(2t)$$

into

$$y' - 3y = 6\cos(2t)$$

giving

$$\tfrac{132}{13}e^{3t} + \tfrac{36}{13}\sin(2t) + \tfrac{24}{13}\cos(2t) - 3(\tfrac{44}{13}e^{3t} - \tfrac{18}{13}\cos(2t) + \tfrac{12}{13}\sin(2t)) = 6\cos(2t)$$

$$\Leftrightarrow \tfrac{78}{13}\cos(2t) = 6\cos(2t)$$

which is correct.

We also check the initial condition by substituting $t=0$, $y=2$ into

$$y = \tfrac{44}{13}e^{3t} - \tfrac{18}{13}\cos(2t) + \tfrac{12}{13}\sin(2t)$$

We find

$$2 = \tfrac{44}{13}e^{0} - \tfrac{18}{13}\cos(0) + \tfrac{12}{13}\sin(0)$$

$$\Leftrightarrow 2 = \tfrac{44}{13} - \tfrac{18}{13}$$

$$\Leftrightarrow 2 = \tfrac{26}{13}$$

which is true.

20.5 SOLVING SECOND-ORDER LTI SYSTEMS

We can solve a second-order system in a manner similar to first-order systems. To solve

$$ay'' + by' + cy = f(t)$$

we solve the homogeneous equation

$$ay'' + by' + cy = 0$$

The solution to this equation is called the complementary function.

We then find any particular solution and the sum of the complementary function, and a particular solution gives us the general solution to the differential equation. We can again list the steps involved. To solve $ay'' + by' + cy = f(t)$:

Step 1: Find the complementary function
The solution of the homogeneous equation

$$ay'' + by' + cy = 0$$

is found by assuming solutions of the form $y = Ae^{\lambda t}$. Hence

$$y' = A\lambda e^{\lambda t} \quad \text{and} \quad y'' = A\lambda^2 e^{\lambda t}$$

and we find

$$aA\lambda^2 e^{\lambda t} + bA\lambda e^{\lambda t} + cAe^{\lambda t} = 0$$

giving the auxiliary equation

$$a\lambda^2 + b\lambda + c = 0$$

The auxiliary equation is a quadratic equation. It has solutions

$$\lambda_1, \lambda_2 = \frac{-b \pm \sqrt{b^2 - 4ac}}{2a}$$

There are three possible types of solution.

Case (1): λ_1 and λ_2 are real and distinct. Then the solution is

$$y = Ae^{\lambda_1 t} + Be^{\lambda_2 t}$$

Case (2): λ_1 and λ_2 are complex. We set $\lambda_1 = k + j\omega_0$ and $\lambda_2 = k - j\omega_0$ so that $k = -b/2a$ and $\omega_0 = \sqrt{4ac - b^2}/2a$. Then the complex exponentials can be written in terms of cosines and sines and the solution becomes

$$y = e^{kt}(A\cos(\omega_0 t) + B\sin(\omega_0 t))$$

Case (3): The roots are equal. Then $\lambda_1 = \lambda_2 = k = -b/2a$, and the solution is

$$y = (At + B)e^{kt}$$

Step 2: Find a particular solution
The particular solution is any solution of the equation

$$ay'' + by' + cy = f(t)$$

As for the first-order system we expect the system output, after the effects of any initial conditions have died out, to mimic the system input. Again, to find a particular solution we try a trial solution, as given in Table 20.1. The 'guess' at the particular solution is substituted into the differential equation and the unknown coefficients can be determined. If this guess does not produce a solution, then Laplace transform methods can be used, as in Chapter 21.

Step 3
The general solution to the equation is given by the sum of the complementary function plus a particular solution.

Step 4
The final stage is to use the initial conditions or boundary conditions to find the appropriate values of the arbitrary constants A and B.

Example 20.8 Solve the differential equation

$$5y'' + 6y' + 5y = 6\cos(t)$$

where $y(0) = 0$ and $y'(0) = 0$.

SOLUTION

Step 1
To find the complementary function we solve the homogeneous equation $5y'' + 6y' + 5y = 0$.
Trying solutions of the form $y = Ae^{\lambda t}$ leads to the auxiliary equation

$$5\lambda^2 + 6\lambda + 5 = 0$$

Notice that a quick way to get the auxiliary equation is to 'replace', y'' by λ^2, y' by λ and y by 1. The auxiliary equation has solutions

$$\lambda = \frac{-6 \pm \sqrt{36 - 100}}{10}$$

$$= \frac{-6 \pm \sqrt{-64}}{10}$$

$$= 0.6 \pm \mathrm{j}0.8$$

Comparing this with $\lambda = k \pm \mathrm{j}\omega_0$ gives $k = -0.6$, $\omega_0 = 0.8$. This means that the complementary function has the form

$$y = e^{kt}(A\cos(\omega_0 t) + B\sin(\omega_0 t))$$

(given as Case (2) in the general method), with $k = -0.6$ and $\omega_0 = 0.8$. This gives

$$y = e^{-0.6t}(A\cos(0.8t) + B\sin(0.8t))$$

Step 2
As $f(t) = 6\cos(t)$, from Table 20.1 we decide to try a particular solution of the form $y = c\cos(t) + d\sin(t)$. Then

$$y' = -c\sin(t) + d\cos(t)$$

and

$$y'' = -c\cos(t) - d\sin(t)$$

Substituting in

$$5y'' + 6y' + 5y = 6\cos(t)$$

we find

$$5(-c\cos(t) - d\sin(t)) + 6(-c\sin(t) + d\cos(t)) + 5(c\cos(t) + d\sin(t)) = 6\cos(t)$$

$$\Leftrightarrow (-5c + 6d + 5c)\cos(t) + (-5d - 6c + 5d)\sin(t) = 6\cos(t)$$

$$\Leftrightarrow 6d\cos(t) - 6c\sin(t) = 6\cos(t)$$

As we want this to be an identity we equate the coefficients of $\cos(t)$ and the coefficients of $\sin(t)$ and get the two equations

$$6d = 6 \Leftrightarrow d = 1$$

$$6c = 0 \Leftrightarrow c = 0$$

Hence a particular solution is $y = \sin(t)$.

Step 3
The general solution is given by the sum of the complementary function and a particular solution, so

$$y = e^{-0.6t}(A \cos(0.8t) + B \sin(0.8t)) + \sin(t)$$

Step 4
Use the given initial conditions to find values for the constants A and B. Substituting $y = 0$ when $t = 0$ into

$$y = e^{-0.6}t(A \cos(0.8t) + B \sin(0.8t)) + \sin(t)$$

gives

$$0 = e^0(A \cos(0) + B \sin(0)) + \sin(0)$$

$$\Leftrightarrow 0 = A$$

To use the other condition, that $y' = 0$ when $t = 0$, we need to differentiate the general solution to find an expression for y'. Differentiating (using the product rule):

$$y = e^{-0.6t}(A \cos(0.8t) + B \sin(0.8t)) + \sin(t)$$

$$\Rightarrow y' = -0.6e^{-0.6t}(A \cos(0.8t) + B \sin(0.8t))$$

$$+ e^{-0.6t}(-0.8A \sin(0.8t) + 0.8B \cos(0.8t)) + \cos(t)$$

and using $y' = 0$ when $t = 0$ gives

$$0 = -0.6A + 0.8B + 1$$

We have already found $A = 0$ from the first condition, so

$$B = -\tfrac{1}{0.8} = -1.25$$

Therefore the solution is

$$y = -1.25e^{-0.6t} \sin(0.8t) + \sin(t)$$

Check

$$y = -1.25e^{-0.6t} \sin(0.8t) + \sin(t)$$

$$y' = 0.75e^{-0.6t} \sin(0.8t) - e^{-0.6t} \cos(0.8t) + \cos(t)$$

$$y'' = -0.45e^{-0.6t} \sin(0.8t) + 0.6e^{-0.6t} \cos(0.8t) + 0.6e^{-0.6t} \cos(0.8t) + 0.8e^{-0.6t} \sin(0.8t) - \sin(t)$$

Substitute into

$$5y'' + 6y' + 5y = 6 \cos(t)$$

giving

$$5(-0.45e^{-0.6t} \sin(0.8t) + 0.6e^{-0.6t} \cos(0.8t) + 0.6e^{-0.6t} \cos(0.8t) + 0.8e^{-0.6t} \sin(0.8t) - \sin(t))$$

$$+ 6(0.75e^{-0.6t} \sin(0.8t) - e^{-0.6t} \cos(0.8t) + \cos(t)) + 5(-1.25e^{-0.6t} \sin(0.8t) + \sin(t))$$

$$= 6 \cos(t)$$

$$\Leftrightarrow e^{-0.6t} \sin(0.8t)(1.75 + 4.5 - 6.25) + e^{-0.6t} \cos(0.8t)(6 - 6) + \sin(t)(-5 + 5) + 6 \cos(t)$$

$$= 6 \cos(t)$$

which is true for all values of t.
 We can also check that the solution satisfies the given initial conditions.

Example 20.9 Find the current in an initially relaxed LRC circuit with $R=5\Omega$, $L=4$ µH and $C=1$ µF, where the a.c. voltage source is given by $v(t)=20\sin(10t)$.

SOLUTION The differential equation is given in Sec. 20.2 as

$$L\frac{d^2i}{dt^2}+R\frac{di}{dt}+\frac{i}{C}=\frac{dv}{dt}$$

As $v=20\sin(10t)$, then

$$\frac{dv}{dt}=200\cos(10t)$$

and substituting in the given values of L, R and C we get

$$4\times10^{-6}\frac{d^2i}{dt^2}+5\frac{di}{dt}+\frac{i}{10^{-6}}=200\cos(10t)$$

$$4\times10^{-6}\frac{d^2i}{dt^2}+5\frac{di}{dt}+10^6i=200\cos(10t)$$

Step 1
Solve the homogeneous equation

$$4\times10^{-6}\frac{d^2i}{dt^2}+5\frac{di}{dt}+10^6i=0$$

This has auxiliary equation

$$4\times10^{-6}\lambda^2+5\lambda+10^6=0$$

with solutions

$$\Leftrightarrow\lambda=\frac{-5\pm\sqrt{25-4(4\times10^{-6}\times10^6)}}{2\times10^{-6}}\Leftrightarrow\lambda=\frac{-5\pm\sqrt{9}}{2\times10^{-6}}$$

$$\Leftrightarrow\lambda=-4\times10^6\vee\lambda=-10^6$$

The complementary function is

$$i=Ae^{-10^6t}+Be^{-4\times10^6t}$$

Step 2
Find a particular solution. As the forcing function is a cosine function we try

$$i=c\cos(10t)+d\sin(10t)$$

$$\frac{di}{dt}=-10c\sin(10t)+10d\cos(10t)$$

$$\frac{d^2i}{dt^2}=-100c\cos(10t)-100d\sin(10t)$$

Substituting into

$$4\times10^{-6}\frac{d^2i}{dt^2}+5\frac{di}{dt}+10^6i=200\cos(10t)$$

$$4 \times 10^{-6}(-100c \cos(10t) - 100d \sin(10t)) + 5(-10c \sin(10t) + 10d \cos(10t))$$
$$+ 10^6(c \cos(10t) + d \sin(10t)) = 200 \cos(10t)$$
$$\Leftrightarrow -0.0004c \cos(10t) - 0.0004d \sin(10t) - 50c \sin(10t) + 50d \cos(10t)$$
$$+ 10^6 c \cos(10t) + 10^6 d \sin(10t)$$

Equating the coefficients of the sine and cosine terms to make this an identity gives

$$-0.0004c + 50d + 10^6 c = 200$$

$$-0.0004d - 50c + 10^6 d = 0$$

Solving these equations we find $d \simeq 10^{-8}$ and $c \simeq 0.0002$, so

$$i = 0.0002 \cos(10t) + 10^{-8} \sin(10t)$$

Step 3
Hence the general solution is

$$i = Ae^{-10^6 t} + Be^{-4 \times 10^6 t} + 0.0002 \cos(10t) + 10^{-8} \sin(10t)$$

Step 4
Given that $i = 0$ and $di/dt = 0$ when $t = 0$ initially, we can substitute into

$$i = Ae^{-10^6 t} + Be^{-4 \times 10^6 t} + 0.0002 \cos(10t) + 10^{-8} \sin(10t)$$

and

$$\frac{di}{dt} = -10^6 Ae^{-10^6 t} - 4B10^6 e^{-4 \times 10^6} - 0.002 \sin(10t) + 10^{-7} \cos(10t)$$

to find equations for A and B

$$A + 4B = -10^{-13}$$

$$A + B = -0.0002$$

which we solve to find $A \simeq -0.000\,267$ and $B \simeq 0.000\,066\,7$. So the solution to the initial value problem is approximately

$$i = -0.000\,267 e^{-10^6 t} + 0.000\,066\,7 e^{-4 \times 10^6 t} + 0.0002 \cos(10t) + 10^{-8} \sin(10t)$$

The transient and steady state solution–system stability

From the systems modelled in Sec. 20.2 we can see that the term by' in the equation

$$ay'' + by' + cy = f(t)$$

comes from the damping factor in the system. For electrical systems this is provided by the resistance. b must be a positive quantity in any of these systems, as are the other coefficients.

The auxiliary equation is $a\lambda^2 + b\lambda + c = 0$ with solutions

$$\lambda_1 \lambda_2 = \frac{-b \pm \sqrt{b^2 - 4ac}}{2a}$$

and this gives three possibilities for the complementary function.

Case (1): Real distinct roots: $y = Ae^{\lambda_1 t} + Be^{\lambda_2 t}$. Notice that λ_1 and λ_2 must be negative when a, b and c are positive and therefore the complementary function will die out as $t \to \infty$. In this case the system is said to be overdamped.

Case (2): Complex roots: $y = e^{kt}(A \cos(\omega_0 t) + B \sin(\omega_0 t))$, where $k = -b/2a$, which is negative, so e^{kt} is a negative exponential term. This then represents dying oscillations. The system is said to be underdamped.

Case (3): The roots are equal; then $k = -b/2a$ and $y = (At + B)e^{kt}$.

Again this has a negative exponential part causing the complementary function to tend to zero as $t \to \infty$. This case is referred to as critical damping.

Because the contribution from the complementary function dies out as $t \to \infty$ it is referred to as the transient solution. Graphs of the form of the transients, for positive initial displacement and zero initial velocity, are shown in Fig. 20.7(a), (b) and (c) for Case (1), Case (2) and Case (3), respectively.

The other part of the solution, where we consider the effect of the forcing function, is called the steady state solution. For the solution found in Example 20.8, where

$$y = -1.25e^{-0.6t} \sin(0.8t) + \sin(t)$$

we find that $-1.25e^{-0.6t} \sin(0.8t)$ is the transient and $\sin(t)$ is the steady state solution. If we consider the system after some time has passed, then the transient will effectively be zero and we are left with $y = \sin(t)$, the steady state solution.

We have said that in any truly linear system, as represented, for instance, by our models in Sec. 20.2, the constants must be positive with positive damping. However, we may wish to analyse non-linear systems by using a locally linear approximation. In this situation they may exhibit unstable behaviour, where the 'transients', instead of dying out, display positive exponential behaviour and grow very large. This is then referred to as an unstable system.

We can analyse systems in the following way.

Stable system: A system is stable if all the solutions to the auxiliary equation have negative real parts. A system with some purely imaginary solutions to the auxiliary equation can also be considered to be stable, although the complementary function does not die out as $t \to \infty$ but represents sustained oscillations.

Unstable system: A system is unstable if any solutions to the auxiliary equation have positive real parts or, for systems of higher order, if there is a repeated, purely imaginary, solution.

Damped oscillations of a mechanical system – resonance

For damped oscillations of a spring we have the differential equation

$$m\frac{d^2 x}{dt^2} + r\frac{dx}{dt} + kx = f(t)$$

as shown in Sec. 20.2, where m is the mass, r is the damping constant and k is the spring constant. We shall look at the steady state solution in the case where there is a single-frequency input. We are interested in the magnitude of the response in this case in order to analyse the problem of resonance.

For simplicity, we can regard the single-frequency input as a complex exponential $f(t) = F_0 e^{j\omega t}$, where F_0 is a constant. Having found the response to the complex exponential, then as $F_0 e^{j\omega t} = F_0 \cos(\omega t) + jF_0 \sin(\omega t)$ we can find the response to a cosine function by taking the real part of the output, or to a sine function by taking the imaginary part of the output (this is an application of the linearity property).

We have the equation $m\ddot{x} + r\dot{x} + kx = F_0 e^{j\omega t}$ and we want to find the steady state solution.

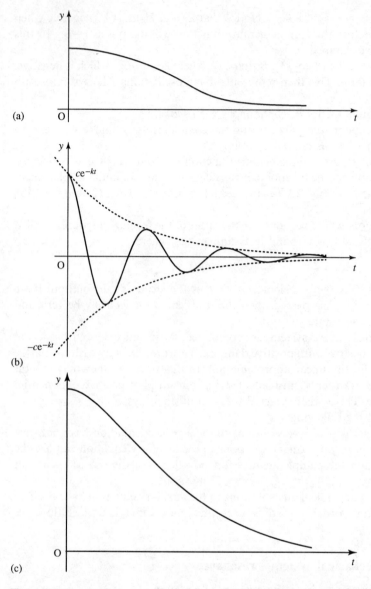

(a)

(b)

(c)

Figure 20.7 The complementary function, found by solving the homogenous equation, provides the transient solution. The graphs shown are for positive initial displacement and zero velocity. (a) The graph of the transient for the overdamped case where $y = Ae^{\lambda_1 t} + Be^{\lambda_2 t}$, λ_1 λ_2 negative. (b) Graph of the underdamped case, where the transient consists of dying oscillations: $y = e^{kt}(A \cos(\omega_0 t) + B \sin(\omega_0 t))$ (k is negative). (c) The critically damped case where $y = (At + B)e^{kt}$, k negative.

Try a solution of the form $x = ce^{j\omega t}$. Then

$$\dot{x} = cj\omega e^{j\omega t}$$

$$\ddot{x} = -c\omega^2 e^{j\omega t}$$

and substituting into the differential equation we find

$$-mc\omega^2 e^{j\omega t} + crj\omega e^{j\omega t} + kce^{j\omega t} = F_0 e^{j\omega t}$$

Dividing both sides by $e^{j\omega t}$ (which is non-zero) we get

$$c(-m\omega^2 + jr\omega + k) = F_0$$

Hence

$$c = \frac{F_0}{-m\omega^2 + jr\omega + k}$$

giving the steady state solution as

$$x = \frac{F_0 e^{j\omega t}}{-m\omega^2 + jr\omega + k}$$

We are interested in the magnitude of the oscillations, which is given by

$$\left| \frac{F_0}{-m\omega^2 + jr\omega + k} \right|$$

We can find this by taking the magnitude of the top and bottom lines. From Chapter 14 we know that the magnitude of a complex number $z = x + jy$ is given by $r = \sqrt{x^2 + y^2}$. Hence we find that the amplitude of the oscillations in response to a single frequency of input of angular frequency ω is given by

$$\frac{F_0}{\sqrt{(m\omega^2 - k)^2 + r^2\omega^2}}$$

This was plotted in Fig. 17.10 for the case $m = k = 1$ and various values of r (there called c). If the damping, r, is quite small then we can set it to be effectively 0. In this case

$$|x| = \begin{cases} \dfrac{F_0}{k - \omega^2 m} & \text{for } \omega^2 < \dfrac{k}{m} \\[3mm] \dfrac{F_0}{\omega^2 m - k} & \text{for } \omega^2 > \dfrac{k}{m} \end{cases}$$

There is an infinite discontinuity at $\omega = \sqrt{k/m}$. We see that $\sqrt{k/m}$ represents the natural angular frequency of the system in the case where r is 0. We can show this by examining the solution to the homogeneous equation in the case where $r = 0$.

$mx'' + kx = 0$ has auxiliary equation

$$m\lambda^2 + k = 0$$

$$\Leftrightarrow \lambda^2 = -\frac{k}{m}$$

Setting $\lambda = j\omega_0$ we find that $\omega_0^2 = k/m$ and the complementary function is $x = A\cos(\omega_0 t) + B\sin(\omega_0 t)$. We can see from this limiting case that if the damping constant is small and the oscillations of the forcing function are near to the natural oscillations of the underdamped system then we have a situation of resonance, where the magnitude of the response can become very large. This effect led to the destruction of some physical systems before the phenomenon of resonance was well understood.

20.6 SOLVING SYSTEMS OF DIFFERENTIAL EQUATIONS

To solve a system of differential equations we can combine the equations into a single differential equation using substitution. In this case we can use the method as outlined in Sec. 20.5. Alternatively, we can solve the system directly using matrices.

We follow the same pattern as in Sec. 20.4 for first-order systems. We solve the homogeneous equation to find the complementary function and then find a particular solution. The sum of these two terms gives a general solution to the system.

Example 20.10 Find the displacement of the spring after time t described by the system of differential equations

$$\frac{d}{dt}\binom{x_1}{x_2} = \begin{pmatrix} 0 & 1 \\ -k/m & -r/m \end{pmatrix}\binom{x_1}{x_2} + \binom{0}{1/m}f(t)$$

in the case where the mass on the spring is 1, the damping constant $r=5$, the spring constant $k=4$, the forcing function $f(t)=t$ and the initial extension of the spring is 0, with 0 initial velocity.

SOLUTION Substituting the given values for m, r, k and f we have the system

$$\frac{d}{dt}\binom{x_1}{x_2} = \begin{pmatrix} 0 & 1 \\ -4 & -5 \end{pmatrix}\binom{x_1}{x_2} + \binom{0}{1}f(t)$$

Step 1
Solve the homogeneous system with $f(t)=0$, giving

$$\frac{d\mathbf{x}}{dt} = \mathbf{A}\mathbf{x} \quad \text{where} \quad \mathbf{A} = \begin{pmatrix} 0 & 1 \\ -4 & -5 \end{pmatrix}$$

Try solutions of the form $\mathbf{x} = \mathbf{v}e^{\lambda t}$, where $\mathbf{v}$ is a constant vector. Then $d\mathbf{x}/dt = \lambda\mathbf{v}e^{\lambda t}$. Substitute for $\mathbf{x}$ and $d\mathbf{x}/dt$ into $d\mathbf{x}/dt = A\mathbf{x}$, giving

$$\lambda\mathbf{v}e^{\lambda t} = \mathbf{A}\mathbf{v}e^{\lambda t}$$

As $e^{\lambda t} \neq 0$, then

$$\lambda\mathbf{v} = \mathbf{A}\mathbf{v}$$

$$\Leftrightarrow \mathbf{A}\mathbf{v} = \lambda\mathbf{v}$$

which we recognize as the eigenvalue problem of Chapter 19. In this case, however, we shall allow the possibility of complex eigenvalues.

The solutions for λ are found by solving

$$|\mathbf{A} - \lambda\mathbf{I}| = 0 \quad \text{as} \quad \mathbf{A} = \begin{pmatrix} 0 & 1 \\ -4 & -5 \end{pmatrix}$$

This gives

$$\begin{vmatrix} -\lambda & 1 \\ -4 & -5-\lambda \end{vmatrix} = 0$$

$$\Leftrightarrow \lambda(5+\lambda) + 4 = 0$$

$$\Leftrightarrow \lambda^2 + 5\lambda + 4 = 0$$

$$\Leftrightarrow \lambda = \frac{-5 \pm \sqrt{25-16}}{2}$$

$$\Leftrightarrow \lambda = \frac{-5 \pm 3}{2} \Leftrightarrow \lambda = -4 \vee \lambda = -1$$

Now we find the eigenvector $\mathbf{v}$ to go with each eigenvalue by solving $(\mathbf{A} - \lambda\mathbf{I})\mathbf{v} = 0$:

$$\begin{pmatrix} -\lambda & 1 \\ -4 & -5-\lambda \end{pmatrix}\begin{pmatrix} v_1 \\ v_2 \end{pmatrix} = \begin{pmatrix} 0 \\ 0 \end{pmatrix}$$

For $\lambda = -4$:

$$\begin{pmatrix} 4 & 1 \\ -1 & -1 \end{pmatrix}\begin{pmatrix} v_1 \\ v_2 \end{pmatrix} = 0$$

Multiplying out, this gives

$$4v_1 + v_2 = 0$$

$$-4v_1 - v_2 = 0$$

These equations are dependent; solving either one of them we find

$$4v_1 = -v_2$$

$$v_2 = -4v_1$$

The vector is given by $(v_1, -4v_1)$. As we are interested only in the direction of the vector, we can set $v_1 = 1$, giving the eigenvector as $(1, -4)$.

For $\lambda = -1$ we get

$$\begin{pmatrix} 1 & 1 \\ -4 & -4 \end{pmatrix}\begin{pmatrix} v_1 \\ v_2 \end{pmatrix} = \begin{pmatrix} 0 \\ 0 \end{pmatrix}$$

Multiplying out we get

$$v_1 + v_2 = 0$$

$$-4v_1 - 4v_2 = 0$$

These equations are dependent; solving either one we get

$$v_2 = -v_1$$

The vector is given by $(v_1, -v_1)$. As we are interested only in the direction of the vector, we can set $v_1 = 1$, giving the eigenvector as $(1, -1)$.

Therefore we have the complementary function for $\mathbf{x}$ as

$$\mathbf{x} = a\begin{pmatrix} 1 \\ -4 \end{pmatrix}e^{-4t} + b\begin{pmatrix} 1 \\ -1 \end{pmatrix}e^{-t}$$

where a and b are arbitrary constant scalars.

Step 2
Find a particular solution to the equation. For

$$\frac{d\mathbf{x}}{dt} = \mathbf{A}\mathbf{x} + \mathbf{b}t$$

the forcing function is $\mathbf{b}t$, so we try a particular solution of the form $\mathbf{x} = \mathbf{c}_1 t + \mathbf{c}_0$, where $\mathbf{c}_0$ and $\mathbf{c}_1$ are constant vectors. We see that this is similar to the choice of trial solution suggested in Table 20.1, but the constant terms are now constant vectors.

$$\mathbf{x} = \mathbf{c}_1 t + \mathbf{c}_0$$

$$\Rightarrow \frac{d\mathbf{x}}{dt} = \mathbf{c}_1$$

Substitute this trial solution into

$$\frac{dx}{dt} = Ax + bt$$

giving

$$c_1 = A(c_1 t + c_0) + bt$$

We can equate vector coefficients of t and the constant terms on both sides, giving

$$0 = Ac_1 + b$$

$$c_1 = Ac_0$$

We solve these two matrix equations:

$$0 = Ac_1 + b$$

$$\Leftrightarrow Ac_1 = -b$$

Premultiply by A^{-1} (if it exists) giving

$$c_1 = -A^{-1}b$$

As

$$A = \begin{pmatrix} 0 & 1 \\ -4 & -5 \end{pmatrix}$$

then

$$A^{-1} = \frac{1}{4}\begin{pmatrix} -5 & -1 \\ 4 & 0 \end{pmatrix} = \begin{pmatrix} -1.25 & -0.25 \\ 1 & 0 \end{pmatrix}$$

and

$$b = \begin{pmatrix} 0 \\ 1 \end{pmatrix}$$

so

$$c_1 = -\begin{pmatrix} -1.25 & -0.25 \\ 1 & 0 \end{pmatrix}\begin{pmatrix} 0 \\ 1 \end{pmatrix} = \begin{pmatrix} 0.25 \\ 0 \end{pmatrix}$$

To find c_0 we can use

$$c_1 = Ac_0$$

We get

$$c_0 = A^{-1}c_1$$

$$= \begin{pmatrix} -1.25 & 0.25 \\ 1 & 0 \end{pmatrix}\begin{pmatrix} 0.25 \\ 0 \end{pmatrix} = \begin{pmatrix} -0.3125 \\ 0.25 \end{pmatrix}$$

Therefore a particular solution $x = c_1 t + c_0$ is given by

$$x = \begin{pmatrix} 0.25 \\ 0 \end{pmatrix}t + \begin{pmatrix} -0.3125 \\ 0.25 \end{pmatrix}$$

Step 3

The general solution is the sum of the complementary function and a particular solution. This gives

$$x = a\begin{pmatrix} 1 \\ -4 \end{pmatrix}e^{-4t} + b\begin{pmatrix} 1 \\ -1 \end{pmatrix}e^{-t} + \begin{pmatrix} 0.25 \\ 0 \end{pmatrix}t + \begin{pmatrix} -0.3125 \\ 0.25 \end{pmatrix}$$

Step 4

We can apply the initial conditions. We are told that at $t=0$ both the displacement and velocity are zero. These are the state variables x_1 and x_2. Therefore we have

$$\mathbf{x} = \begin{pmatrix} 0 \\ 0 \end{pmatrix}$$

when

$$t = 0$$

Substituting into the solution we find

$$\begin{pmatrix} 0 \\ 0 \end{pmatrix} = a\begin{pmatrix} 1 \\ -4 \end{pmatrix} + b\begin{pmatrix} 1 \\ -1 \end{pmatrix} + \begin{pmatrix} -0.3125 \\ 0.25 \end{pmatrix}$$

$$0 = a + b - 0.3125$$

$$0 = -4a - b + 0.25$$

Solving these two equations gives

$$a \simeq -0.02083$$

$$b \simeq 0.3333$$

and therefore the particular solution to this initial value problem is

$$\mathbf{x} \simeq 0.020\,83\begin{pmatrix} 1 \\ -4 \end{pmatrix}e^{-4t} + 0.3333\begin{pmatrix} 1 \\ -1 \end{pmatrix}e^{-t} + \begin{pmatrix} 0.25 \\ 0 \end{pmatrix}t + \begin{pmatrix} -0.3125 \\ 0.25 \end{pmatrix}$$

Check

$$\frac{d\mathbf{x}}{dt} = -0.08332\begin{pmatrix} 1 \\ -4 \end{pmatrix}e^{-4t} - 0.3333\begin{pmatrix} 1 \\ -1 \end{pmatrix}e^{-t} + \begin{pmatrix} 0.25 \\ 0 \end{pmatrix}$$

Substitute into

$$\frac{d\mathbf{x}}{dt} = \begin{pmatrix} 0 & 1 \\ -4 & -5 \end{pmatrix}\mathbf{x} + \begin{pmatrix} 0 \\ 1 \end{pmatrix}t$$

giving

$$-0.083\,32\begin{pmatrix} 1 \\ -4 \end{pmatrix}e^{-4t} - 0.3333\begin{pmatrix} 1 \\ -1 \end{pmatrix}e^{-t} + \begin{pmatrix} 0.25 \\ 0 \end{pmatrix}$$

$$= \begin{pmatrix} 0 & 1 \\ -4 & -5 \end{pmatrix}\begin{pmatrix} 0.02083e^{-4t} + 0.3333e^{-t} + 0.25t - 0.3125 \\ -0.08332e^{-4t} - 0.3333e^{-t} + 0.25 \end{pmatrix} + \begin{pmatrix} 0 \\ 1 \end{pmatrix}t$$

Multiplying out and simplifying:

$$\begin{pmatrix} -0.08332e^{-4t} - 0.3333e^{-t} + 0.25 \\ 0.3333e^{-4t} + 0.3333e^{-t} \end{pmatrix} = \begin{pmatrix} -0.08332e^{-4t} - 0.3333e^{-t} + 0.25 \\ 0.3333e^{-4t} + 0.3333e^{-t} \end{pmatrix}$$

which is true for all t.

20.7 DIFFERENCE EQUATIONS

Discrete systems are designed using digital adders, multipliers and shift registers, and are represented mathematically by difference equations. Difference equations are used in the design of digital filters and also to model continuous systems approximately.

The system has an input, which is a sequence of values $f_0, f_1, f_2, \ldots$, and the output is a sequence $y_0, y_1, y_2, \ldots$. If y_n depends solely on the values of the input then

$$y_n = b_0 f_n + b_1 f_{n-1} + b_2 f_{n-2} + \ldots$$

This is a non-recursive system, as we do not need previous knowledge of y to determine y_n. It is called a finite impulse response (FIR) system (the meaning of 'impulse response' will be explained in Chapter 21). If, however, the output y depends on the previous state of the system as well as the input, then we have a recursive system, modelled by a difference equation (also called a recurrence relation). This is called an infinite impulse response (IIR) system.

We shall look at linear time-invariant discrete systems. We shall consider the solution of a second-order difference equation

$$ay_n + by_{n-1} + cy_{n-2} = f_n$$

The method of solution, as for differential equations, is to solve the homogeneous system and find a particular solution. The sum of these gives the general solution to the difference equation. We employ the initial conditions to find the value of the arbitary constants.

Example 20.11 Solve the difference equation ($n \geqslant 2$):

$$6y_n - 5y_{n-1} + y_{n-2} = 1$$

given $y_0 = 4.5$ and $y_1 = 2$.

SOLUTION
Step 1
Solve the homogeneous difference equation

$$6y_n - 5y_{n-1} + y_{n-2} = 0$$

We know from Chapters 14 and 18 that $y_{n+1} = ry_n$ defines a geometric series $y_n = ar^{n-1}$, where a is the first term of the sequence and r is the common ratio. If we start at the zeroth term then we have $y_n = a_0 r^n$. We therefore try a solution of this form for the homogeneous equation and substitute $y_n = a\lambda^n$ into $6y_n - 5y_{n-1} + y_{n-2} = 0$. This gives

$$6a\lambda^n - 5a\lambda^{n-1} + a\lambda^{n-2} = 0$$

Assuming that a and λ^{n-2} are non-zero we can divide by $a\lambda^{n-2}$ to give

$$6\lambda^2 - 5\lambda + 1 = 0$$

This is called the auxiliary equation, and solving this we find

$$\lambda = \frac{5 \pm \sqrt{25 - 24}}{12} \Leftrightarrow \lambda = \tfrac{1}{3} \vee \lambda = \tfrac{1}{2}$$

Hence the general solution of the homogeneous equation is

$$y_n = a(\tfrac{1}{3})^n + b(\tfrac{1}{2})^n$$

As in the case of differential equations there are three possible types of solution, depending on whether the roots for λ are real and distinct, complex, or equal, where

$$\lambda_1, \lambda_2 = \frac{-b \pm \sqrt{b^2 - 4ac}}{2a}$$

Case (1): λ_1 and λ_2 are real and distinct:

$$y_n = a(\lambda_1)^n + b(\lambda_2)^n$$

Case (2): λ_1 and λ_2 are complex. Then we write λ_1, λ_2 in exponential form: $\lambda_1 = re^{j\theta}$ and $\lambda_2 = re^{-j\theta}$. Then the solution can be expressed as

$$y = r^n(a \cos(n\theta) + b \sin(n\theta))$$

Case (3): $\lambda_1 = \lambda_2 = \lambda$ are equal real roots and the solution is

$$y_n = (a + bn)\lambda^n$$

Step 2

Find a particular solution. We use a trial solution which depends on the form of the input sequence f_n as suggested by Table 20.2.

Table 20.2 A table of trial solutions to be used to find a particular solution of a difference equation. The form of the trial solution is suggested by the form of the input sequence f_n. The coefficients $c_0, c_1, \ldots$ and c, d are to be determined

Input function	Trial solution
Power series: $f_n = n^k$ (k an integer)	$y = c_0 + c_1 n + c_2 n^2 + \ldots + c_k n^k$
Exponential function: $f_n = a^n$	$y = c a^n$
Sine or cosine function: $f_n = \cos(\omega n)$ or $\sin(\omega n)$	$y = c \cos(\omega n) + d \sin(\omega n)$

In this case the input is a constant, so we try a constant output. The trial solution is $y_n = c$. Substituting this into $6y_n - 5y_{n-1} + y_{n-2} = 1$ gives the equation for c as

$$6c - 5c + c = 1$$
$$\Leftrightarrow 2c = 1$$
$$c = \tfrac{1}{2}$$

A particular solution for y is $y_n = \tfrac{1}{2}$.

Step 3
The general solution is the sum of the solution of the homogeneous equation and a particular solution, giving

$$y_n = a(\tfrac{1}{3})^n + b(\tfrac{1}{2})^n + \tfrac{1}{2}$$

Step 4
Solve for the initial conditions $y_0 = 4.5$ and $y_1 = 2$. Substituting $n = 0$ and $y_0 = 4.5$ in the solution

$$y_n = a(\tfrac{1}{3})^n + b(\tfrac{1}{2})^n + \tfrac{1}{2}$$

gives

$$4.5 = a + b + \tfrac{1}{2} \Leftrightarrow a + b = 4$$

Substituting $n = 1$ and $y_1 = 4.5$ gives

$$2 = \frac{a}{3} + \frac{b}{2} + \tfrac{1}{2} \Leftrightarrow 2a + 3b = 9$$

Solving for the constants a and b gives

$$\begin{matrix} a + b = 4 \\ 2a + 3b = 9 \end{matrix} \Leftrightarrow \begin{matrix} 2a + 2b = 8 \\ 2a + 3b = 9 \end{matrix} \Rightarrow -b = -1$$

Substituting $b = 1$ into $a + b = 4$:

$$a + 1 = 4 \Leftrightarrow a = 3$$

$a = 3$ and $b = 1$, so the solution is

$$y_n = 3(\tfrac{1}{3})^n + (\tfrac{1}{2})^n + \tfrac{1}{2}$$

System stability

We can see that the solution of the above example, as in the continuous case, is made up of a transient, in this case, $3(\tfrac{1}{3})^n + (\tfrac{1}{2})^n$, and a steady state response, which in this case is $\tfrac{1}{2}$. The transient tends to zero as $n \to \infty$. This will be so for any system as long as the roots to the auxiliary equation are such that the modulus of λ is less than 1. If the roots have modulus of exactly 1 then the solution of the homogeneous equation does not die away as $n \to 0$ but neither does it tend to infinity. Therefore we can identify a stable system as one with roots of the auxiliary equation such that $|\lambda| \leqslant 1$.

20.8 SUMMARY

1. Dynamic physical processes involve variables which are interdependent and constantly changing. When modelling the system we will obtain relationships between variables, many of which will be related through derivatives. We can combine these to make a single differential equation relating the system input to the system output or we can choose state variables and express the relationship through a system of differential equations.

2. Linear time-invariant systems (LTI) are represented by linear differential equations with constant coefficients. A second-order equation is of the form

$$a\frac{d^2y}{dt^2} + b\frac{dy}{dt} + cy = f(t)$$

$f(t)$ represents the system input and y the system output.

3. If a function $y = g(t)$ is a solution to a differential equation in y then if, in the differential equation, we replace y and by $g(t)$ and all the derivatives of y by the corresponding derivatives of $g(t)$, the result should be an identity for t. That is, it should be true for all values of t.

4. To solve linear differential equations with constant coefficients we solve the homogeneous equation (with $f(t) = 0$), thus finding the complementary function. This leads to the auxiliary equation which, for a second-order system, is

$$a\lambda^2 + b\lambda + c = 0$$

This gives three possibilities for the complementary function, depending on the roots for λ.

$$\lambda_1, \lambda_2 = \frac{-b \pm \sqrt{b^2 - 4ac}}{2a}$$

   Case (1): Real distinct roots: $y = Ae^{\lambda_1 t} + Be^{\lambda_2 t}$.
   Case (2): Complex roots: $y = e^{kt}(A\cos(\omega_0 t) + B\sin(\omega_0 t))$.
   Case (3): The roots are equal. Then $\lambda = k = -b/2a$ and $y = (At + B)e^{kt}$.

We then find a particular solution to the equation by using a trial solution, as suggested in Table 20.1. The sum of the complementary function and a particular solution gives the general solution of the differential equation. The initial conditions are used to find the arbitary constants A and B.

5. For a truly linear system the complementary function dies out as $t \to \infty$ and is called the transient solution. The other part of the solution, in response to the input $f(t)$, is called the steady state solution. When analysing a system as locally linear, we say that the system is unstable if the roots of the complementary function have a positive real part. In these circumstances the complementary function $\to \infty$ as $t \to \infty$. Higher order systems are also unstable if there is a repeated purely imaginary root.

6. Resonance occurs when, in an underdamped system, the forcing frequency approaches the natural frequency of the system.

7. Systems of differential equations may be solved using a matrix method.

8. Discrete systems may be represented by difference equations (also called recurrence relations). A second-order linear system is represented by $ay_n + by_{n-1} + cy_{n-2} = f_n$. To solve this equation we find the solution to the homogeneous system (setting $f_n = 0$) which leads to the auxiliary equation $a\lambda^2 + b\lambda + c = 0$. This leads to three possibilities for the complementary function depending on the roots for λ.

$$\lambda_1, \lambda_2 = \frac{-b \pm \sqrt{b^2 - 4ac}}{2a}$$

   Case (1): λ_1 and λ_2 are real and distinct:

$$y_n = a(\lambda_1)^n + b(\lambda_2)^n$$

   Case (2): λ_1 and λ_2 are complex. Then we write λ_1 and λ_2 in exponential form, $\lambda_1 = re^{j\theta}$ and $\lambda_2 = re^{-j\theta}$, and the solution can be expressed as

$$y = r^n(a\cos(n\theta) + b\sin(n\theta))$$

Case (3): $\lambda_1 = \lambda_2 = \lambda$ are equal real roots and the solution is

$$y_n = (a + bn)\lambda^n$$

The system is stable if $|\lambda| \leqslant 1$.

We then find a particular solution by attempting a trial solution, as suggested in Table 20.2. The general solution of the difference equation is given by the sum of the complementary function and a particular solution.

20.9 EXERCISES

20.1 For the following differential equations:

(i) State whether the equation is linear.
(ii) Give the order of the equation.
(iii) Show that the given function represents a solution to the differential equation.

(a) $\dfrac{t}{4}\dfrac{dy}{dt} = y,\, y = ct^4$

(b) $\dfrac{d^2y}{dt^2}(t - \tfrac{1}{2}) - \dfrac{dy}{dt} = 0,\, y = t^2 - t$

(c) $t\dfrac{d^2y}{dt^2} + t\dfrac{dy}{dt} + y = 4,\, y = 3te^{-t} + 4$

(d) $y\dfrac{dy}{dt} = t,\, y^2 = t^2 + c,\, y \geqslant 0$

(e) $3y^2\dfrac{d^2y}{dt^2} + 6y\left(\dfrac{dy}{dt}\right)^2 = 2,\, y^3 = t^2$

20.2 A linear differential equation

$$\dfrac{dy}{dt} + 4y = f(t)$$

is found to have a particular solution $y = 2t^2 - t + 1$ when $f(t) = 8t^2 + 3$ and a particular solution $y = 0.8 \cos(3t) + 0.6 \sin(3t)$ when $f(t) = 5 \cos(3t)$.

(a) Suggest a particular solution when $f(t) = 10 \cos(3t) + 4t^2 + 3/2$ and show by substitution that your solution is correct.
(b) Suggest a particular solution when $f(t) = 5 \cos(3(t - 10))$ and show by substitution that your solution is correct.

20.3 Show that the differential equation

$$\cos(t)\dfrac{dy}{dt} - \sin(t)y = f(t)$$

has a solution

$$y = \dfrac{t^2}{\cos(t)}$$

when $f(t) = 2t$, and a solution

$$y = \tan(t)$$

when $f(t) = \cos(t)$.

(a) Use the linearity property to find a solution when $f(t) = 2(t - \cos(t))$.
(b) By considering a time-shifted input of one of the functions for $f(t)$ given, show that

$$\cos(t)\dfrac{dy}{dt} - \sin(t)y = f(t)$$

cannot represent a time-invariant system.

20.4 Solve the following differential equations with the given initial conditions:

(a) $\dfrac{dy}{dt} - 3y = 0$, $y(0) = 1$

(b) $4\dfrac{dy}{dt} - y = 4$, $y(0) = -2$

(c) $\dfrac{dy}{dt} + 5y = \sin(12t)$, $y(0) = 0$

(d) $3\dfrac{dy}{dt} + 2y = e^{-t}$, $y(0) = 3$

(e) $\dfrac{dy}{dt} - 6t = 0$, $y(1) = 6$

(f) $\frac{1}{2}\dfrac{dy}{dt} + 6y - 3\sin(5t) = 2\cos(5t)$, $y(0) = 0$

20.5 A spring of length l has a mass m attached to it and a dashpot damping the motion. It is subject to a force $f(t)$ forcing the motion, as in Fig. 20.2. The extension x of the spring obeys the differential equation

$$m\frac{d^2x}{dt^2} + r\frac{dx}{dt} + kx = f(t)$$

Solve for x given the following information:

(a) $m = 2$, $r = 7$, $k = 5$, $f(t) = t^2$, $x(0) = 0$, $\dfrac{dx}{dt}(0) = 0.16$

(b) $m = 1$, $r = 4$, $k = 4$, $f(t) = 4\cos(6t)$, $x(0) = 0$, $\dfrac{dx}{dt}(0) = 0$

(c) $m = 1$, $r = 4$, $k = 5$, $f(t) = e^{-t}$, $x(0) = 0$, $\dfrac{dx}{dt}(0) = 0$

20.6 An LRC circuit, as in Fig. 20.3, obeys the equation

$$L\frac{d^2q}{dt^2} + R\frac{dq}{dt} + \frac{q}{C} = v(t)$$

where q is the charge on the capacitor, $v(t)$ is the applied voltage, L is the inductance, R the resistance and C the capacitance. Find a steady state solution for q and hence calculate the voltages across the capacitor, resistor and inductor, given by

$$v_c = \frac{q}{C}, \quad v_R = R\frac{dq}{dt} \quad \text{and} \quad v_L = L\frac{d^2q}{dt^2}$$

in each of the following cases:

(a) $R = 120\ \Omega$, $L = 0.06\ \text{H}$, $C = 0.0001\ \text{F}$, $v(t) = 130\cos(1000t)$
(b) $R = 1\ \text{k}\Omega$, $L - 0.001\ \text{H}$, $C = 1\ \mu\text{F}$, $v(t) - t$

20.7 An RC circuit is subjected to a single-frequency input of angular frequency ω and magnitude v_i.

(a) Find the steady state solution of the equation

$$R\frac{dq}{dt} + \frac{q}{C} = v_i e^{j\omega t}$$

and hence find the magnitude of

(i) The voltage across the capacitor $v_c = q/C$.
(ii) The voltage across the resistor $v_R = R\,dq/dt$.

(b) Using the impedance method of Chapter 16 confirm your results to part (a) by calculating

(i) The voltage across the capacitor v_c.
(ii) The voltage across the resistor v_R in response to a single frequency of angular frequency ω and magnitude v_i.

(c) For the case where $R = 1\ \text{k}\Omega$ and $C = 1\ \mu\text{F}$ find the ratio $|v_c/v_i|$ and fill in the table below:

ω	10	10^2	10^3	10^4	10^5	10^6
$|v_c/v_i|$						

Explain why the table results show that an RC circuit can be used as a high cut filter and find the value of the high cut frequency, defined as $f_{hc} = \omega_{hc}/2\pi$ such that

$$\left|\frac{v_c}{v_i}\right| = \frac{1}{\sqrt{2}}$$

20.8 A spring of length l has a mass m attached to it and a dashpot damping the motion. It is subject to a force $f(t)$ forcing the motion as in Fig. 20.2. The extension x_1 of the spring and velocity x_2 obey the system of differential equations

$$\frac{d}{dt}\begin{pmatrix} x_1 \\ x_2 \end{pmatrix} = \begin{pmatrix} 0 & 1 \\ -k/m & -r/m \end{pmatrix}\begin{pmatrix} x_1 \\ x_2 \end{pmatrix} + \begin{pmatrix} 0 \\ 1/m \end{pmatrix} f(t)$$

By solving this system of equations, find the extension of the spring after time t in the following cases

(a) $m=1$, $r=12$, $k=11$, $f(t)=e^{-2t}$ with initial extension 0 and 0 initial velocity.
(b) $m=2$, $r=13$, $k=20$, $f(t)=10t$, $x_1(0)=0=x_2(0)$.

20.9 Solve the following difference equations:

(a) $2y_n - y_{n-1} = -3$, $y_0 = 1$
(b) $4y_n + y_{n-1} = n$, $y_0 = 2$
(c) $y_n - 1.2y_{n-1} + 0.36y_{n-2} = 4$, $y_0 = 3$, $y_1 = 2$
(d) $y_n + 0.6y_{n-1} = 0.16y_{n-2} + (0.5)^n$, $y_0 = 1$, $y_1 = 2$
(e) $2y_n + y_{n-1} = \cos\left(\dfrac{\pi n}{2}\right)$, $y_0 = 1$

LAPLACE AND z TRANSFORMS

21.1 INTRODUCTION

In this chapter we shall present a quick review of Laplace transform methods for continuous and piecewise continuous systems and z transform methods for discrete systems. The widespread application of these methods means that engineers need increasingly to be able to apply these techniques in multitudes of situations.

Laplace transforms are used to reduce a differential equation to a simple equation in s-space and a system of differential equations to a system of linear equations. We discover that in the case of zero initial conditions, we can solve the system by multiplying the Laplace transform of the input function by the transfer function of the system. Furthermore, if we are only interested in the steady state response and we have a periodic input then we can find the response by simply adding the response to each of the frequency components of the input signal. In Chapter 22 we therefore look at finding the frequency components of a periodic function by finding its Fourier series.

Similar ideas apply to z transform methods applied to difference equations representing discrete systems.

The system transfer function is the Laplace transform of the impulse response function of the system. The impulse response is its response to an impulse function, also called a delta function. Inputting an impulse function can be thought of as something like giving the system a short kick to see what happens next. The impulse function is an idealized kick, as it lasts for no time at all and has energy of exactly 1. Because of these requirements we find that the delta function is not a function at all, in the sense that we defined functions in Chapter 7. We find that it is the most famous example of a generalized function.

21.2 THE LAPLACE TRANSFORM: DEFINITION

The Laplace transform $F(s)$ of the function $f(t)$ defined for $t > 0$ is

$$F(s) = \int_0^\infty e^{-st} f(t)\, dt$$

The Laplace transform is a function of s, where s is a complex variable. Because the definition of the Laplace transform involves an integral to ∞ it is usually necessary to limit possible values of s so that the integral converges (i.e. does not tend to ∞). Also, for many functions the Laplace transform does not exist at all.

$\mathscr{L}\{\ \}$ is the symbolic representation of the process of taking a Laplace transform (LT). Therefore

$$\mathscr{L}\{f(t)\} = F(s)$$

Example 21.1 Find from the definition $\mathscr{L}\{e^{3t}\}$.

SOLUTION

$$\mathscr{L}\{e^{3t}\} = \int_0^\infty e^{3t} e^{-st}\, dt = \int_0^\infty e^{(3-s)t}\, dt = \left[\frac{e^{(3-s)t}}{3-s}\right]_0^\infty$$

Now $e^{(3-s)t} \to 0$ as $t \to \infty$ only if the real part of $3-s$ is negative, i.e. $\text{Re}(s) > 3$. Then the upper limit of the integration gives 0. Hence

$$\mathscr{L}\{e^{3t}\} = \frac{1}{s-3}$$

where $\text{Re}\{s\} > 3$.

Example 21.2 Find, from the definition, $\mathscr{L}\{\cos(at)\}$.

SOLUTION We need to find the integral

$$I = \int_0^\infty e^{-st} \cos(at)\, dt$$

We employ integration by parts twice until we find an expression which involves the original integral, I. We are then able to express I in terms of a and s.

Integrating by parts, using the formula $\int u\, dv = uv - \int v\, du$, where $u = \cos(at)$, $dv = e^{-st}\, dt$, $du = -a \sin(at)\, dt$ and $v = e^{-st}/(-s)$ gives

$$I = \left[\frac{-e^{-st}}{s} \cos(at)\right]_0^\infty - \int_0^\infty \frac{e^{-st}}{s} a \sin(at)\, dt$$

The expression $[-e^{-st}\cos(at)]_0^\infty$ only has a finite value if $e^{-st} \to 0$ as $t \to \infty$. This will only be so if $\text{Re}(s) > 0$. With that proviso we get

$$I = \frac{1}{s} - \frac{a}{s}\int_0^\infty e^{-st} \sin(at)\, dt$$

integrating the remaining integral on the right-hand side, again by parts, where this time $u = \sin(at)$, $dv = e^{-st}\, dt$, $du = a\cos(at)\, dt$ and $v = e^{-st}/(-s)$ gives

$$I = \frac{1}{s} - \left[\frac{-a}{s^2} e^{-st} \sin(at)\right]_0^\infty - \frac{a^2}{s^2}\int_0^\infty e^{-st} \cos(at)\, dt$$

$$= \frac{1}{s} - \frac{a^2}{s^2}\int_0^\infty e^{-st} \cos(at)\, dt$$

We see that the remaining integral is the integral we first started with, which we called I. Hence

$$I = \frac{1}{s} - \frac{a^2}{s^2} I$$

and solving this for I gives

$$I\left(1 + \frac{a^2}{s^2}\right) = 1/s$$

$$I = \frac{s}{s^2 + a^2}$$

So we have that

$$\mathcal{L}\{\cos(at)\} = \frac{s}{s^2 + a^2}, \qquad \text{where } \mathrm{Re}(s) > 0$$

21.3 THE UNIT STEP FUNCTION AND THE (IMPULSE) DELTA FUNCTION

The unit step function and delta function are very useful in systems theory.

The unit step function

The unit step function is defined as follows:

$$u(t) = \begin{cases} 1 & t \geqslant 0 \\ 0 & t < 0 \end{cases}$$

and its graph is as shown in Fig. 21.1.

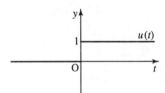

Figure 21.1 The graph of the unit step function $u(t)$.

The function is used to represent an idealized switch. It switches on at time $t = 0$. If it is multiplied by any other function then it acts to switch that function on at $t = 0$. That is, the function

$$f(t)u(t) = \begin{cases} 0 & \text{for } t < 0 \\ f(t) & \text{for } t \geqslant 0 \end{cases}$$

An example for $f(t) = t^2$ is given in Fig. 21.2.

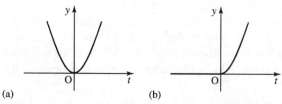

(a)   (b)

Figure 21.2 Any function when multiplied by the unit step function is 'switched on' at $t = 0$. (a) The function $y = t^2$. (b) The function $y = t^2 u(t)$.

As the Laplace transform only involves the integral from $t=0$, then all functions can be thought of as multiplied by the unit step function because $f(t)=f(t)u(t)$ as long as $t>0$.

Shifting the unit step function

As we saw in Chapter 8, the graph of $f(t-a)$ can be found by taking the graph of $f(t)$ and moving it a units to the right. There is an example of a shifted unit step function in Fig. 21.3. Notice that the unit step function 'switches on' where the argument of u is zero. Multiplying any function by the shifted unit step function changes where it is switched on. That is,

$$f(t)u(t-a)=\begin{cases}0 & t<a \\ f(t) & t\geq a\end{cases}$$

The example of $\sin(\omega t)u(t-1)$ is shown in Fig. 21.4.

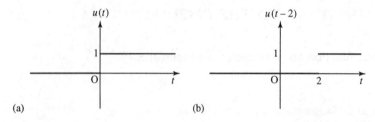

(a)   (b)

Figure 21.3 (a) The graph of $y=u(t)$. (b) The graph of $y=u(t-2)$ is found by shifting the graph of $u(t)$ two units to the right.

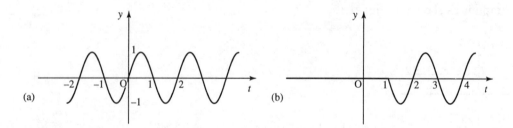

(a)   (b)

Figure 21.4 (a) The graph of $\sin(\omega t)$. (b) The graph of $u(t-1)\sin(\omega t)$. Notice that this function is zero for $t<1$ and equal to $\sin(\omega t)$ for $t>1$.

The delta function

The impulse function, or delta function, is a mathematical representation of a kick. It is an idealized kick which lasts for no time at all and has energy of exactly 1.

The delta function, $\delta(t)$, is an example of a generalized function. A generalized function can be defined in terms of a sequence of functions. One way of defining $\delta(t)$ is as the limit of a rectangular pulse function, with area 1, as it halves in width and doubles in height. This sequence of functions is pictured in Fig. 21.5. Although the height of the pulse is tending to infinity, the area of the pulse remains 1.

Two important properties of the delta function are

1. $\delta(t-a)=0$ for $t\neq a$

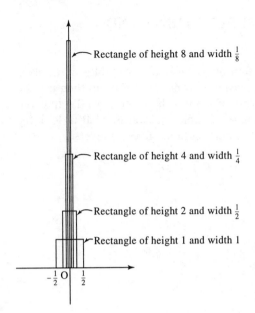

Rectangle of height 8 and width $\frac{1}{8}$

Rectangle of height 4 and width $\frac{1}{4}$

Rectangle of height 2 and width $\frac{1}{2}$

Rectangle of height 1 and width 1

Figure 21.5 The sequence of rectangular pulse functions of area 1. Starting from a pulse of height 1 and width 1 we double the height and halve the width.

2. $\displaystyle\int_{-\infty}^{\infty} \delta(t)\, dt = 1$

The second property expresses the fact that the area enclosed by the delta function is 1.

The unit step function, $u(t)$, has no derivative at $t=0$. Because of the sharp edges present in its graph and its jump discontinuity it is impossible to define a single tangent at that point. However, if we also consider the unit step function as a generalized function (by taking the limit of nice smooth, continuous curves as they approach the shape of the unit step function) we are able to define its derivative, which turns out to be the delta function. This gives the third property:

3. $\displaystyle\frac{du}{dt} = \delta(t)$

Symbolic representation of $\delta(t)$

$\delta(t)$ can be represented on a graph by an arrow of height 1. The height represents the weight of the delta function. A shifted delta function $\delta(t-a)$ is represented by an arrow at $t=a$. Examples are given in Fig. 21.6.

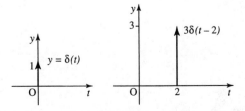

Figure 21.6 Symbolic representations of delta functions. (a) $\delta(t)$. (b) $3\delta(t-2)$.

21.4 LAPLACE TRANSFORMS OF SIMPLE FUNCTIONS AND PROPERTIES OF THE TRANSFORM

Rather than find a Laplace transform from the definition we usually use tables of Laplace transforms. A table of common Laplace transforms is given in Table 21.1. We can then use the properties of the transform to find Laplace transforms of many other functions. To find the inverse transform, represented by $\mathcal{L}^{-1}\{\ \}$, we use the same table backwards. That is, look for the function of s in the right-hand column, and the left-hand column gives its inverse.

$$F(s) = \int_0^\infty f(t)e^{-st}\,dt \qquad f(t) = \mathcal{L}^{-1}\{F(s)\}$$

Table 21.1 Common Laplace transforms

$f(t)$	$F(s) = \mathcal{L}\{f(t)\}$	
$u(t)$	$\dfrac{1}{s}$	$\mathrm{Re}(s) > 0$
$\delta(t)$	1	
$\dfrac{t^{n-1}}{(n-1)!}$	$\dfrac{1}{s^n}$	$\mathrm{Re}(s) > 0$
e^{-at}	$\dfrac{1}{s+a}$	$\mathrm{Re}(s) > -a$
$\dfrac{1}{a}\sin(at)$	$\dfrac{1}{s^2+a^2}$	$\mathrm{Re}(s) > 0$
$\cos(at)$	$\dfrac{s}{s^2+a^2}$	$\mathrm{Re}(s) > 0$
$\dfrac{1}{a}\sinh(at)$	$\dfrac{1}{s^2-a^2}$	$\mathrm{Re}(s) > |a|$
$\cosh(at)$	$\dfrac{s}{s^2-a^2}$	$\mathrm{Re}(s) > |a|$

Properties of the Laplace transform

In all of the following, $F(s) = \mathcal{L}\{f(t)\}$.

1. Linearity:

$$\mathcal{L}\{af_1(t) + bf_2(t)\} = aF_1(s) + bF_2(s)$$

where a and b are constants.

2. First translation (or shift rule):

$$\mathcal{L}\{e^{at}f(t)\} = F(s-a)$$

3. Second translation:

$$\mathcal{L}\{f(t-a)u(t-a)\} = e^{-as}F(s)$$

4. Change of scale:

$$\mathcal{L}\{f(at)\} = \frac{1}{a}F\left(\frac{s}{a}\right)$$

5. Laplace transforms of derivatives:

$$\mathcal{L}\{f'(t)\} = sF(s) - f(0)$$

$$\mathcal{L}\{f''(t)\} = s^2 F(s) - sf(0) - f'(0)$$

and

$$\mathcal{L}\{f^{(n)}(t)\} = s^n F(s) - s^{n-1} f(0) - \cdots - sf^{(n-2)}(0) - f^{(n-1)}(0)$$

6. Integrals:

$$\mathcal{L}\left\{\int_0^t f(\tau)\,d\tau\right\} = \frac{F(s)}{s}$$

7. Convolution:

$$\mathcal{L}\{f * g\} = \mathcal{L}\left\{\int_0^t f(\tau)g(t-\tau)\,d\tau\right\} = F(s)G(s)$$

8. Derivatives of the transform:

$$\mathcal{L}\{t^n f(t)\} = (-1)^n F^{(n)}(s)$$

where

$$F^{(n)}(s) = \frac{d^n F(s)}{ds^n}$$

Example 21.3: Linearity Find

$$\mathcal{L}\{3t^2 + \sin(2t)\}$$

SOLUTION From Table 21.1

$$\mathcal{L}\left\{\frac{t^2}{2}\right\} = \frac{1}{s^3}$$

and

$$\mathcal{L}\left\{\frac{\sin(2t)}{2}\right\} = \frac{1}{s^2 + 4}$$

$$\mathcal{L}\{3t^2 + \sin(2t)\} = 6\mathcal{L}\left\{\frac{t^2}{2}\right\} + 2\mathcal{L}\left\{\frac{\sin(2t)}{2}\right\} = \frac{6}{s^3} + \frac{2}{s^2 + 4}$$

using

$$\mathcal{L}\{af_1(t) + bf_2(t)\} = aF_1(s) + bF_2(s)$$

Example 21.4: Linearity and the inverse transform Find

$$\mathcal{L}^{-1}\left\{\frac{2}{s+4} + \frac{4s}{s^2 + 9}\right\}$$

SOLUTION From Table 21.1,

$$\mathcal{L}^{-1}\left\{\frac{1}{s+4}\right\} = e^{-4t}$$

and

$$\mathcal{L}^{-1}\left\{\frac{s}{s^2+9}\right\}=\cos(3t)$$

Therefore

$$\mathcal{L}^{-1}\left\{\frac{2}{s+4}+\frac{4s}{s^2+9}\right\}=2e^{-4t}+4\cos(3t)$$

Example 21.5: First translation Find

$$\mathcal{L}\{t^2\,e^{-3t}\}$$

SOLUTION As

$$\mathcal{L}\{t^2\}=\frac{2}{s^3}$$

using

$$\mathcal{L}\{e^{at}f(t)\}=F(s-a)$$

and then multiplying by e^{-3t} in the t domain will shift the function $F(s)$ by 3:

$$\mathcal{L}\{t^2\,e^{-3t}\}=\frac{2}{(s+3)^3}$$

Example 21.6: First translation – inverse transform Find

$$\mathcal{L}^{-1}\left\{\frac{s+2}{(s+2)^2+9}\right\}$$

SOLUTION As

$$\mathcal{L}^{-1}\left\{\frac{s}{s^2+9}\right\}=\cos(3t)$$

and as

$$\frac{s+2}{(s+2)^2+9}$$

is $s/(s^2+9)$ translated by 2, using the first translation rule this will multiply in the t domain by e^{-2t}, so

$$\mathcal{L}^{-1}\left\{\frac{s+2}{(s+2)^2+9}\right\}=\cos(3t)e^{-2t}$$

Example 21.7: Second translation Find the Laplace transform of

$$f(t)=\begin{cases}\sin(3t-2) & t\geqslant\frac{2}{3}\\ 0 & t<\frac{2}{3}\end{cases}$$

SOLUTION $f(t)$ can be expressed using the unit step function:

$$f(t) = \sin(3t - 2)u(t - \tfrac{2}{3}) = \sin(3(t - \tfrac{2}{3}))u(t - \tfrac{2}{3})$$

Using

$$\mathcal{L}\{f(t-a)u(t-a)\} = e^{-as}F(s)$$

and as

$$\mathcal{L}\{\sin(3t)\} = \frac{3}{s^2 + 9}$$

then

$$\sin(3(t - \tfrac{2}{3}))u(t - \tfrac{2}{3}) = \frac{3e^{-2s/3}}{s^2 + 9}$$

Example 21.8: Second translation – inverse transform Find

$$\mathcal{L}^{-1}\left\{\frac{e^{-s}}{(s+2)^2}\right\}$$

SOLUTION As

$$\mathcal{L}^{-1}\left\{\frac{1}{(s+2)^2}\right\} = te^{-2t} \qquad \text{(using first translation)}$$

then the factor of e^{-s} will translate in the t domain, $t \to t - 1$. Then

$$\mathcal{L}^{-1}\left\{\frac{e^{-s}}{(s+2)^2}\right\} = (t-1)e^{-2(t-1)}u(t-1)$$

by second translation.

Example 21.9: Change of scale Given

$$\mathcal{L}\{\cos(t)\} = \frac{s}{s^2 + 1}$$

find $\mathcal{L}\{\cos(3t)\}$ using the change of scale property of the Laplace transform:

$$\mathcal{L}\{f(at)\} = \frac{1}{a} F\left(\frac{s}{a}\right)$$

SOLUTION As

$$\mathcal{L}\{\cos(t)\} = \frac{s}{s^2 + 1} = F(s)$$

to find $\cos(3t)$ we put $a = 3$ into the change of scale property

$$\mathcal{L}\{f(at)\} = \frac{1}{a} F\left(\frac{s}{a}\right)$$

giving

$$\mathcal{L}\{\cos(3t)\} = \tfrac{1}{3}\frac{s/3}{(s/3)^2 + 1} = \frac{s}{s^2 + 9}$$

Example 21.10: Derivatives Given

$$\mathcal{L}\{\cos(2t)\} = \frac{s}{s^2 + 4}$$

find $\mathcal{L}\{\sin(2t)\}$ using the derivative rule.

SOLUTION If $f(t) = \cos(2t)$ then $f'(t) = -2\sin(2t)$.

$$\mathcal{L}\{\cos(2t)\} = \frac{s}{s^2 + 4}$$

and $f(0) = \cos(0) = 1$. From the derivative rule, $\mathcal{L}\{f'(t)\} = sF(s) - f(0)$,

$$\mathcal{L}\{-2\sin(2t)\} = s\left(\frac{s}{s^2 + 4}\right) - 1 = \frac{s^2}{s^2 + 4} - 1 = \frac{s^2 - s^2 - 4}{s^2 + 4}$$

$$= \frac{-4}{s^2 + 4}$$

By linearity,

$$\mathcal{L}\{\sin(2t)\} = \frac{1}{-2}\left(\frac{-4}{s^2 + 4}\right) = \frac{2}{s^2 + 4}$$

Example 21.11: Integrals Using

$$\mathcal{L}\left\{\int_0^t f(\tau)\,d\tau\right\} = \frac{F(s)}{s}$$

and

$$\mathcal{L}\{t^2\} = \frac{2}{s^3}$$

find

$$\mathcal{L}\left\{\frac{t^3}{3}\right\}$$

SOLUTION As

$$\int_0^t \tau^2\,d\tau = \left[\frac{\tau^3}{3}\right]_0^t = \frac{t^3}{3}$$

and

$$\mathcal{L}\{t^2\} = \frac{2}{s^3}$$

then

$$\mathcal{L}\left\{\int_0^\tau \tau^2\,d\tau\right\} = \frac{2/s^3}{s} = \frac{2}{s^4} \Rightarrow \mathcal{L}\left\{\frac{t^3}{3}\right\} = \frac{1}{s}\left(\frac{2}{s^3}\right) = \frac{2}{s^4}$$

Example 21.12: Convolution Find

$$\mathcal{L}^{-1}\left\{\frac{1}{(s-2)(s-3)}\right\}$$

using the convolution property:

$$\mathcal{L}\left\{\int_0^t f(t)g(t-\tau)\,d\tau\right\} = F(s)G(s)$$

SOLUTION

$$\frac{1}{(s-2)(s-3)} = \frac{1}{s-2}\frac{1}{s-3}$$

Therefore, call

$$F(s) = \frac{1}{s-2}, \qquad G(s) = \frac{1}{s-3}$$

$$f(t) = \mathcal{L}^{-1}\left\{\frac{1}{s-2}\right\} = e^{2t}$$

$$g(t) = \mathcal{L}^{-1}\left\{\frac{1}{s-3}\right\} = e^{3t}$$

Then by the convolution rule

$$\mathcal{L}^{-1}\left\{\frac{1}{(s-2)(s-3)}\right\} = \int_0^\tau e^{2\tau}e^{3(t-\tau)}\,d\tau$$

This integral is an integral over the variable τ. t is a constant as far as the integration process is concerned. We can use the properties of powers to separate out the terms in τ and the terms in t, giving

$$\mathcal{L}^{-1}\left\{\frac{1}{(s-2)(s-3)}\right\} = e^{3t}\int_0^t e^{-\tau}\,d\tau = e^{3t}\left[\frac{e^{-\tau}}{-1}\right]_0^t$$

$$= e^{3t}(-e^{-t}+1)$$

$$= -e^{2t}+e^{3t}$$

Example 21.13: Derivatives of the transform Find $\mathcal{L}\{t\sin(3t)\}$.

SOLUTION Using the derivatives of the transform property

$$\mathcal{L}\{t^n f(t)\} = (-1)^n F^{(n)}(s)$$

we have $f(t) = \sin(3t)$ and

$$\mathcal{L}\{\sin(3t)\} = \frac{3}{s^2+9} = F(s)$$

from Table 21.1. As $f(g)$ is multiplied by t in the expression that we want to find the transform of, then we write (with $n=1$):

$$\mathcal{L}\{t^1 f(t)\} = (-1)^1 F^{(1)}(s) = -F'(s)$$

So

$$\mathcal{L}\{t\sin(3t)\} = -\frac{\mathrm{d}}{\mathrm{d}s}\left(\frac{3}{s^2+9}\right) = \frac{6s}{(s^2+9)^2}$$

Example 21.14: Derivatives of the transform and the inverse transform Find

$$\mathcal{L}^{-1}\left\{\frac{s}{(s^2+4)^2}\right\}$$

SOLUTION We notice that

$$\frac{s}{(s^2+4)^2} = -\frac{1}{2}\frac{\mathrm{d}}{\mathrm{d}s}\left(\frac{1}{s^2+4}\right)$$

Using the derivatives of the transform property

$$\mathcal{L}\{t^n f(t)\} = (-1)^n F^{(n)}(s)$$

and setting $n=1$ we have

$$\mathcal{L}^{-1}\left\{(-1)\frac{\mathrm{d}F}{\mathrm{d}s}\right\} = tf(t)$$

and as

$$\mathcal{L}^{-1}\left\{\frac{1}{s^2+4}\right\} = \tfrac{1}{2}\sin(2t)$$

then

$$\mathcal{L}^{-1}\left\{\frac{s}{(s^2+4)^2}\right\} = \tfrac{1}{2}t(\tfrac{1}{2}\sin(2t)) = \frac{t}{4}\sin(2t)$$

Using partial fractions to find the inverse transform

Partial fractions can be used to find the inverse transform of expressions like

$$\frac{11s+7}{s^2-1}$$

by expressing $F(s)$ as a sum of fractions with a simple factor in the denominator.

Example 21.15 Find the inverse Laplace transform of

$$\frac{11s+7}{s^2-1}$$

SOLUTION Factorize the denominator. This is equivalent to finding the values of s for which the denominator is 0, because if the denominator has factors s_1 and s_2 we know that we can write it as $c(s-s_1)(s-s_2)$, where c is some number. In this case we could solve for $s^2-1=0$. However, it is not difficult to spot that $s^2-1=(s-1)(s+1)$. The values for which the denominator is zero are called the **poles** of the function.

$$\frac{11s+7}{s^2-1}=\frac{11s+7}{(s-1)(s+1)}$$

We assume that there is an identity such that

$$\frac{11s+7}{s^2-1}=\frac{A}{s-1}+\frac{B}{s+1}$$

Multiply both sides of the equation by $(s-1)(s+1)$:

$$11s+7=A(s+1)+B(s-1)$$

Substitute $s=1$; then $11+7=2A\Rightarrow A=9$.
Substitute $s=-1$; then $-11+7=-2B\Rightarrow B=2$.
So

$$\mathcal{L}^{-1}\left\{\frac{11s+7}{s^2-1}\right\}=\mathcal{L}^{-1}\left\{\frac{9}{(s-1)}+\frac{2}{s+1}\right\}=9e^t+2e^{-t}$$

A quick formula for partial fractions

There is a quick way of getting the partial fractions expansion, called the 'cover up' rule which works in the case where all of the roots of the denominator of $F(s)$ are distinct. If $F(s)=P(s)/Q(s)$, then write $Q(s)$ in terms of its factors:

$$Q(s)=c(s-s_1)(s-s_2)(s-s_3)\ldots(s-s_r)\ldots(s-s_n)$$

where $s_1\ldots s_n$ are its distinct roots. Then we can find the constant A_r etc. for the partial fraction expansion from covering up each of the factors of Q in turn and substituting $s=s_r$ in the rest of the expression.

$$F(s)=\frac{P(s)}{c(s-s_1)(s-s_2)(s-s_3)\ldots(s-s_r)\ldots(s-s_n)}$$

Then

$$F(s)=\frac{A_1}{(s-s_1)}+\frac{A_2}{(s-s_2)}+\cdots+\frac{A_r}{(s-s_r)}+\cdots+\frac{A_n}{(s-s_n)}$$

where

$$A_r=\frac{P(s_r)}{c(s_r-s_1)(s_r-s_2)(s_r-s_3)\ldots(s_r-s_r)\ldots(s_r-s_n)}$$

In this case

$$F(s)=\frac{11s+7}{s^2-1}=\frac{11s+7}{(s-1)(s+1)}$$

$$=\frac{A}{(s-1)}+\frac{B}{(s+1)}$$

Then A is found by substituting $s=1$ into

$$\frac{11s+7}{(s-1)(s+1)}=\frac{11+7}{2}=9$$

and B is found by substituting $s=-1$ into

$$\frac{11s+7}{(s-1)(s+1)}=\frac{-11+7}{-2}=2$$

This gives the partial fraction expansion, as before, as

$$F(s)=\frac{9}{s+1}+\frac{2}{s-1}$$

This method can also be used for complex poles, for instance

$$\frac{1}{(s^2+4)(s+3)}=\frac{1}{(s+3)(s-j2)(s+j2)}$$

The roots of the denominator are $-3, j2$ and $-j2$, so we get for the partial fraction expansion

$$\frac{1}{(s^2+4)(s+3)}=\frac{1}{13(s+3)}+\frac{1}{(-8+j12)(s-j2)}+\frac{1}{(-8-j12)(s+j2)}$$

The inverse transform can be found of this directly, or the last two terms can be combined to give the expansion as

$$\frac{1}{(s^2+4)(s+3)}=\frac{1}{13(s+3)}+\frac{-16s+48}{208(s^2+4)}=\frac{1}{13(s+3)}+\frac{3-s}{13(s^2+4)}$$

Repeated poles

If there are repeated factors in the denominator, e.g.

$$\frac{4}{(s+1)^2(s-2)}$$

then try a partial fraction expansion of the form

$$\frac{4}{(s+1)^2(s-2)}=\frac{A}{(s+1)^2}+\frac{B}{s+1}+\frac{C}{s-2}$$

The 'cover up' rule cannot be used to find the coefficients A and B.

21.5 SOLVING LINEAR DIFFERENTIAL EQUATIONS WITH CONSTANT COEFFICIENTS

The scheme for solving differential equations is as outlined below. The Laplace transform transforms the linear differential equation with constant coefficients to an algebraic equation in s. This can be solved and then the inverse transform of this solution gives the solution to the original differential equation.

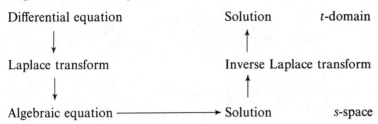

Differential equation → Laplace transform → Algebraic equation → Solution (s-space) → Inverse Laplace transform → Solution (t-domain)

Example 21.16 A d.c. voltage of 3 V is applied to an RC circuit with $R=2000\,\Omega$ and $C=0.001$ F, where $q(0)=0$. Find the voltage across the capacitor as a function of t.

SOLUTION From Kirchhoff's voltage law we get the differential equation

$$2000\frac{dq}{dt}+\frac{q}{0.001}=3$$

$$\frac{dq}{dt}+0.5q=0.0015$$

Taking Laplace transforms of both sides of the equation, where $Q(s)=\mathcal{L}\{q(t)\}$, then

$$sQ(s)-q(0)+0.5Q(s)=\frac{0.0015}{s}$$

Here we have used the derivative property of the Laplace transform $\mathcal{L}\{q'(t)\}=sQ(s)-q(0)$.
Solve the algebraic equation

$$(s+0.5)Q=\frac{0.0015}{s}$$

$$\Leftrightarrow Q=\frac{0.0015}{s(s+0.5)}$$

We need to find the inverse Laplace transform of this function of s, so we expand using partial fractions, giving

$$Q=\frac{0.003}{s}-\frac{0.003}{s+0.5}$$

$$q=\mathcal{L}^{-1}\left\{\frac{0.003}{s}-\frac{0.003}{s+0.5}\right\}=0.003-0.003e^{-0.5t}$$

The voltage across the capacitor is

$$\frac{q(t)}{c}=\frac{0.003-0.003e^{-0.5t}}{0.001}=3(1-e^{-0.5t})$$

Example 21.17 Solve the following differential equation using Laplace transforms:

$$\frac{d^2x}{dt^2}+4\frac{dx}{dt}+3x=e^{-3t}$$

given $x(0)=0.5$ and $dx(0)/dt=-2$.

SOLUTION Transform the differential equation

$$\frac{d^2x}{dt^2} + 4\frac{dx}{dt} + 3x = e^{-3t}$$

to get

$$\left(s^2X(s) - sx(0) - \frac{dx}{dt}(0)\right) + 4(sX(s) - x(0)) + 3X(s) = \frac{1}{s+3}$$

Here we have used

$$\mathcal{L}\left\{\frac{d^2x}{dt^2}\right\} = s^2X(s) - sx(0) - \frac{dx}{dt}(0)$$

and

$$\mathcal{L}\left\{\frac{dx}{dt}\right\} = sX(s) - x(0)$$

Substitute $x(0) = 0.5$ and $dx(0)/dt = -2$:

$$s^2X(s) - \tfrac{1}{2}s + 2 + 4sX(s) - 2 + 3X(s) = \frac{1}{s+3}$$

$$\Leftrightarrow X(s)(s^2 + 4s + 3) = \frac{1}{s+3} + \tfrac{1}{2}s$$

$$\Leftrightarrow X(s) = \frac{1}{(s+3)(s^2+4s+3)} + \frac{s}{2(s^2+4s+3)}$$

Use partial fractions:

$$\frac{1}{(s+3)(s^2+4s+3)} = \frac{1}{(s+3)(s+3)(s+1)}$$

$$\frac{1}{(s+3)(s+3)(s+1)} = \frac{A}{(s+3)} + \frac{B}{(s+3)^2} + \frac{C}{s+1}$$

$$\Rightarrow 1 = A(s+3)(s+1) + B(s+1) + C(s+3)^2$$

Substituting $s = -3$:

$$1 = -2B \Leftrightarrow B = -\tfrac{1}{2}$$

Substituting $s = -1$:

$$1 = 4C \Rightarrow C = \tfrac{1}{4}$$

Substituting $s = 0$:

$$1 = 3A + B + 9C$$

Substituting $B = -\tfrac{1}{2}$, $C = \tfrac{1}{4}$ gives

$$1 = 3A - \tfrac{1}{2} + \tfrac{9}{4} \Leftrightarrow A = -\tfrac{1}{4}$$

Therefore

$$\frac{1}{(s+3)^2(s+1)} = -\frac{1}{4(s+3)} - \frac{1}{2(s+3)^2} + \frac{1}{4(s+1)}$$

We can use the 'cover up' rule to find

$$\frac{s}{2(s+3)(s+1)} = \frac{3}{4(s+3)} - \frac{1}{4(s+1)}$$

so

$$X(s) = -\frac{1}{4(s+3)} - \frac{1}{2(s+3)^2} + \frac{1}{4(s+1)} + \frac{3}{4(s+3)} - \frac{1}{4(s+1)} = \frac{-1}{2(s+3)^2} + \frac{1}{2(s+3)}$$

Taking the inverse transform, we find

$$x(t) = -\tfrac{1}{2}te^{-3t} + \tfrac{1}{2}e^{-3t} = \frac{e^{-3t}}{2}(1-t)$$

21.6 LAPLACE TRANSFORMS AND SYSTEMS THEORY

The transfer function and impulse response function

An important role is played in systems theory by the impulse response function, the Laplace transform of which is called the transfer function (or system function). We remember from Chapter 20 that a linear, time-invariant system is represented by a linear differential equation with constant coefficients. We will stick to second-order equations, although the results can be generalized to any order.

An LTI system can be represented by

$$a\frac{d^2y}{dt^2} + b\frac{dy}{dt} + cy = f(t)$$

If we take $f(t)$ as the delta function $\delta(t)$, then we get

$$a\frac{d^2y}{dt^2} + b\frac{dy}{dt} + cy = \delta(t)$$

By definition of the impulse response function we consider all initial conditions to be 0.

Taking the Laplace transform we get

$$as^2Y(s) + bsY(s) + cY(s) = 1$$

$$Y(s)(as^2 + bs + c) = 1$$

$$Y(s) = \frac{1}{as^2 + bs + c}$$

Notice that the poles of this function are found by solving $as^2 + bs + c = 0$, which we recognize as the auxiliary equation or characteristic equation from Chapter 20.

This function, the Laplace transform of the impulse response function, is called the transfer function, and is usually denoted by $H(s)$, so we have

$$H(s) = \frac{1}{as^2 + bs + c}$$

and $\mathscr{L}^{-1}\{H(s)\} = h(t)$, where $h(t)$ is the impulse response function.

Note that the impulse response function describes the behaviour of the system after it has been given an idealized kick.

Example 21.18 Find the transfer function and impulse response of the system described by the following differential equation:

$$3\frac{dy}{dt} + 4y = f(t)$$

SOLUTION To find the transfer function, replace $f(t)$ by $\delta(t)$ and take the Laplace transform of the resulting equation assuming zero initial conditions:

$$3\frac{dy}{dt} + 4y = \delta(t)$$

Taking the Laplace transform of both sides of the equation we get

$$3(sY - y(0)) + 4Y = 1$$

As $y(0) = 0$, then

$$Y = \frac{1}{3s + 4} = H(s)$$

To find the impulse response function we take the inverse transform of the transfer function to find

$$h(t) = \mathcal{L}^{-1}\left\{\frac{1}{3s + 4}\right\} = e^{-4t/3}$$

We now discover that we can find the system response to any input function $f(t)$, with zero initial conditions, by using the transfer function.

Response of a system with zero initial conditions to any input $f(t)$

Assuming all initial conditions are zero, that is $y'(0) = 0$ and $y(0) = 0$, then the equation

$$a\frac{d^2y}{dt^2} + b\frac{dy}{dt} + cy = f(t)$$

transforms to become

$$(as^2 + bs + c)Y(y) = F(s)$$

$$\Leftrightarrow Y(s) = \frac{F(s)}{as^2 + bs + c}$$

We have just discovered that the transfer function, $H(s)$, is given by

$$H(s) = \frac{1}{as^2 + bs + c}$$

Then we find that

$$Y(s) = F(s)H(s)$$

In order to find the response of the system to the function $f(t)$ we take the inverse transform of this expression. We are able to use the convolution property of Laplace transforms, which states:

$$\mathcal{L}\left\{\int_0^t f(\tau)g(t-\tau)\,d\tau\right\} = F(s)G(s)$$

or equivalently

$$\mathcal{L}^{-1}\{F(s)G(s)\} = \int_0^t f(\tau)g(t-\tau)\,d\tau$$

The integral $\int_0^t f(\tau)g(t-\tau)\,d\tau$ is called the convolution of f and g and can be expressed as $f(t) * g(t)$.

So we find that the response of the system with zero initial conditions to any input function $f(t)$ is given by the convolution of $f(t)$ with the system's impulse response:

$$y(t) = \mathcal{L}^{-1}\{F(s)H(s)\} = f(t) * h(t)$$

we can use this result to solve two types of problem, given zero initial conditions. If we know the impulse response function of the system then we can find the system response to any input $f(t)$ either by convolving in the time domain, $y(t) = f(t) * h(t)$, or by finding the Laplace transform of $f(t)$, $F(s)$ and finding the transfer function of the system, $\mathcal{L}\{h(t)\} = H(s)$, and then finding $Y(s) = H(s)F(s)$ and taking the inverse transform of this to find $y(t)$. This type of problem is called a convolution problem. The other type of problem is that given any output $y(t)$ and given the input $f(t)$ then we can deduce the impulse response of the system. This we do by finding $F(s) = \mathcal{L}\{f(t)\}$, $Y(s) = \mathcal{L}\{y(t)\}$ and then $H(s) = Y(s)/F(s)$, thus giving the transfer function. To find the impulse response we find $h(t) = \mathcal{L}^{-1}\{H(s)\}$.

Example 21.19 The impulse response of a system is known to be $h(t) = e^{3t}$. Find the response of the system to an input of $f(t) = 6\cos(2t)$ given zero initial conditions.

SOLUTION

Method 1
We can take Laplace transforms and use $Y(s) = H(s)F(s)$. In this case

$$h(t) = e^{3t} \Leftrightarrow H(s) = \mathcal{L}\{e^{3t}\} = \frac{1}{s-3}$$

$$f(t) = 6\cos(2t) \Leftrightarrow F(s) = \mathcal{L}\{6\cos(2t)\} = \frac{6s}{4+s^2}$$

Hence

$$Y(s) = H(s)F(s) = \frac{6s}{(s-3)(4+s^2)}$$

We want to find $y(t)$ so we use partial fractions:

$$\frac{6s}{(s-3)(4+s^2)} = \frac{6s}{(s-3)(s+j2)(s-j2)}$$

$$= \frac{18}{13(s-3)} + \frac{3}{(j2-3)(s-j2)} - \frac{3}{(j2+3)(s+j2)} \quad \text{(using the 'cover up' rule)}$$

$$= \frac{18}{13(s-3)} - \frac{3(6s-8)}{13(s^2+4)}$$

$$= \frac{18}{13(s-3)} - \frac{18s}{13(s^2+4)} + \frac{12}{13}\frac{2}{(s^2+4)}$$

We can now take the inverse transform to find the system response:

$$y(t) = \mathscr{L}^{-1}\{Y(s)\} = \mathscr{L}^{-1}\left\{\frac{18}{13(s-3)} - \frac{18s}{13(s^2+4)} + \frac{12}{13}\frac{2}{(s^2+4)}\right\}$$

$$= \frac{18}{13}e^{3t} - \frac{18}{13}\cos(2t) + \frac{12}{13}\sin(2t)$$

Alternative method

Find $y(t)$ by taking the convolution of $f(t)$ with the impulse response function:

$$y(t) = f(t) * h(t) = (6\cos(2t) * (e^{3t}))$$

By the definition of convolution

$$f(t) * h(t) = \int_0^t 6\cos(2\tau)e^{3(t-\tau)}\, d\tau$$

as this is a real integral, we can use the trick of writing $\cos(2\tau) = \mathrm{Re}(e^{j2\tau})$ to make the integration easier. So we find

$$I = \int_0^t 6e^{j2\tau}e^{3(t-\tau)}\, d\tau = 6e^{3t}\int_0^t e^{\tau(j2-3)}\, d\tau$$

$$= 6e^{3t}\left[\frac{e^{\tau(j2-3)}}{j2-3}\right]_0^t$$

$$= 6e^{3t}\left(\frac{e^{t(j2-3)}}{j2-3} - \frac{1}{j2-3}\right)$$

$$= \frac{6e^{3t}(-j2-3)(e^{-3t}(\cos(2t)+j\sin(2t))-1)}{4+9}$$

Taking the real part of this result we get the system response as

$$\int_0^t 6\cos(2\tau)e^{3(t-\tau)}\, d\tau = \tfrac{6}{13}(-3\cos(2t) + 2\sin(2t)) + \tfrac{18}{13}e^{3t}$$

$$= -\tfrac{18}{13}\cos(2t) + \tfrac{12}{13}\sin(2t) + \tfrac{18}{13}e^{3t}$$

which confirms the result of the first method.

Example 21.20 A system at rest has a constant input of $f(t) = 3$ applied at $t = 0$. The output is found to be $u(t)(\tfrac{3}{2} - \tfrac{3}{2}e^{-2t})$. Find the impulse response of the system.

SOLUTION Since $y(t) = u(t)(\tfrac{3}{2} - \tfrac{3}{2}e^{-2t})$, we have

$$Y(s) = \frac{3}{2s} - \frac{3}{2(s+2)} = \frac{3s+6-3s}{2s(s+2)} = \frac{3}{s(s+2)}$$

and $f(t) = 3$, so

$$F(s) = \frac{3}{s}$$

Hence

$$H(s)=\frac{Y(s)}{F(s)}=\frac{3/(s(s+2))}{3/s}=\frac{1}{s+2}$$

giving

$$h(t)=\mathscr{L}^{-1}\left\{\frac{1}{s+2}\right\}=e^{-2t}$$

Hence the impulse response is $h(t)=e^{-2t}$.

The frequency response

In this section we shall seek to establish a relationship between the transfer function and the steady state response to a single frequency input.

Consider the response of the system

$$\ddot{y}+3\dot{y}+2y=f(t)$$

to a single sinusoidal input $e^{j\omega t}=\cos(\omega t)+j\sin(\omega t)$ with $y(0)=2$ and $\dot{y}(0)=1$. Taking Laplace transforms we find

$$s^2Y-2s-1+3(sY-2)+2Y=\frac{1}{s-j\omega}$$

$$Y(s^2+3s+2)=2s+7+\frac{1}{s-j\omega}$$

$$Y=\frac{2s+7}{s^2+3s+2}+\frac{1}{(s-j\omega)(s^2+3s+2)}$$

Remember

$$H(s)=\frac{1}{s^2+3s+2}=\frac{1}{(s+2)(s+1)}$$

$$Y=\frac{2s+7}{(s+2)(s+1)}+\frac{1}{(s-j\omega)(s+2)(s+1)}$$

The first term in this expression is a result of the non-zero initial conditions. We are particularly interested in the second term in the expression for $Y(s)$, which we notice may be written as $H(s)/(s-j\omega)$.

Using the 'cover up' rule to write $Y(s)$ in partial fractions, then

$$Y(s)=\frac{5}{s+1}-\frac{3}{s+2}+\frac{1}{(-1-j\omega)(s+1)}+\frac{1}{(2+j\omega)(s+2)}+\frac{H(j\omega)}{s-j\omega}$$

Taking inverse transforms we find

$$y(t)=5e^{-t}-3e^{-2t}+\frac{1}{-1-j\omega}e^{-t}+\frac{1}{2+j\omega}e^{-2t}+H(j\omega)e^{j\omega t}$$

The first two terms in this expression are caused by the non-zero initial values and decay exponentially. The next two terms also decay with increasing t and are a result of the abrupt turning on of the input at $t=0$. Thus for a stable system, all four terms are part of the transient solution which dies out as t increases. The fifth and final term is the sinusoidal input $f(t) = e^{j\omega t}$ multiplied by $H(j\omega)$. $H(j\omega)$ is a complex constant. This is the steady state response of the system to a single sinusoidal input, which we have shown in this case is given by $Y(t) = H(j\omega)e^{j\omega t}$.

We can then see that for a single sinusoidal input the steady state response is found by substituting $s = j\omega$ into the transfer function for the system and multiplying the resulting complex constant by the sinusoidal input. In other words, the steady state response is a scaled and phase-shifted version of the input. We can find the response to a sine or cosine input by

$$y(t) = \text{Re}(H(j\omega)e^{j\omega t}) \qquad \text{for } f(t) = \cos(\omega t)$$

and

$$y(t) = \text{Im}(H(j\omega)e^{j\omega t}) \qquad \text{for } f(t) = \sin(\omega t)$$

Alternatively, we can find the response to $e^{j\omega t}$ and to $e^{-j\omega t}$ and use the fact that

$$\cos(\omega t) = \tfrac{1}{2}(e^{j\omega t} + e^{-j\omega t})$$

$$\sin(\omega t) = \frac{1}{2j}(e^{j\omega t} - e^{-j\omega t})$$

to write

$$\text{for } f(t) = \cos(\omega t), \ y(t) = \tfrac{1}{2}(H(j\omega)e^{j\omega t} + H(-j\omega)e^{-j\omega t})$$

$$\text{for } f(t) = \sin(\omega t), \ y(t) = \frac{1}{2j}(H(j\omega)e^{j\omega t} - H(-j\omega)e^{-j\omega t})$$

All of these results make use of the principle of superposition for linear systems.

We have seen in this section that the response to a simple sinusoid can be characterized by multiplying the input by a complex constant, $H(j\omega)$, where ω is the angular frequency of the input. The function $H(j\omega)$ is called the frequency response function. It is this result that motivates us towards the desirability of expressing all signals in terms of cosines and sines of single frequencies – a technique known as Fourier analysis. We shall look at Fourier analysis for periodic functions in Chapter 22.

Example 21.21 A system transfer function is known to be

$$H(s) = \frac{1}{3s+1}$$

Find the steady state response to the following:
(a) $f(t) = e^{j2t}$
(b) $f(t) = 3\cos(2t)$

SOLUTION
(a) The steady state response to a single frequency $e^{j\omega t}$ is given by $H(j\omega)e^{j\omega t}$. Here $f(t) = e^{j2t}$, so in this case $\omega = 2$ and $H(s)$ is given as $1/(3s+1)$. Hence we get the steady state response as

$$H(j2)e^{j2t} = \frac{1}{3(j2)+1}e^{j2t} = \frac{e^{j2t}}{1+j6} = \frac{(1-j6)e^{j2t}}{37}$$

(b) Using $\tfrac{1}{2}(H(j\omega)e^{j\omega t} + H(-j\omega)e^{-j\omega t})$ as the response to $\cos(\omega t)$ and substituting for H and

$\omega = 2$ gives

$$\frac{1}{2}\left(\frac{(1-j6)e^{j2t}}{37} + \frac{(1+j6)}{37}e^{-j2t}\right) = \tfrac{1}{74}((1-j6)(\cos(2t)+j\sin(2t)) + (1+j6)(\cos(2t)-j\sin(2t))$$

$$= \tfrac{1}{37}(\cos(2t)+6\sin(2t))$$

Hence the steady state response to an input of $3\cos(2t)$ is $\tfrac{3}{37}(\cos(2t)+6\sin(2t))$.

21.7 z TRANSFORMS

z transforms are used to solve discrete systems in a manner similar to that for Laplace transforms. We take z transforms of sequences. We shall assume that our sequences begin with the zeroth term and have terms for positive n. $f_0, f_1, f_2, \ldots, f_n, \ldots$, is an input sequence to the system. However, when considering the initial conditions for a difference equation it is convenient to assign them to $y_{-j}, \ldots, y_{-2}, y_{-1}$ etc., where j is the order of the difference equation. So, in that case, we shall allow some elements in the sequence with negative subscript.

Our output sequence will be of the form $y_{-j}, \ldots, y_{-2}, y_{-1}, y_0, y_1, y_2, \ldots, y_n, \ldots$ where the difference equation describing the system only holds for $n \geqslant 0$.

z transform definition

The z transform of a sequence $f_0, f_1, f_2, \ldots, f_n, \ldots$ is given by

$$F(z) = \sum_{n=0}^{\infty} f_n z^{-n}$$

As this is an infinite sum it will not always converge. The set of values of z for which it exists is called the region of convergence. The sequence $f_0, f_1, f_2, \ldots, f_n, \ldots$ is a function of an integer; however, its z transform is a function of a complex variable z. The operation of taking the z transform of the sequence f_n is represented by $\mathscr{Z}\{f_n\} = F(z)$.

Example 21.22 Find the z transform of the finite sequence 1, 0, 0.5, 3.

SOLUTION We multiply the terms in the sequence by z^{-n}, where $n = 0, 1, 2, \ldots$, and then sum the terms, giving

$$F(z) = 1 + 0z^{-1} + 0.5z^{-2} + 3z^{-3}$$

$$= 1 + \frac{0.5}{z^2} + \frac{3}{z^3}$$

Example 21.23 Find the z transform of the geometric sequence $a_0 r^n$ where $n = 0, 1, \ldots$.

SOLUTION

$$F(z) = \sum_{n=0}^{\infty} a_0 r^n z^{-n} = \sum_{n=0}^{\infty} a_0 \left(\frac{r}{z}\right)^n$$

Writing this out we get

$$F(z) = a_0 + a_0\left(\frac{r}{z}\right) + a_0\left(\frac{r}{z}\right)^2 + a_0\left(\frac{r}{z}\right)^3 + \cdots$$

From this we can see that we have another geometric progression with zeroth term a_0 and common ratio r/z. Hence we can sum to infinity provided $|r/z| < 1$, giving

$$F(z) = \frac{a_0}{1 - (r/z)} = \frac{a_0 z}{z - r} \quad \text{where} \quad \left|\frac{r}{z}\right| < 1 \quad \text{or} \quad |z| > |r|$$

We can see that in the case of infinite sequences there will be a region of convergence for the z transform.

The impulse function and the step function

The impulse function or delta function for a discrete system is the sequence

$$\delta_n = \begin{cases} 1 & n = 0 \\ 0 & n \neq 0 \end{cases}$$

The step function is the function

$$u_n = \begin{cases} 1 & n \geq 0 \\ 0 & n < 0 \end{cases}$$

and the shifted unit step function is

$$u_{n-j} = \begin{cases} 1 & n \geq j \\ 0 & n < j \end{cases}$$

As we are mainly considering sequences defined for $n \geq 0$ we could consider that all of the sequences are multiplied by the step function, u_n. That is, they are all 'switched on' at $n = 0$.

Rather than always using the definition we will usually make use of a table of well-known transforms and properties of the z transform in order to discover the transform of various sequences. A table of z transforms is given in Table 21.2.

Table 21.2 z transforms

f_n	$F(z)$	
u_n	$\dfrac{z}{z-1}$	$|z| > 1$
δ_n	1	
n	$\dfrac{z}{(z-1)^2}$	$|z| > 1$
r^n	$\dfrac{z}{z-r}$	$|z| > |r|$
$\cos(\theta n)$	$\dfrac{z(z - \cos(\theta))}{z^2 - 2z\cos(\theta) + 1}$	$|z| > 1$
$\sin(\theta n)$	$\dfrac{z \sin(\theta)}{z^2 - 2z\cos(\theta) + 1}$	$|z| > 1$
$e^{j\theta n}$	$\dfrac{z}{z - e^{j\theta}}$	$|z| > 1$

Properties of the z transform

For the following,

$$\mathscr{L}\{f_n\} = \sum_{n=0}^{\infty} f_n z^{-n} = F(z), \ \mathscr{L}\{g_n\} = \sum_{n=0}^{\infty} g_n z^{-n} = G(z)$$

1. Linearity:

$$\mathscr{L}\{af_n + bg_n\} = aF(z) + bG(z)$$

2. Left shifting property:

$$\mathscr{L}\{f_{n+k}\} = z^k F(z) - \sum_{i=0}^{k} z^{k-i} f_i$$

3. Right shifting property
   (Although usually we assume $f_n = 0$, for $n < 0$ we use f_{-1}, f_{-2} to specify the initial conditions when solving difference equations using z transforms.)

$$\mathscr{L}\{f_{n-1}\} = z^{-1}\mathscr{L}\{f_n\} + f_{-1}$$

$$\mathscr{L}\{f_{n-2}\} = z^{-2}\mathscr{L}\{f_n\} + f_{-2} + z^{-1}f_{-1}$$

$$\mathscr{L}\{f_{n-k}\} = z^{-k}\mathscr{L}\{f_n\} + \sum_{i=0}^{k-1} f_{i-k}z^{-i}$$

4. Change of scale:

$$\mathscr{L}\{a^n f_n\} = F\left(\frac{z}{a}\right)$$

   where a is a constant.
5. Convolution:

$$\mathscr{L}\left\{\sum_{k=0}^{n} g_k f_{n-k}\right\} = G(z)F(z)$$

The convolution of f and g can be written as

$$g * f = \sum_{k=0}^{n} g_k f_{n-k}$$

   where g_n and f_n are sequences defined for $n \geq 0$.
6. Derivatives of the transform

$$\mathscr{L}\{nf_n\} = -z\frac{dF}{dz}(z)$$

Example 21.24: Linearity
Find the z transform of $3n + 2 \times 3^n$.

SOLUTION From the linearity property

$$\mathscr{L}\{3n + 2 \times 3^n\} = 3\mathscr{L}\{n\} + 2\mathscr{L}\{3^n\}$$

and from Table 21.2

$$\mathscr{Z}\{n\} = \frac{z}{(z-1)^2} \quad \text{and} \quad \mathscr{Z}\{3^n\} = \frac{z}{z-3}$$

(r^n with $r=3$). Therefore

$$\mathscr{Z}\{3n + 2 \times 3^n\} = \frac{3z}{(z-1)^2} + \frac{2z}{(z-3)}$$

Example 21.25: Linearity and the inverse transform Find the inverse z transform of

$$\frac{2z}{z-1} + \frac{3z}{z-2}$$

SOLUTION From tables:

$$\mathscr{Z}^{-1}\left\{\frac{z}{z-1}\right\} = u_n$$

$$\mathscr{Z}^{-1}\left\{\frac{z}{z-2}\right\} = 2^n \quad (r=2)$$

So

$$\mathscr{Z}^{-1}\left\{\frac{2z}{z-1} + \frac{3z}{z-2}\right\} = 2u_n + 3 \times 2^n$$

Example 21.26: Change of scale Find the inverse z transform of

$$\frac{z}{(z-2)^2}$$

SOLUTION

$$\frac{z}{(z-2)^2} = \frac{\frac{1}{2}(z/2)}{((z/2)-1)^2}$$

From Table 21.2,

$$\mathscr{Z}^{-1}\left\{\frac{z}{(z-1)^2}\right\} = n$$

Using the change of scale property and linearity:

$$\mathscr{Z}^{-1}\left\{\frac{\frac{1}{2}(z/2)}{((z/2)-1)^2}\right\} = \frac{1}{2}n(2)^n = n2^{n-1}$$

Example 21.27: Convolution Find the inverse z transform of

$$\frac{z}{z-1}\frac{z}{z-4}$$

SOLUTION Note that

$$\mathscr{L}^{-1}\left\{\frac{z}{z-1}\right\}=u_n \qquad \text{and} \qquad \mathscr{L}^{-1}\left\{\frac{z}{z-4}\right\}=4^n$$

Hence, using convolution:

$$\mathscr{L}^{-1}\left\{\frac{z}{z-1}\frac{z}{z-4}\right\}=u_n * 4^n = \sum_{k=0}^{n} u_k 4^{n-k}$$

Writing out this sequence for $n=0, 1, 2, 3,\dots$:

$$1 \quad, (1+4), 1+4+16, 1+4+16+64, \dots$$
$$(n=0) \ (n=1) \quad (n=2) \qquad (n=3)$$

We see that the nth term is a geometric series with $n+1$ terms and first term 1 and common ratio 4. From the formula for the sum of n terms of a geometric progression, $S_n = a(r^n-1)/(r-1)$, where a is the first term, r is the common ratio and n is the number of terms we get

$$\frac{4^{n+1}-1}{4-1} = \frac{4^{n+1}-1}{3}$$

So we have found

$$\mathscr{L}^{-1}\left\{\frac{z}{z-1}\frac{z}{z-4}\right\} = \frac{4^{n+1}-1}{3}$$

Example 21.28: Derivatives of the transform Using

$$\mathscr{L}\{n\} = \frac{z}{(z-1)^2}$$

find $\mathscr{L}\{n^2\}$.

SOLUTION Using the derivative of the transform property

$$\mathscr{L}\{n^2\} = \mathscr{L}\{nn\} = -z\frac{d}{dz}\mathscr{L}\{n\}$$

$$= -z\frac{d}{dz}\left(\frac{z}{(z-1)^2}\right)$$

As

$$\frac{d}{dz}\left(\frac{z}{(z-1)^2}\right) = \frac{(z-1)^2 - 2z(z-1)}{(z-1)^4} = \frac{z-1-2z}{(z-1)^3}$$

$$= \frac{-z-1}{(z-1)^3}$$

we obtain

$$\mathscr{L}\{n^2\} = -z\left(\frac{-z-1}{(z-1)^3}\right) = \frac{z(z+1)}{(z-1)^3}$$

Using partial fractions to find the inverse transform

Example 21.29 Find

$$\frac{z^2}{(z-1)(z-0.5)}$$

SOLUTION Notice that most of the values of the transform in Table 21.2 have a factor of z in the numerator. We write

$$\frac{z^2}{(z-1)(z-0.5)} = z\left(\frac{z}{(z-1)(z-0.5)}\right)$$

We use the 'cover up' rule to write

$$\frac{z}{(z-1)(z-0.5)} = \frac{1}{0.5(z-1)} - \frac{1}{z-0.5}$$

$$= \frac{2}{z-1} - \frac{1}{z-0.5}$$

So

$$\frac{z^2}{(z-1)(z-0.5)} = \frac{2z}{z-1} - \frac{z}{z-0.5}$$

and using Table 21.2 we find

$$\mathscr{Z}^{-1}\left\{\frac{2z}{z-1} - \frac{z}{z-0.5}\right\} = 2u_n - (0.5)^n$$

21.8 SOLVING LINEAR DIFFERENCE EQUATIONS WITH CONSTANT COEFFICIENTS USING z TRANSFORMS

The scheme for solving difference equations is very similar to that for solving differential equations using Laplace transforms and is outlined below. The z transform transforms the linear difference equation with constant coefficients to an algebraic equation in z. This can be solved and then the inverse transform of this solution gives the solution to the original difference equation.

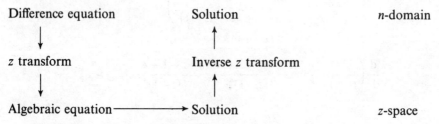

Example 21.30 Solve the difference equation $y_n + 2y_{n-1} = 2u_n$ for $n \geq 0$, given $y_{-1} = 1$.

SOLUTION We take the z transform of both sides of the difference equation

$$y_n + 2y_{n-1} = 2u_n$$

and using the right shift property to find

$$\mathscr{L}\{y_{n-1}\}=z^{-1}Y(z)+y_{-1}$$

this gives

$$Y(z)+2(z^{-1}Y(z)+y_{-1})=\frac{2z}{z-1}$$

As $y_{-1}=1$,

$$Y(z)(1+2z^{-1})=\frac{2z}{z-1}-2$$

$$Y(z)=\frac{2z^2}{(z-1)(z+2)}-\frac{2z}{z+2}$$

To take the inverse transform we need to express the first term using partial fractions. Using the 'cover up' rule we get

$$\frac{2z}{(z-1)(z+2)}=\frac{2}{3(z-1)}+\frac{4}{3(z+2)}$$

So

$$Y(z)=\frac{2z}{3(z-1)}+\frac{4z}{3(z+2)}-\frac{2z}{(z+2)}=\frac{2z}{3(z-1)}-\frac{2z}{3(z+2)}$$

Taking inverse transforms we find

$$y_n=\tfrac{2}{3}u_n-\tfrac{2}{3}(-2)^n$$

Check
To check that we have the correct solution we can substitute in a couple of values for n and see that we get the same value from the difference equation as from the explicit formula found.
From the explicit formula, $n=0$ gives

$$y_0=\tfrac{2}{3}u_0-\tfrac{2}{3}(-2)^0=\tfrac{2}{3}-\tfrac{2}{3}=0$$

From the difference equation, $y_n+2y_{n-1}=2u_n$, where $y_{-1}=1$, $n=0$ gives

$$y_0+2y_{-1}=2u_0$$

Substituting $y_{-1}=1$ gives $y_0=0$, as before.
From the explicit formula, $n=1$ gives

$$y_1=\tfrac{2}{3}u_1-\tfrac{2}{3}(-2)^1=\tfrac{2}{3}+\tfrac{4}{3}=2$$

From the difference equation, $n=1$ gives

$$y_1+2y_0=2u_1$$

Substituting $y_0=0$ gives $y_1=2$, confirming the result of the explicit formula.

Example 21.31 Solve the difference equation

$$6y_n-5y_{n-1}+y_{n-2}=(0.25)^n \qquad n\geqslant0$$

given $y_{-1}=1$, $y_{-2}=0$.

SOLUTION We take the z transform of both sides of the difference equation $6y_n - 5y_{n-1} + y_{n-2} = (0.25)^n$ and use the right shift property to find

$$\mathcal{Z}\{y_{n-1}\} = z^{-1}Y(z) + y_{-1}$$

$$\mathcal{Z}\{y_{n-2}\} = z^{-2}Y(z) + z^{-1}y_{-1} + y_{-2}$$

This gives

$$6Y(z) - 5(z^{-1}Y(z) + y_{-1}) + z^{-2}Y(z) + y_{-2} + z^{-1}y_{-1} = \frac{z}{z - 0.25}$$

Substituting the initial conditions $y_{-1} = 1$ and $y_{-2} = 0$ and collecting the terms involving $Y(z)$ we get

$$Y(z)(6 - 5z^{-1}Y(z) + z^{-2}) - 5 + z^{-1} = \frac{z}{z - 0.25}$$

$$\Leftrightarrow Y(z) = \frac{z}{(z - 0.25)(6 - 5z^{-1} + z^{-2})} + \frac{5}{6 - 5z^{-1} + z^{-2}} - \frac{z^{-1}}{6 - 5z^{-1} + z^{-2}}$$

$$= z\left(\frac{z^2}{(z - 0.25)(6z^2 - 5z + 1)}\right) + z\left(\frac{5z}{(3z - 1)(2z - 1)}\right) - z\left(\frac{1}{(3z - 1)(2z - 1)}\right)$$

Using the 'cover up' rule we write each of these terms as partial fractions:

$$\frac{z^2}{(z - 0.25)(3z - 1)(2z - 1)} = \frac{z^2}{6(z - \frac{1}{4})(z - \frac{1}{3})(z - \frac{1}{2})} = \frac{1}{2(z - \frac{1}{4})} - \frac{4}{3}\frac{1}{(z - \frac{1}{3})} + \frac{1}{(z - \frac{1}{2})}$$

$$\frac{5z}{6(z - \frac{1}{3})(z - \frac{1}{2})} = \frac{-5}{3(z - \frac{1}{3})} + \frac{5}{2(z - \frac{1}{2})}$$

$$\frac{1}{(3z - 1)(2z - 1)} = \frac{1}{6(z - \frac{1}{3})(z - \frac{1}{2})} = \frac{-1}{z - \frac{1}{3}} + \frac{1}{z - \frac{1}{2}}$$

giving

$$Y(z) = \frac{z}{2(z - \frac{1}{4})} - \frac{4z}{3(z - \frac{1}{3})} + \frac{z}{z - \frac{1}{2}} - \frac{5z}{3(z - \frac{1}{3})} + \frac{5z}{2(z - \frac{1}{2})} + \frac{z}{z - \frac{1}{3}} - \frac{z}{z - \frac{1}{2}}$$

$$\Leftrightarrow Y(z) = \frac{z}{2(z - \frac{1}{4})} - \frac{2z}{z - \frac{1}{3}} + \frac{5}{2(z - \frac{1}{2})}$$

Taking the inverse transform:

$$y_n = \tfrac{1}{2}(\tfrac{1}{4})^n - 2(\tfrac{1}{3})^n + \tfrac{5}{2}(\tfrac{1}{2})^n$$

21.9 z TRANSFORMS AND SYSTEMS THEORY

The transfer function and impulse response function

As before, when considering Laplace transforms, we find that an important role is played in systems theory of discrete systems by the impulse response function, which in this case is a sequence, h_n, the z transform of which is called the transfer function (or system function). We remember from Chapter 20 that a linear, time-invariant system is represented by a linear difference equation with constant coefficients, i.e. a second order LTI system can be represented by

$$ay_n + by_{n-1} + cy_{n-2} = f_n$$

If we take f_n as the delta function δ_n, then we get

$$ay_n + by_{n-1} + cy_{n-2} = \delta_n$$

By definition of the impulse response function we consider all initial conditions to be 0, i.e. $y_{-1} = 0$, $y_{-2} = 0$, taking the z transform we get

$$aY(z) + bz^{-1}Y(z) + cz^{-2}Y(z) = 1$$

$$Y(z)(a + bz^{-1} + cz^{-2}) = 1$$

$$Y(z) = \frac{1}{a + bz^{-1} + cz^{-2}}$$

$$= \frac{z^2}{az^2 + bz + c}$$

Notice that the poles of this function are found by solving $az^2 + bz + c = 0$, which we recognize as the auxiliary equation or characteristic equation from Chapter 20.

This function, the z transform of the impulse response function, is called the transfer function and is usually denoted by $H(z)$, so we have

$$H(z) = \frac{z^2}{az^2 + bz + c}$$

and $\mathscr{Z}^{-1}\{H(z)\} = h_n$, where h_n is the impulse response function.

Example 21.32 Find the transfer function and impulse response of the system described by the following difference equation:

$$3y_n + 4y_{n-1} = f_n$$

SOLUTION To find the transfer function, replace f_n by δ_n and take the z transform of the resulting equation, assuming zero initial conditions:

$$4y_n + 3y_{n-1} = \delta_n$$

taking the z transform of both sides of the equation we get

$$4Y + 3(z^{-1}Y + y_{-1}) = 1$$

As $y_{-1} = 0$ then

$$Y = \frac{z}{4z + 3} = H(z) = \frac{z/4}{z + \frac{3}{4}}$$

To find the impulse response sequence we take the inverse transform of the transfer function to find:

$$h_n = \mathscr{Z}^{-1}\left\{\frac{z/4}{z + \frac{3}{4}}\right\} = \tfrac{1}{4}(-\tfrac{3}{4})^n$$

We can now see why, in the design of digital filters this is referred to as an infinite impulse response (IIR) filter. This impulse response has a non-zero value for all n. Hence it represents an IIR system.

We now discover that we can find the system response to any input sequence f_n, with zero initial conditions, by using the transfer function.

Response of a system with zero initial conditions to any input f_n

Assuming all initial conditions are zero, that is $y_{-1} = 0$ and $y_{-2} = 0$, then the equation

$$ay_n + by_{n-1} + cy_{n-2} = f_n$$

transforms to become

$$aY(z) + bz^{-1}Y(z) + cz^{-2}Y(z) = F(z) \Leftrightarrow y(z) = \frac{F(z)z^2}{az^2 + bz + c}$$

we have just discovered that the transfer function, $H(z)$, is given by

$$H(z) = \frac{z^2}{az^2 + bz + c}$$

Then we find that

$$Y(z) = F(z)H(z)$$

In order to find the response of the system to the function f_n we take the inverse transform of this expression. We are able to use the convolution property of z transforms, which states:

$$\mathscr{Z}\left\{\sum_{k=0}^{n} f_k g_{n-k}\right\} = F(z)G(z)$$

or equivalently

$$\mathscr{Z}^{-1}\{F(z)G(z)\} = \sum_{k=0}^{n} f_k g_{n-k}$$

So we find that the response of the system with zero initial conditions to any input sequence f_n is given by the convolution of f_n with the system's impulse response:

$$y_n = \mathscr{Z}^{-1}\{F(z)H(z)\} = f_n * h_n$$

we can use this result to solve two types of problem, given zero initial conditions. If we know the impulse response function of the system then we can find the system response to any input f_n either by convolving the two sequences:

$$y_n = \sum_{k=0}^{n} f_k h_{n-k} = f_n * h_n$$

or by finding the z transform of f_n, $F(z)$, finding the transfer function of the system $\mathscr{Z}\{h_n\} = H(z)$, then finding $Y(z) = H(z)F(z)$ and taking the inverse transform of this to find y_n. This type of problem is called a convolution problem. The other type of problem is that given any out y_n and given the input f_n we can deduce the impulse response of the system. This we do by finding $F(z) = \mathscr{Z}\{f_n\}$, $Y(z) = \mathscr{Z}\{y_n\}$ and then $H(z) = Y(z)/F(s)$, thus giving the transfer function. To find the impulse response we find $h_n = \mathscr{Z}^{-1}\{H(z)\}$.

Example 21.33 The impulse response of a system is known to be $h_n = 3(0.5)^n$. Find the response of the system to an input of $f_n = 2u_n$ given zero initial conditions.

SOLUTION

Method 1
We can take z transforms and use

$$Y(z) = H(z)F(z)$$

In this case

$$h_n = 3(0.5)^n \Leftrightarrow H(z) = \mathscr{Z}\{3(0.5)^n\} = \frac{3z}{z-0.5}$$

$$f(t) = 2u_n \Leftrightarrow F(z) = \mathscr{Z}\{2u_n\} = \frac{2z}{z-1}$$

Hence

$$Y(z) = H(z)F(z) = \frac{6z^2}{(z-0.5)(z-1)}$$

We want to find y_n, so we use partial fractions:

$$\frac{6z}{(z-0.5)(z-1)} = \frac{-6}{z-0.5} + \frac{12}{z-1}$$

(using the 'cover up' rule). We can now take the inverse transform to find the system response:

$$y_n = \mathscr{Z}^{-1}\{Y(z)\} = \mathscr{Z}^{-1}\left\{\frac{-6z}{z-0.5} + \frac{12z}{z-1}\right\}$$

$$= -6(0.5)^n + 12u_n$$

Alternative method
Find y_n by taking the convolution of f_n with the impulse response function:

$$y_n = f_n * h_n = (2u_n) * (3(0.5)^n)$$

By the definition of convolution:

$$y_n = \sum_{k=0}^{n} 2u_k(3(0.5)^{n-k}) = \sum_{k=0}^{n} 6(0.5)^{n-k}$$

Writing out some of these terms we get

$$\underset{n=0}{6}\ ,\ \underset{n=1}{6(0.5+1)},\ \underset{n=2}{6((0.5)^2+0.5+1)},\ \underset{n=3}{6((0.5)^3+(0.5)^2+0.5+1)}, \ldots$$

We see that each term in the sequence is a geometric progression with first term 6 and common ratio 0.5 and $n+1$ terms. Hence

$$y_n = \frac{6(1-(0.5)^{n+1})}{1-0.5} = 12(1-(0.5)^{n+1}) = 12 - 12(0.5)(0.5)^n$$

$$= 12 - 6(0.5)^n$$

which confirms the result of the first method.

The frequency response

As in the case of Laplace transforms and continuous systems we find we are able to establish a relationship between the transfer function and the steady state response to a single frequency input.

The steady state response of the system to a sequence representing a single frequency input, $e^{j\omega n}$, is found to be $Y(z) = H(e^{j\omega})e^{j\omega n}$.

We can then see that for a single sinusoidal input the steady state response is found by substituting $z = e^{j\omega}$ into the transfer function for the system and multiplying the resulting complex constant by the sinusoidal input. In other words, the steady state response is a scaled and phase-shifted version of the input. The function $H(e^{j\omega})$ is called the frequency response function for a discrete system.

Example 21.34 A system transfer function is known to be

$$H(z) = \frac{z}{z + 0.2}$$

Find the steady state response to the following:
(a) $f_n = e^{j2n}$
(b) $f(t) = \cos(2n)$

SOLUTION
(a) The steady state response to a single frequency $e^{j\omega n}$ is given by $H(e^{j\omega})e^{j\omega n}$. $f_n = e^{j2n}$, so in this case $\omega = 2$ and $H(z)$ is given as $z/(z + 0.2)$. Hence we get the steady state response as

$$\frac{e^{j2}}{e^{j2} + 0.2}e^{j2n}$$

(b) Using $\frac{1}{2}(H(e^{j\omega})e^{j\omega n} + H(e^{-j\omega})e^{-j\omega n})$ as the response to $\cos(\omega n)$ and substituting for H and $\omega = 2$ gives

$$y_n = \frac{1}{2}\left(\frac{e^{j2}e^{j2n}}{e^{j2} + 0.2} + \frac{e^{-j2}}{e^{-j2} + 0.2}e^{-j2n}\right)$$

Expressing this over a real denominator and simplifying gives

$$\frac{\cos(2n)(1 + 0.2\cos(2)) + 0.2\sin(2)\sin(2n)}{1.4 + 0.4\cos(2)}$$

Hence the steady state response to an input of $3\cos(2n)$ is

$$\frac{3\cos(2n)(1 + 0.2\cos(2)) + 0.6\sin(2)\sin(2n)}{1.4 + 0.4\cos(2)}$$

21.10 SUMMARY

1. The Laplace transform $F(s)$ of the function $f(t)$ defined for $t \geqslant 0$ is

$$F(s) = \int_0^\infty e^{-st}f(t)\,dt$$

The Laplace transform is a function of s, where s is a complex variable. Because the integral definition of the Laplace transform involves an integral to ∞ it is usually necessary to limit possible values of s so that the integral above converges (i.e. does not tend to ∞).

2. The impulse function, $\delta(t)$, also called a delta function, is the most famous example of a generalized function. The impulse function represents an idealized kick, as it lasts for no time at all and has energy of exactly 1.

3. Laplace transforms are usually found by using a table of transforms (as in Table 21.1) and also by using properties of the transform, some of which are listed in Section 21.4.

4. Laplace transforms are used to reduce a differential equation to a simple equation in s-space. This can then be solved and the inverse transform used to find the solution to the differential equation.

5. The transfer function of the system, $H(s)$, is the Laplace transform of its impulse response function with zero initial conditions, $h(t)$. The Laplace transform of the response to any input function with zero initial conditions can be found by multiplying the Laplace transform of the input function by the transfer function of the system, $Y(s) = H(s)F(s)$.

6. The steady state response to a single frequency input $e^{j\omega t}$ is $H(j\omega)e^{j\omega t}$. $H(j\omega)$ is called the frequency response function.

7. The z transform of a sequence $f_0, f_1, f_2, \ldots, f_n, \ldots$ is given by

$$F(z) = \sum_{n=0}^{\infty} f_n z^{-n}$$

As this is an infinite summation it will not always converge. The set of values of z for which it exists is called the region of convergence.

8. The discrete impulse function or delta function, is defined by

$$\delta_n = \begin{cases} 1 & n=0 \\ 0 & n \neq 0 \end{cases}$$

9. z transforms are usually found by using a table of transforms (as in Table 21.2) and also by using properties of the transform, some of which are listed in Sec. 21.7.

10. z transforms are used to reduce a difference equation to a simple equation in z-space. This can then be solved and the inverse transform used to find the solution to the difference equation.

11. The transfer function of a discrete system, $H(z)$, is the z transform of its impulse response function with zero initial conditions, h_n. The z transform of the response to any input function with zero initial conditions can be found by multiplying the Laplace transform of the input function by the transfer function of the system, $Y(z) = H(z)F(z)$.

12. The steady state response of a discrete system to a single frequency input $e^{j\omega n}$ is $H(e^{j\omega})e^{j\omega n}$ $H(e^{j\omega})$ is called the frequency response function.

21.11 EXERCISES

21.1 Using the definition of the Laplace transform,

$$\mathcal{L}\{f(t)\} = \int_0^\infty e^{-st} f(t) \, dt$$

show the following. In each case, specify the values of s for which the transform exists.

(a) $\mathcal{L}\{2e^{4t}\} = \dfrac{2}{s-4}$

(b) $\mathcal{L}\{3e^{-2t}\} = \dfrac{3}{s+2}$

(c) $\mathcal{L}\{5t-3\} = \dfrac{5}{s^2} - \dfrac{3}{s}$

(d) $\mathcal{L}\{3\cos(5t)\} = \dfrac{3s}{s^2+25}$

21.2 Find Laplace transforms of the following using Table 21.1:

(a) $5\sin(3t)$ (b) $\cos\left(\dfrac{t}{2}\right)$ (c) $\dfrac{t^4}{3}$ (d) $\tfrac{1}{2}e^{-5t}$

21.3 Find inverse Laplace transforms of the following using tables:

(a) $\dfrac{1}{s-4}$ (b) $\dfrac{3}{s+1}$ (c) $\dfrac{4}{s}$ (d) $\dfrac{2}{s^2+3}$

(e) $\dfrac{4s}{9s^2+4}$ (f) $\dfrac{1}{s^3}$ (g) $\dfrac{5}{s^4}$

21.4 Find Laplace transforms of the following using the properties of the transform:

(a) te^{-3t} (b) $4\sin(t)e^{-2t}$ (c) $t\sinh(4t)$

(d) $u(t-1)$ (e) $f(t)=\begin{cases} \sin(t-\pi) & t \geqslant \pi \\ 0 & t < \pi \end{cases}$

21.5 Find inverse Laplace transforms of the following using the properties of the transform:

(a) $\dfrac{1}{(s-2)^2+9}$ (b) $\dfrac{s+4}{(s+4)^2+1}$ (c) $\dfrac{e^{-2s}}{s+1}$ (d) $\dfrac{s-1}{s^2-2s+10}$

21.6 Find inverse Laplace transforms of the following using partial fractions:

(a) $\dfrac{1}{s(s+1)}$ (b) $\dfrac{1}{s(s^2+2)}$ (c) $\dfrac{s}{(s^2-4)(s+1)}$ (d) $\dfrac{4}{(s^2+1)(s-3)}$

21.7 In each case solve the given differential equation using Laplace transforms:

(a) $y'+5y=0$, where $y(0)=3$
(b) $y'+y=t$, where $y(0)=0$
(c) $y''+4y=9t$, where $y(0)=0$, $y'(0)=7$
(d) $y''-3y'+2y=4e^{2t}$, where $y(0)=-3$, $y'(0)=5$
(e) $y''+y=t$, where $y(0)=1$, $y'(0)=-2$
(f) $y''+y'+2y=4$, $y(0)=0$, $y'(0)=0$

21.8 A capacitor of capacitance C in an RC circuit, as in Fig. 21.7, is charged so that initially its potential is V_0. At $t=0$ it begins to discharge. Its charge q is then described by the differential equation

$$R\frac{dq}{dt}+\frac{q}{C}=0$$

Using Laplace transforms, find the charge on the capacitor at time t after the switch was closed.

Figure 21.7 An RC circuit for Exercise 21.8.

21.9 Find the response of a system with zero initial conditions to an input of $f(t)=2e^{-3t}$, given that the impulse response of the system is $h(t)=\tfrac{1}{2}(e^{-t}-e^{-2t})$.

21.10 (a) The impulse response of a system is given by $h(t)=3e^{-4t}$. Find the system's step response, i.e. the response of the system to an input of the step function, $u(t)$.

(b) Use the result that $Y(s)=H(s)F(s)$, where Y is the Laplace transform of the system output, $F(s)$ is the Laplace transform of the input and $H(s)$ is the system transfer function, to show that the step response can be found by

$$y_u(t)=\mathcal{L}^{-1}\left\{\frac{H(s)}{s}\right\}$$

(c) Given that the step response of a system is

$$-\tfrac{4}{3}u(t)-e^{-2t}-\tfrac{2}{3}e^{-3t}$$

then

(i) Find the system's transfer function.
(ii) Find its response to an input of e^{-t}.

21.11 A system has a known impulse response of $h(t)=e^{-t}\sin(2t)$. Find the input function $f(t)$ that would produce an output of

$$y(t)=-0.16u(t)+0.4t+0.16e^{-t}\cos(2t)-0.04\sin(2t)e^{-t}$$

given zero initial conditions.

21.12 A system has transfer function

$$H(s)=\frac{1}{(s+2)^2+4}$$

(a) Find its steady state response to a single frequency input of e^{j5t}.
(b) Find the steady state response to inputs of $\cos(5t)$ and $\sin(5t)$.

21.13 Using the definition of the z transform,

$$\mathcal{Z}\{f_n\}=\sum_{n=0}^{\infty}f_n z^{-n}=F(z)$$

show the following. In each case specify the values of z for which the transform exists.

(a) $\mathcal{Z}\{3\delta_n\}=3$ (b) $\mathcal{Z}\{6u_n\}=\dfrac{6z}{z-1}$

(c) $\mathcal{Z}\{3^n\}=\dfrac{z}{z-3}$

21.14 Find z transforms of the following using Table 21.2:

(a) $2u_n+\tfrac{1}{2}n$ (b) $\cos(3n)+2\sin(3n)$
(c) $4(0.2)^n-6(2)^n$ (d) $2e^{j4n}$

21.15 Find inverse z transforms of the following using Table 21.2:

(a) $\dfrac{2z}{z-1}+2$ (b) $\dfrac{z}{2(z-1)^2}+\dfrac{1}{1+3/z}$

(c) $\dfrac{2z}{2z-3}+\dfrac{4z}{2z+3}$ (d) $\dfrac{z^2+z(\sin(1)-\cos(1))}{z^2-2z\cos(1)+1}$

21.16 Find z transforms of the following using the properties of the transform:

(a) $u_{n+2}3^{n+2}$ (b) $(\tfrac{1}{2})^n n$ (c) ne^{jn} (d) n^3

21.17 Find inverse z transforms of the following using the properties of the transform:

(a) $\dfrac{z}{(z-4)^2}$ (b) $\dfrac{z}{z-2e^{j4}}$ (c) $\dfrac{z}{(z-0.4)^2}$ (d) $\dfrac{z}{z-1}\dfrac{2z}{z-2}$

21.18 Find inverse z transforms using partial fractions:

(a) $\dfrac{z^2}{(z-1)(z+1)}$ (b) $\dfrac{z^3}{(z-1)(z^2-2)}$ (c) $\dfrac{z^2}{(z-0.1)(5z-2)}$

(d) $\dfrac{z^2}{(z-1)^2(z+2)}$ (e) $\dfrac{z^2}{z^2+1}$

21.19 In each case solve the given difference equation using z transforms, $n \geqslant 0$:

(a) $y_n + 5y_{n-1} = 0$, where $y_{-1} = 3$
(b) $y_n + y_{n-1} = n$, where $y_{-1} = 0$
(c) $y_n + 4y_{n-1} = 9$, where $y_{-1} = 1$
(d) $y_n - 3y_{n-1} + 2y_{n-2} = 4 \times 2^n$, where $y_{-1} = -3$, $y_{-2} = 5$
(e) $y_n + y_{n-2} = n$, where $y_{-1} = 0$, $y_{-2} = 0$
(f) $10y_n - 3y_{n-1} - y_{n-2} = 4$, where $y_{-1} = -1$, $y_{-2} = 2$

21.20 Find the response of a system with zero initial conditions to an input of $f_n = 2\,(0.3)^n$, given that the impulse response of the system is $h_n = (0.1)^n + (-0.5)^n$.

21.21 (a) The impulse response of a discrete system is given by $h_n = (0.8)^n$. Find the system's step response, i.e. the response of the system to an input of the step function, u_n.

(b) Use the result that $Y(z) = H(z)F(z)$, where Y is the z transform of the system output, $F(z)$ is the z transform of the input and $H(z)$ is the system transfer function, to show that the step response can be found by

$$y_n = \mathscr{Z}^{-1}\left\{\frac{H(z)z}{z-1}\right\}$$

(c) Given that the step response of a discrete system is

$$\tfrac{1}{24}(0.2)^n + \tfrac{10}{48}u_n$$

then:
(i) Find the system's transfer function.
(ii) Find its response to an input of $6(0.5)^n$.

21.22 A system has a known impulse response of $h_n = (0.5)^n$. Find the input function f_n that would produce an output of

$$2(0.5)^n + 2n - 2u_n$$

given zero initial conditions.

21.23 A system has transfer function

$$H(z) = \frac{z}{(10z-3)}$$

(a) Find its steady state response to a single-frequency input of e^{j5n}.
(b) Find the steady state response to an input of $\cos(5n)$ and $\sin(5n)$.

22.1 INTRODUCTION

Fourier analysis is the theory behind frequency analysis of signals. This chapter is concerned with the Fourier analysis of periodic, piecewise continuous functions. A periodic function can be represented by a Fourier series. A non-periodic function can be represented by its Fourier transform, which we shall not be concerned with here. Discrete functions may be represented by a discrete Fourier transform, which we also shall not look at in this book.

Any periodic signal is made up of the sum of single frequency components. These componnts consist of a fundamental frequency component, multiples of the fundamental frequency, called the harmonics, and a bias term, which represents the average offset from zero. There are three ways of representing this information, which are equivalent. We can represent the frequency components as the sum of sine and cosine terms, or by considering the amplitude and phase of each component, or we can represent them using a complex Fourier series. The use of the complex Fourier series simplifies the calculation.

Having found the Fourier components we can use the system's frequency response function, as found in the previous chapter, to find the steady state response to any periodic signal.

22.2 PERIODIC FUNCTIONS

In Chapter 11 we discussed the property of periodicity of the trigonometric functions. A periodic function is one whose graph can be translated to the right or left by an amount, called the period, such that the new graph fits exactly on top of the original graph. The fundamental period, also called the cycle, is the minimum non-zero amount the graph needs to be shifted in order to fit over the original graph.

A periodic function, with period τ, satisfies $f(t + \tau) = f(t)$ for all values of t. Examples of periodic functions are given in Fig. 22.1.

The fundamental frequency of a periodic function is the number of cycles in an interval of unit length, $f = 1/\tau$. The fundamental angular frequency is then given by $\omega_0 = 2\pi f = 2\pi/\tau$. A periodic function need only be defined in one cycle, as the periodicity property will then define it

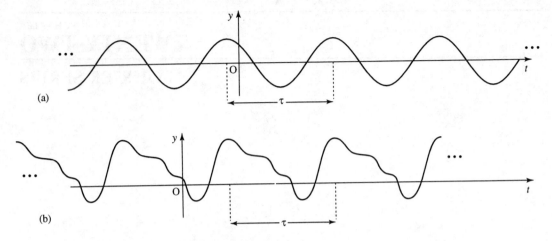

(a)

(b)

Figure 22.1 Some periodic functions with their fundamental periods marked.

everywhere. For example, the graph of the periodic square wave

$$f(t) = \begin{cases} \frac{1}{2} & \text{for } 0 < t < 1 \\ -\frac{1}{2} & \text{for } 1 < t < 2 \end{cases}$$

is drawn in Fig. 22.2. First we draw the section of the graph as given in the definition, and then shift the section along by the period, in this case of 2, and copy the section. By repeatedly shifting and copying in this way, both to the left and right, we get the graph as shown.

As we mentioned in the introduction there are three ways of expressing the Fourier series: as a sine and cosine series, in amplitude and phase form or in complex form.

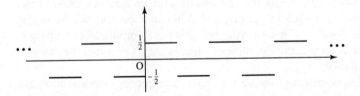

Figure 22.2 The periodic square wave defined by

$$f(t) = \begin{cases} \frac{1}{2} & \text{for } 0 < t < 1 \\ -\frac{1}{2} & \text{for } 1 < t < 2 \end{cases}$$

22.3 SINE AND COSINE SERIES

If $f(t)$ is periodic with period $\tau = 2\pi/\omega_0$, then the Fourier series for f is given by

$$f(t) = \tfrac{1}{2}a_0 + \sum_{n=1}^{\infty} a_n \cos(n\omega_0 t) + b_n \sin(n\omega_0 t)$$

where the coefficients are given by

$$a_0 = \frac{2}{\tau} \int_{-\frac{\tau}{2}}^{\frac{\tau}{2}} f(t)\, dt$$

$$a_n = \frac{2}{\tau} \int_{-\frac{\tau}{2}}^{\frac{\tau}{2}} f(t) \cos (n\omega_0 t) \, dt$$

$$b_n = \frac{2}{\tau} \int_{-\frac{\tau}{2}}^{\frac{\tau}{2}} f(t) \sin (n\omega_0 t) \, dt$$

The steps for finding the Fourier series are:

Step 1: Plot the periodic function $f(t)$.
Step 2: Determine its fundamental period τ and its fundamental angular frequency $\omega_0 = 2\pi/\tau$.
Step 3: Evaluate a_0, a_n and b_n as given above.
Step 4: Write down the resulting Fourier series.

Example 22.1 Find the Fourier series for

$$f(t) = \begin{cases} \frac{1}{2} & \text{for } 0 < t < 1 \\ -\frac{1}{2} & \text{for } 1 < t < 2 \end{cases}$$

SOLUTION
Step 1
We have already plotted the graph, as shown in Fig. 22.2.

Step 2
The fundamental period of this is $\tau = 2$ so that $\omega_0 = 2\pi/2 = \pi$

Step 3
Calculate a_0, a_n and b_n. We find

$$a_0 = \frac{2}{\tau} \int_{-\frac{\tau}{2}}^{\frac{\tau}{2}} f(t) \, dt = \frac{2}{2} \int_{-1}^{1} f(t) \, dt = \int_{-1}^{0} -\tfrac{1}{2} \, dt + \int_{0}^{1} \tfrac{1}{2} \, dt = [-\tfrac{1}{2}t]_{-1}^{0} + [\tfrac{1}{2}t]_{0}^{1} = -\tfrac{1}{2} + \tfrac{1}{2} = 0$$

$$a_n = \frac{2}{\tau} \int_{-\frac{\tau}{2}}^{\frac{\tau}{2}} f(t) \cos (n\omega_0 t) \, dt = \frac{2}{2} \int_{-1}^{1} f(t) \cos (n\pi t) \, dt$$

$$= \int_{-1}^{0} -\tfrac{1}{2} \cos (n\pi t) \, dt + \int_{0}^{1} \tfrac{1}{2} \cos (n\pi t) \, dt = \left[-\tfrac{1}{2} \frac{\sin (n\pi t)}{n\pi} \right]_{-1}^{0} + \left[\tfrac{1}{2} \frac{\sin (n\pi t)}{n\pi} \right]_{0}^{1}$$

$$= -\tfrac{1}{2} \frac{\sin (0)}{n\pi} + \frac{\sin (-n\pi)}{n\pi} + \tfrac{1}{2} \frac{\sin (n\pi)}{n\pi} - \tfrac{1}{2} \frac{\sin (0)}{n\pi} = -0 + 0 + 0 - 0 = 0$$

$$b_n = \frac{2}{\tau} \int_{-\frac{\tau}{2}}^{\frac{\tau}{2}} f(t) \sin (n\omega_0 t) \, dt = \frac{2}{2} \int_{-1}^{1} f(t) \sin (n\pi t) \, dt$$

$$= \int_{-1}^{0} -\tfrac{1}{2} \sin (n\pi t) \, dt + \int_{0}^{1} \tfrac{1}{2} \sin (n\pi t) \, dt$$

$$= \left[\tfrac{1}{2} \frac{\cos (n\pi t)}{n\pi} \right]_{-1}^{0} + \left[-\tfrac{1}{2} \frac{\cos (n\pi t)}{n\pi} \right]_{0}^{1}$$

$$= \frac{1}{2n\pi} (\cos (0) - \cos (-n\pi)) + \frac{1}{2\pi n} (-\cos (n\pi) + \cos (0))$$

$$= \frac{1}{2n\pi}(1 - (-1)^n) + \frac{1}{2n\pi}(-(-1)^n + 1)$$

$$= \frac{1}{n\pi}(1 - (-1)^n)$$

since $\cos(n\pi) = (-1)^n$.

Step 4
The fourier series for $f(t)$ is

$$f(t) = \tfrac{1}{2}a_0 + \sum_{n=1}^{\infty} a_n \cos(n\omega_0 t) + b_n \sin(n\omega_0 t)$$

in this case giving

$$f(t) = \frac{1}{\pi} \sum_{n=1}^{\infty} \frac{1 - (-1)^n}{n} \sin(n\pi t)$$

Note that the even values of n all give zero coefficients as $1 - (-1)^n = 0$ for n even. Odd values give $2/n\pi$. In this case we can change the variable for the summation, using $n = 2m - 1$, which is always odd. This gives

$$f(t) = \frac{1}{\pi} \sum_{m=1}^{\infty} \frac{2}{2m - 1} \sin((2m - 1)\pi t)$$

It is interesting to plot graphs of the first few partial sums that we obtain from this series. In Fig. 22.3 we have plotted the graph given by the terms up to $n = 3$, $n = 5$ and $n = 7$:

$$S_3 = \frac{2}{\pi} \sin(\pi t) + \frac{2}{3\pi} \sin(3\pi t)$$

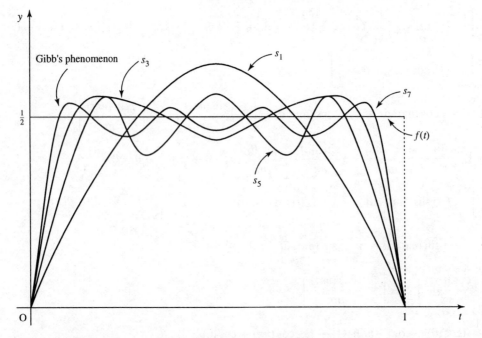

Figure 22.3 Partial sums of the Fourier series for the square wave.

$$S_5 = \frac{2}{\pi}\sin(\pi t) + \frac{2}{3\pi}\sin(3\pi t) + \frac{2}{5\pi}\sin(5\pi t)$$

$$S_7 = \frac{2}{\pi}\sin(\pi t) + \frac{2}{3\pi}\sin(3\pi t) + \frac{2}{5\pi}\sin(5\pi t) + \frac{2}{7\pi}\sin(7\pi t)$$

You will notice that there is an overshoot of the value near the jump discontinuities. This is an example of Gibb's phenomenon. However many terms we take in the partial sum this overshoot remains significant, at about 10 per cent of the function value. It is also interesting to see the value that the Fourier series takes at the discontinuous points, e.g. $t = 1$. Substituting $t = 1$ into S_3, S_5 and S_7 gives 0. This is half-way between the values of f at either side of the point $t = 1$.

Example 22.2 Find the Fourier series for the periodic function defined in the interval $0 < t < 1$ by

$$f(t) = \begin{cases} t & \text{for } 0 < t < \frac{1}{2} \\ 0 & \text{for } \frac{1}{2} < t < 1 \end{cases}$$

SOLUTION
Step 1
We plot the graph as shown in Fig. 22.4.

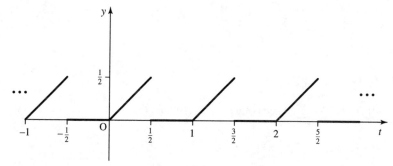

Figure 22.4 The graph for Example 22.2.

Step 2
The fundamental period of this is $\tau = 1$ so that $\omega_0 = 2\pi/\tau = 2\pi$.

Step 3
Calculate a_0, a_n and b_n. We find

$$a_0 = \frac{2}{1}\int_{-\frac{1}{2}}^{\frac{1}{2}} f(t)\,dt = 2\int_0^{\frac{1}{2}} t\,dt = [t^2]_0^{\frac{1}{2}} = \frac{1}{4} - 0 = \frac{1}{4}$$

$$a_n = \frac{2}{\tau}\int_{-\frac{1}{2}}^{\frac{1}{2}} f(t)\cos(\omega_0 nt)\,dt = 2\int_{-\frac{1}{2}}^{\frac{1}{2}} f(t)\cos(2\pi nt)\,dt = 2\int_0^{\frac{1}{2}} t\cos(2\pi nt)\,dt$$

To find this integral we perform integration by parts. The formula, as given in Chapter 13, is $\int u\,dv = uv - \int v\,du$. In this case we choose $u = t$, $dv = \cos(2\pi nt)\,dt$, $du = dt$ and $v = \sin(2\pi nt)/2\pi n$.

$$a_n = 2\left(\left[t\frac{\sin(2\pi nt)}{2\pi n}\right]_0^{\frac{1}{2}} - \int_0^{\frac{1}{2}}\frac{\sin(2\pi nt)}{2\pi n}\,dt\right)$$

As $\sin(\pi n) = 0$ for all n we get

$$a_n = 2\left(0 - \left[-\frac{\cos(2\pi nt)}{4\pi^2 n^2}\right]_0^{\frac{1}{2}}\right) = 2\left(\frac{\cos(\pi n)}{4\pi^2 n^2} - \frac{1}{4\pi^2 n^2}\right)$$

$\cos(\pi n)$ is -1 when n is odd and 1 when n is even. This means that $\cos(\pi n) = (-1)^n$. Then

$$a_n = 2\frac{((-1)^n - 1)}{4\pi^2 n^2} = \frac{(-1)^n - 1}{2\pi^2 n^2}$$

Next:

$$b_n = \frac{2}{\tau}\int_{-\frac{1}{2}}^{\frac{1}{2}} f(t)\sin(2\pi nt)\,dt = 2\int_0^{\frac{1}{2}} t\sin(2\pi nt)\,dt$$

Integrating by parts, using the formula $\int u\,dv = uv - \int v\,du$, and choosing $u = t$, $dv = \sin(2\pi nt\,dt)$, $du = dt$ and $v = -\cos(2\pi nt)/2\pi n$, we obtain

$$b_n = 2\left(\left[t\left(-\frac{\cos(2\pi nt)}{2\pi n}\right)\right]_0^{\frac{1}{2}} + \int_0^{\frac{1}{2}}\frac{\cos(2\pi nt)}{2\pi n}\,dt\right)$$

$$= 2\left(-\frac{\cos(\pi n)}{4\pi n} + \left[\frac{\sin(2\pi nt)}{4\pi^2 n^2}\right]_0^{\frac{1}{2}}\right)$$

$$= 2\left(-\frac{(-1)^n}{4\pi n}\right) = -\frac{(-1)^n}{2\pi n}$$

Step 4
The Fourier series for $f(t)$ is

$$f(t) = \tfrac{1}{2}a_0 + \sum_{n=1}^{\infty} a_n\cos(n\omega_0 t) + b_n\sin(n\omega_0 t)$$

in this case giving

$$f(t) = \tfrac{1}{8} + \sum_{n=1}^{\infty}\frac{(-1)^n - 1}{2\pi^2 n^2}\cos(2\pi nt) + \sum_{n=1}^{\infty}\frac{(-1)^{n+1}}{2\pi n}\sin(2\pi nt)$$

Signal bias: the d.c. (direct current) component

The term $\tfrac{1}{2}a_0$ is called the bias term, or the d.c. component (a name adopted from electronic signals), as it corresponds to the average value of the function over a single period.

22.4 FOURIER SERIES OF SYMMETRIC PERIODIC FUNCTIONS

We looked at even functions and odd functions in Chapter 8.

Even functions

We find that even functions, which have the property that $f(-t) = f(t)$, have all $b_n = 0$ in their Fourier series. They are represented by cosine terms only. This is not surprising, as the cosine is an even function and the sine function is odd. We would expect that an even function would be expressed in terms of other even functions. Another simplification in this case is that

$$a_n = \frac{2}{\tau} \int_{-\frac{\tau}{2}}^{\frac{\tau}{2}} f(t) \cos\left(n\omega_0 t\right) dt = \frac{2}{\tau} \int_{-\frac{\tau}{2}}^{0} f(t) \cos\left(n\omega_0 t\right) dt + \frac{2}{\tau} \int_{0}^{\frac{\tau}{2}} f(t) \cos\left(n\omega_0 t\right) dt$$

$$= \frac{4}{\tau} \int_{0}^{\frac{\tau}{2}} f(t) \cos\left(n\omega_0 t\right) dt$$

and it is therefore only necessary to integrate over a half cycle. To summarize, for an even function

$$a_0 = \frac{4}{\tau} \int_{0}^{\frac{\tau}{2}} f(t) \, dt$$

$$a_n = \frac{4}{\tau} \int_{0}^{\frac{\tau}{2}} f(t) \cos\left(n\omega_0 t\right) dt$$

$$b_n = 0 \qquad \text{all } n$$

Odd functions

Odd functions, where $f(-t) = -f(t)$ have all $a_n = 0$ and only have sine terms in their Fourier series. We need also only to consider the half cycle, because

$$b_n = \frac{2}{\tau} \int_{-\frac{\tau}{2}}^{\frac{\tau}{2}} f(t) \sin\left(n\omega_0 t\right) = \frac{2}{\tau} \int_{-\frac{\tau}{2}}^{0} f(t) \sin\left(n\omega_0 t\right) dt + \frac{2}{\tau} \int_{0}^{\frac{\tau}{2}} f(t) \sin\left(n\omega_0 t\right) dt$$

$$= \frac{4}{\tau} \int_{0}^{\frac{\tau}{2}} f(t) \sin\left(n\omega_0 t\right) dt$$

To summarize, for an odd function

$$a_n = 0 \qquad \text{all } n$$

$$b_n = \frac{4}{\tau} \int_{0}^{\frac{\tau}{2}} f(t) \sin\left(n\omega_0 t\right) dt$$

Half-wave symmetry

There is another sort of symmetry that has an important effect on the Fourier series representation. This is called half-wave symmetry. A function with half-wave symmetry obeys $f(t + \tau/2) = -f(t)$. that is the graph of the function in the second half of the period is the same as the graph of the function in the first half turned upside down. A function with half-wave symmetry has no even harmonics. This can be shown by considering one of the even terms where $n = 2m$. Then

$$b_{2m} = \frac{2}{\tau} \int_{-\frac{\tau}{2}}^{\frac{\tau}{2}} f(t) \sin\left(2m\omega_0 t\right) dt = \frac{2}{\tau} \int_{-\frac{\tau}{2}}^{0} f(t) \sin\left(2m\omega_0 t\right) dt + \frac{2}{\tau} \int_{0}^{\frac{\tau}{2}} f(t) \sin\left(2m\omega_0 t\right) dt$$

Substitute $t' = t - \tau/2$ in the second term, so that $dt' = dt$, giving

$$\frac{2}{\tau} \int_{-\frac{\tau}{2}}^{0} f\left(t' + \frac{\tau}{2}\right) \sin\left(2m\omega_0\left(t' + \frac{\tau}{2}\right)\right) dt'$$

As $\tau = 2\pi/\omega_0$,

$$\sin\left(2m\omega_0\left(t' + \frac{\tau}{2}\right)\right) = \sin\left(2m\omega_0 t' + 2m\frac{\omega_0 2\pi}{\omega_0 2}\right)$$

$$= \sin(2m\omega_0 t' + 2m\pi) = \sin(2m\omega_0 t')$$

As $f(t' + \tau/2) = -f(t')$, the second term in b_{2m} becomes

$$\frac{2}{\tau}\int_{-\frac{\tau}{2}}^{0} -f(t')\sin(2m\omega_0 t')\,dt'$$

which cancels the first term, giving $b_{2m} = 0$.

A similar argument shows that the coefficients of the cosine terms for even n are also zero. In this case also it is only necessary to consider the half-cycle, as

$$\frac{2}{\tau}\int_{-\frac{\tau}{2}}^{\frac{\tau}{2}} f(t)\sin(n\omega_0 t)\,dt = \frac{4}{\tau}\int_{0}^{\frac{\tau}{2}} f(t)\sin(n\omega_0 t)\,dt \qquad n \text{ odd}$$

To summarize, for a function with half-wave symmetry

$$a_n = \frac{4}{\tau}\int_{0}^{\frac{\tau}{2}} f(t)\cos(n\omega_0 t)\,dt \qquad n \text{ odd}$$

$$b_n = \frac{4}{\tau}\int_{0}^{\frac{\tau}{2}} f(t)\sin(n\omega_0 t)\,dt \qquad n \text{ odd}$$

$$a_n = b_n = 0 \qquad n \text{ even}$$

An even function, an odd function and a function with half-wave symmetry are shown in Fig. 22.5.

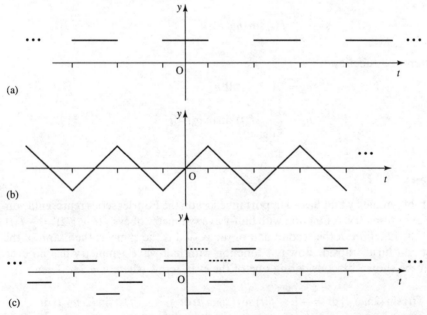

Figure 22.5 (a) An even function satisfies $f(-t) = f(t)$, that is reflecting the graph in the y-axis results in the same graph. This function has only cosine terms in its Fourier series. (b) An odd function satisfies $f(-t) = -f(t)$, that is reflecting the graph in the y-axis results in an upside down version of the same graph. This function has only sine terms in its Fourier series. (c) A function with half-wave symmetry satisfies $f(t + (\tau/2)) = -f(t)$ that is the graph of the function in the second half of the period is the same as the graph of the function in the first half reflected in the x-axis. This function has no even harmonics.

22.5 AMPLITUDE AND PHASE REPRESENTATION OF A FOURIER SERIES

In Chapter 15, when considering using vectors to represent single-frequency waves, we saw that terms like $c \cos(\omega t) - d \sin(\omega t)$ can be represented by a single cosine term $A \cos(\omega t + \phi)$, where A is the amplitude and ϕ is the phase. The terms A and ϕ can be found by expressing the vector (c, d) in polar form. We can employ this idea to represent the Fourier series in amplitude and phase form. This can be very useful because, for instance, a filter may be designed to attenuate frequencies outside of the desired pass band. This requirement species its amplitude characteristics. The phase characteristics may then be considered separately.

The Fourier series becomes

$$f(t) = \tfrac{1}{2}a_0 + \sum_{n=1}^{\infty} c_n \cos(n\omega_0 t + \phi_n)$$

where

$$c_n \cos(n\omega_0 t + \phi_n) = a_n \cos(n\omega_0 t) + b_n \sin(n\omega_0 t)$$

From the trigonometric identity for $\cos(A + B)$ we can expand the left-hand side of the above expression to get

$$c_n \cos(n\omega_0 t) \cos(\phi_n) - c_n \sin(n\omega_0 t) \sin(\phi_n) = a_n \cos(n\omega_0 t) + b_n \sin(n\omega_0 t)$$

Equating terms in $\cos(n\omega_0 t)$ and $\sin(n\omega_0 t)$ we get

$$c_n \cos(\phi_n) = a_n$$

$$-c_n \sin(\phi_n) = b_n$$

giving

$$c_n = \sqrt{a_n^2 + b_n^2} \quad \text{and} \quad \phi_n = -\tan^{-1}\left(\frac{b_n}{a_n}\right) \qquad (\pm \pi \text{ if } a_n \text{ is negative})$$

We see that $c_n \underline{/\phi_n}$ can be found by expressing $(a_n, -b_n)$ in polar form.

The amplitudes can be plotted against the frequency f (or angular frequency ω), giving the amplitude spectrum, and the phases can be plotted, giving the phase spectrum. The amplitude gives information about the distribution of energy among the different frequencies.

For the unbiased square wave which we considered in Example 22.1 we found

$$f(t) = \frac{1}{\pi} \sum_{n=1}^{\infty} \frac{1 - (-1)^n}{n} \sin(n\pi t) = \frac{1}{\pi} \sum_{m=1}^{\infty} \frac{2}{2m - 1} \sin((2m - 1)\pi t)$$

$$a_m = 0$$

$$b_m = \frac{2}{(2m - 1)\pi}$$

$$c_m = \sqrt{a_m^2 + b_m^2} = \frac{2}{\pi(2m - 1)}$$

As there is only a sine term, the phase is given by $-\pi/2$. So we get

$$f(t) = \sum_{m=1}^{\infty} \frac{2}{\pi(2m - 1)} \cos\left((2m - 1)\pi t - \frac{\pi}{2}\right)$$

where $n = 2m - 1$.

Plotting these amplitudes and phases we get Fig. 22.6. The phases are not defined for frequencies with zero amplitude, but we can consider them as 0.

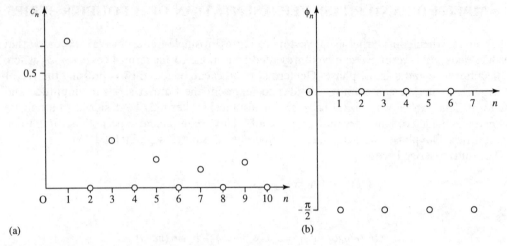

(a)

(b)

Figure 22.6 (a) The amplitude and (b) phase spectra of the square wave given in Fig. 22.1.

22.6 FOURIER SERIES IN COMPLEX FORM

The most useful form of the Fourier series is the amplitude and phase form of the previous section. c_n and ϕ_n can often be calculated more simply by considering the complex form of the Fourier series. We can find this by taking the expression from the previous section:

$$f(t) = \tfrac{1}{2}a_0 + \sum_{n=1}^{\infty} c_n \cos(n\omega_0 t + \phi_n)$$

and expressing the cosine terms as the sums of complex exponentials:

$$c_n \cos(n\omega_0 t + \phi_n) = \frac{c_n}{2}\left(e^{j(n\omega_0 t + \phi_n)} + e^{-j(n\omega_0 t + \phi_n)}\right)$$

$$= \frac{c_n}{2}e^{j\phi_n}e^{jn\omega_0 t} + \frac{c_n}{2}e^{-j\phi_n}e^{-jn\omega_0 t}$$

Then setting

$$\alpha_n = \frac{c_n}{2}e^{j\phi_n} \quad \text{and} \quad \alpha_{-n} = \frac{c_n}{2}e^{-j\phi_n}$$

we get

$$f(t) = \tfrac{1}{2}a_0 + \sum_{n=1}^{\infty} \alpha_n e^{jn\omega_0 t} + \alpha_{-n}e^{-jn\omega_0 t}$$

$$\Rightarrow f(t) = \sum_{n=-\infty}^{n=\infty} \alpha_n e^{jn\omega_0 t}$$

where $\alpha_0 = \tfrac{1}{2}a_0$,

$$\alpha_n = \frac{c_n}{2}e^{j\phi_n} \quad \text{and} \quad \alpha_{-n} = \frac{c_n}{2}e^{-j\phi_n}$$

As c_n is real, $|\alpha_n| = |\alpha_{-n}| = c_n/2$, and $\alpha_n^* = \alpha_{-n}$.

Therefore α_n ($n > 0$) is a complex coefficient of the Fourier series with amplitude $c_n/2$ and phase ϕ_n.

The complex form is generally the most convenient form of the Fourier series because in the expression

$$f(t) = \sum_{n=-\infty}^{\infty} \alpha_n e^{jn\omega_0 t}$$

the complex Fourier components can be found from

$$\alpha_n = \frac{1}{\tau} \int_{-\frac{\tau}{2}}^{\frac{\tau}{2}} f(t) e^{-jn\omega_0 t}\, dt$$

and hence involves performing only a single integration.

This form of the Fourier series gives apparent negative frequencies, but for any real function of time the coefficients of negative frequencies have equal amplitude to the equivalent positive frequencies and negative phase. Thus only the positive frequency coefficients need be given to totally specify the function.

From the complex form we can easily find the amplitude and phase spectra. If we write the coefficients α_n in exponential form, $|\alpha_n|e^{j\theta_n}$, we find $c_n = 2|\alpha_n|$ and $\theta_n = \phi_n$, $n \geqslant 0$.

We can also easily find the sine and cosine form of the Fourier series by using

$$a_n = 2|\alpha_n| \cos(\phi_n) = 2\, \mathrm{Re}(\alpha_n)$$

$$b_n = -2|\alpha_n| \sin(\phi_n) = -2\, \mathrm{Im}(\alpha_n)$$

Example 22.3 Find the complex Fourier series for the function defined in the interval $0 < t < 1$ by

$$f(t) = \begin{cases} t & \text{for } 0 < t < \frac{1}{2} \\ 0 & \text{for } \frac{1}{2} < t < 1 \end{cases}$$

SOLUTION

Step 1
We have already plotted the graph as shown in Fig. 22.4.

Step 2
The fundamental period of this is $\tau = 1$ so that $\omega_0 = 2\pi$.

Step 3
Calculate α_n. We find

$$\alpha_n = \frac{1}{\tau} \int_{-\frac{\tau}{2}}^{\frac{\tau}{2}} f(t) e^{-jn\omega_0 t}\, dt = \int_0^{\frac{1}{2}} t e^{-j2\pi n t}\, dt$$

To find this integral we perform integration by parts, using $\int u\, dv = uv - \int v\, du$. In this case we choose $u = t$, $dv = e^{-j2\pi n t}\, dt$, $du = dt$ and

$$v = \frac{e^{-j2\pi n t}}{-j2\pi n} = j\frac{e^{-j2\pi n t}}{2\pi n}$$

giving

$$\alpha_n = \int_0^{\frac{1}{2}} te^{-j2\pi nt} = \left[j\frac{te^{-j2\pi nt}}{2\pi n} \right]_0^{\frac{1}{2}} - \int_0^{\frac{1}{2}} \frac{je^{-j2\pi nt}}{2\pi n} \, dt$$

$$= j\frac{e^{-j\pi n}}{4\pi n} - \left[j\frac{e^{-j2\pi nt}}{(2\pi n)(-j2\pi n)} \right]_0^{\frac{1}{2}} = \frac{je^{-j\pi n}}{4\pi n} + \frac{e^{-j\pi n} - 1}{4\pi^2 n^2}$$

Since $e^{-j\pi n} = (-1)^n$, this means

$$\alpha_n = \frac{(-1)^n - 1}{4\pi^2 n^2} + j\frac{(-1)^n}{4\pi n}$$

Step 4

The complex Fourier series for $f(t)$ is

$$f(t) = \sum_{n=-\infty}^{n=\infty} \alpha_n e^{jn\omega t}$$

in this case giving

$$f(t) = \sum_{n=-\infty}^{\infty} \left(\frac{(-1)^n - 1}{4\pi^2 n^2} + j\frac{(-1)^n}{4\pi n} \right) e^{-jn\omega t}$$

To find the trigonometric form from this we can use

$$a_n = 2\,\mathrm{Re}(\alpha_n) = \frac{(-1)^n - 1}{2\pi^2 n^2}$$

$$b_n = -2\,\mathrm{Im}(\alpha_n) = \frac{-(-1)^n}{2\pi_n}$$

which agrees with our previous result found in Example 22.2.

22.7 SUMMARY

1. If $f(t)$ is periodic with period τ, $\omega_0 = 2\pi/\tau$ and the Fourier series for f is given by

$$f(t) = \tfrac{1}{2}a_0 + \sum_{n=1}^{\infty} a_n \cos(n\omega_0 t) + b_n \sin(n\omega_0 t)$$

where the coefficients are given by

$$a_0 = \frac{2}{\tau} \int_{-\frac{\tau}{2}}^{\frac{\tau}{2}} f(t)\, dt$$

$$a_n = \frac{2}{\tau} \int_{-\frac{\tau}{2}}^{\frac{\tau}{2}} f(t) \cos(n\omega_0 t)\, dt$$

$$b_n = \frac{2}{\tau} \int_{-\frac{\tau}{2}}^{\frac{\tau}{2}} f(t) \sin(n\omega_0 t)\, dt$$

This is the sine and cosine form (trigonometric form) of the Fourier series.

2. An even function, such that $f(-t) = f(t)$, has $b_n = 0$, and a_n can be found by integrating over half the cycle. An odd function, such that $f(-t) = -f(t)$, has $a_n = 0$, and the terms b_n can

be found by integrating over the half-cycle. A function with half-wave symmetry, $f(t + \tau/2) = -f(t)$ has no even terms in the expansion. The terms a_n and b_n can be found by integrating over a half-cycle only.

3. The amplitude and phase form of the Fourier series is

$$f(t) = \tfrac{1}{2}a_0 + \sum_{n=1}^{\infty} c_n \cos(n\omega_0 t + \phi_n)$$

$c_n \,\underline{/\phi_n}$ can be found by expressing $(a_n, -b_n)$ (as defined above) in polar form: $a_n = c_n \cos(\phi_n)$, $b_n = -c_n \sin(\phi_n)$.

4. The complex form of the Fourier series is given by

$$f(t) = \sum_{n=-\infty}^{n=\infty} \alpha_n e^{jn\omega_0 t}$$

where

$$\alpha_n = \frac{1}{\tau}\int_{-\frac{\tau}{2}}^{\frac{\tau}{2}} f(t)e^{-jn\omega_0 t}\, dt$$

The coefficients α_n are related to a_n and b_n by

$$a_n = 2\,\mathrm{Re}(\alpha_n),\ b_n = -2\,\mathrm{Im}(\alpha_n),\ \alpha_n = \tfrac{1}{2}(a_n - jb_n)$$

and to the amplitude and phase form by $c_n = 2|\alpha_n|$ and $\phi_n = \arg(\alpha_n)$ $(n \geqslant 0)$.

22.8 EXERCISES

For the periodic functions in Exercises 22.1–22.7, find the Fourier series in trigonometric and complex form:

$$f(t) = \tfrac{1}{2}a_0 + \sum_{n=1}^{\infty} a_n \cos(n\omega_0 t) + b_n \sin(n\omega_0 t)$$

and

$$f(t) = \sum_{n=-\infty}^{\infty} \alpha_n e^{jn\omega_0 t}$$

and check that the results are equivalent in each case.

22.1 $f(t) = \begin{cases} 1 & 0 < t < 1 \\ 0 & 1 < t < 4 \end{cases}$

22.2 $f(t) = \begin{cases} t^2 & 0 < t < 1 \\ 0 & 1 < t < 2 \end{cases}$

22.3 $f(t) = \cos^2(t)$

22.4 $f(t) = \begin{cases} t & 0 < t < 1 \\ 0 & 1 < t < 2 \end{cases}$

22.5 $f(t) = \begin{cases} t^2 & 0 < t < 1 \\ 1 - (t - 1)^2 & 1 < t < 2 \end{cases}$

22.6 $f(t) = \begin{cases} \cos(2\pi t) & 0 < t < 1 \\ 0 & 1 < t < 2 \end{cases}$

22.7 $f(t) = \begin{cases} t - 0.5 & 0 < t < 1 \\ 1.5 - t & 1 < t < 2 \end{cases}$

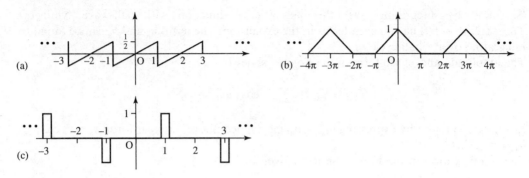

Figure 22.7 Graphs for Exercise 22.8.

22.8 Which of the periodic functions whose graphs are shown in Fig. 22.7 are even, odd or have half-wave symmetry? What consequences will such properties have for the Fourier series in trigonometric form?

22.9 Find the fundamental period and give the amplitude and phase spectra of the Fourier series representation of the following periodic functions:

(a) $f(t) = \sin(3t) - \sin(t)$

(b) $g(t) = \frac{1}{2}\cos(2t) + 2 - \frac{1}{2}\sin(2t)$

(c) $f(t) = \sum_{n=-\infty}^{\infty} \frac{1}{n^2} e^{jnt} \qquad n \neq 0$

22.10 Find and plot the amplitude and phase spectra of your results from Exercises 22.1, 22.3 and 22.7.

22.11 Consider the functions in Exercises 22.1, 22.2 and 22.4 and show that at the position of the jump discontinuities, $t = t_d$, the Fourier series converges to the value

$$\frac{f(t_d^+) + f(t_d^-)}{2}$$

where

$$f(t_d^+) = \lim_{t \to t_d^+} f(t)$$

and

$$f(t_d^-) = \lim_{t \to t_d^-} f(t)$$

You may assume that:

$$\sum_{n=1}^{\infty} \frac{(-1)^{n-1}}{2n-1} = \frac{\pi}{4}$$

$$\sum_{n=1}^{\infty} \frac{1}{n^2} = \frac{\pi^2}{6}$$

and

$$\sum_{n=1}^{\infty} \frac{(-1)^{n-1}}{n^2} = \frac{\pi^2}{12}$$

22.12 The steady state response to a signal $e^{j\omega t}$ is given by $H(j\omega)e^{j\omega t}$. Find the steady state response to the function $f(t) = \cos^2(t)$ from a system with transfer function $1/(s+2)$.

GRAPH THEORY

23.1 INTRODUCTION

A graph consists of points, called vertices, and lines connecting them, called edges. They can be used to represent many diverse situations, e.g. five cities and five roads connected as in Fig. 23.1. The same graph could represent some people and the edge connections could represent those people who do business with each other.

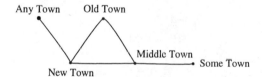

Figure 23.1 Five cities with five roads connecting them represented by a graph.

Sometimes the relationship is 'one-way', for instance in the case of a network of one-way streets, or a graph representing a circuit where the arrows on the edges represent the current flow. In this case the graph is called a directed graph, shortened to 'digraph'.

We will look at a few definitions and applications of different types of graph and their matrix representations (which we also looked at in Chapter 19). We also look at applications to the solving of routing problems through networks.

23.2 DEFINITIONS

Graph

A graph G consists of a finite set of vertices $V(G)$, which cannot be empty and a finite set of edges, $E(G)$, which connect pairs of vertices. The number of vertices of G is called the order of G. The number of edges in G is represented by $|E|$ and the number of vertices by $|V|$. The graph in Fig. 23.2 has $|V| = 6$ and $|E| = 9$.

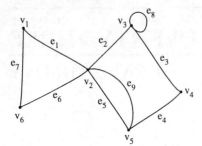

Figure 23.2 A graph with six vertices and nine edges.

Incidence, adjacency and neighbours

Two vertices are adjacent if they are joined by an edge. In Fig. 23.2, v_3 is adjacent to v_4 and is also said to be a neighbour of v_4. The edge which joins the vertices is said to be incident to them. That is, e_3 is incident to v_3 and v_4.

Multiple edges, loops and simple graphs

Two or more edges joining the same pair of vertices are multiple edges. An edge joining a vertex to itself is called a loop. In Fig. 23.2, edges e_5 and e_9 are multiple edges and e_8 is a loop. A graph containing no multiple edges or loops is called a simple graph. There is an example of a simple graph in Fig. 23.3.

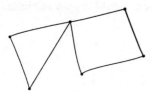

Figure 23.3 A simple graph.

Isomorphism

Graphs are isomorphic to each other if one can be obtained from a redrawing of the other one. Some isomorphic graphs are shown in Fig. 23.4.

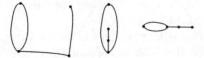

Figure 23.4 Some isomorphic graphs.

Weighted graph

A weighted graph has a number assigned to each of its edges, called its weight. The weight can be used to represent distances or capacities. A weighted graph is shown in Fig. 23.5.

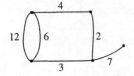

Figure 23.5 A weighted graph.

Digraphs

A digraph is a directed graph, i.e. instead of edges in the definition of graphs we have arcs joining pairs of vertices in a specified order. A digraph is shown in Fig. 23.6(a).

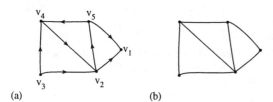

(a)   (b)   **Figure 23.6** (a) A digraph and (b) its underlying graph.

Underlying graph

The underlying graph of a digraph D is the graph obtained by replacing each arc by an (undirected) edge, as in Fig. 23.6(b).

Degree

The number of times edges are incident to a vertex v is called its degree, denoted by $d(v)$. In Fig. 23.2 vertex v_5 has degree 3 and vertex v_3 has degree 4.

A vertex of a digraph has an in-degree $d_-(v)$ and an out-degree, $d_+(v)$. In Fig. 23.6(a), v_5 has in-degree of 2 and out-degree of 1. Vertex v_1 has in-degree 1 and out-degree 1.

The sum of the values of the degrees, $d(v)$, over all the vertices of a graph totals twice the number of edges:

$$\sum_i d(v_i) = 2|E|$$

where $|E|$ is the number of edges.

Checking this result for Fig. 23.2 we find that the degrees are $d(v_1) = 2$, $d(v_2) = 5$, $d(v_3) = 4$, $d(v_4) = 2$, $d(v_5) = 3$ and $d(v_6) = 2$. Summing these gives 18, which is the same as twice the number of edges.

For a digraph we get

$$\sum_i d_-(v_i) = |A|$$

and

$$\sum_i d_+(v_i) = |A|$$

where $|A|$ is the number of arcs.

Subgraph

A subgraph of G is a graph, H, whose vertex set is a subset of G's vertex set, and similarly its edge set is a subset of the edge set of G.

$$V(H) \subseteq V(G)$$

$$E(H) \subseteq E(G)$$

If it spans all of the vertices of G, i.e. $V(H) = V(G)$, then H is called a spanning subgraph of G. A graph G, a subgraph and a spanning subgraph are shown in Fig. 23.7.

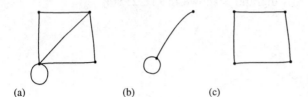

(a) (b) (c)

Figure 23.7 (a) A graph G. (b) A subgraph of G. (c) A spanning subgraph of G.

Complete graph

A simple graph in which every pair of two vertices is adjacent is called a complete graph. K_n is the complete graph with n vertices. K_n has $\frac{1}{2}n(n-1)$ edges. Some examples of complete graphs are shown in Fig. 23.8.

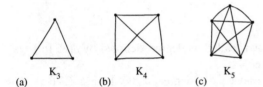

(a) K_3 (b) K_4 (c) K_5

Figure 23.8 Complete graphs: (a) K_3; (b) K_4; (c) K_5.

Bipartite graph

This is a graph whose vertex set can be partitioned into two sets in such a way that each edge joins a vertex of the first set to a vertex of the second set. Examples are given in Fig. 23.9.

A complete bipartite graph is a bipartite simple graph in which every vertex in the first set is adjacent to every vertex in the second set. $K_{m,n}$ is the complete bipartite graph with m vertices in the first set and n vertices in the second set. Some examples of complete bipartite graphs are given in Fig. 23.9(b) and (c).

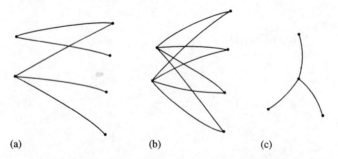

(a) (b) (c)

Figure 23.9 Bipartite graphs. (b) and (c) are the complete bipartite graphs $K_{2,4}$ and $K_{3,1}$.

Walks, paths and circuits

A sequence of edges of the form

$$v_s v_i, \ v_i v_j, \ v_j v_k, \ v_k v_l, \ v_l v_t$$

is a walk of from v_s to v_t. If these edges are all distinct then the walk is called a trail, and if the vertices are also distinct then the walk is called a path. A walk or trail is closed if $v_s = v_t$. A closed walk in which all the vertices are distinct except v_s and v_t is called a cycle or circuit. Examples are given in Fig. 23.10.

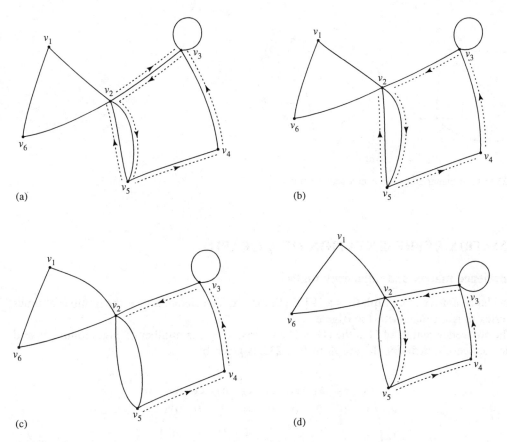

Figure 23.10 (a) A graph G with a walk marked. (b) A graph G with a trail marked. (c) A graph G with a path marked. (d) A graph G with a circuit marked.

Connected graph

A graph G is connected if there is a path from any one of its vertices to any other vertex. A disconnected graph is said to be made up of components. All the graphs we have drawn so far have been connected. In Fig. 23.11(a) there is a disconnected graph with two components and in Fig. 23.11(b) a disconnected graph with three components.

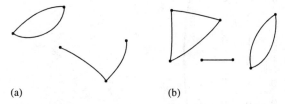

Figure 23.11 (a) A disconnected graph with two components. (b) A disconnected graph with three components.

Planar graphs

A planar graph is one which can be drawn in the plane without any edges intersecting except at vertices to which they are both incident. There is an example of a planar graph in Fig. 23.12(a) and a non-planar graph in Fig. 23.12(b).

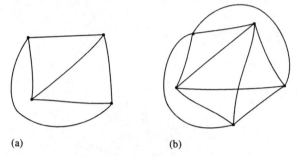

(a)   (b)

Figure 23.12 (a) A planar graph. (b) A non-planar graph.

23.3 MATRIX REPRESENTATION OF A GRAPH

The incidence matrix and adjacency matrix

The incidence matrix of a graph G is a $|V| \times |E|$ matrix. The element $a_{ij} =$ the number of times that vertex v_i is incident with the edge e_j.

The adjacency matrix of G is the $|V| \times |V|$ matrix. $a_{ij} =$ the number of edges joining v_i and v_j. The incidence matrix for the graph in Fig. 23.2 is given by

$$
\begin{array}{c c}
& \begin{array}{ccccccccc} e_1 & e_2 & e_3 & e_4 & e_5 & e_6 & e_7 & e_8 & e_9 \end{array} \\
\begin{array}{c} v_1 \\ v_2 \\ v_3 \\ v_4 \\ v_5 \\ v_6 \end{array} &
\left(\begin{array}{ccccccccc}
1 & 0 & 0 & 0 & 0 & 0 & 1 & 0 & 0 \\
1 & 1 & 0 & 0 & 1 & 1 & 0 & 0 & 1 \\
0 & 1 & 1 & 0 & 0 & 0 & 0 & 2 & 0 \\
0 & 0 & 1 & 1 & 0 & 0 & 0 & 0 & 0 \\
0 & 0 & 0 & 1 & 1 & 0 & 0 & 0 & 1 \\
0 & 0 & 0 & 0 & 0 & 1 & 1 & 0 & 0
\end{array}\right)
\end{array}
$$

and the adjacency matrix by

$$
\begin{array}{c c}
& \begin{array}{cccccc} v_1 & v_2 & v_3 & v_4 & v_5 & v_6 \end{array} \\
\begin{array}{c} v_1 \\ v_2 \\ v_3 \\ v_4 \\ v_5 \\ v_6 \end{array} &
\left(\begin{array}{cccccc}
0 & 1 & 0 & 0 & 0 & 1 \\
1 & 0 & 1 & 0 & 2 & 1 \\
0 & 1 & 1 & 1 & 0 & 0 \\
0 & 0 & 1 & 0 & 1 & 0 \\
0 & 2 & 0 & 1 & 0 & 0 \\
1 & 1 & 0 & 0 & 0 & 0
\end{array}\right)
\end{array}
$$

23.4 TREES

A tree is a connected graph with no cycles. A forest is a graph with no cycles, but which may or may not be connected (i.e. a forest is a graph whose components are trees). In Fig. 23.13(a) there is a tree and in Fig. 23.13(b) there is a forest.

If T is a tree with at least two vertices, then it has the three properties (T1), (T2) and (T3):

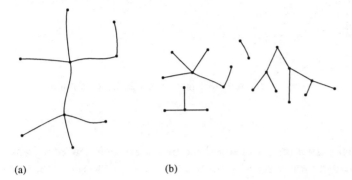

(a)   (b)

Figure 23.13 (a) A tree is a connected graph with no cycles. (b) A forest has no cycles, but may or may not be connected.

(T1) There is exactly one path from any vertex v_i in T to any other vertex v_j

(T2) The graph obtained from T by removing any edge has two components, each of which is a tree.

(T3) $|E| = |V| - 1$

Trees have many applications, particularly rooted trees. Decision trees are used to represent the possible decisions at each stage of a problem or algorithm. Probability trees can be used to analyse conditional probabilities, which we shall see in the next chapter. Another application is to the parsing of a sentence. The tree in Fig. 23.14 represents the sentence 'Fortune favours the brave'. The vertices, other than the terminal vertices, represent grammatical categories and the terminal vertices represent the words of the sentence. The same sort of idea can be used to analyse allowed constructs of statements in programming languages and the syntax of arithmetic expressions.

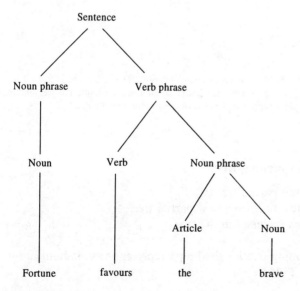

Figure 23.14 A parsing tree for the sentence 'Fortune favours the brave'.

Spanning trees

A spanning tree of a graph G is a tree T which is a spanning subgraph of G. That is, T has the same vertex set as G. Examples of graphs with spanning trees marked are given in Fig. 23.15.

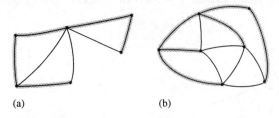

(a)   (b)   **Figure 23.15** Graphs with spanning trees shaded.

How to grow a spanning tree

Take any vertex v of G as an initial partial tree. Add edges one by one so each new edge joins a new vertex to the partial tree. If there are n vertices in the graph G then the spanning tree will have n vertices and $n - 1$ edges.

Minimum spanning tree

Supposing we have a group of offices which need to be connected by a network of communication lines. The offices may communicate with each other directly or through another office. In order to decide on which offices to build links between we firstly work out the cost of all possible connections. This will then give us a weighted complete graph, as shown in Fig. 23.16. The minimum spanning tree is then the spanning tree that represents the minimum cost.

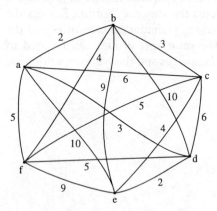

Figure 23.16 A weighted complete graph. The vertices represent offices and the edges represent possible communication links. The weights on the edges represent the cost of construction of the link.

The greedy algorithm for the minimum spanning tree

1. Choose any start vertex to form the initial partial tree (v_i).
2. Add the cheapest edge, e_i, to a new vertex, to form a new partial tree.
3. Repeat Step 2 until all vertices have been included in the tree.

Example 23.1 Find the minimum spanning tree for the graph representing communication links between offices as shown in Fig. 23.16.

SOLUTION We start with any vertex and choose the one marked a. Add the edge ab, which is the cheapest edge of those incident to a. Add a new edge in order to form a partial tree and choose bc, which is one of the cheapest remaining edges incident with either a or b. Now we add edge ad, which is the cheapest remaining edge of those incident with a or b or c which forms a partial tree. Continuing in this manner we find the minimum spanning tree, as shown in Fig. 23.17. The total cost of the communication links in our solution is found to be $2 + 3 + 3 + 2 + 4 = 14$.

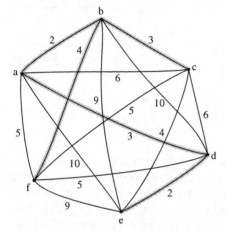

Figure 23.17 The graph of Fig. 23.16 with its minimum spanning tree marked.

23.5 THE SHORTEST PATH PROBLEM

The weights on a graph may represent delays on a communication network or travel times along roads. A practical problem that we may wish to solve is to find the shortest path from any two vertices. The algorithm is illustrated in Example 23.2.

Example 23.2 The weighted graph in Fig. 23.18 represents a communication network with weights indicating the delays associated with each edge. Find the minimum delay path from s to t.

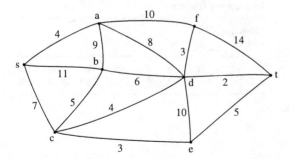

Figure 23.18 The graph representing a communications network for Example 23.2.

SOLUTION
Stage 1
To find the path we begin at the start vertex s. s is the reference vertex for stage 1. Label all the adjacent vertices with the lengths of the paths using only one edge. Mark all the other

vertices with a very high number (bigger than the sum of all the weights in the graph). In this case we choose 100. This is shown in Fig. 23.19. At the same time start to fill in a table, as in Table 23.1.

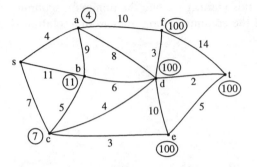

Table 23.1 The lengths of paths using one edge from s

Reference vertex	a	b	c	d	e	f	t
s	4	11	7	100	100	100	100

Figure 23.19 Stage 1 of solving the shortest path problem of Example 23.2.

Stage 2

Choose the vertex with the smallest label that has not already been a reference vertex, so that this becomes the new reference vertex. In this case we choose vertex a. Once a vertex has been used as a reference vertex it gains a permanent label and for this reason we circle the label in bold, as in Fig. 23.20 and circle the label on the table, as in Table 23.2. This label represents the length of the shortest path from s to the new reference vertex. Consider any vertex adjacent to the new reference vertex and calculate the length of the path via a to this vertex. If this is less than the current temporary label on the vertex then change the label on the vertex. For instance, b is adjacent to vertex a. The length of the path to b via a is $4 + 9 = 13$. This is not less than the current label of 11, so we keep the current label. However, looking at vertex d, which is adjacent to a, the length of the path via a is $4 + 8 = 12$, which is shorter than the current label of d; hence we change the current label. Considering also the vertices b and f, we end up at the end of Stage 2 with the labels as given in Fig. 23.20 and the table as shown in Table 23.2.

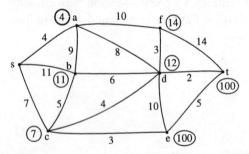

Table 23.2 The lengths of paths using up to two edges from s

Reference vertex	a	b	c	d	e	f	t
s	④	11	7	100	100	100	100
a		11	7	12	100	14	100

Figure 23.20 Stage 2 of solving the shortest path problem. Vertex a now has a permanent label, shown in bold.

Stage 3

Choose the vertex with the smallest label that has not already been a reference vertex; this becomes the new reference vertex. In this case we choose vertex c, so this gains a permanent label which we circle in bold as in Fig. 23.21; we also circle the label on the table, as in Table 23.3. This label represents the length of the shortest path from s to c. Consider any vertex adjacent to c which does not have a permanent label, and calculate the length of the

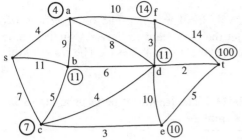

Figure 23.21 Stage 3 of solving the shortest path problem.

Table 23.3 The lengths of paths using up to three edges from s

Reference vertex	a	b	c	d	e	f	t
s	④	11	7	100	100	100	100
a		11	⑦	12	100	14	100
c		11		11	10	14	100

path via c to this vertex. If this is less than the current temporary label on the vertex then change the label on the vertex. For instance d is adjacent to vertex c. The length of the path to d via c is $7 + 4 = 11$. This is less than the current label of 12, so we change the label. However, looking at vertex b, which is adjacent to c, the length of the path via c is $7 + 5 = 12$, which is longer than the current label of 11, so we keep the current label. Considering also the vertex e we end up, at the end of Stage 3, with the labels as given in Fig. 23.21 and the table as shown in Table 23.3.

Stage 4
Choose e as the new reference vertex and circle it in bold. Compare paths via e to the labels on any adjacent vertices and relabel the vertices if the paths via e are found to be shorter. The result of this stage is shown in Fig. 23.22 and Table 23.4.

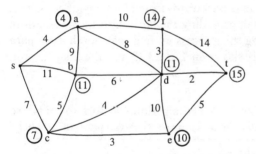

Figure 23.22 Stage 4 of solving the shortest path problem.

Table 23.4 The lengths of paths using up to four edges from s

Reference vertex	a	b	c	d	e	f	t
s	④	11	7	100	100	100	100
a		11	⑦	12	100	14	100
c		11		11	⑩	14	100
e		11		11		14	15

Stage 5
Choose b as the new reference vertex (we could choose d instead but this would make no difference to the final result) and circle its label in bold. Compare paths via b to the labels on any adjacent vertices and relabel the vertices if the paths via b are found to be shorter. The result of this stage is shown in Fig. 23.23 and Table 23.5.

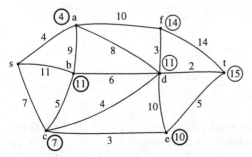

Figure 23.23 Stage 5 of solving the shortest path problem.

Table 23.5 The lengths of paths using up to five edges from s

Reference vertex	a	b	c	d	e	f	t
s	④	11	7	100	100	100	100
a		11	⑦	12	100	14	100
c		11		11	⑩	14	100
e	⑪		11			14	15
b			11			14	15

Stage 6

Choose d as the new reference vertex and circle its label in bold. The only vertices left without permanent labels are t and f. We find that the path via d to t gives a smaller value than the current label of 15 and hence we change the label to $11 + 2 = 13$. The result of this stage is shown in Fig. 23.24 and Table 23.6.

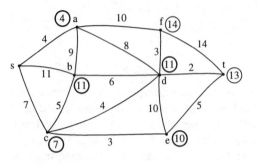

Figure 23.24 Stage 6 of solving the shortest path problem.

Table 23.6 The lengths of paths using up to six edges from s

Reference vertex	a	b	c	d	e	f	t
s	④	11	7	100	100	100	100
a		11	⑦	12	100	14	100
c		11		11	⑩	14	100
e		⑪		11		14	15
b			⑪			14	15
d						14	13

Stage 7

The remaining vertex with the smallest label is t. We therefore give t the permanent label of 13. As soon as t receives a permanent label we can stop the algorithm, as this must now represent the length of the shortest path from s to t which is the result we set out to find. To find the actual path that has this length we move backwards from t looking for consistent labels. To get to t we must have gone through d, as adding the label on d to the length of edge dt gives the correct value of 13. Continuing this process we find that the shortest path is scdt and is of length 13. The final solution is shown in Fig. 23.25 and the final table in Table 23.6.

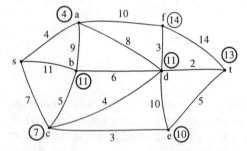

Figure 23.25 The solution to the shortest path problem. Vertex t now has a permanent label, showing that the length of the shortest path from s to t is 13. By working backwards from t to s we find that the actual path must be scdt.

Table 23.7 The algorithm to find the shortest path stops as soon as t receives a permanent label

Reference vertex	a	b	c	d	e	f	t
s	④	11	7	100	100	100	100
a		11	⑦	12	100	14	100
c		11		11	⑩	14	100
e		⑪		11		14	15
b			⑪			14	15
d						14	⑬

23.6 NETWORKS AND MAXIMUM FLOW

A network is a digraph with a weight function. The weight function represents the number of links between the vertices, e.g. the units of traffic (channels) in a communication network, or the maximum capacity of a network of one-way streets in a busy town. This weight function value

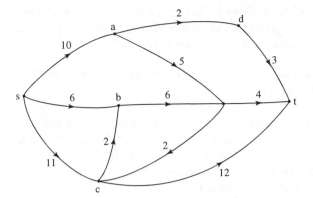

Figure 23.26 A network. The weights on the arcs represent the capacity of the arc. s is the source and t is the sink.

is called the capacity of the arc. An example of a network is shown in Fig. 23.26. s is the source and t is the sink. The capacity of an arc a_k will be referred to as $c(a_k)$. We wish to find the maximum possible flow from s the source to the sink t. That is, we want to find a flow function $f(a_k)$ assigning a flow to each of the arcs which gives this maximum flow.

Flow function

A valid flow function must be such that

1. $f(a_k) \leqslant c(a_k)$
2. The out-flow at any vertex equals the in-flow except at the source or the sink.

An arc is saturated if $f(a_k) = c(a_k)$. The total flow is then the flow out of s (which equals the flow into t).

Finding a maximum flow

1. Assign an initial flow to each edge (this will probably be 0 flow on each arc).
2. Find a flow-augmenting path from s to t. Any arc along this flow-augmenting path may be either in the forward direction, i.e. in the same direction as the path, or in the backward direction, i.e. in the opposite direction to the path.

   If the arc along the path is in the forward direction then it must have some 'extra' capacity, such that $c(a_k) > f(a_k)$. Then the extra capacity is represented by $\Delta_k = c(a_k) - f(a_k)$.

   If the arc along the path is in the backward direction then $f(a_k)$ must be greater than 0, and we assign the possible increase in flow $\Delta_k = f(a_k)$.

   We can then calculate the amount that this path can augment the flow by as $\Delta = \text{minimum}$ (Δ_k), where we consider all the values of Δ_k along the path.

3. We can now change the value of the flow function along the flow-augmenting path to $f(a_k) + \Delta$ (for arcs in the forward direction along the path) or $f(a_k) - \Delta$ (for arcs in the backward direction along the path). We then repeat the second stage and look for another flow-augmenting path. If no more flow-augmenting paths exist then we have found the maximum flow. This algorithm is due to Ford and Fulkerson. It can take a long while to terminate on complicated networks and better algorithms have been developed (e.g. the Dinic algorithm, which is not given here).

We can check that we have in fact found the maximum flow in the network by finding the minimum cut. The max-flow, min-cut theorem states that the maximum value of a flow from s to t in a network is equal to the minimum capacity of a cut separating s and t.

Example 23.3 Find the maximum flow from s to t in the network shown in Fig. 23.27(a). The weights on the arcs represent their capacities.

SOLUTION We begin by identifying a flow-augmenting path, s, e, t. As initially we consider the flow function to be zero, the spare capacities along the arcs are 12 and 14, the minimum of which is 12. We therefore assign a flow of 12 along the path to get Fig. 23.27(b).

We spot another flow-augmenting path, s, c, d, t, with a possible flow of 10 (the minimum of 14, 19 and 10). We add 10 to the flow function along this path to give Fig. 23.27(c).

We identify the flow-augmenting path s, a, b, t, with a possible flow of 8 (the minimum of the values of the spare capacities of 14, 8 and 15). We add 8 to the flow function along this path to give Fig. 23.27(d).

Another flow-augmenting path is s, a, c, d, b, t, with a possible flow of 6 (the minimum of the values of the spare capacities of 6, 14, 9, 6, 7). We add 6 to the flow function along this path to give Fig. 23.27(e).

There are no more entirely forward paths but we can see one that involves an arc in the backward direction that has a non-zero flow function. This path is s, c, a, e, t, where ca is in the backward direction. We take the minimum of the spare capacity in the forward direction along with the value of the flow in the backward direction. This gives 2 as the minimum of 4, 6, 3 and 2. We add 2 to the flows on the arcs in the forward direction and

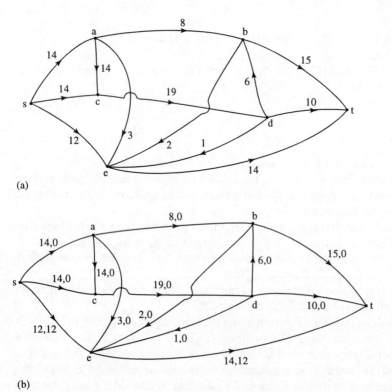

(a)

(b)

Figure 23.27 (a) The network for Example 23.3. The solution is presented in (b)–(f). (f) also shows the minimum cut, which confirms that the maximum flow of 38 has been found.

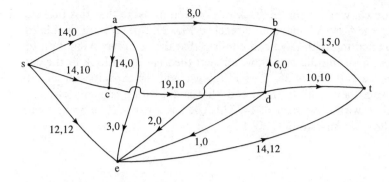

(c)

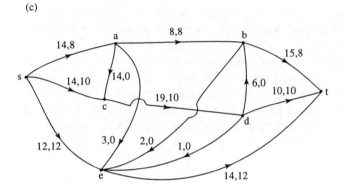

(d)

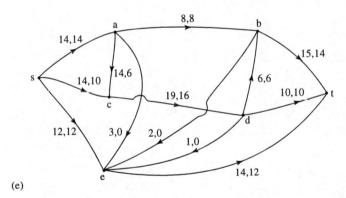

(e)

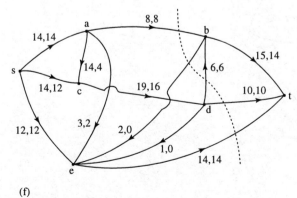

(f)

Figure 23.27 *continued*

GRAPH THEORY **531**

subtract 2 from the flow on ca, which is in the backward direction. Note that this process is equivalent to redirecting the flow. We have now reached Fig. 23.27(f). We are unable to identify any more flow-augmenting paths. We can confirm that this is in fact a solution to the maximum flow problem by using the max-flow, min-cut theorem. If we look at the cut marked we see that this has a capacity of 38. To find this we add the capacities of all the arcs which cross the cut from the 's end' to the 't end' that is db, with a capacity of 6, dt, with a capacity of 10, and et, with a capacity of 14. This cut has a capacity of 38 in total, which is the value of the flow we have found.

23.7 SUMMARY

1. A graph consists of a set of vertices, V, and a set of edges, E. A digraph is a directed graph with arcs instead of edges. Various definitions for different types of graphs are given in Sec. 23.2.
2. Graphs can be represented by an incidence matrix or an adjacency matrix.
3. Various optimization algorithms have been demonstrated in this chapter, including the greedy algorithm for the minimum spanning tree and the solution to the minimum path problem. For these two examples the weights on the graph represent lengths, costs or delays. We have also looked at one solution to the maximum flow problem. For this problem the weights on the arcs of a digraph represent the maximum capacity of the arc and the problem is to maximize the flow through the network.

23.8 EXERCISES

23.1 From the graphs in Fig. 23.28 identify any that are isomorphic.

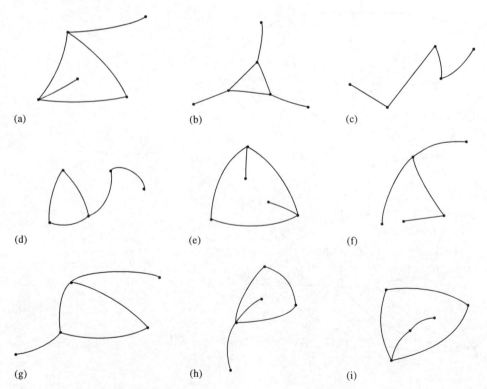

Figure 23.28 Graphs for Exercise 23.1.

23.2 The sequence of complete graphs $K_1, K_2, K_3, K_4, \ldots$ can be drawn by adding a vertex to the previous member of the sequence and drawing in all the edges necessary to make the new graph complete. Using this method, show that the number of edges in the complete graph K_n can be found from the series $1 + 2 + 3 + \ldots + n - 1$.

23.3 For what values of n is the complete graph K_n planar? For what values of m, n is the complete bipartite graph K_{mn} planar?

23.4 (a) Find the incidence and adjacency matrices for the graphs in Fig. 23.29.

(b) Show that in each case the column sum of the incidence matrix is 2 and explain this property.

(c) What are the column sums of A and what property of the graph to they represent?

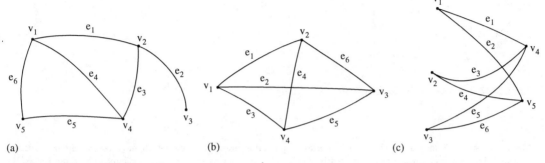

(a) (b) (c)

Figure 23.29 Graphs for Exercise 23.4.

23.5 Find a minimum spanning tree for the weighted graph in Fig. 23.30. Is there more than one possible minimum spanning tree for this graph?

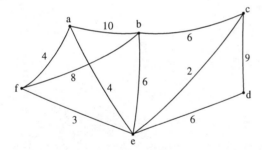

Figure 23.30 Graph for Exercise 23.5.

23.6 Find the shortest path from s to t in the weighted graph shown in Fig. 23.31.

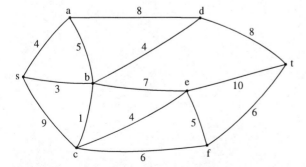

Figure 23.31 Graph for Exercise 23.6.

23.7 Find the maximum flow in the network shown in Fig. 23.32. The number next to each arc is its capacity. Use the max-flow, min-cut theorem to check you have found the maximum flow.

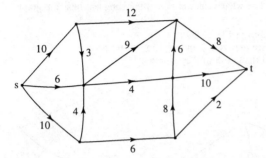

Figure 23.32 The network for Exercise 23.7.

23.8 In the network in Fig. 23.33, the arcs represent communication channels which have the maximum capacity indicated. The vertices represent switching centres, which also have a maximum capacity as indicated on the vertex. By replacing each of the vertices by two vertices and a connecting arc with a capacity equal to that of the original vertex, find the maximum flow from s to t. Use the max-flow, min-cut theorem to check that you have found the maximum flow.

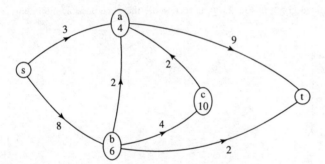

Figure 23.33 The network for Exercise 23.8.

FOUR

PROBABILITY AND STATISTICS

TWENTY FOUR
PROBABILITY AND STATISTICS

24.1 INTRODUCTION

Consider a system of computers and communication networks supporting a bank's cash dispensing machines. The machines provide instant money on the street on production of a cash card and a personal identification code, with the only prerequisite being that of a positive bank balance.

The communication network must be reliable – that is, not subject too often to breakdowns. The reliability of the system, or each part of the system, will depend on each component part.

A very simple system with only two components can be configured in series or in parallel. If the components are in series then the system will fail if one component fails (see Fig. 24.1). If the components are in parallel, then only one component need function (Fig. 24.2), and we have built in redundancy just in case one of the components fails.

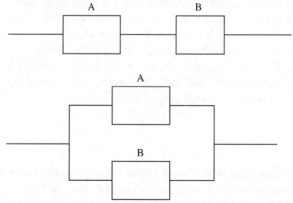

Figure 24.1 Component A and B must function for the system to function.

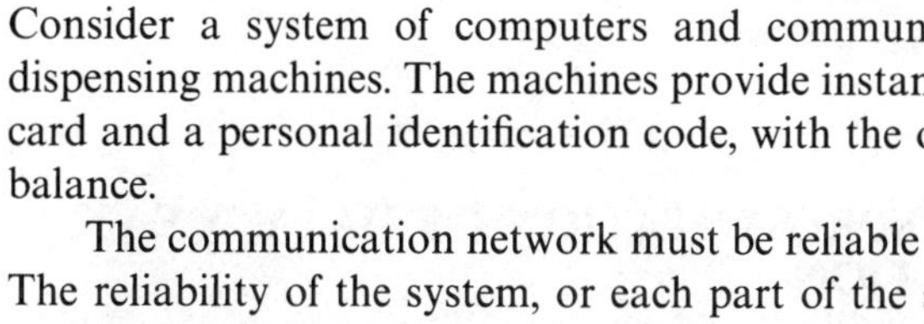

Figure 24.2 Either component A or component B must function for the system to function.

A network could be made up of tens, hundreds or even thousands of components. It is important to be able to estimate the reliability of a complex system and therefore the rate of replacement of components necessary and the overall annoyance level of our customers should the system fail.

The reliability will depend on the reliability of each one of the components, but it is impossible to say exactly how long even one single component will last.

We would like to be able to answer questions like:

1. What is the likely lifetime of any one element of the system?
2. How can we estimate the reliability of the whole system by considering the interaction of its component parts?

This is just one example of the type of problem for which we need the ideas of probability and statistics. Others are a factory producing 1000 electronic components each day, where we wish to be able to estimate how many defective components are produced each day, or a company with eight telephone lines, where we wish to say whether that is a sufficient number, in general, for the number of calls likely to be received at any one time.

All of these questions are those that need the ideas of probability and statistics. Statistics is used to analyse data and produce predictions about it. Probability is the theory upon which statistical models are built.

For the engineering situations in previous chapters we have considered that we know all the factors that determine how a system will behave, that is, we have modelled the system deterministically. This is not very realistic. Any real situation will have some random element. Some problems, such as the one we have just introduced, contain a large random element, so that it is difficult to predict the behaviour of each part of the system. However, it is possible to say, for instance, what the average overall behaviour would be.

24.2 POPULATION AND SAMPLE, REPRESENTATION OF DATA, MEAN, VARIANCE AND STANDARD DEVIATION

In statistics we are generally interested in a 'population' which is too large for us to measure completely. For instance, we might be interested in all the light bulbs that are produced by a certain manufacturer. The manufacturer may claim that the light bulbs have a lifetime of 1500 h. Clearly this cannot be exact, but we might be satisfied to agree with the manufacturer if most of the bulbs have a lifetime greater than 1500. If the factory produces half a million light bulbs per year then we do not have the time to test them all. In this case we test a 'sample'. The larger the sample, the more accurate will be the comparison with the whole population.

We have recorded the lifetimes of a sample of light bulbs in Table 24.1. We can build up a table to give information about our sample. Columns 8 and 9 are given for comparison with statistical modelling and columns 4, 6 and 7 are to help with the calculations. The following describes each column of Table 24.1, beginning with a list of the 'raw' data.

Column 1: The class intervals

Find the minimum and the maximum of the data. The difference between these gives the range of the data. Choose a number of class intervals (usually up to about 20) so that the class range can be chosen as some multiple of 10, 100, 1000 etc. (like the class range of 100 in Table 24.1). The data range divided by the class range and then rounded up to the integer above gives the number of classes. The class intervals are chosen so that the lowest class minimum is less than the minimum data value. In our example, the lowest class minimum has been chosen as 900. Add the class range of 100 to find the class interval, e.g. 900–1000. Carry on adding the class range to find the next class interval until you pass the maximum value.

Table 24.1 Frequency distribution of the lifetimes of a sample of light bulbs showing the method of calculating the mean and standard deviation. N represents the number of classes, n the number in the sample

(1)	(2)	(3)	(4)	(5)	(6)	(7)	(8)	(9)
Lifetime (h)	Class mid-point x_i	Class frequency f_i	$f_i x_i$	Cumulative frequency F_i	$(x_i - \bar{x})^2$	$f_i(x_i - \bar{x})^2$	Relative frequency f_i/n	Relative cumulative frequency (F_i/n)
900–1000	950	2	1 900	2	536 849	1 073 698	0.002	0.002
1000–1100	1050	0	0	2	400 309	0	0.002	0.002
1100–1200	1150	5	5 750	7	283 769	1 418 845	0.005	0.007
1200–1300	1250	23	28 750	30	187 229	4 306 267	0.023	0.03
1300–1400	1350	47	63 450	77	110 689	5 202 383	0.047	0.077
1400–1500	1450	103	149 350	180	54 149	5 577 347	0.103	0.18
1500–1600	1550	160	248 000	340	17 609	2 817 440	0.16	0.34
1600–1700	1650	190	313 500	530	1 069	203 110	0.19	0.53
1700–1800	1750	165	288 750	695	4 529	747 285	0.165	0.695
1800–1900	1850	164	303 400	859	27 989	4 590 196	0.164	0.859
1900–2000	1950	92	179 400	951	71 449	6 573 308	0.092	0.951
2000–2100	2050	49	100 450	1000	134 909	6 610 541	0.049	1
		1000	1 682 700			39 120 420	1	

$$\sum_i f_i = n \qquad \bar{x} = \frac{1\,682\,700}{1000} \qquad F_N = n \qquad \sigma^2 = \frac{1}{n}\sum_i f_i(x_i - \bar{x})^2 \qquad \sum_i \frac{f_i}{n} = 1 \qquad \frac{F_N}{n} = 1$$

$$\bar{x} = \frac{1}{n}\sum_i f_i x_i \qquad\qquad \sigma^2 = \frac{39\,120\,420}{1000}$$

$$\sigma \approx 198$$

Column 2: The class mid-point

The class mid-point is found by the maximum value in the class interval plus the minimum value in the class interval divided by 2. For the interval 1300 to 1400, the class mid-point is $(1400 + 1300)/2 = 1350$.

The class mid-point is taken as a representative value for the class. That is, for the sake of simplicity we treat the data in the class 1300 to 1400 as though there were 47 values at 1350. This is an approximation that is not too serious if the classes are not too wide.

Column 3: Class frequency and the total number in the sample

The class frequency is found by counting the number of data values that fall within that class range. For instance, there are 47 values between 1300 and 1400. You have to decide whether to include values that fall exactly on the class boundary in the class above or the class below. It does not usually matter which you choose as long as you are consistent within the whole data set.

The total number of values in the sample is given by the sum of the frequencies for each class interval.

Column 4: $f_i x_i$ and the mean of the sample, $\bar{x}$

This is the product of the frequency and the class mid-point. Remembering that the class mid-point is taken as a representative value for the class, then $f_i x_i$ gives the sum of all the values in that class. For instance, in the eighth class, with class mid-point 1650, there are 190 values. If we say that each of the values is approximately 1650 then the total of all the values in that class is $1650 \times 190 = 313\,500$. Summing this column gives the approximate total value for the whole sample $= 1\,682\,700$ h. If we only used one light bulb at a time and changed it when it failed, then the 1000 light bulbs would last approximately $1\,682\,700$ hours. The mean of the sample is this total divided by the number in the sample, 1000, giving $1\,682\,700/1000 = 1682.7$ h. The mean is a measure of the central tendency. That is, if you wanted a simple number to sum up the lifetime of these light bulbs you would say that the average life was 1683 h.

Column 5: The cumulative frequency

The cumulative frequency gives the number that fall into the current class interval or any class interval that comes before it. You could think of it as 'the number so far' function. To find the cumulative frequency for a class, take the number in the current class and add on the previous cumulative frequency for the class below, e.g. for 1900–2000 we have a frequency of 92. The cumulative frequency for 1800 – 1900 is 859. Add $859 + 92$ to get the cumulative frequency of 951. That is 951 lightbulbs in the sample have a lifetime below 2000 h. Notice that the cumulative frequency of the final class must equal the total number in the sample. This is because the final class must include the maximum value in the sample, and all the others have lifetimes less than this.

Column 6: $(x_i - \bar{x})^2$

This column is one used in the process of calculating the standard deviation (see column 7). $x_i - \bar{x}$ represents the amount that the class mid-point differs from the mean of the sample. If we want to measure how spread out around the mean the data is, then this would seem like a useful number. However, adding up $(x_i - \bar{x})$ for the whole sample will just give zero, as the positive and negative values will cancel each other out. Hence we take the square, so that all the numbers are positive. For the sixth class interval the class mid-point is 1450. Subtracting the mean, 1682.7, gives -232.7. Squaring this gives $54\,149.2$. This value is the square of the deviation from the mean, or just the squared deviation.

Column 7: $f_i(x_i - \bar{x})^2$, the variance and the standard deviation

For each class we multiply the frequency by the squared deviation, calculated in column 6. This gives an approximation to the total squared deviation for that class. For the sixth class we multiply the squared deviation of $54\,149$ by the frequency 103 to get $5\,577\,347$. The sum of this column gives the total squared deviation from the mean for the whole sample. Dividing by the number of sample points gives an idea of the average squared deviation. This is called the variance. It is found by summing column 7 and dividing by 1000, the number in the sample, giving a variance of $39\,120$. A better measure of the spread of the data is given by the square root of this number, called the standard deviation and usually represented by σ. Here, $\sqrt{39\,120} \simeq 198$.

Column 8: The relative frequency

If instead of 1000 data values in the sample we had 2000 or 10 000 or 500 we might expect that the proportion falling into each class interval would remain roughly the same. This proportion of the total number is called the relative frequency and is found by dividing the frequency by the total number in the sample. Hence for the third class, 1100–1200, we find $5/1000 = 0.005$.

Column 9: The cumulative relative frequency

By the same argument as above, we would expect the proportion with a lifetime of less than 1900 h to be roughly the same whatever the sample size. This cumulative relative frequency can be found by dividing the cumulative frequency by the number in the sample. Notice that the cumulative relative frequency for the final class interval is 1. That is, the whole of the sample has a lifetime of less than 2000 h.

The data can be represented on a histogram, as in Fig. 24.3. This gives a simple pictoral representation of the data. The right-hand side has a scale equal to the left-hand scale divided by the number in the sample. These readings therefore give the relative frequencies, as given in the Table 24.1.

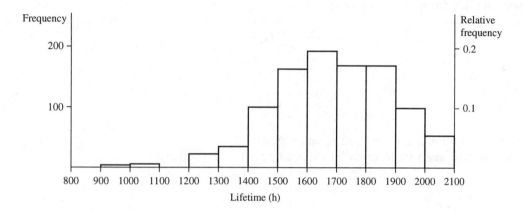

Figure 24.3 Histogram of the lifetimes of a sample of light bulbs.

We may want to sum up our findings with a few simple statistics. To do this we use a measure of the central tendency and the dispersion of the data, which are calculated as already described above. These are summarized below.

Central tendency: the mean

The most commonly used measure of the central tendency is the mean, $\bar{x}$:

$$\bar{x} = \frac{1}{n}\sum_i f_i x_i = \sum_i \frac{f_i x_i}{n}$$

where x_i is a representative value for the class and f_i is the class frequency. The summation is over all classes.

If the data has not been grouped into classes then the mean is found by summing all the individual data values and dividing by the number in the sample:

$$\bar{x} = \frac{1}{n}\sum_i x_i$$

where the summation is now over all sample values.

Dispersion: the standard deviation

The variance is the mean square deviation given by

$$v = \sigma^2 = \frac{1}{n} \sum_i f_i(x_i - \bar{x})^2 = \sum_i \frac{f_i}{n}(x_i - \bar{x})^2$$

where $\bar{x}$ is the sample mean, x_i is a representative value for the class, f_i is the class frequency and the summation is over all classes. This value is the same as

$$v = \sigma^2 = \frac{1}{n} \sum f_i x_i^2 - \bar{x}^2$$

which can sometimes be quicker to calculate.

For data not grouped into classes we have

$$v = \sigma^2 = \frac{1}{n} \sum_i (x_i - \bar{x})^2$$

or

$$v = \sigma^2 = \frac{1}{n} \sum_i x_i^2 - \bar{x}^2$$

where the summation is now over all sample values.

The standard deviation is given by the square root of the variance:

$$\sigma = \sqrt{\frac{1}{n} \sum_i f_i(x_i - \bar{x})^2}$$

We have already calculated the average lifetime and the standard deviation of the lifetimes of the sample of light bulbs. We repeat the calculation below.

The mean is given by

class frequency ⟶ class midpoint

$$(2 \times 950 + 0 \times 1050 + 5 \times 1150 + 23 \times 1250 + 47 \times 1350 + 103 \times 1450 + 160 \times 1550$$
$$+ 190 \times 1650 + 165 \times 1750 + 164 \times 1850 + 92 \times 1950 + 49 \times 2050)/1000 = 1682.7$$

The variance is given by

class frequency ⟶ class midpoint ⟶ sample mean

$$(2 \times (950 - 1682.7)^2 + 0 \times (1050 - 1682.7)^2 + 5 \times (1150 - 1682.7)^2 + 23 \times (1250 - 1682.7)^2$$
$$+ 47 \times (1350 - 1682.7)^2 + 103 \times (1450 - 1682.7)^2 + 160 \times (1550 - 1682.7)^2$$
$$+ 190 \times (1650 - 1682.7)^2 + 165 \times (1750 - 1682.7)^2 + 164 \times (1850 - 1682.7)^2$$
$$+ 92 \times (1950 - 1682.7)^2 + 49 \times (2050 - 1682.7)^2)/1000 = 39\,120\,420/1000 \simeq 39\,120.4$$

Hence the standard deviation is

$$\sqrt{39\,120.4} \simeq 198$$

Can we agree with the manufacturer's claim that the light bulbs have a lifetime of 1500 h? The number with a lifetime of less than 1500 h is 180 (the cumulative frequency up to 1500 h). This represents only 0.18, less than 20 per cent of the sample. However, this is probably an unreasonable number to justify the manufacturer's claim as nearly 20 per cent of the customers will get light bulbs that are not as good as advertised. We might accept a small number, say 5 per cent, failing to live up to a manufacturer's promise. Hence it would be better to claim a lifetime of around 1350 h, with a relative cumulative frequency of between 0.03 and 0.077 – approximately 0.053, or just over 5 per cent that would fail to live up to the manufacturer's claim.

Example 24.1 A sample of 2000 Ω resistors were tested and their resistances were found as below:

1997	1998	2004	2002	1999	2000	2001	2002	1999	1998	1997	1999	2001
2003	2005	1996	2000	2000	1998	1999	1998	2001	2003	2002	1996	2002
1995	2000	2000	1999	1997	2004	2001	1999	1999	1998	2002	2003	2002
2001	1999	1998	1997	2002	2001	2000	1999	2001	1999	1998	2000	1999
1998	2000	2002	1999	2001	2000	2001	2002	2004	1996	2000	2002	2000
1998	2000	2005	2000	1999	2000	2000	2001	1999	2001	1997	2002	2001
2004	2000	2000	1999	1998	2001	1998	2000	2002	1998	1998	1999	2002
1999	2001	2002	2006	2000	2001	2001	2002	2003				

Group the data in class intervals and represent it using a table and a histogram. Calculate the mean resistance and the standard deviation.

SOLUTION We follow the method of building up the table as described previously.

Column 1: Class intervals
To decide on class intervals, look at the range of values presented. The minimum value is 1995 and the maximum is 2006. A reasonable number of class intervals would be around 10, and in this case, as the data is presented to the nearest whole number, then the smallest possible class range we could choose would be 1 Ω. A 1 Ω class range would give 12 classes, and this seems a reasonable choice. If we choose the class mid-points to be the integer values, then we get class intervals of 1994.5–1995.5, 1995.5–1996.5 etc. We can now produce Table 24.2.
The other columns are filled in as follows.

Column 2
To find the frequency, count the number of data values in each class interval. The sum of the frequencies gives the total number in the sample; in this case 100.

Column 3
The cumulative frequency is given by adding up the frequencies so far. The number of resistances up to 1995.5 is 1. As there are 3 in the interval 1995.5–1996.5 then add $1 + 3 = 4$ to get the number up to a resistance of 1996.5. The number up to 1997.6 is given by $4 + 5 = 9$ etc.
Columns 4, 6 and 7 help us calculate the mean and standard deviation. To find column 4, multiply the frequencies (column 3) by the class mid-points (column 2) to get $f_i x_i$.
The sum of this column is an estimate of the total if we added all the values in the sample together. Therefore, dividing by the number of items in the sample gives the mean:

$$\text{Mean} = (1 \times 1995 + 3 \times 1996 + 5 \times 1997 + 13 \times 1998 + 17 \times 1999$$

$$+ 19 \times 2000 + 16 \times 2001 + 15 \times 2002 + 4 \times 2003 + 4 \times 2004$$

$$+ 2 \times 2005 + 1 \times 2006)/100 = 200\,015/100 = 2000.15$$

To find the standard deviation we calculate the variance first. The variance is the mean squared deviation. Find the difference between the mean and each of the class intervals and

Table 24.2 Frequency distribution of resistances of a sample of resistors. N represents the number of classes, n the number in the sample

(1) Resistance (Ω)	(2) Class mid-point x_i	(3) Class frequency f_i	(4) $f_i x_i$	(5) Cumulative frequency F_i	(6) $(x_i - \bar{x})^2$	(7) $f_i(x_i - \bar{x})^2$	(8) Relative frequency f_i/n	(9) Relative cumulative frequency F_i/n
1994.5–1995.5	1995	1	1 995	1	26.52	26.52	0.01	0.01
1995.5–1996.5	1996	3	5 998	4	17.22	51.66	0.03	0.04
1996.5–1997.5	1997	5	9 985	9	9.92	49.6	0.05	0.09
1997.5–1998.5	1998	13	25 974	22	4.62	60.06	0.13	0.22
1998.5–1999.5	1999	17	33 983	39	1.32	22.44	0.17	0.39
1999.5–2000.5	2000	19	38 000	58	0.02	0.38	0.19	0.58
2000.5–2001.5	2001	16	32 016	74	0.72	11.52	0.16	0.74
2001.5–2002.5	2002	15	30 030	89	3.42	51.3	0.15	0.89
2002.5–2003.5	2003	4	8 012	93	8.12	32.48	0.04	0.93
2003.5–2004.5	2004	4	8 016	97	14.82	59.28	0.04	0.97
2004.5–2005.5	2005	2	4 010	99	24.52	47.04	0.02	0.99
2005.5–2006.5	2006	1	2 006	100	34.22	34.22	0.01	1
		100	200 015			446.5	1	

$$\sum_i f_i = n \qquad \bar{x} = \frac{200\,015}{100} \qquad F_N = n \qquad \sigma^2 = \frac{1}{n}\sum_i f_i(x_i - \bar{x})^2 \qquad \sum_i \frac{f_i}{n} = 1 \qquad \frac{F_N}{n} = 1$$

$$\bar{x} = \frac{1}{n}\sum_i f_i x_i \qquad\qquad = \frac{446.5}{100}$$

$$\bar{x} = 2000.15 \qquad\qquad \sigma = 2.11$$

square it. This gives column 6. Multiply by the class frequency to get column 7. Finally, add them all up and divide by the total number. This gives the variance:

$$\sigma^2 = \frac{446.5}{100} = 4.465$$

and the standard deviation:

$$\sigma = \sqrt{4.465} \simeq 2.11$$

The relative frequency (given in Column 8) is the fraction in each class interval, that is, the frequency divided by the total number. For example, the relative frequency in the class interval 2000.5–2001.5 is $16/100 = 0.16$.

The cumulative relative frequency is the cumulative frequency divided by the total number n. The cumulative frequency up to 2000.5 is 58. Dividing this by the total number gives 0.58 for the relative cumulative frequency.

We can sum up this sample by saying that the mean resistance is 2000.15 with a standard deviation of 2.11. These simple figures can be used to sum up how closely the claim that they are resistors of 2000 Ω can be justified. We can also picture the frequency distribution using a histogram, as in Fig. 24.4.

Returning to the lifetimes of the light bulbs, we might want to ask what average lifetime and standard deviation might lead to less than 5 per cent of the light bulbs having a lifetime less than 1500 hours.

To answer this sort of problem we need to build up a theory of statistical models and use probability theory.

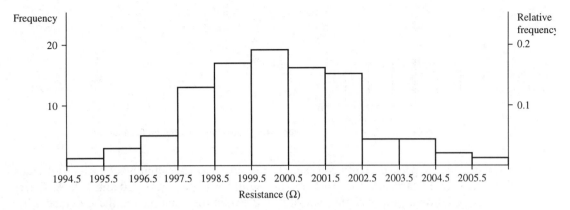

Figure 24.4 Histogram of the frequency distribution of the resistances given in Table 24.2.

24.3 RANDOM SYSTEMS AND PROBABILITY

We are dealing with complicated systems with a number of random factors affecting their behaviour, such as those that we have already seen: the production of resistors or the lifetimes of light bulbs. We cannot determine the exact behaviour of the system but using its frequency distribution we can estimate the probabilities of certain events.

Relative frequency and probability

The probability of an event is related to its relative frequency. If I chose a light bulb from the sample in Sec. 24.2, then the probability that its lifetime is between 1400–1500 is 0.103. This is the same as the relative frequency for that class interval, i.e. the number in the class interval divided by the total number of light bulbs in the sample. Here we have assumed that we are no more likely to pick any one light bulb than any other, i.e. each outcome is equally likely. The histogram of the relative frequencies (Fig. 24.3 using the right-hand scale) gives the probability distribution function (or simply the probability function) for the lifetimes of the sample of light bulbs.

As an introduction to probability, examples are often quoted involving throwing dice or dealing cards from a pack of playing cards. These are used because the probabilities of events are easy to justify and not because of any particular predilection on the part of mathematics authors to a gambling vocation!

Example 24.2 A die has the numbers 1 to 6 marked on its sides. Draw a graph of the probability distribution function for the outcome from one throw of the die, assuming it is fair.

SOLUTION If we throw the die 10 000 times we would expect the number of times each number appeared face up to be roughly the same. Here we have assumed that the die is fair; that is, any one number is as likely to be thrown as any other. The relative frequencies would be approximately $\frac{1}{6}$. The probability distribution function therefore is a flat function with a value of $\frac{1}{6}$ for each of the possible outcomes of 1, 2, 3, 4, 5 and 6. This is pictured in Fig. 24.5.

Notice two important things about the probabilities in the probability distribution for the die:

1. Each probability is less than 1.
2. The sum of all the probabilities is one:

$$\tfrac{1}{6} + \tfrac{1}{6} + \tfrac{1}{6} + \tfrac{1}{6} + \tfrac{1}{6} + \tfrac{1}{6} = 1$$

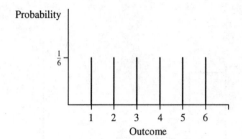

Figure 24.5 Probability distribution for a fair die.

Example 24.3 A pack of cards consists of four suits: hearts, diamonds, spades and clubs. Each suit has 13 cards; that is, cards for the numbers 1 (the ace) to 10 and a Jack, Queen and King. Draw a graph of the probability distribution function for the outcome when dealing one card from the pack, where only the suit is recorded. Assume the card is replaced each time and the pack is perfectly shuffled.

SOLUTION If a card is selected, the suit is recorded (hearts, spades, clubs or diamonds), the card is placed back in the pack and the pack is shuffled. If this is repeated (say 10 000 times), then we might expect that each suit will occur as often as any other, that is, $\frac{1}{4}$ of the time. The probability distribution function again is a flat distribution, and has the value of $\frac{1}{4}$ for each of the four possible outcomes. The probability distribution for suits is given in Fig. 24.6.

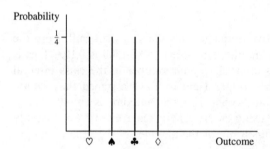

Figure 24.6 The probability distribution for suits from a pack of cards.

Notice again that in Example 24.3 each probability is less than 1 and that the sum of all the probabilities is one:

$$\tfrac{1}{4} + \tfrac{1}{4} + \tfrac{1}{4} + \tfrac{1}{4} = 1$$

We have seen that a probability distribution can be represented using a graph. Each item along the x-axis has an associated probability. Probability is a function defined on some set. The set is called the sample space, S, and contains all possible *outcomes* of the random system. The probability distribution function is often abbreviated to p.d.f.

Some definitions

A **trial** is a single observation on the random system, e.g. one throw of a die or one measurement of a resistance in the example in Sec. 24.2.

The **sample space** is the set of all possible outcomes, e.g. for the die it is the set $\{1,2,3,4,5,6\}$, and for the resistance problem it is the set of all possible measured resistances. This set may be discrete or continuous.

An *event* is a set of outcomes. For instance, A is the event of throwing less than 4, and B is the event of throwing a number greater than or equal to 5.

An event is a subset of the sample space S.

Notice that in the case of a continuous sample space an event is also a continuous set, represented by an interval of values. For example, C is the event that the resistance lies in the interval 2000 ± 1.5.

Probability

The way that probability is defined is slightly different for the case of a discrete sample space or a continuous sample space.

Discrete sample space

The outcome of any trial is uncertain. However, in a large number of trials the proportion showing a particular outcome approaches a certain number. We call this the probability of that outcome. The probability distribution function, or simply probability function, gives the value of the probability that is associated with each outcome. The probability function obeys two important conditions:

1. All probabilities are less than or equal to 1, i.e. $0 \leqslant p(x) \leqslant 1$, where x is any outcome in the sample space.
2. The probabilities of the individual outcomes sum to 1, i.e.

$$\sum p(x) = 1$$

Continuous sample space

Consider a continuous sample space. We assign probabilities to intervals. We find the probability of, for instance, the resistance being in the interval 2000 ± 0.5, and hence we assign probabilities to events and not to individual outcomes. In this case, the function which gives the probabilities is called the probability density function and we can find the probability of some event by integrating the probability density function over the interval. For instance, if we have a probability density function $f(x)$, where x can take values from a continuous sample space, then the probability of x being in the interval a to b is given by, for example,

$$P(a < x < b) = \int_{a}^{b} f(x) \, dx$$

The probability density function obeys the condition

$$\int_{-\infty}^{\infty} f(x) \, dx = 1$$

That is, the total area under the graph of the probability density function must be 1.

Equally likely events

If all outcomes are equally likely, then the probability of an event E from a discrete sample set is given by

$$\frac{\text{The number of outcomes in E}}{\text{The total number of outcomes in S}}$$

That is, the probability of E is equal to the proportion of the whole sample space that is in E when each of the outcomes is equally likely.

Example 24.4. What is the probability on one throw of a die getting a number less than 3?

SOLUTION Give this event the name A. Then $A=\{1,2\}$ and $S=\{1,2,3,4,5,6\}$.

The number in A is 2 and the number in S is 6. Therefore, as each outcome is equally likely, $p(A)$, the probability of A, is $\frac{2}{6}=\frac{1}{3}$.

This particular result leads to another way of representing probability – by using area.

If we use a rectangle to represent the set of all possible outcomes, S, then an event is a subset of S, i.e. one section of the rectangle, and if all outcomes are equally likely then its probability can be pictured by the proportion of S that is in A. We will put a dotted line around the picture representing the set A to indicate the number in A or the area of A.

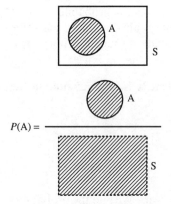

$$P(A) = \underline{\qquad\qquad\qquad\qquad}$$

Figure 24.7 $A \subseteq S$ (A is a subset of S), where all the outcomes in S are equally likely. The probability of A can be pictured as the proportion of S than is in A: the ratios of the areas.

This way of picturing the probability of an event can help in remembering some of the probabilities of combined events.

24.4 ADDITION LAW OF PROBABILITY

Disjoint events

Disjoint events are events with no outcomes in common. They cannot both happen simultaneously.

Example 24.5 A is the event that a card chosen from a playing pack is under 6 (counting ace as low, i.e. = 1) and B is the event of choosing a picture card (Jack, Queen or King). Find the probability that a card chosen from the pack is under 6 or is a picture card.

SOLUTION

$A=\{1\heartsuit,2\heartsuit,3\heartsuit,4\heartsuit,5\heartsuit,1\diamondsuit,2\diamondsuit,3\diamondsuit,4\diamondsuit,5\diamondsuit,1\spadesuit,2\spadesuit,3\spadesuit,4\spadesuit,5\spadesuit,1\clubsuit,2\clubsuit,3\clubsuit,4\clubsuit,5\clubsuit\}$

$B=\{J\heartsuit,Q\heartsuit,K\heartsuit,J\diamondsuit,Q\diamondsuit,K\diamondsuit,J\spadesuit,Q\spadesuit,K\spadesuit,J\clubsuit,Q\clubsuit,K\clubsuit\}$

A and B are disjoint. As each outcome is considered to be equally likely, the probabilities are easy to find:

$$p(A)=(\text{Number in A})/(\text{Number in S})$$

There are 52 cards, and therefore 52 possible outcomes in S, so

$$p(A) = 20/52 = 5/13$$

Also, $p(B) = $ (Number in B)/(Number in S) $= 12/52 = 3/13$.

We want to find the probability of A or B happening, that is, the probability that the card is either under 6 or is a picture card:

$$A \cup B = \{1\heartsuit, 2\heartsuit, 3\heartsuit, 4\heartsuit, 5\heartsuit, 1\diamondsuit, 2\diamondsuit, 3\diamondsuit, 4\diamondsuit, 5\diamondsuit, 1\spadesuit, 2\spadesuit, 3\spadesuit, 4\spadesuit, 5\spadesuit, 1\clubsuit, 2\clubsuit, 3\clubsuit,$$
$$4\clubsuit, 5\clubsuit, J\heartsuit, Q\heartsuit, K\heartsuit, J\diamondsuit, Q\diamondsuit, K\diamondsuit, J\spadesuit, Q\spadesuit, K\spadesuit, J\clubsuit, Q\clubsuit, K\clubsuit\}$$

We can see that:

$$p(A \cup B) = (\text{Number in } A \cup B)/(\text{Number in S}) = 32/52 = 8/13$$

$$\frac{\text{Number in A} + \text{Number in B}}{\text{Number in S}} = \frac{\text{Number in A}}{\text{Number in S}} + \frac{\text{Number in B}}{\text{Number in S}}$$

$$p(A \cup B) = p(A) + p(B)$$

$$\tfrac{8}{13} = \tfrac{5}{13} + \tfrac{3}{13}$$

We can also see this by using the idea of area to picture it, as in Fig. 24.8.

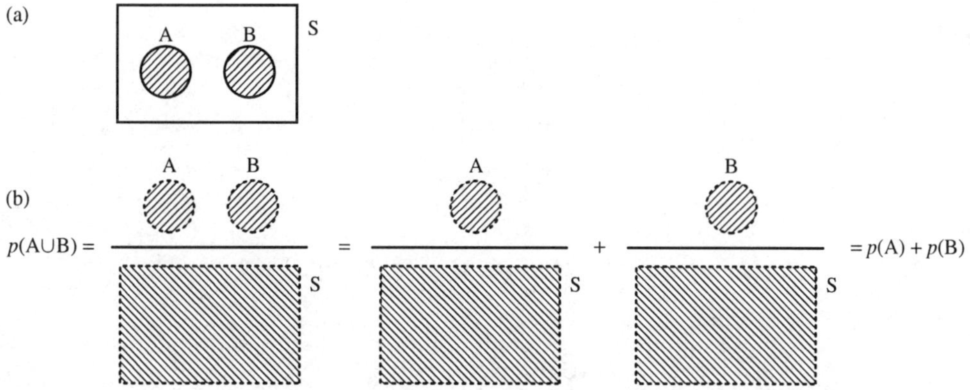

Figure 24.8 The two events A and B are disjoint; that is, they have no outcomes in common. On a Venn diagram, as in (a), they do not intersect. The probability of either A or B can be found by $p(A \cup B) = p(A) + p(B)$, as in (b).

We can also consider the probability of A not happening, that is, the probability of the complement of A, which we represent by A'. As A and A' are disjoint:

$$p(A') + p(A) = 1$$

$$p(A') = 1 - p(A)$$

This is pictured in Fig. 24.9.

Non-disjoint events

Example 24.6 Consider one throw of a die:

$$A = \{a | a \text{ is an even number}\}$$

$$B = \{b | b \text{ is a multiple of 3}\}$$

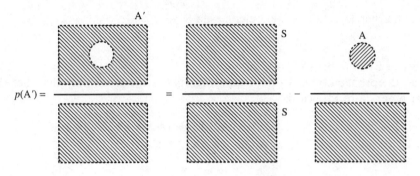

Figure 24.9 $p(A')=1-p(A)$.

Find the probability that the result of one throw of the die is either an even number or a multiple of 3.

SOLUTION

$$A=\{a|a \text{ is an even number}\}$$
$$B=\{b|b \text{ is a multiple of 3}\}$$

Then

$$A=\{2,4,6\}$$
$$B=\{3,6\}$$
$$A\cup B=\{2,3,4,6\}$$

and

$$p(A)=\tfrac{3}{6}=\tfrac{1}{2}, \qquad p(B)=\tfrac{2}{6}=\tfrac{1}{3}, \qquad p(A\cup B)=\tfrac{4}{6}=\tfrac{2}{3}$$

and

$$p(A\cup B)\neq p(A)+p(B)$$

Looking at the problem using the idea of areas in the set, we can see from Fig. 24.10 that the rule becomes

$$p(A\cup B)=p(A)+p(B)-p(A\cap B)$$

24.5 REPEATED TRIALS, OUTCOMES AND PROBABILITIES

Suppose we start to consider more complicated situations, like throwing a die twice. Then the outcomes can be found by considering all the possible outcomes of throwing the die the first time combined with all the possible outcomes of throwing the die the second time. As there are six possible outcomes for the first throw and six possible outcomes for the second throw, there are $6\times6=36$ outcomes of throwing the die twice.

If we would like to find the probability that the first throw is a 5 and the second throw is a 5 or 6 then we can do this by listing all the 36 outcomes and finding the proportion that fall into our event.

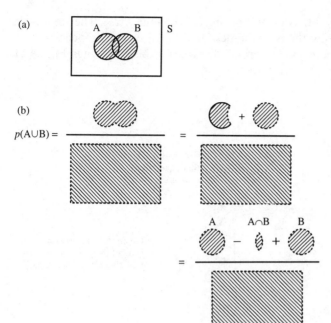

Figure 24.10 (a) $A \cup B$. (b) $p(A \cup B) = p(A) + p(B) - p(A \cap B)$.

The set S of all possible outcomes has 36 elements:

$$S = \{(1,1),(1,2),(1,3),(1,4),(1,5),(1,6),$$
$$(2,1),(2,2),(2,3),(2,4),(2,5),(2,6),$$
$$(3,1),(3,2),(3,3),(3,4),(3,5),(3,6),$$
$$(4,1),(4,2),(4,3),(4,4),(4,5),(4,6),$$
$$(5,1),(5,2),(5,3),(5,4),(5,5),(5,6),$$
$$(6,1),(6,2),(6,3),(6,4),(6,5),(6,6)\}$$

and each has an equally likely outcome.

E = the first throw is a five and the second is a 5 or 6. Hence

$$E\{(5,5),(5,6)\}$$

$$p(E) = \frac{\text{Number in E}}{\text{Number in S}} = \tfrac{2}{36} = \tfrac{1}{18}$$

This sort of problem can be pictured more easily using a probability tree.

24.6 REPEATED TRIALS AND PROBABILITY TREES

Repeatedly tossing a coin

The simplest sort of trial to consider is one with only two outcomes, for instance tossing a coin. Each trial has the outcome of head or tail, and each is equally likely.

We can picture repeatedly tossing the coin by drawing a probability tree. The tree works by drawing all the outcomes and writing the probabilities for the trial on the branches.

Let us consider tossing a coin three times. The probability tree for this is shown in Fig. 24.11. There are various rules and properties can can notice.

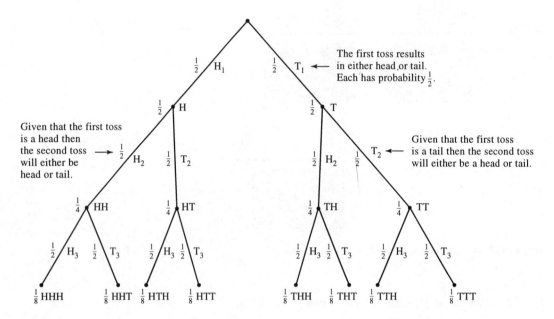

Figure 24.11 A probability tree for three tosses of a coin.

Rules of probability trees and repeated trials

1. The probabilities of outcomes associated with each of the vertices can be found by multiplying all the probabilities along the branches leading to it from the top of the tree. HTH has the probability $\frac{1}{2} \times \frac{1}{2} \times \frac{1}{2} = \frac{1}{8}$.
2. At each level of the tree, the sum of all the probabilities on the vertices must be 1.
3. The probabilities along branches out of a single vertex must sum to 1. For example, after getting a head on the first trial we have a probability of $\frac{1}{2}$ of getting a head on the second trial and a probability of $\frac{1}{2}$ of getting a tail. Together they sum to 1.

The fourth property is one that is only true when the various repeated trials are independent, that is, the result of the first trial has no effect on the possible result of the second trial etc.

4. For independent trials, the tree keeps repeating the same structure with the same probabilities associated with the branches. Here the event of getting a head on the third toss is independent of the event of getting a head on the first or second toss.

Using the tree we can find various probabilities, as we see in Example 24.7.

Example 24.7 What is the probability on three tosses of the coin that exactly two of them will be heads?

SOLUTION Count up all the ways that we could have two heads and one tail, looking at the foot of the probability tree in Fig. 24.11. We find three possibilities:

$$HHT, \ HTH, \ THH$$

Each of these has probability of $\frac{1}{8}$. Therefore the probability of exactly two heads is $\frac{3}{8}$.

Picking balls from a bag without replacement

We have 20 balls in a bag. Ten are red and ten are black. A ball is picked out of the bag, its colour recorded and then it is *not* replaced in the bag. There are two possible outcomes of each trial, red (R) or black (B). Let us consider three trials and their associated outcomes in a probability tree, as in Fig. 24.12.

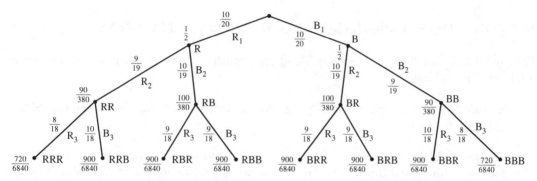

Figure 24.12 A probability tree for three trials of picking a ball out of a bag without replacement.

To find the probabilities we consider how many balls remain in each case. If the first ball chosen is red, then there are only 19 balls left, of which 9 are red and 10 are black. Therefore the probability of picking a red ball on the second trial is only 9/19 and the probability of picking black is 10/19.

The first three rules given in the last example apply.

1. The probabilities of outcomes associated with each of the vertices can be found by multiplying all the probabilities along the branches leading to it from the top of the tree. BRB has the probability

$$\frac{10}{20} \times \frac{10}{19} \times \frac{9}{18} = \frac{900}{6840} = \frac{1}{76}$$

2. At each level of the tree the sum of all the probabilities on the vertices must be 1. For example, at the second level we have

$$\frac{90}{380} + \frac{100}{380} + \frac{100}{380} + \frac{90}{380} = \frac{9+10+19+9}{38} = \frac{38}{38} = 1$$

3. The probabilities along branches out of a single vertex must sum to 1. For example, after picking red on the first trial we have a probability of 9/19 of picking red and 10/19 of picking black. Together they sum to 1.

The fourth property is no longer true, as the various repeated trials are not independent. The result of the first trial has an effect on the possible result of the second trial etc.

Using the tree we can find the probability of various events, as in Example 24.8.

Example 24.8 Of 20 balls in a bag, 10 are red and 10 are black. A ball is picked out of the bag, its colour recorded and then it is not replaced in the bag. What is the probability that of the first three balls chosen, exactly two will be red?

SOLUTION We can use the probability tree in Fig. 24.12 to solve this problem. We look at the foot of the tree, which gives all the possible outcomes after three balls have been selected. The ways of getting two reds are RRB, RBR, BRR, and the associated probabilities are

$$\frac{900}{6840}, \frac{900}{6840}, \frac{900}{6840} \quad \text{or} \quad \frac{1}{76}, \frac{1}{76}, \frac{1}{76}$$

Summing these gives the probability of exactly two reds being chosen out of the three as $\frac{3}{76}$.

24.7 CONDITIONAL PROBABILITY AND PROBABILITY TREES

The probabilities we have been writing along the branches of the probability tree are called conditional probabilities.

Example 24.9 What is the probability that the first throw of a die will be a 5 and the second throw will be a 5 or 6?

SOLUTION A is the set of those outcomes with first throw a five and B is the set of those outcomes with second throw a 5 or 6. We can use a probability tree in the following way. After the first throw of the die, either A is true or not, i.e. we have only two possibilities, A or A'. After that, we are interested in whether B happens or not. Again, we either get B or B'. We get the probability tree as in Fig. 24.13. Here $p(B|A)$ means the probability of B given A; similarly $p(B|A')$ means the probability of B given (not A). We can fill in the probabilities using our knowledge of the fair die. The probability of A is $\frac{1}{6}$. The second throw of the die is unaffected by the first throw of the die; therefore the $p(B) = p$(throwing a 5 or a 6 on one throw of the die) $= \frac{2}{6} = \frac{1}{3}$.

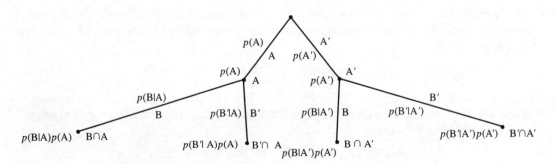

Figure 24.13 A probability tree showing conditional probabilities.

Working out the other probabilities gives the probability tree in Fig. 24.14. The probability that the first throw of a die will be a 5 and the second throw will be a 5 or 6 is $p(B \cap A)$, given from the tree in Fig. 24.14 as $\frac{1}{6} \times \frac{1}{3} = \frac{1}{18}$.

Notice that the probability we have calculated is the intersection of the two events A and B, i.e. we calculated the probability that both occurred. When multiplying the probabilities on

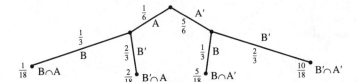

Figure 24.14 The probability tree for Example 24.9.

the branches of the probability tree we are using the following:

$$p(A \cap B) = p(A)p(B|A)$$

Furthermore, because of independence we have used the fact that $p(B|A) = p(B)$. That is, the probability of B does not depend on whether A has happened or not: B is independent of A. We therefore have the following important result.

Multiplication law of probability

$$p(A \cap B) = p(A)p(B|A)$$

This law applies for any two events A and B. It is the law used in finding the probabilities of the vertices of the probability trees.

Example 24.10 It is known that 10 per cent of a selection of 100 electrical components are faulty. What is the probability that the first two components selected are faulty if the selection is made without replacement?

SOLUTION The probability we are looking for is

$$p(\text{first faulty} \cap \text{second faulty}) = p(\text{first faulty})p(\text{second faulty}|\text{first faulty})$$

Here there are only two possibilities at each stage, faulty or not faulty. The probability tree is as shown in Fig. 24.15. Notice that there are only 99 components left after the first trial and whether the first was faulty or not changed the probability that the second is faulty or otherwise. Each branch of the tree, after the first layer, represents a conditional probability.

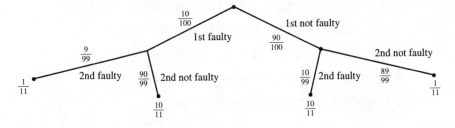

Figure 24.15 The probability tree for Example 24.10.

The answer to our problem is therefore

$$p(\text{first faulty} \cap \text{second faulty}) = p(\text{first faulty})p(\text{second faulty}|\text{first faulty}) = \tfrac{10}{100} \times \tfrac{9}{99} = \tfrac{1}{10}$$

Condition of independence

If two events A and B are independent, then

$$p(B|A) = p(B)$$

Notice that the multiplication law changes in this case, as described below.

Multiplication law of probability for independent events

$$p(A \cap B) = p(A)p(B)$$

There is one other much-quoted law of probability that completes all the basic laws from which probabilities can be worked out. That is Bayes's theorem and it comes from the multiplication law.

Bayes's theorem

As $p(A \cap B) = p(A)p(B|A)$, then as $A \cap B = B \cap A$ we can also write $p(B \cap A) = p(B)p(A|B)$. Putting the two together gives $p(A)p(B|A) = p(B)p(A|B)$ or

$$p(B|A) = \frac{p(A|B)p(B)}{p(A)}$$

Bayes's theorem is important because it gives a way of swapping conditional probabilities which may be useful in diagnostic situations where not all of the conditional probabilities can be found directly.

> **Example 24.11** In a certain town there are only two brands of hamburgers available, Brand A and Brand B. It is known that people who eat Brand A hamburger have a 30 per cent probability of suffering stomach pains and those who eat Brand B hamburger have a 25 per cent probability of suffering similarly. Twice as many people eat Brand B as eat Brand A hamburgers. However no one eats both varieties. Suppose that one day you meet someone suffering from stomach pains who has just eaten a hamburger. What is the probability that they have eaten Brand A and what is the probability that they have eaten Brand B?
>
> SOLUTION First we define the sample space S, and the other simple events.
>
> $\qquad$ S = people who have just eaten a hamburger
>
> $\qquad$ A = people who have eaten a Brand A hamburger
>
> $\qquad$ B = people who have eaten a Brand B hamburger
>
> $\qquad$ C = people who are suffering stomach pains
>
> We are given that
>
> $$p(A) = \tfrac{1}{3}$$
> $$p(B) = \tfrac{2}{3}$$
> $$p(C|A) = 0.3$$
> $$p(C|B) = 0.25$$

Note also that $S = A \cup B$.

As those who have stomach pains have either eaten brand A or brand B, but not both, and as $S = A \cup B$ and $A \cap B = \varnothing$:

$$p(C) = p(C \cap S) = p(C \cap A) + p(C \cap B) = p(C|A)p(A) + p(C|B)p(B) = 0.3 \times \tfrac{1}{3} + 0.25 \times \tfrac{2}{3} = \tfrac{8}{30}$$

Then

$$p(A|C) = \frac{p(C|A)p(A)}{p(C)} = \frac{0.3 \times \tfrac{1}{3}}{8/30} = \tfrac{3}{8}$$

and

$$p(B|C) = \frac{p(C|A)p(A)}{p(C)} = \frac{0.25 \times \tfrac{2}{3}}{8/30} = \tfrac{5}{8}$$

Hence if they have a stomach pain the probability that they have eaten brand A is $\tfrac{3}{8}$ and the probability that they have eaten brand B is $\tfrac{5}{8}$.

24.8 APPLICATION OF THE PROBABILITY LAWS TO THE PROBABILITY OF FAILURE OF AN ELECTRICAL CIRCUIT

Components in series

See Figure 24.16. We denote 'The probability that A does not fail' as $p(A)$. Then the probability that S fails is $p(S') = 1 - p(S)$. As they are in series, then S will function if both A and B function: $S = A \cap B$. Therefore

$$p(S) = p(A \cap B)$$

As A and B are independent, $p(S) = p(A)p(B)$ and $p(S') = 1 - p(A)p(B)$.

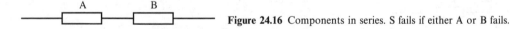

Figure 24.16 Components in series. S fails if either A or B fails.

Example 24.12 An electrical circuit has three components in series (see Fig. 24.17). One has a probability of failure within the time of operation of the system of $\tfrac{1}{5}$. The second has been found to function on 99 per cent of occasions and the third component has proved to be very unreliable, with one failure for every three successful runs. What is the probability that the system will fail on a single operation run?

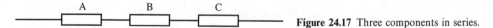

Figure 24.17 Three components in series.

SOLUTION Using the method of reasoning above we call the components A, B and C. The information we have is

$$p(A') = \tfrac{1}{5}$$

$$p(B) = 0.99$$

$$p(C'){:}p(C) = 1{:}3$$

The probability we would like to find is $p(S')$, the probability of failure of the system, and we know that the system is in series, so $S = A \cap B \cap C$. That is, all the components must function for the system to function.

We can find the probability that the system will function and subtract it from 1 to find the probability of failure:

$$p(S') = 1 - p(S)$$

As all the components are independent,

$$p(A \cap B \cap C) = p(A)p(B)p(C)$$

We can find $p(A)$, $p(B)$ and $p(C)$ from the information we are given.

$$p(A) = 1 - p(A') = 1 - \tfrac{1}{5} = 0.8$$

$p(B)$ is given as 0.99.

Given that $p(C'){:}p(C) = 1{:}3$ then C fails one in every $1 + 3$ occasions, so $p(C') = \tfrac{1}{4} = 0.25$ and $p(C) = 1 - p(C') = 0.75$.

We can now find $p(A)p(B)p(C) = 0.8 \times 0.99 \times 0.75 = 0.594$. This is the probability that the system will function, so the probability that it will fail is given by $1 - 0.594 = 0.406$.

Components in parallel

If a system S consists of two components A and B in parallel, as in Fig. 24.18, S fails if both A and B fail. Again, denote 'The probability that A does not fail' as $p(A)$. Then the probability that S fails is $p(S') = 1 - p(S)$. As they are in parallel, S will function if either A and B functions: $S = A \cup B$. Therefore

$$p(S) = p(A \cup B)$$

as A and B are independent, but not disjoint. They are not disjoint because both A and B can function simultaneously. From the addition law of probabilities given in Sec. 24.4,

$$p(A \cup B) = p(A) + p(B) - p(A \cap B)$$

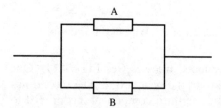

Figure 24.18 Two components in parallel. S fails if both A and B fail.

As A and B are independent,

$$p(A \cap B) = p(A)p(B)$$

Therefore

$$p(S) = p(A \cup B) = p(A) + p(B) - p(A)p(B)$$

$$p(S') = 1 - (p(A) + p(B) - p(A)p(B)) = 1 - p(A) - p(B) + p(A)p(B)$$

As this is a long-winded expression it may be more useful to look at the problem the other way round. S fails only if both A and B fail:

$$S' = A' \cap B'$$

So

$$p(S') = p(A' \cap B') = p(A')p(B')$$

as A and B are independent. Finally, $p(S') = p(A')p(B')$. This is a simpler form which we may use in preference to the previous expression we found. It must be equivalent to our previous result, so we should just check that by substituting $p(A') = 1 - p(A)$ and $p(B') = 1 - p(B)$. Hence

$$p(S') = p(A')p(B')$$

$$p(S') = (1 - p(A))(1 - p(B))$$

$$p(S') = 1 - p(A) - p(B) + p(A)p(B)$$

which is the same as we had before.

Let us try a mixed example with some components in series and others in parallel.

Example 24.13 An electrical circuit has three components, two in parallel, components A and B, and one in series, component C. They are arranged as in Fig. 24.19.

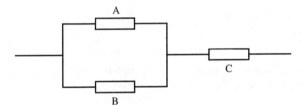

Figure 24.19 Two components in parallel and one in series (Example 24.12).

Components A and B are identical components with a $\frac{2}{3}$ probability of functioning and C has a probability of failure of 0.1%. Find the probability that the system fails.

SOLUTION Abbreviating $p(A$ functions$)$ to $p(A)$, the information we have is

$$p(A) = \tfrac{2}{3}$$

$$p(B) = \tfrac{2}{3}$$

$$p(C') = 0.001 \Rightarrow p(C) = 0.999$$

The probability we would like to find is $p(S')$, the probability of failure of the system, and we know that the system will function if (either A or B functions) and C functions. So

$$S = (A \cup B) \cap C$$

From the multiplication law for independent events we have:

$$p((A \cup B) \cap C) = p(A \cup B)p(C)$$

From the addition law for non-disjoint events (A and B are not disjoint because both A and B can occur) we have:

$$p(A \cup B) = p(A) + p(B) - p(A)p(B)$$

$$p(A \cup B) = \tfrac{2}{3} + \tfrac{2}{3} - \tfrac{2}{3} \times \tfrac{2}{3} = \tfrac{8}{9} = 0.8889$$

So

$$p(S) = p(A \cup B)p(C) = 0.8889 \times 0.999 = 0.888$$

Hence the probability that the system functions is 0.888.

24.9 STATISTICAL MODELLING

Suppose, as in Example 24.2, that we have tested 5000 resistors and recorded the resistance of each one to an accuracy of 0.01 Ω. We may wish to decide quickly whether a manufacturer's claim of producing 2000 Ω resistors is correct. One way of doing this is to divide our resistors into class intervals and draw up a table and a histogram. We can then count the percentage of resistors which are outside acceptable limits and assume that the population behaviour is the same as the sample behaviour. Hence if 98 per cent of the resistors lie between 2000 Ω±0.1% we may be quite happy.

A quicker way of doing this is to use a statistical model. That is, we can guess what the histogram would look like based on our past experience. To use such a model we only need to know the population mean, μ, and the population standard deviation, σ.

In the rest of this chapter we shall look at three possible ways of modelling data, the normal distribution, the exponential distribution (both continuous models) and the binomial distribution, a discrete model.

24.10 THE NORMAL DISTRIBUTION

The normal distribution is symmetric about its mean. It is bell-shaped and the fatness of the bell depends on its standard deviation. Examples are given in Figs 24.20 and 24.21.

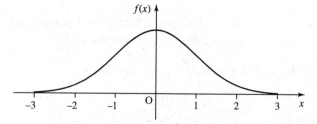

Figure 24.20 A normal distribution with $\mu=0$ and $\sigma=1$ (called the standard normal distribution $N(0,1)$).

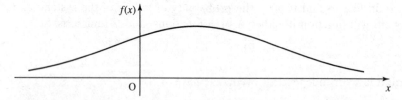

Figure 24.21 Normal distribution with $\mu=2$ and $\sigma=3$.

The normal distribution is very important because of the following points.

1. Many practical distributions approximate to the normal distribution.

Look at the histograms of lifetimes given in Fig. 24.3 and of resistances given in Fig. 24.4 and you will see that they resemble the normal distribution. Another common example is the distribution of errors. If you were to get a large group of students to measure the diameter of a washer to the nearest 0.1 mm then a histogram of the results would give an approximately normal distribution. This is because the errors in the measurement are normally distributed.

2. The central limit theorem

If we take a large number of samples from a population and calculate the sample means, then the distribution of the sample means will behave like the normal distribution for all populations (even those populations which are not distributed normally). This is as a result of what is called the central limit theorem. There is a project exploring the behaviour of sample means given in the Appendix.

3. Many other common distributions become like the normal distribution in special cases.

For instance, the binomial distribution, which we shall look at in Sec. 24.12, can be approximated by the normal distribution when the number of trials is very large.

Finding probabilities from a continuous graph

Before we look at the normal distribution in more detail we need to find out how to relate the graph of a continuous function to our previous idea of probability. In Sec. 24.3 we identified the probability of a class with its relative frequency in a frequency distribution. That was all right when we had already divided the various sample values into classes. The problem with the normal distribution is that it has no such divisions along the x-axis and no individual class heights, just a nice smooth curve. To overcome this problem we define the probability of the outcome lying in some interval of values as the area under the graph of the probability density function between those two values, as shown in Fig. 24.22.

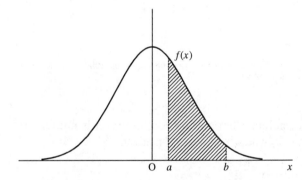

Figure 24.22 A continuous probability density function and the probability that the outcome lies in the interval $a \leqslant x \leqslant b$.

As we found in Chapter 13, the area under a curve is given by the integral; therefore for a continuous probability distribution, $f(x)$, we define

$$p(x \text{ lies between } a \text{ and } b) = \int_a^b f(x)\,dx$$

The cumulative distribution function gives us the probability of this value or any previous value (it is like the cumulative relative frequency). For a continuous distribution this becomes 'the area so far' function; the integral from the lowest possible value that can occur in the distribution up to the current value.

The cumulative distribution up to a value a is represented by

$$F(a) = \int_{-\infty}^a f(x)\,dx$$

and is the total area under the graph of the probability density function up to a; this is shown in Fig. 24.23.

We can also use the cumulative distribution function to represent probabilities of a certain interval. The area between two values can be found by subtracting two values of the cumulative distribution function, as in Fig. 24.24.

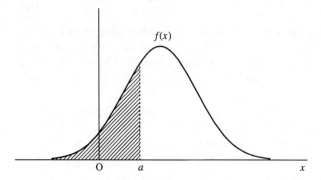

Figure 24.23 The value of the cumulative distribution $F(a)$ marked as an area on the graph of the probability density function $f(x)$ of a continuous distribution.

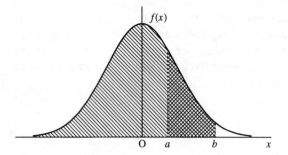

Figure 24.24 The probability $p(a < x < b)$ can be found by the difference between two values of the cumulative distribution function $F(b) - F(a)$. Compare the difference in areas with Fig. 24.22.

However, there is a problem with the normal distribution function, in that it is not easy to integrate! The probability density function for x, where x is $N(\mu, \sigma^2)$ is given by

$$f(x) = \frac{1}{\sqrt{2\pi}\sigma} e^{-\frac{(x-\mu)^2}{2\sigma^2}}$$

It is only integrated by using numerical methods. Hence the values of the integrals can only be tabulated. The values that we have tabulated are the areas in the tail of the standardized normal distribution; that is $\int_u^\infty f(z)\, dz$ where $f(z)$ is the probability distribution with 0 mean ($\mu = 0$), and standard deviation of 1 ($\sigma = 1$). This is pictured in Fig. 24.25 and tabulated in Table 24.3. In order to use these values we need to use the ideas of transformation of graphs from Chapter 8 to transform any normal distribution into its standardized form.

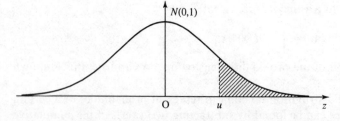

Figure 24.25 The area in the tail of the standardized normal curve, $\int_u^\infty f(z)\, dz$, tabulated in Table 24.3.

Table 24.3 Areas in the tail of the standardized normal distribution. The table gives $P(z>u)$ where z is a variable with distribution $N(0,1)$

u	0.00	0.01	0.02	0.03	0.04	0.05	0.06	0.07	0.08	0.09
0.0	0.50000	0.49601	0.49202	0.48803	0.48405	0.48006	0.47608	0.47210	0.46812	0.46414
0.1	0.46017	0.45620	0.45224	0.44828	0.44433	0.44038	0.43644	0.43251	0.42858	0.42465
0.2	0.42074	0.41683	0.41294	0.40905	0.40517	0.40129	0.39743	0.39358	0.38974	0.38591
0.3	0.38209	0.37828	0.37448	0.37070	0.36693	0.36317	0.35942	0.35569	0.35197	0.34827
0.4	0.34458	0.34090	0.33724	0.33360	0.32997	0.32636	0.32276	0.31918	0.31561	0.31207
0.5	0.30854	0.30503	0.30153	0.29806	0.29460	0.29116	0.28774	0.28434	0.28096	0.27760
0.6	0.27425	0.27093	0.26763	0.26435	0.26109	0.25785	0.25463	0.25143	0.24825	0.24510
0.7	0.24196	0.23885	0.23576	0.23270	0.22965	0.22663	0.22363	0.22065	0.21770	0.21476
0.8	0.21186	0.20897	0.20611	0.20327	0.20045	0.19766	0.19489	0.19215	0.18943	0.18673
0.9	0.18406	0.18141	0.17879	0.17619	0.17361	0.17106	0.16853	0.16602	0.16354	0.16109
1.0	0.15866	0.15625	0.15386	0.15151	0.14917	0.14686	0.14457	0.14231	0.14007	0.13786
1.1	0.13567	0.13350	0.13136	0.12924	0.12714	0.12507	0.12302	0.12100	0.11900	0.11702
1.2	0.11507	0.11314	0.11123	0.10935	0.10749	0.10565	0.10383	0.10204	0.10027	0.09853
1.3	0.09680	0.09510	0.09342	0.09176	0.09012	0.08851	0.08691	0.08534	0.08379	0.08226
1.4	0.08076	0.07927	0.07780	0.07636	0.07493	0.07353	0.07215	0.07078	0.06944	0.06811
1.5	0.06681	0.06552	0.06426	0.06301	0.06178	0.06057	0.05938	0.05821	0.05705	0.05592
1.6	0.05480	0.05370	0.05262	0.05155	0.05050	0.04947	0.04846	0.04746	0.04648	0.04551
1.7	0.04457	0.04363	0.04272	0.04182	0.04093	0.04006	0.03920	0.03836	0.03754	0.03673
1.8	0.03593	0.03515	0.03438	0.03362	0.03288	0.03216	0.03144	0.03074	0.03005	0.02938
1.9	0.02872	0.02807	0.02743	0.02680	0.02619	0.02559	0.02500	0.02442	0.02385	0.02330
2.0	0.02275	0.02222	0.02169	0.02118	0.02068	0.02018	0.01970	0.01923	0.01876	0.01831
2.1	0.01786	0.01743	0.01700	0.01659	0.01618	0.01578	0.01539	0.01500	0.01463	0.01426
2.2	0.01390	0.01355	0.01321	0.01287	0.01255	0.01222	0.01191	0.01160	0.01130	0.01101
2.3	0.01072	0.01044	0.01017	0.00990	0.00964	0.00939	0.00914	0.00889	0.00866	0.00842
2.4	0.00820	0.00798	0.00776	0.00755	0.00734	0.00714	0.00695	0.00676	0.00657	0.00639
2.5	0.00621	0.00604	0.00587	0.00570	0.00554	0.00539	0.00523	0.00508	0.00494	0.00480
2.6	0.00466	0.00453	0.00440	0.00427	0.00415	0.00402	0.00391	0.00379	0.00368	0.00357
2.7	0.00347	0.00336	0.00326	0.00317	0.00307	0.00298	0.00289	0.00280	0.00272	0.00264
2.8	0.00256	0.00248	0.00240	0.00233	0.00226	0.00219	0.00212	0.00205	0.00199	0.00193
2.9	0.00187	0.00181	0.00175	0.00169	0.00164	0.00159	0.00154	0.00149	0.00144	0.00139
3.0	0.00135	0.00131	0.00126	0.00122	0.00118	0.00114	0.00111	0.00107	0.00104	0.00100
3.1	0.00097	0.00094	0.00090	0.00087	0.00084	0.00082	0.00079	0.00076	0.00074	0.00071
3.2	0.00069	0.00066	0.00064	0.00062	0.00060	0.00058	0.00056	0.00054	0.00052	0.00050
3.3	0.00048	0.00047	0.00045	0.00043	0.00042	0.00040	0.00039	0.00038	0.00036	0.00035
3.4	0.00034	0.00032	0.00031	0.00030	0.00029	0.00028	0.00027	0.00026	0.00025	0.00024
3.5	0.00023	0.00022	0.00022	0.00021	0.00020	0.00019	0.00019	0.00018	0.00017	0.00017
3.6	0.00016	0.00015	0.00015	0.00014	0.00014	0.00013	0.00013	0.00012	0.00012	0.00011
3.7	0.00011	0.00010	0.00010	0.00010	0.00009	0.00009	0.00008	0.00008	0.00008	0.00008
3.8	0.00007	0.00007	0.00007	0.00006	0.00006	0.00006	0.00006	0.00005	0.00005	0.00005
3.9	0.00005	0.00005	0.00004	0.00004	0.00004	0.00004	0.00004	0.00004	0.00003	0.00003

The standardized normal curve

The standardized normal curve is obtained from the normal curve by the substitution $z=(x-\mu)/\sigma$, and it converts the original distribution into one with zero mean and standard deviation 1. This is useful because we can use the table of values for z given in Table 24.3 to perform calculations.

Finding the probability that x lies between a given range of values

Suppose that we have decided that a sample of resistors has a mean of 10.02 and a standard deviation of 0.06. What percentage lies inside an acceptable tolerance of 10 ± 0.1?

We want to find the area under the normal curve $N(10.02, 0.06)^2$ between $x = 9.9$ and $x = 10.1$, i.e. the shaded area in Fig. 24.26.

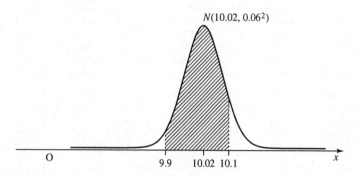

$N(10.02, 0.06^2)$

O   9.9   10.02 10.1 x

Figure 24.26 The area under the normal curve of mean 10.02 and standard deviation 0.06 between 9.9 and 10.1.

First, convert the x values to z values:

$$x = 9.9 \Rightarrow z = \frac{9.9 - 10.02}{0.06} = -2$$

$$x = 10.1 \Rightarrow z = \frac{10.1 - 10.02}{0.06} = 1.3333$$

So we now want to find the shaded area for z values (which will be the same area as above), shown in Fig. 23.27.

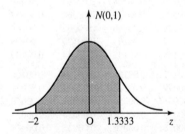

$N(0,1)$

−2   O   1.3333 z

Figure 24.27 The area under the standardized normal curve of mean 0 and standard deviation 1 between $z = -2$ and $z = 1.333$. The area is equivalent to that shown in Fig. 24.26.

In order to use Table 24.3 we need to express the problem in terms of the proportion that lies outside of the tolerance limits, as in Fig. 24.28. We use the table of the standardized normal distribution to find the proportion less than $z = -2$. As the curve is symmetric this will be the same as the proportion greater than $z = 2$. From the tables, this gives 0.022 75.

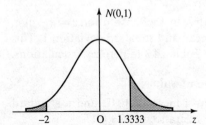

$N(0,1)$

−2   O   1.3333 z

Figure 24.28 The area outside the tolerance limits as given in Fig. 24.27.

The proportion greater than $z = 1.33$ from the tables is 0.091 76. Hence the proportion which lies outside our limits is $0.022\,75 + 0.091\,76 = 0.114\,51$.

As the total area is 1, the proportion within the limits is $1 - 0.114\,51 = 0.885\,49$.

24.11 THE EXPONENTIAL DISTRIBUTION

The exponential distribution is also called the failure rate function, as it can be used to model the rate of failure of components.

Consider a set of 1000 light bulbs, a similar make to those tested in Sec. 24.1. However, now consider a batch of bulbs at random which have already been in use for some unknown time. They are therefore of mixed ages. Measuring the time of failure we get Table 24.4. This data is represented in a histogram in Fig. 24.29 (giving the frequencies and relative frequencies) and in Fig. 24.30, giving the cumulative frequencies and relative cumulative frequencies. Notice that Fig. 24.29 looks like a dying exponential. This is not unreasonable, as we might expect failure rates to be something like the problem of radioactive decay of Chapter 14, i.e. a dying exponential.

Table 24.4 Time to failure of a sample of light bulbs

Time of failure (h)	Class mid-point	Frequency	Cumulative frequency	$f_i x_i$
0–200	100	260	260	26 000
200–400	300	194	454	58 200
400–600	500	154	608	77 000
600–800	700	100	708	70 000
800–1000	900	80	788	72 000
1000–1200	1100	60	848	66 000
1200–1400	1300	38	886	49 400
1400–1600	1500	33	919	49 500
1600–1800	1700	23	942	39 100
1800–2000	1900	14	956	26 600
2000–2200	2100	12	968	25 200
2200–2400	2300	10	978	23 000
2400–2600	2500	9	987	22 500
2600–2800	2700	13	1000	33 800
		1000		638 300

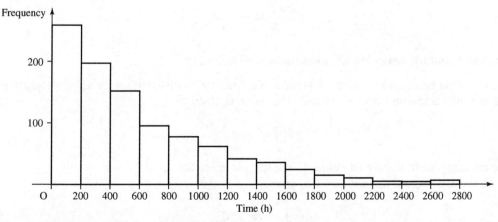

Figure 24.29 Histogram of frequencies given in Table 24.4.

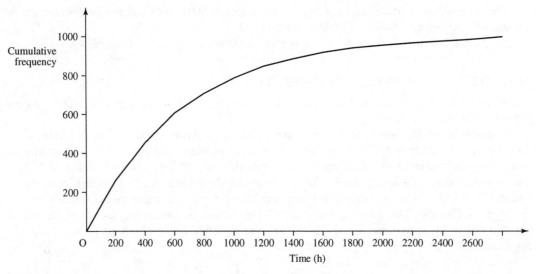

Figure 24.30 Cumulative frequency of time to failure of a sample of 1000 light bulbs.

We could think of it in a similar way to a population problem. The proportion that have failed after time t is given by the cumulative distribution function F. The proportion that are still functioning is therefore $1 - F$. The increase in the total proportion of failures is given by the failure rate multiplied by the number still functioning. If λ is the failure rate, this gives

$$\frac{\mathrm{d}F}{\mathrm{d}t} = \lambda(1 - F)$$

This differential equation can be solved to give

$$F = 1 - A\mathrm{e}^{-\lambda t}$$

Using the fact that at time 0 there are no failures, we find $A = 1$. This gives the cumulative distribution of the exponential distribution as

$$F = 1 - \mathrm{e}^{-\lambda t}$$

where λ is the failure rate, i.e. the proportion that will fail in unit time. The probability distribution can be found from the cumulative distribution by differentiating, giving

$$f = \frac{\mathrm{d}F}{\mathrm{d}t} = \lambda\mathrm{e}^{-\lambda t}$$

Mean and standard deviation of a continuous distribution

We can find the mean and standard deviation of a continuous distribution by using integration to replace the summation over all values. The mean is given by

$$\mu = \int xf(x)\,\mathrm{d}x$$

where the integration is over all values in the sample space for x.

For the exponential distribution this gives:

$$\mu = \int_0^\infty x\lambda\mathrm{e}^{-\lambda x}\,\mathrm{d}x$$

which can be found, using integration by parts (Chapter 13), to be $1/\lambda$. The standard deviation is given by

$$\sigma = \sqrt{\int (x-\mu)^2 f(x)\,dx}$$

where the integration is over all values of x. For the exponential distribution this gives

$$\sigma = \sqrt{\int_0^\infty \left(x - \frac{1}{\lambda}\right)^2 \lambda e^{-\lambda x}\,dx}$$

Again, using integration by parts, we obtain $\sigma = 1/\lambda$.

So we see that the mean is $1/\lambda$, as is the standard deviation for the exponential distribution.

Comparison of the data with the model

We can now compare a statistical model with the data given in Table 24.4. To do this we calculate the cumulative frequencies for the maximum value in each of the class intervals. The mean of the sample is 638.3. We calculated that the mean of the exponential distribution is given by $1/\lambda$, the inverse of the failure rate.

$$638.3 = 1/\lambda \Rightarrow \lambda \simeq 1.567 \times 10^{-3}$$

Using $F(x) = 1 - e^{-\lambda t}$:

$$F(200) = 1 - e^{-1.567 \times 10^{-3} \times 200} \simeq 0.269$$
$$P(0 < x < 200) = F(200) - F(0) = 0.269$$
$$F(400) = 1 - e^{-1.567 \times 10^{-3} \times 400} \simeq 0.466$$
$$P(200 < x < 400) = F(400) - F(200) = 0.197$$

etc. giving the values as in Table 24.5. The model's predictions agree quite well with the data. To find the model's predicted frequencies and cumulative frequencies we multiply by the number in the sample, 1000.

Table 24.5 Time to failure of a sample of light bulbs compared with values obtained by modelling with the exponential distribution

Time of failure (h)	Class mid-point	Frequency	Cumulative frequency	$f_i x_i$	$F(x)$	Probabilities	Cumulative frequency	Frequency
		Data				Model predictions		
0–200	100	260	260	26 000	0.269	0.269	269	269
200–400	300	194	454	58 200	0.466	0.197	466	197
400–600	500	154	608	77 000	0.609	0.143	609	143
600–800	700	100	708	70 000	0.714	0.105	714	105
800–1000	900	80	788	72 000	0.791	0.077	719	77
1000–1200	1100	60	848	66 000	0.847	0.056	847	56
1200–1400	1300	38	886	49 400	0.888	0.041	888	41
1400–1600	1500	33	919	49 500	0.918	0.03	918	30
1600–1800	1700	23	942	39 100	0.940	0.022	940	22
1800–2000	1900	14	956	26 600	0.956	0.016	956	16
2000–2200	2100	12	968	25 200	0.968	0.012	968	12
2200–2400	2300	10	978	23 000	0.977	0.009	977	9
2400–2600	2500	9	987	22 500	0.983	0.006	983	6
2600–2800	2700	13	1000	33 800	0.988	0.005	988	5

638 300

24.12 THE BINOMIAL DISTRIBUTION

Consider a random system with a sequence of trials, the trials being such that:

1. Each trial has two possible outcomes, e.g. (non-defective, defective) which we assign the outcomes of 1 (success) and 0 (failure).
2. On each trial $p(1)=\theta$ and $p(0)=1-\theta$ are the same.
3. The outcomes of the n trials are mutually independent.

$p_n(r)=$ the outcome of r successes in n trials and

$$p_n(r)=\binom{n}{r}\theta^r(1-\theta)^{n-r}$$

where

$$\binom{n}{r}={}^nC_r=\frac{n!}{(n-r)!r!}=\frac{n(n-1)\ldots(n-r+1)}{r!}$$

Setting $\alpha=1-\theta$, the probability of r successes in n trials is given by the rth term in the binomial expansion:

$$(\theta+\alpha)^n=\alpha^n+n\theta\alpha^{n-1}+\frac{n(n-1)}{2!}\theta^2\alpha^{n-2}+\ldots+\frac{n(n-1)\ldots(n-r+1)}{r!}\theta^r\alpha^{n-r}\ldots\theta^n$$

Hence the term binomial distribution.

Example 24.14 In five tosses of a coin, find the probability of obtaining three heads.

SOLUTION Assign the outcome of obtaining a head to 1 and tail to zero. Assume that the coin is fair and therefore $\theta=\frac{1}{2}$, $1-\theta=\frac{1}{2}$. The probability of obtaining three heads in five tosses of a coin is given by the binomial probability:

$$p_5(3)=\frac{5!}{3!2!}\theta^3(1-\theta)^2=\frac{5\times4}{2!}(\tfrac{1}{2})^3(\tfrac{1}{2})^2=0.3125$$

Mean and variance of a single trial

The mean of a discrete distribution can be found by using $\mu=\sum xp(x)$, and the variance is

$$\sigma^2=\sum(x-\mu)^2p(x)$$

where the summation is over the sample space.

We can use these to find the mean and variance of a single trial with only two outcomes, success or failure. The outcome of success we have given the value 1 and occurs with probability θ, and the outcome of failure has the value 0 with probability $1-\theta$. Then the mean is given by

$$1\times\theta+(1-\theta)\times0=\theta$$

The variance of a single trial is given by

$$(1-\theta)^2\theta+(0-\theta)^2(1-\theta)=\theta-2\theta^2+\theta^3+\theta^2-\theta^3=\theta(1-\theta)$$

$$\uparrow\quad\uparrow\quad\uparrow\quad\uparrow\quad\uparrow\quad\quad\uparrow$$
$$x\quad\mu\quad p(1)\quad x\quad\mu\quad\quad p(0)$$

The standard deviation is the square root of the variance, giving

$$\sigma = \sqrt{\theta(1-\theta)}$$

The mean and standard deviation of the binomial distribution

The expressions involving a summation over the entire sample space can be used to find the mean and standard deviation of the binomial distribution, but they take a bit of manipulation to find. Instead, we can take a short cut and use the fact that each trial is independent. The mean of the union of n trials is given by the sum of the means of the n trials. Similarly (for independent trials only) the variance of the union of the n trials is given by the sum of the variances.

Therefore the mean of the binomial distribution for n trials is given by number of trials × mean for a single trial $= n\theta$. The variance is given by $n\theta(1-\theta)$, and therefore the standard deviation is $\sqrt{n\theta(1-\theta)}$.

Example 24.15 A file of data is stored on a magnetic tape with a parity bit stored with each byte (eight bits) making nine bits in all. The parity bit is set so that the nine bits add up to an even number. The parity bit allows errors to be detected, but not corrected. However, if there are two errors in the nine bits then the errors will go undetected. Three errors will be detected, but four errors will go undetected etc. A very poor magnetic tape was tested for the reproduction of 1024 bits and 16 errors were found. If on one record on the tape there are 4000 groups of 9 bits, estimate how many bytes will have undetected errors.

SOLUTION Call 1 the outcome a bit being in error and 0 that it is correct. We are given that in 1024 (n) trials there were 16 errors. Taking 16 as the mean over 1024 trials and using $1024\theta = 16$:

$$\theta = \frac{16}{1024} = \frac{1}{64}$$

Errors go undetected if there are 2, 4, 6 etc. The probability of two errors in nine bits is given by

$$P_2(9) = \binom{9}{2}\theta^2(1-\theta)^7 = \frac{9!}{7!2!}\left(\frac{1}{64}\right)^2\left(\frac{63}{64}\right)^7 = 0.007\,871\,7$$

Multiplying by the number of data bytes (4000) gives approximately 31 undetected errors. The probability of four errors will obviously be much less. We check how small:

$$P_4(9) = \binom{9}{4}\theta^4(1-\theta)^5 = \frac{9!}{4!5!}\left(\frac{1}{64}\right)^2\left(\frac{63}{64}\right)^7 = 0.000\,000\,8$$

This probability is too small to show up on only 4000 bytes. As the probabilities of six or eight errors are even smaller they can safely be ignored.

The probable number of undetected errors is 31.

24.13 SUMMARY

1. The mean of a sample of data can be found by using

$$\bar{x} = \frac{1}{n}\sum_i x_i$$

where the summation is over all sample values and n is the number of values in the sample.

If the sample is divided into class intervals, then

$$\bar{x} = \frac{1}{n} \sum_i f_i x_i$$

where x_i is a representative value for the class and the summation is over all classes.

2. The standard deviation of a sample of data can be found by using

$$\sigma = \sqrt{\frac{1}{n} \sum_i (x_i - \bar{x})^2}$$

where the summation is over all sample values, n is the number of values in the sample and $\bar{x}$ is the sample mean. If the sample is divided into class intervals then

$$\sigma = \sqrt{\frac{1}{n} \sum_i f_i (x_i - \bar{x})^2}$$

where x_i is a representative value for the class, f_i is the class frequency, $\bar{x}$ is the sample mean and the summation is over all classes. The square of the standard deviation is called the variance.

3. The cumulative frequency is found by summing the values of the current class and all previous classes. It is the 'number so far'.

4. The relative frequency of a class is found by dividing the frequency by the number of values in the data sample – this gives the proportion that fall into that class. The cumulative relative frequency is found by dividing the cumulative frequency by the number in the sample.

5. In probability theory the set of all possible outcomes of a random experiment is called the sample space. The probability distribution function for a discrete sample space is a function of the outcomes which obeys the conditions:

$$0 \leqslant p(x) \leqslant 1$$

$$\sum_i p(x_i) = 1$$

6. An event is a subset of the sample space. The probability of an event in a discrete sample space (if all outcomes are equally likely) is

$$p(E) = \frac{\text{The number of outcomes in the event}}{\text{The number of outcomes in the sample space}}$$

7. The addition law of probability is given by $P(A \cup B) = P(A) + P(B) - P(A \cap B)$ for non-disjoint events. If $A \cap B = \varnothing$ this becomes $P(A \cup B) = P(A) + P(B)$ for disjoint events.

8. Multiplication law of probabilities: $p(A \cap B) = p(A)p(B|A)$ if events A and B are not independent.

9. The definition of independence is that B is independent of A if the probability of B does not depend on A: $p(B|A) = p(B)$ if B is independent of A. In this case, the multiplication law becomes $p(A \cap B) = p(A)p(B)$ for A and B independent events.

10. Bayes's theorem is

$$p(B|A) = \frac{p(A|B)p(B)}{p(A)}$$

11. The normal or Gaussian distribution is a bell-shaped distribution. Many things, particularly involving error distributions, have a probability distribution that is approximately normal.

12. To calculate probabilities using a normal distribution we use areas of the standard normal

distribution in table form, so the variable must be standardized by using the transformation

$$z = \frac{x - \mu}{\sigma}$$

where μ is the mean and σ is the standard deviation of the distribution.
13. The exponential distribution is used to model times to failure.
14. The probability density function of the exponential distribution is $f(x) = \lambda e^{-\lambda x}$ and the cumulative density function is given by $F(x) = 1 - e^{-\lambda x}$.
15. The mean and standard deviation of a continuous distribution can be found by

$$\mu = \int x f(x)\, dx$$

$$\sigma = \sqrt{\int (x - \mu)^2 f(x)\, dx}$$

where the integration is over the sample space.

For the exponential distribution these give $\mu = 1/\lambda$, i.e. the mean time of failure is the reciprocal of the failure rate. Also, $\sigma = 1/\lambda$.
16. The binomial distribution is a discrete distribution which models repeated trials where the outcome is either success or failure and each trial is independent of the others.
17. The mean of a discrete distribution is given by $\mu = \sum_i x_i p(x_i)$ and the variance $\sigma^2 = \sum_i (x_i - \mu)^2 p(x_i)$. The mean of the binomial distribution, for n trials, is $n\theta$, the variance is $n\theta(1 - \theta)$ and the standard deviation $\sqrt{n\theta(1 - \theta)}$.

24.14 EXERCISES

24.1 An integrated circuit design includes a capacitor of 100 pF (picofarads). After manufacture, 80 samples are tested and the following capacitances found for the nominal 100 pF capacitor (the data is expressed in pF):

90	100	115	80	113	114	99	105	90	99	106	103	95	105	95	101	87
91	101	102	103	90	96	97	99	86	105	107	93	118	94	113	92	110
104	104	95	93	95	85	96	99	98	83	97	96	98	84	102	109	98
111	119	110	108	102	100	101	104	105	93	104	97	83	98	91	85	92
100	91	101	103	101	86	120	96	101	102	112	119					

Express the data in table form and draw a histogram. Find the mean and standard deviation.

24.2 What is the probability of throwing a number over 4 on one throw of a die?

24.3 What is the probability of throwing either a number less than 4 or a 6 on one throw of a die?

24.4 What is the probability of throwing either an odd number or a number over 4 on one throw of a die?

24.5 What is the probability of drawing any Heart or a King from a well-shuffled pack of cards?

24.6 On two throws of a die, what is the probability of a 6 on the first throw followed by an even number on the second throw?

24.7 Throwing two dice, what is the probability that the sum of the dice is 7?

24.8 What is the probability that the first two cards dealt from a pack will be clubs?

24.9 A component in a communication network has a one per cent probability of failure over a 24 h period. To guard against failure an identical component is fitted in parallel with an automatic switching device should the original component fail. If that also has a one per cent probability of failure, what is the probability that despite this precaution the communication will fail?

24.10 Find the reliability of the system, S, in Fig. 24.31. Each component has its reliability marked in the figure. Assume that each of the components is independent of the others.

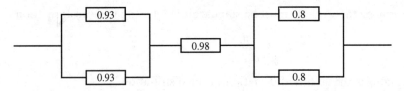

Figure 24.31 A system for Exercise 24.10.

24.11 A ball is chosen at random out of a bag containing two black balls and three red balls, and then a selection is made from the remaining four balls. Assuming all outcomes are equally likely, find the probability that a red ball will be selected:

(a) the first time
(b) the second time
(c) both times

24.12 A certain brand of compact disc (CD) player has an unreliable integrated circuit (IC) which fails to function on one per cent of the models as soon as the player is connected. On 20 per cent of these occasions the light displays fail and the buttons fail to respond, so that it appears exactly the same as if the power connection is faulty. No other component failure causes that symptom. However, two per cent of people who buy the CD player fail to fit the plug correctly, in such a way that they also experience a complete loss of power. A customer rings the supplier of the CD player saying that the light displays and buttons are not functioning on the CD. What is the probability that the fault is due to the IC failing as opposed to the poorly fitted plug?

24.13 If a population which is normally distributed has mean 6 and standard deviation 2, then find the proportion of values greater than the following:

(a) 9   (b) 10   (c) 12   (d) 7

24.14 If a population which is normally distributed, has mean 3 and standard deviation 4, find the proportion of values less than the following:

(a) 1   (b) −5   (c) −1   (d) 0

24.15 If a population which is normally distributed has mean 10 and standard deviation 3, find the proportion of values which satisfy the following:

(a) $x > 3$ (b) $x < 12$ (c) $x > 9$ (d) $x < 11$ (e) $9 < x < 11$ (f) $3 < x < 12$

24.16 A car battery has a mean life of 4.2 years and a standard deviation of 1.3 years. It is guaranteed for 3 years. Estimate the percentage of batteries that will need replacing under the guarantee.

24.17 A certain component has a failure rate of 0.3 per hour. Assuming an exponential distribution calculate the following:

(a) The probability of failure in a 4 h period.
(b) The probability of failure in a 30 min period.
(c) The probability that a component functions for 1 h and then fails to function in the second hour.
(d) Of a group of five components the probability that exactly two fail in an hour.

24.18 A certain town has 50 cash dispenser machines, but due to inaccessibility is only visited for repairs once a week, after which all the machines are working. The failure rate of the machines is approximately 0.05 per 24 hours.

A town-councillor makes a public complaint in the newspaper that on average at least one in 10 of the machines does not function. Assuming an exponential model, calculate the number that are not functioning 1 day, 2 days,..., 7 days after the day of the visit. Take the mean of these results to assess whether the councillor is correct.

24.19 A bag contains two red balls and eight green balls. A ball is repeatedly chosen at random from the bag, its colour recorded and then replaced. Find the following probabilities:

(a) The first three picked were green.
(b) In a selection of five there were exactly two red balls.
(c) There were no more than three red balls out of the first 10.

APPENDIX
PROJECTS AND INVESTIGATIONS

A1 AN INVESTIGATION OF THE NUMBER e

An important equation which describes many physical situations is

$$\frac{dy}{dt} = ky \qquad k \text{ is a constant}$$

That is, the derivative of the function is equal to a constant times the function value for all values of t. Examples of models which lead to this differential equation are given in Chapter 14. This is an investigation of functions which obey this equation. For the purposes of the investigation we assume no previous knowledge of the number e.

A1.1 Draw the graph of $y = 2^t$ with $-3 < t < 3$ (use a scale of 2 cm = 1 unit on both axes).

A1.2 Measure the gradient of the tangent to the graph for various values of t (e.g. at intervals of 0.5) and clearly show on your graph how you have measured the tangent gradients.

Using your measured values of the gradient of the tangent, plot an approximate graph of dy/dt against t.

A1.3 Examine the ratio

$$\frac{dy/dt}{y}$$

for various t values. Is this ratio approximately constant?

A1.4 Repeat steps 1 to 3 for $y = 3^t$ (use a t range of $-3 \leqslant t \leqslant 2$).

A1.5 If we were to suppose that a number exists, which we could call e, such that the function $y = e^t$ was such that $dy/dt = 1y$, then how could you use your results to justify the assumption that $2 < e < 3$?

A1.6 Investigate a graphical, or any other method, of estimating e to one decimal place, assuming that all you know is that $y = e^t$ gives $dy/dt = y$.

A2 NUMERICAL INTEGRATION

This is an investigation of numerical integration and the relationship between the error and the step size used.

A2.1 Use the software provided to draw a graph of the function $f(t) = t(t-2)(t+3)$ for $-4 \leqslant t \leqslant 4$.

A2.2 Perform numerical integration using the trapezoidal rule with a starting value of -3 and a step size of 0.1, i.e. calculate an approximation to

$$A(t) = \int_{-3}^{t} t'(t'-2)(t'+3)\, dt'$$

and plot the approximate graph of $A(t)$ for $-3 \leqslant t \leqslant 4$.

A2.3 Record in a table the approximate values of

$$A(t) = \int_{-3}^{t} t'(t'-2)(t'+3)\, dt'$$

for $t = -2, -1.5, -1, -0.5, 0, 0.5, 1, 1.5, 2, 2.5, 3, 3.5, 4$.

A2.4 Compare your graph of $A(t)$ with the graph of

$$g(t) = \frac{t^4}{4} + \frac{t^3}{3} - 3t^3 + 2$$

In what ways are the graphs of $g(t)$ and $A(t)$ similar? How and why are they disimilar?

A2.5 Find

$$\int_{-3}^{0} t'(t'-2)(t'+3)\, dt'$$

by calculating the exact integral.

A2.6 Using the software provided, calculate the approximate value of

$$\int_{-3}^{0} t'(t'-2)(t'+3)\, dt'$$

using a step size of $h = 1.5, 1, 0, 0.5, 0.3, 0.15, 0.1, 0.05, 0.03, 0.015$ and 0.01.

A2.7 For each value of h used in A2.6, calculate:

(a) The error.
(b) The relative error.

A2.8 Using the error values obtained in A2.7, draw a graph of error values, ε, against h. Comment on the relationship between ε and h shown in the graph.

A2.9 Draw a graph of the relative error, ε_r, against h. Using the graph (or otherwise) suggest a step size h such that the integral could be assumed to be correct to six significant figures. Explain the reasons for your choice and check your result by performing the calculation with your suggested value of h.

A2.10 Is it possible to obtain a general formula for the error in terms of the step size h whatever the function $f(t)$ we are integrating? What would you expect for Simpson's rule – would the error be larger or smaller than the error in this case?

A3 SEQUENCES TO SOLVE EQUATIONS

A3.1 Using the software provided, draw a graph of the function

$$y = t^3 - 5t - 1$$

A3.2 From the graph estimate the three solutions to the equation $t^3 - 5t - 1 = 0$.

A3.3 Using the software provided and the Newton–Raphson method of solving equations, i.e. the recurrence relation

$$t \leftarrow t - \frac{f(t)}{f'(t)}, \qquad \text{where } f(t) = t^3 - 5t - 1$$

attempt to find the three solutions of the equation, $t^3 - 5t - 1 = 0$, to 2, 4, 6 and 8 significant figures. How many iterations of the algorithm were required in each case? Comment on what happens when the method fails to converge.

A3.4 The equation $t^3 - 5t - 1 = 0$ can be rearranged in the following way:

$$t^3 - 5t - 1 = 0$$

$$\Leftrightarrow t^3 - 1 = 5t$$

$$\Leftrightarrow t = \tfrac{1}{5}(t^3 - 1)$$

and the recurrence relation

$$t \leftarrow \tfrac{1}{5}(t^3 - 1)$$

Then provides another method for numerical solution of the equation.

Now use this recurrence relation to attempt to find the three solutions of the equation $t^3 - 5t - 1 = 0$ to 2, 4, 6 and 8 significant figures. How many iterations of the algorithm were required in each case? Again comment on what happens when the method fails to converge. Compare the rate of convergence of this method with that of the Newton–Raphson method.

A3.5 A fixed point method, $t \leftarrow g(t)$, is known to converge if $|g'(t)| < 1$ when t is near a solution, t^*, such that $t^* = g(t^*)$. Use this to explain those cases where the two methods above succeeded or failed to converge.

A4 AN INTRODUCTION TO SYSTEMS OF LINEAR EQUATIONS

A4.1 Using three sets of axes draw graphs of the following systems of equations (two equations on each set of axes):

(a) $2x + 3y = 2$
$\quad\ 6x - y = 2$

(b) $9x - 6y = 12$
$\quad\ -6x + 4y = -8$

(c) $3x + y = -1$
$\quad\ -6x + 2y = 3$

A4.2 By using the graphs, list all the solutions of each set of simultaneous equations given in Sec. A4.1.

A4.3 Check your graphical solutions by an algebraic solution to each set of equations.

A4.4 Can you draw any general conclusions concerning systems of linear equations from the graphical approach? Can you also do this from the algebraic approach?

A4.5 Another method of analysing systems of equations is to use determinants. A 2×2 determinant is defined in the following manner:

$$\begin{vmatrix} a & b \\ c & d \end{vmatrix} = ad - cb$$

i.e. it takes an arrangement of four numbers and results in a single number. (Notice that swapping the order of the rows in the determinant merely changes the sign of the determinant; similarly swapping the columns.)

In a system of two linear equations there are basically three sets of determinants (ignoring swapping rows or columns), e.g. system A4.1(a):

$$2x + 3y = 2$$

$$6x - y = 2$$

contains the determinants:

$$\begin{vmatrix} 2 & 3 \\ 6 & -1 \end{vmatrix} \quad \begin{vmatrix} 2 & 2 \\ 6 & 2 \end{vmatrix} \quad \begin{vmatrix} 2 & 3 \\ 2 & -1 \end{vmatrix}$$

A4.6 Calculate the three determinants present in each of the three systems of equations given in A4.1.

A4.7 Using the results of A4.6, investigate a way of analysing systems of equations using determinants.

A4.8 Invent three other systems of equations which display similar properties to A4.1(a), A4.1(b) and A4.1(c) (but are not equal systems) and check your graphical, algebraic and determinant method of analysing the systems of equations.

A4.9 Generalize your results for a general system of equations as below:

$$ax + by = e$$

$$cx + dy = f$$

where a, b, c, d, e and f are real numbers.

Give a way of deciding whether a system of equations is consistent or inconsistent and whether it is an indeterminate system.

A4.10 Investigate a system of three equations and two unknowns, e.g.

$$2x + 3y = 1$$

$$4x + y = 2$$

$$3x - 4y = 0$$

Can you solve this system of equations? Are there some such systems that you can or cannot solve? Generalize your results if possible.

A4.11 Suggest any other interesting lines (or planes!) of investigation.

A5 A NON-LINEAR ELECTRICAL CIRCUIT

A two-terminal electrical direct current device has the voltage–current characteristic shown in Fig. A5.1, with the equation $v = f(i)$. The device is connected in series with a d.c. voltage source of 50 V and resistor of resistance 250 Ω. The current i in the resulting circuit can be modelled by the equation:

$$50 = 250i + f(i)$$

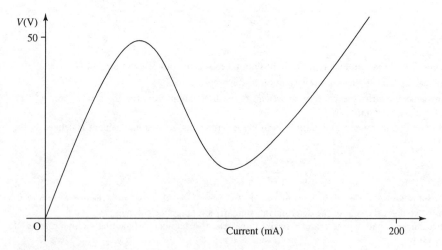

Figure A5.1 The voltage-current characteristic of a non-linear device.

A5.1 Show by graphical means or otherwise that the possible steady state currents are approximately 35, 80 and 125 mA.

A5.2 When an attempt is made to operate the circuit at the equilibrium values it is found that it is only stable at two of the predicted values of the current, 35 and 125 mA. If an attempt is made to operate at 80 mA the circuit settles down to operate at either 35 or 125 mA.

To improve the model we assume that there is always some residual inductance in an electrical circuit and that for small inductance, L, the v–i characteristic still holds. The circuit is now modelled by the equation

$$L\frac{di}{dt} + 250i + f(i) = 50$$

Suppose that i_s is a possible steady state current; that is,

$$50 = 250i_s + f(i_s)$$

and set $i = i_s + i_p$, where i_p is small. Then a Taylor first-order approximation would give

$$f(i) \approx f(i_s) + i_p f'(i_s)$$

for a region near one of the steady state values. Substitute these assumptions into our equation and use this to explain the unstable behaviour at $i = 80$ mA.

A5.3 Assuming that at time, $t = 0$, $i = 80$, suggest method(s) to obtain an approximate solution to the differential equation

$$L\frac{di}{dt} + 250i + f(i) = 50$$

using a suitable (small) value for the residual inductance L. Using your method(s) give the approximate time before the circuit has settled to operating at either 35 or 125 mA.

A6 PROBABILITY AND THE BINOMIAL DISTRIBUTION

There is a box of 20 green balls and 5 red balls. One is picked out of the box at random, recorded and then placed back in the box. If this process is repeated we can represent this using a probability tree.

After performing two trials we have the probability tree as in Fig. A6.1. This data can also be represented as in Table A6.1. If the order is not considered to be important we can represent the outcomes by the number of greens (the number of 'successes') and hence get Table A6.2. This gives us the probability distribution function $P_2(r)$ (the probability of r successes in two trials). It can also be represented in a graph, as in Fig. A6.2.

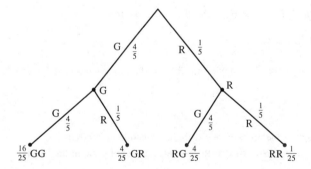

Figure A6.1 A probability tree for two trials.

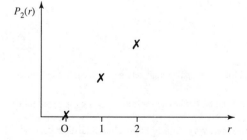

Figure A6.2 A graph of $P_2(r)$.

A6.1 Draw the probability tree for five trials, marking all the probabilities.

A6.2 Produce the table of outcomes as in Table A6.1 and in Table A6.2 for five trials.

Table A6.1

Outcome	Probability
GG	0.64
GR	0.16
RG	0.16
RR	0.04

Table A6.2

Outcome	Probability
2	0.64
1	0.32
0	0.04

A6.3 Draw the graph of the probability distribution function for five trials.

A6.4 Use your results to answer the following questions:

(a) What is the probability of selecting 4 green balls out of 5?
(b) What is the probability of selecting no green balls out of 5?
(c) What is the most probable number of green balls selected out of five?
(d) What is the probability that two or more balls are red out of the five?
(e) Given that the first four balls selected are green, what is the probability that the fifth ball selected is red?

A6.5 n represents the number of trials and r is the number of green balls after n trials. Given that

$$\binom{n}{r} = \frac{n!}{(n-r)!r!}$$

is the number of combinations of r objects chosen from n, briefly justify (if possible) the following conclusion for our green/red ball problem

$$P_n(r) = \binom{n}{r}\left(\frac{4}{5}\right)^r\left(\frac{1}{5}\right)^{n-r}$$

A6.6 Use the expression given in A6.5 to draw a graph of the probability distribution function for 10 trials.

A6.7 An electronics company has 10 telephone lines. It is discovered that the average number of engaged lines at any instant is 2. Calculate the probability that:

(a) Exactly 2 lines are engaged.
(b) More than 5 lines are engaged.

A6.8 There are three assumptions used in the binomial distribution model:

(a) Each trial is independent.
(b) There are two possible outcomes of each trial (success/failure).
(c) The probability of success on any one trial is a constant.

Suggest two other examples of the use of the binomial distribution (in particular related to communication or other engineering problems).

A6.9 Suppose that in our original problem (A6.1) we had not replaced the ball after each selection. Draw the probability tree for three trials.

Looking at the assumptions of the binomial distribution of A6.8, explain why this is not now a binomial distribution.

A6.10 Give one communication or other engineering example where the outcome of each trial is success/failure but is not suitable for modelling using the binomial distribution.

A7 THE DISTRIBUTION OF THE SAMPLE MEAN

This is an investigation into the distribution of the sample mean. In order to perform the investigation a lot of data needs to be collected, so it is preferable if organized among a group of students. If students work in pairs, then sufficient data can be collected in less than 20 min. Alternatively, a computer simulation could be used.

The result of this investigation illustrates one application of the central limit theorem. That is, the sum of a large number of random variables approaches the normal distribution, even when the original random variables are not themselves normally distributed. We also attempt to derive the relationship between the distribution of the sample means and the population mean and standard deviation.

The distribution of the sample mean is important in quality assurance. Suppose a factory claims to produce 30 mm nails and accepts a standard deviation of 2 mm. Each day a certain sample of the production is measured. The mean of the sample should be 'something near' 30 mm, but how near is acceptable if the hypothesis that the population is of mean 30 mm and standard deviation is 2 mm is to be considered reasonable? After this investigation you should be able to answer this question.

The example of a die is used because (a) it is very simple to collect lots of data quickly and (b) the population distribution (outcomes of throwing a die) is nowhere near normal, and yet the sample means display nearly normal behaviour for quite small numbers in the sample.

A7.1 Draw the probability distribution for the outcomes of throwing a fair die. Find the mean and the standard deviation. This is the population distribution.

A7.2 Throw a die four (sample size $n=4$) times and find the mean of the four throws. Record the mean and repeat this process about 50 times. You now have 50 data values, each one found from the mean of four numbers.

A7.3 Draw a histogram of the distribution of the sample means found in Sec. A7.2. Find the mean of this data and its standard deviation that is, the mean, $\bar{x}_n$ and the standard deviation, σ_n of the sample means, where the number in the sample is $n=4$).

A7.4 Repeat A7.2 and A7.3 for sample sizes of 5, 6, 7, 8, 9 and 10.

A7.5 Comment on the shape of the histograms found in each case.

A7.6 Draw a graph of the standard deviation of the sample means, σ_n, against n (the number in the sample) and draw a log–log graph of the same data. Estimate the relationship between σ_n and n.

A7.7 Write out your conclusions about the distribution of the sample mean and suggest how this could be used in the quality assurance problem given in the introduction.

ANSWERS TO EXERCISES

CHAPTER 1

1.1 (a) $a \times b = b \times a$; true   (b) $a \times (b+c) = a \times b + a \times c$; true
(c) $a + (b+c) = (a+b) + c$; true   (d) $1/(1/a) = a$; true

1.3 (a) $^-5$ (b) $^+10$ (c) x (d) $1/x$

1.4 (a) $3a - 11b$ (b) $-3a + 4b - c - 10$

1.5 (a) No   (b) No   (c) Yes

1.6 Yes; $x = 1$

1.7 No

1.8 x

1.9 (a) 20   (b) 39   (c) 60   (d) -6

1.10 (a) $5q$ (b) $-5y$ (c) $y^2 + 5y + 6$ (d) $5a^2 + 10ab + 5b^2$

1.11 (a) True   (b) True   (c) False   (d) True   (e) False   (f) False

1.12 (a) 3   (b) $-\frac{1}{7}$ (c) a (d) $y - x$ (e) $x - y$

1.13 (a) 4   (b) -4 (c) $1\frac{5}{9}$ (d) $1/a$ (e) a (f) $\dfrac{1}{a-4}$ (g) $\dfrac{1}{a^2}$

1.14 (a) $-5y$ (b) $5a - 4$

CHAPTER 2

2.1 (a) 42   (b) 11.523

2.2 $2a + \dfrac{1}{b}$

2.3 $\dfrac{10b(a+3)}{a+2}$

2.4 (1) $\left(\dfrac{ac}{bd}\right)$, $(a)\left(\dfrac{c}{bd}\right)$, $\left(\dfrac{1}{b}\right)\left(\dfrac{ac}{d}\right)$, $(c)\left(\dfrac{a}{bd}\right)$, $\left(\dfrac{1}{d}\right)\left(\dfrac{ac}{b}\right)$, $\left(\dfrac{a}{b}\right)\left(\dfrac{c}{d}\right)$, $(ac)\left(\dfrac{1}{bd}\right)$, $\left(\dfrac{a}{d}\right)\left(\dfrac{c}{b}\right)$

2.5 (a) $\dfrac{1+y}{x+1}$, defined for $x \neq -1$

   (b) $\dfrac{1+b+c}{d+e+f}$, defined for $d+e+f \neq 0$

2.6 $-3/x$

2.7 (a) $\dfrac{2x-1}{x(x-1)}$ (b) $\dfrac{a(a-d)+b(a-c)+c(a-c)(a-d)}{(a-b)(a-c)(a-d)}$

2.8 $\dfrac{2x(x+1)}{4(2x-1)}$

2.9 (a) $4\frac{5}{6}$ (b) $4\frac{45}{56}$ (c) $2\frac{7}{8}$

2.10 (a) $\dfrac{y+5}{(y+1)(y+3)}$ (b) $\dfrac{5}{y-8}$ (c) $\dfrac{y-2}{y(y+1)(y+2)}$ (d) $\dfrac{-5}{6y}$ (e) $\dfrac{1}{(y+2)(y+3)}$

2.11 (a) $\frac{1}{4}$ (b) $\dfrac{1}{3y}$ (c) $\dfrac{y+1}{y}$ (d) $\dfrac{y(y-3)}{3y+1}$ (e) $\dfrac{3y-2}{y^2(y+1)}$

2.12 (a) False   (b) True   (c) False   (d) True

2.13 295.4 µH

2.14 0.000 484

CHAPTER 3

3.1 (a) $x=-2$ (b) $y=4\frac{3}{4}$ (c) $z=5$ (d) $a=2$ (e) $x=162$
(f) $a=4$ (g) $p=5$ (h) $x=12$ (i) $p=20$ (j) $x=-21$
(k) $x=-36$ (l) $p=-\frac{1}{2}$ (m) $y=\frac{3}{4}$ (n) $x=\frac{2}{3}$

3.2 (a) $y=3$ or -20 (b) $z=5$ or -2 (c) $p=5$ or 3
(d) $x=5$ or -4 (e) $y=6$ or 2 (f) $y=1$ or -6
(g) $z=4$ or 3 (h) $x=-0.349$ or -2.151 (i) $y=4.562$ or 0.439
(j) $z=1.193$ or -4.193 (k) $q=2.303$ or -1.303 (l) $x=2.618$ or 0.382
(m) $y=3.815$ or -6.815

3.3 $u=\dfrac{vf}{v-f}$

3.4 $t=\dfrac{-u\pm\sqrt{u^2+2as}}{a}$;   (a) $t=7.838$ s   (b) $t=12.583$ s

3.5 $T_2=\dfrac{Qx}{t\lambda A}+T_1$

CHAPTER 4

4.2 (a) $-\frac{3}{4}$ (b) $\frac{8}{3}$ (c) $\frac{9}{16}$ (d) 3.674

4.3 (a) $21\frac{1}{3}$ (b) 3 (c) 0.794

4.4 (a) $a^{\frac{31}{2}}$ (b) 3^5 (c) 2^{20} (d) 2^2 (e) not possible
(f) $(3a)^4$ (g) 2^8 (h) 10^4 (i) $y^{5/6}$ (j) not possible

4.5 (a) $\log(x^6)$ (b) $\log(2)$ (c) $\log(x)$

4.6 (a) 0.862 (b) 2.262 (c) 4.983

4.7 (a) $x=2$ or -2 (b) $x=-3$ (c) no solutions   (d) $x=16$
(e) $x=3125$ (f) $x=1$ (g) $x=8$ (h) $x=0$ (i) $x=100$

4.8 $k=2$, $n=3.033$

CHAPTER 5

5.1 (a) 2.57 (b) $105\,879.2$ (c) 0.0 (d) 39.00

5.2 (a) 23.8 (b) $10\,457\,000$ (c) 1.0 (d) $0.000\,000\,057$

5.3 (a) 1.65, 1.55, 3% (b) 0.3555, 0.3545, 0.14%
(c) $0.000\,046\,5$, $0.000\,045\,5$, 1% (d) $2\,150\,000\,000$, $2\,050\,000\,000$, 2.4%

5.4 275, 265 or 270.5, 269.5 km

5.5 (a) 5.253, 4.947 Ω   (b) 0.006 130 5, 0.006 069 5 s
(c) 13.545, 12.255 kg   (d) 606 320, 559 680 m

5.6 (a) 12.05 cm, 0.3 cm, 1.2%   (b) 5.255 N, 0.6 N, 5.7%
(c) 0.1246 s, 0.05 s, 20%

5.7 (a) 1.2×10^7 (b) 7.869×10^{-7} (c) 4.6789×10^1 (d) 9.0005×10^0

5.8 (a) 50.9 (3 s.f.)   (b) 6.5 (2 s.f.)   (c) 18.26 (4 s.f.)   (d) 0.27 (2 s.f.)

5.9 7.14 Ω, 0.61 Ω

5.10 4.3×10^{-3} Ω (2 s.f.)

5.11 (0.010 097 Ω (6 s.f.)

CHAPTER 6

6.1 (a) 10.11   (b) 6.91   (c) 5.23

6.2 (a) 20.36°   (b) 41.14°   (c) 45°

6.3 (a) 7.07   (b) 11.76   (c) 3.27

6.4 29.85 m, 2.99 m

6.5 20.02 m

6.6 6.49 N

6.7 294.15I, 88.97°

CHAPTER 7

7.1 (a) {b}   (b) {a,b,c,d,e,f}   (c) {a,b}   (d) {c,d,e}
(e) {c,d,e}   (f) {b,c,d,e}   (g) {b,c,d,e}   (h) {a,b,c,d,f,g}
(i) {a,b,c,d,f,g}

7.3 (a) {0,1,2,3,4}   (b) {3,4,5,6,7,8,9}   (c) {0,1,2,3,4,5,6,7,8,9}
(d) {5,6,7,8,9}   (e) {5,6,7,8,9}

7.4 (a) False   (b) False   (c) True   (d) False

7.5 (a) A   (b) A∪B   (c) A∩B

7.6 (a) $\{(x,y)|0 \leqslant x \leqslant 79 \text{ and } 0 \leqslant y \leqslant 24\}$
(b) (i) $\{(x,y)|0 \leqslant x \leqslant 79 \text{ and } 13 \leqslant y \leqslant 24\}$
(ii) $\{(x,y)|0 \leqslant x \leqslant 79 \text{ and } 0 \leqslant y \leqslant 24 \text{ and } y \geqslant \frac{79}{24}x\}$

7.7 1

7.9 (a) (i) 3   (ii) 3   (iii) $\frac{3}{5}$ (iv) $2\frac{5}{6}$ (v) $\frac{9}{125}$ (vi) 2
(vii) $2\frac{1}{4}$ (viii) 3
(b) (i) $f \circ g : x \mapsto \frac{2}{3}x^2 - 1$ (ii) $g \circ f : x \mapsto \frac{1}{3}(2x-1)^2$
(iii) $h \circ g : x \mapsto 9/x^2$ (iv) $f^{-1} : x \mapsto \dfrac{x+1}{2}$ (v) $h^{-1} : x \mapsto 3/x$
(d) (i) 1   (ii) $\frac{9}{25}$ (iii) $16\frac{1}{3}$

CHAPTER 8

8.1 (a) 3   (b) 2   (c) −5   (d) $\frac{1}{2}$

8.2 (a) 3   (b) −5   (c) $\frac{4}{7}$ (d) 0

8.3 $y = -5x + 11$

8.4 (a) $y = -3x + 2$ (b) $y = 2x$ (c) $y = -x + 0.7$

8.5 (a) 2, −2   (b) $\frac{1}{2}$, −1   (c) 3   (d) 4, −4   (e) 2, −3
(f) −3, −4   (g) 4, −3

8.7 (a) Even   (b) Neither   (c) Odd

8.8 (a) Even   (b) Even   (c) Neither   (d) Odd   (e) Even   (f) Even

8.9 (a) Yes   (b) No   (c) Yes

8.10 (a) $t \leqslant 3.3$ (b) $x > -\frac{2}{7}$ (c) $y \leqslant -\frac{8}{5}$ (d) $t < -3$

8.11 (a) $x > -11$ (b) $t \geqslant 2$ (c) $u > 2$ or $u < -5$

8.12 (a) $-3 \leqslant x \leqslant 3$ (b) $-1 < x < \frac{3}{2}$ or $x > 5$ (c) $-7 \leqslant t \leqslant 3$
(d) $-3\frac{1}{2} \leqslant w \leqslant 2\frac{1}{2}$

8.13 (a) $A = 119$, $k = -0.2$ (b) $A = 1$, $k = 3.1$

8.14 $\gamma = 1.52$, $c = 1.38$, volume $= 1.23$ m^3

CHAPTER 9

9.1 (a) F   (b) F   (c) T   (d) T   (e) F   (f) F   (g) T   (h) T
(i) T for all t (j) T for $t = 3$, F for all other values of t (k) F

9.2 (a) 2   (b) 0, 5   (c) $-2\frac{1}{5}, \frac{1}{5}$ (d) $3\frac{1}{4}, \frac{1}{2}$ (e) $t \leqslant \frac{1}{10}$ (f) $x < -\frac{1}{2}$

9.3 (a) $x \geqslant 2 \vee x < -1$ (b) $t > 1$ (c) $x > 3$

9.4 (a) $\Leftarrow$ (b) $\Leftrightarrow$ (c) No implication   (d) $\Leftrightarrow$ (e) $\Rightarrow$ (f) $\Leftrightarrow$
(g) $\Leftarrow$ (h) $\Leftrightarrow$ (i) No implication   (j) $\Rightarrow$ (k) $\Leftrightarrow$

9.5 (a) F   (b) F   (c) T   (d) T   (e) F

9.6 (a) $x/3 \notin \mathbb{Z}$ (b) $3 \leqslant y \leqslant 60$ (c) $w/2 \in \mathbb{Z} \wedge w > 20$ (d) $|t - t_{n-1}| < 0.001$

9.7 (a) (0,6), (1,4), (2,3), (3,1), (4,0)   (b) 5 m

9.8 13.15 ms^{-1}

CHAPTER 10

10.2 (b) (i) F   (ii) T   (iii) F   (iv) T   (v) T   (vi) T

10.4 (a) a (b) abc (c) a (d) $a\bar{b}$

10.5 (b) (i) 1   (ii) 1   (iii) 0   (iv) 0

10.6 (a) $a(\bar{b} + c)$ (b) $a + b + c$ (c) $a(b + \bar{c})$ (d) $\bar{a}\bar{b}\bar{d} + ab$

10.7 (a) $(\overline{ab + c}).c$ (b) $ab\bar{c} + \bar{c}d + c\bar{d}$

10.8 $r = \bar{c} + cd + \bar{a}b$

CHAPTER 11

11.1 (a) 120°   (b) 720°   (c) 108°   (d) 360°   (e) 90°

11.2 (a) $\pi/4$ (b) $3\pi/4$ (c) $\pi/18$ (d) $5\pi/6$

11.3 (a) $\sqrt{3}/2$ (b) $\sqrt{3}$ (c) $-\frac{1}{2}$ (d) $\sqrt{3}/2$ (e) $\sqrt{3}$

11.5 (a) $-\pi/4$, 2, 2, $\frac{1}{2}$ (b) $-\pi/8$, $\frac{1}{2}$, 4, $\frac{1}{4}$

11.6 (a) 3, $\pi/2$, 4, $\pi/2$ (b) 1, $2\pi/377$, 377, $0.4 - \pi/2$ (c) 40, $\pi/1500$, 3000, -0.8

11.7 (a) 0.5, π, $1/\pi$, 2   (b) 2, $\pi/36$, $36/\pi$, 2   (c) 52, $\pi/40$, $40/\pi$, 80

11.8 (c) 0.4 ms^{-1}

11.9 110 dB

11.10 -5.74 dB

11.12 $0.0\dot{2}\dot{5}/\pi$ Hz

11.14 (a) 1.5 m, 0.75 m, 0.5 m   (b) 442 Hz

11.15 $C = 2\cos(\pi/3)$, $d = -2\sin(\pi/3)$

11.16 (a) $\sin(104°)$ (b) $\sin(4°)$ (c) $\cos(52°)$ (d) $-\cos(63°)$ (e) $\sin(3x)$ (f) $\cos(2x)$

11.17 $\cos(x)\cos(y)\cos(z) - \sin(x)\sin(y)\cos(z) - \sin(x)\cos(y)\sin(z) - \cos(x)\sin(y)\sin(z)$

11.18 (a) $\pi + 0.5236$, $2\pi - 0.5236$, $3\pi + 0.5236$, $4\pi - 0.5236$, $5\pi + 0.5236$, $6\pi - 0.5236$

(b) 0.3218, $\pi + 0.3218$, $2\pi + 0.3218$, $3\pi + 0.3218$, $4\pi + 0.3218$, $5\pi + 0.3218$

(c) 2.4981, $2\pi - 2.4981$, $2\pi + 2.4981$, $4\pi - 2.4981$, $4\pi + 2.4981$, $6\pi - 2.4981$

(d) 0.5236, $\pi - 0.5236$, $\pi + 0.5236$, $2\pi - 0.5236$, $2\pi + 0.5236$, $3\pi - 0.5236$, $3\pi + 0.5236$, $4\pi - 0.5236$, $4\pi + 0.5236$, $5\pi - 0.5236$, $5\pi + 0.5236$, $6\pi - 0.5236$

(e) $\pi/2$, $3\pi/2$, 2.0944, $2\pi - 2.0944$, $5\pi/2$, $7\pi/2$, $2\pi + 2.0944$, $4\pi - 2.0944$, $9\pi/2$, $11\pi/2$, $4\pi + 2.0944$, $6\pi - 2.0944$

(f) No solutions

CHAPTER 12

12.1 (a) 8.667 m s^{-1} (b) $v = \begin{cases} \frac{1}{5}t^2 + 2 & \text{for } 0 \leqslant t < 10 \\ 22 & \text{for } t \geqslant 10 \end{cases}$

(c) (i) 7 m s^{-1} (ii) 22 m s^{-1} (iii) 22 m s^{-1}

(d) 2 m s^{-2} (e) 0 m s^{-2} (f) $a = \begin{cases} \frac{2}{5}t & \text{for } 0 \leqslant t < 10 \\ 0 & \text{for } t \geqslant 10 \end{cases}$

(g) (i) 2 m s^{-2} (ii) 0 m s^{-2} (iii) 0 m s^{-2}

12.2 (1) $6x + 6$ (2) $\frac{1}{2}x^{-1/2} + \frac{1}{2}x^{-3/2}$ (3) $\frac{3}{2}\sqrt{2x} + \dfrac{5}{3x^3}$

(4) $(9x^2 + 1)\cos(3x^3 + x)$ (5) $-12\sin(6x - 2)$

(6) $2x\sec^2(x^2)$ (7) $\dfrac{-2}{(2x-3)^2}$ (8) $24(4x-5)^5$

(9) $\dfrac{-x}{\sqrt{(x^2-1)^3}}$ (10) $\dfrac{-2}{\sqrt{1-(5-2x)^2}}$ (11) $\dfrac{-\sec^2(1/x)}{x^2}$

(12) $\dfrac{x}{\sqrt{x^2+2}}$ (13) $-\frac{3}{2}(x+4)^{-5/2}$ (14) $2\sin(x)\cos(x)$

(15) $-15\cos^2(x)\sin(x)$ (16) $\dfrac{-3\cos(x)}{\sin^4(x)}$ (17) $-10\cos(5x)\sin(5x)$

(18) $3x^2\sqrt{x+1} + \dfrac{\frac{1}{2}x^3}{\sqrt{x+1}}$ (19) $5\cos(x) - 5x\sin(x)$

(20) $12x\sin(x) + 6x^2\cos(x)$ (21) $3\tan(5x) + (15x+5)\sec^2(5x)$

(22) $3x^2\cos^{-1}(x) - \dfrac{x^3}{\sqrt{1-x^2}}$ (23) $\dfrac{3x^2\cos(x) + x^3\sin(x)}{\cos^2(x)}$

(24) $\dfrac{-2\cos(x)}{\sin^3(x)}$ (25) $\dfrac{(2x+10)\cos(x) - 2\sin(x)}{(2x+10)^2}$

(26) $\dfrac{2x\tan(x) - x^2\sec^2(x)}{\tan^2(x)}$ (27) $\dfrac{6x}{\sqrt{x-1}} - \dfrac{3x^2}{2\sqrt{(x-1)^3}}$

(28) $\dfrac{20x}{(5x^2+1)^2}$ (29) $\dfrac{-\cos(x)}{(x+1)^2} - \dfrac{\sin(x)}{x+1}$

(30) $\dfrac{2x}{\sqrt{1-x^4}}$ (31) $\dfrac{(\frac{5}{2}x-2)x\sin(x) + x^2\cos(x)}{\sqrt{x-1}}$

(32) $-4x\cos(x^2)\sin(x^2)$ (33) $\dfrac{5\tan(\sqrt{5x-1})\sec^2(\sqrt{5x-1})}{\sqrt{5x-1}}$

12.3 $-(9.475 \times 10^5 \cos(20\pi t) + 3.553\cos(30\pi t))$

CHAPTER 13

13.1 (a) $\dfrac{x^4}{4} + \dfrac{x^3}{3} + c$ (b) $-2\cos(x) + \tan(x) + c$

(c) $\dfrac{-1}{2x^2}+c$ (d) $x+\dfrac{x^3}{3}+\dfrac{3x^4}{4}+c$

(e) $x-\dfrac{5x^2}{2}+c$ (f) $-\frac{1}{4}\sin(2-4x)+c$

(g) $\dfrac{\sqrt{(2x-1)^3}}{3}+c$ (h) $2\sqrt{x+2}+c$

(i) $\dfrac{(x^2-4)^4}{8}+c$ (j) $\dfrac{\sqrt{(1+x^2)^3}}{3}+c$

(k) $\dfrac{-1}{1+\sin(x)}+c$ (l) $\frac{1}{2}(x^2+x-6)+c$

(m) $\dfrac{-4}{3(x^3-7)}+c$ (n) 0.0207

(o) $x^2\sin(x)+2x\cos(x)-2\sin(x)+c$ (p) 8.633

(q) $(2x-3)^5\left(\dfrac{10x+3}{120}\right)+c$ (r) $-\cos(x)+\dfrac{2\cos^3(x)}{3}-\dfrac{\cos^5(x)}{5}+c$

(s) $\frac{1}{32}\sin(4x)+\frac{1}{4}\sin(2x)+\frac{3}{8}x+c$ (t) $-\frac{1}{16}\cos(8x)+\frac{1}{2}\cos(2x)+c$

13.2 $s=3t-\dfrac{t^2}{2}+5$; when $t=2$ s, $s=9$ m

13.3 $y=-5x$

13.4 $y=x-\frac{2}{3}x^3+1$

13.5 0.0187 A

13.6 $F=\dfrac{1}{\pi}(1-\cos(3\pi t))$

13.8 $28\frac{1}{2}$

13.9 9

13.10 2

13.11 4.76

13.12 1.382

13.13 (a) $0.945\,078\,8$, $0.945\,832\,09$
(b) $0.946\,145\,9$, $0.946\,086\,93$

13.14 (a) $1.106\,746$ (b) $1.098\,942$

13.15 0.1667

CHAPTER 14

14.2 $p_0=1$, $k=\frac{1}{1200}$

14.3 $N_0=5\times10^{-6}$, $k=-4.3\times10^{-4}$

14.4 $A=100$, $k=-0.1$

14.7 (a) 4.14431 (b) $0.995\,05$ (c) $0.568\,82$
(d) Not defined   (e) Not defined

14.8 (a) $2te^{t^2-2}$ (b) $e^{-t}(2\sinh(2t)-\cosh(2t))$

(c) $\dfrac{2x\sinh(x)-(x^2-1)\cosh(x)}{\sinh^2(x)}$ (d) $\dfrac{3x^2-3}{x^3-3x}$

(e) $\dfrac{1}{\ln(2)x}$ (f) $4\ln(a)a^{4t}$ (g) $2^t(\ln(2)t^2+2t)$

(h) $\dfrac{-2}{(e^{t-1})^2}$

14.9 (a) $\frac{1}{4}e^{4t-3}+c$ (b) 0.113 (c) $\frac{1}{4}\cosh(2x^2)+c$

(d) $\dfrac{x^2}{2}\ln(x)-\dfrac{x^2}{4}+c$ (e) $0.718\,28$

(f) $\ln(\cosh(t)) + c$ (g) $\ln(x^2 - 2x - 4) + c$

(h) $2\ln(t-3) - \ln(t-1) + c$ (i) $-0.405\,47$

14.10 $I = -0.01e^{-10t}$, 0.0693 s

14.11 (a) -0.0025 A   (b) -1.684×10^{-5} A   (c) -1.135×10^{-7} A

CHAPTER 15

15.1 (a) $\mathbf{b} - \mathbf{a}$ (b) $\mathbf{a} - \mathbf{b}$ (c) $\mathbf{a} + \mathbf{b}$ (d) $-\mathbf{b}$

(e) $-\mathbf{a}$ (f) $\mathbf{b} - \mathbf{a}$ (g) $\mathbf{a} - \mathbf{b}$ (h) $2\mathbf{a}$

(i) $-2\mathbf{b}$ (j) $\mathbf{a} + \mathbf{b}$

15.2 (a) $(0,5)$ (b) $(2,1)$ (c) $(-2,-1)$

(d) $(2,1)$ (e) $(-2,4)$ (f) $(-1,7)$

(g) $(4,7)$ (h) $(-3,2,3)$ (i) $(30,60,20)$

(j) $(39,30,-4)$ (k) $(12,32,19)$

15.3 (a) $3.162\underline{/1.249}$ (b) $3.162\underline{/-0.322}$

(c) $3.162\underline{/-1.893}$ (d) $7.810\underline{/-0.876}$

15.4 (a) $(-5,0)$ (b) $(-1,0)$ (c) $(0.354,-0.354)$ (d) $(1.5,2.598)$

15.5 (a) $-1.248\cos(3t) + 2.728\sin(3t)$

(b) $2.837\cos(20t) + 9.589\sin(20t)$

15.6 (a) $5\cos(10t + 0.644)$ (b) $10.20\cos(157t - 1.768)$

15.7 (a) $13.040\cos(2t + 2.457)$ (b) 0

(c) $6.325\cos(628t - 2.820)$

15.8 (a) $(0.6,0.8)$ (b) $(\frac{5}{13},\frac{12}{13})$ (c) $(\frac{5}{13},\frac{-12}{13})$

(d) $(0.707,0.707)$ (e) $(0.832,0.555)$ (f) $(1,0)$

(g) $(0,-1)$ (h) $(\frac{1}{3},\frac{2}{3},\frac{2}{3})$ (i) $(0.408,-0.408,0.816)$

(j) $(0.707,0,-0.707)$

15.9 (a) $5\mathbf{i} + 2\mathbf{j}$ (b) $-\mathbf{i} - 2\mathbf{j}$ (c) $-6\mathbf{i} + 2\mathbf{j}$

(d) $-\mathbf{i} + 2\mathbf{j} - 3\mathbf{k}$ (e) $0.2\mathbf{i} - 1.6\mathbf{j} + 3.3\mathbf{k}$

15.10 (a) -3 (b) 3 (c) -3

15.11 (a) 1.305 (b) $\pi/2$ (c) 1.616

15.14 (a) 7 (b) 10 (c) 11

CHAPTER 16

16.1 (b) (i) $4 + \mathrm{j}$ (ii) $-2 + \mathrm{j}6$ (iii) $1 + \mathrm{j}2$

(c) (i) $3 + \mathrm{j}5$ (ii) $5 - \mathrm{j}3$ (iii) 5 (iv) $-\frac{1}{6} - \dfrac{\mathrm{j}}{2}$ (v) $7 + \mathrm{j}6$

16.2 (a) 1 (b) $-\mathrm{j}$ (c) 1

16.3 (a) $34 - \mathrm{j}2$ (b) $-3 - \mathrm{j}4$ (c) $\frac{23}{26} - \mathrm{j}\frac{11}{26}$ (d) $\frac{57}{97} + \mathrm{j}5\frac{95}{97}$

16.4 $\mathrm{j}\frac{3}{2}$

16.5 $x = 3$, $y = 5$

16.6 (a) $1,2$ (b) $1 + \mathrm{j}\sqrt{5}$, $1 - \mathrm{j}\sqrt{5}$ (c) $\frac{1}{6} + \mathrm{j}\dfrac{\sqrt{11}}{6}$, $\frac{1}{6} - \mathrm{j}\dfrac{\sqrt{11}}{6}$

(d) $-\frac{1}{4}, 2$ (e) $\mathrm{j}\sqrt{\frac{3}{2}}$, $-\mathrm{j}\sqrt{\frac{3}{2}}$

16.7 (a) $-1 - \mathrm{j}3$ (b) $b = 2$, $c = 10$

16.8 (a) $4.899\underline{/1.030}$ (b) $6.708\underline{/2.678}$

(c) $6.403\underline{/-2.246}$ (d) $4.899\underline{/-2.601}$

16.9 (a) $-3.536 - \mathrm{j}3.536$ (b) $3.464 - \mathrm{j}2$ (c) $-1.827 + \mathrm{j}0.813$

(d) $3.992 - \mathrm{j}3.010$

16.10 $x = 5\frac{1}{4}$, $y = 4\frac{1}{2}$

16.11 (a) $36\underline{/23\pi/20}$ (b) $4\underline{/7\pi/20}$ (c) $13.626\underline{/2.159}$

(d) $10.968\underline{/-0.539}$ (e) $12\underline{/-3\pi/4}$ (f) $9\underline{/4\pi/5}$

16.12 $16\underline{/-3.083}$

16.13 $Z = 409174\underline{/-0.212}$, $V = 2.046 \times 10^6$ V, relative phase $= -0.212$

16.14 $Y = 3.916 \times 10^{-4}\underline{/-1.561}$, $I = 3.916 \times 10^{-3}$ A, relative phase $= -1.561$

16.15 (a) $313\underline{/32.4°}$ (b) $9.7347 \times 10^{-4}\underline{/106.5°}$

16.16 (a) $4.899e^{1.030j}$ (b) $6.708e^{2.678j}$ (c) $6.403e^{-2.246j}$
(d) $8e^{0.384j}$ (e) $3e^{2.13j}$ (f) $6e^{1.9j}$

16.17 (a) $4\underline{/2}$, $-1.665 + 3.637j$ (b) $1\underline{/-\pi/2}$, $-j$
(c) $2\underline{/\pi}$, -2 (d) $6\underline{/0.858}$, $3.922 + j4.541$
(e) $\frac{1}{2}\underline{/1.283}$, $0.142 + 0.479j$ (f) $3\underline{/11\pi/12}$, $-2.898 + j0.776$
(g) $2.906\underline{/2.017}$, $-1.255 + j2.621$

16.18 (a) $-2, 0$ (b) $-2.633, -1.438$ (c) $0.183, 0.285$
(d) $-1.821, 0.260$ (e) $-3.539, -12.133$

16.19 (a) $36e^{\frac{23}{20}\pi j}$ (b) $4e^{\frac{7}{20}\pi j}$ (c) $13.627e^{2.159j}$
(d) $10.969e^{-0.539j}$ (e) $12e^{-\frac{3}{4}\pi j}$ (f) $4e^{-\frac{23}{20}\pi j}$
(g) $144e^{0j}$

16.20 $27\underline{/1.38}$, $27e^{1.38j}$

16.21 (a) $1, j, -1, -j$
(b) $0.866 + j0.5$, j, $-0.866 + j0.5$, $-0.866 - j0.5$, $-j$, $0.866 - j0.5$
(c) $1.618 + j1.176$, $-0.618 + j1.902$, -2, $-0.618 - j1.902$, $1.618 - j1.176$
(d) $0.437 + j0.757$, -0.874, $0.437 - j0.757$

16.22 $\cos^3(\theta) - 3\cos(\theta)\sin^2(\theta)$

16.23 $\frac{3}{4}\sin(\theta) - \frac{1}{4}\sin(3\theta)$

16.24 $1.174 + j0.540$, $-0.151 + j1.284$, $-1.267 + j0.252$, $-0.631 - j1.128$, $0.877 - j0.949$

16.25 (a) $1 - j$, $-1 - j$ (b) $0.327 + j3.035$, $-0.327 - j0.035$

16.26 (a) 4 (b) 0.5

CHAPTER 17

17.1 (a) $(\frac{5}{2}, -\frac{17}{4})$ is a local minimum
(b) $(\frac{2}{3}, \frac{4}{3})$ is a local maximum
(c) $(\frac{1}{3}, -\frac{2}{9})$ is a local minimum and $(-\frac{1}{3}, \frac{2}{9})$ is a local maximum
(d) $(10, 40)$ is a local minimum and $(-10, -40)$ is a local maximum
(e) $(-4, -126)$ is a local minimum, $(0, 2)$ is a local maximum and $(1, -1)$ is a local minimum

17.2 $1/(2\sqrt{2})$, $-1/(2\sqrt{2})$

17.5 8, $-27/256$

17.6 1.193 m, 2.384 m

17.8 $0.21\omega^2$, $-0.106875\omega^2$

17.9 $1/2$, $1/2(1 + \cos(\theta))$

17.10 $\sqrt{h/3c}$, $66\frac{2}{3}\%$

CHAPTER 18

18.1 (a) 21, 25, 29, $a_{n+1} = a_n + 4$, $a_1 = -3$
(b) 0.25, 0.125, 0.0625, $a_{n+1} = \frac{1}{2}a_n$, $a_1 = 8$
(c) 0, -3, -6, $a_{n+1} = a_n - 3$, $a_1 = 18$
(d) 6, -6, 6, $a_{n+1} = -a_n$, $a_1 = 6$
(e) 2, 0, -2, $a_{n+1} = a_n - 2$, $a_1 = 10$
(f) 29, 37, 46, $a_{n+1} = a_n + n$, $a_1 = 1$
(g) 21, 28, 36, $a_{n+1} = a_n + n + 1$, $a_1 = 1$

18.2 (a) 2, 5, 8, 11, 14   (b) 720, 360, 240, 180, 144
(c) 0, -3, -8, -15, -24 (d) 6, 8, 10, 12, 14
(e) 2, 6, 18, 54, 162   (f) -1, 2, -4, 8, -16
(g) $\frac{1}{2}$, 1, $1\frac{1}{2}$, 2, $2\frac{1}{2}$ (h) 2, 5, 8, 11, 14
(i) 1, 3, 9, 27, 81

18.3 (a) $\sum\limits_{n=0}^{n=10} x^n$ (b) $\sum\limits_{n=1}^{n=8}(-2)^n$ (c) $\sum\limits_{n=1}^{n=6} n^3$ (d) $\sum\limits_{n=1}^{n=8}(-\frac{1}{3})^n$

(e) $\sum\limits_{n=2}^{n=10}\dfrac{1}{n^2}$ (f) $\sum\limits_{n=1}^{n=8}(-4)(\frac{1}{4})^{n-1}$

18.4 (a) 0, 0.1987, 0.3894, 0.5646, 0.7174, 0.8415, 0.9320, 0.9854, 0.9996, 0.9738
(b) 1, 0.9553, 0.8253, 0.6216, 0.3624, 0.0707, -0.2272, -0.5048, -0.7374, -0.9041
(c) 0, 2, 4, 6, 8, 6, 4, 2, 0, -2
(d) 1, 1, 1, 1, -1, -1, -1, -1, 1, 1

18.5 (a) 22, 42, $6+(n-1)4$
(b) 1, -1.5, $3+(n-1)(-0.5)$
(c) 17, 47, $-7+(n-1)6$

18.6 (a) $-10+(n-1)4$, 560   (b) $-5+(n-1)0.5$, -5
(c) $25+(n-1)(-3)$, -70

18.7 $\frac{2}{3}$, $2+(n-1)\frac{2}{3}$, 2, $2\frac{2}{3}$, $3\frac{1}{3}$, 4, $4\frac{2}{3}$, $5\frac{1}{3}$

18.8 32

18.9 (a) 8, 128, 2^{n-1} (b) $\frac{1}{192}$, $\frac{1}{49152}$, $\frac{1}{3}(\frac{1}{4})^{n-1}$
(c) $\frac{1}{3}$, $\frac{1}{243}$, $-9(-\frac{1}{3})^{n-1}$ (d) 29.296875, 71.525574, $15(\frac{5}{4})^{n-1}$

18.10 (a) $\frac{8}{25}5^{n-1}$, 31 249.92   (b) $(-1.3867)(-2.0801)^{n-1}$, 157.3377
(c) $64(-\frac{1}{2})^{n-1}$, 42.5

18.11 8

18.12 17

18.13 (a) 30   (b) 1.996 093 75   (c) $-2.249 961 9$

18.14 (a) $\dfrac{1-z^n}{1-z}$ (b) $\dfrac{1-(-1)^n y^{2n}}{1+y^2}$ (c) $\dfrac{2(1-(2/x)^n)}{1-2/x}$

18.15 (a) Convergent, 4   (b) Not convergent
(c) Convergent, 20.25   (d) Convergent, $\frac{1}{3}$

18.16 (a) $\frac{4}{9}$ (b) $\frac{1}{6}$ (c) $\frac{1}{45}$

18.17 (a) $1+\frac{3}{2}x+\frac{3}{4}x^2+\frac{1}{8}x^3$
(b) $1-4x+6x^2-4x^3+x^4$
(c) x^3-3x^2+3x-1
(d) $1-8y+24y^2-32y^3+16y^4$
(e) $1+8x+28x^2+56x^3+70x^4+56x^5+28x^6+8x^7+x^8$
(f) $8x^3+12x^2+6x+1$
(g) $8a^3+12a^2b+6ab^2+b^3$
(h) $x^7+7x^5+21x^3+35x+35\dfrac{1}{x}+21\dfrac{1}{x^3}+7\dfrac{1}{x^5}+\dfrac{1}{x^7}$
(i) $a^4-8a^3b+24a^2b^2-32ab^3+16b^4$

18.18 (a) 1.331   (b) 0.6561   (c) 8.120 601

18.19 (a) $1+10x+40x^2+80x^3+\ldots$
(b) $1-24x+252x^2-1512x^3+\ldots$
(c) $64-192z+240z^2+160z^3+\ldots$
(d) $1+8x+30x^2+70x^3+\ldots$
(e) $1-6x+15x^2-20x^3+\ldots$
(f) $\frac{1}{32}-\frac{5}{8}x+5x^2-20x^3+\ldots$

18.20 $5\cos^4(\theta)\sin(\theta)-10\cos^2(\theta)\sin^3(\theta)+\sin^5(\theta)$

18.21 (a) 0, 8   (b) -7, -24 (c) -12, 316

18.22 (a) 0.9227   (b) 1.0721   (c) 74.2204

18.23 (a) $1-\dfrac{x^2}{2!}+\dfrac{x^4}{4!}-\dfrac{x^6}{6!}+\ldots$, all x

(b) $1+\dfrac{x^2}{2!}+\dfrac{x^4}{4!}+\dfrac{x^6}{6!}+\dots$, all x

(c) $x-\dfrac{x^2}{2}+\dfrac{x^3}{3}-\dfrac{x^4}{4}+\dots$, $|x|<1$

(d) $1+1.5x+0.75\dfrac{x^2}{2!}-0.375\dfrac{x^3}{3!}+\dots$, $|x|<1$

(e) $1-2x+3x^2-4x^3+\dots$, $|x|<1$

18.24 (a) $1-x^2+\dfrac{x^4}{3}-\dfrac{17}{6!}x^6+\dots$

(b) $x-\dfrac{x^3}{3}+\dfrac{x^5}{5}-\dfrac{x^7}{7}+\dots$

(c) $x+x^2+\dfrac{x^3}{3}-\dfrac{4}{5!}x^5+\dots$

(d) $1-2.5x+2.875x^2-2.8125x^3+\dots$

18.25 (a) 1.025   (b) 0.099 67   (c) 0.03   (d) 0.9713

18.26 (a) 0.1951   (b) 2.005

18.27 (a) $\frac{3}{25}$ (b) 6   (c) 0   (d) 1   (e) 1   (f) -0.2273

18.28 (a) $-0.618\,034$ (b) 1.495 35   (c) 0.450 184

18.29 (a) $|x_n-x_{n-1}|<0.000\,000\,5$

(b) $|x_n-x_{n-1}|<0.000\,000\,5|x_n|$

CHAPTER 19

19.1 (a) $\begin{pmatrix} 2 & 0 & 0 \\ 3 & -4 & 2 \end{pmatrix}$ (b) $\begin{pmatrix} 4 & -2 & 0 \\ 2 & -21 & 4 \end{pmatrix}$

(c) Not possible   (d) $\begin{pmatrix} 4 & 10 \\ 2 & -19 \\ 0 & -4 \end{pmatrix}$ (e) $\begin{pmatrix} 4 \\ 0 \\ -1 \end{pmatrix}$

(f) Not possible   (g) $(8 \quad 10)$ (h) $(8 \quad -7 \quad -4)$

(i) $(24 \quad 30)$ (j) $\begin{pmatrix} 3 & 5 \\ -2 & 3\frac{1}{3} \end{pmatrix}$ (k) $\begin{pmatrix} -\frac{3}{2} & 8 \\ -\frac{5}{2} & \frac{37}{3} \end{pmatrix}$

(l) Not possible   (m) $\begin{pmatrix} 3 & -8 & -1 \\ 8 & 38 & 10 \\ 3 & 30 & 19 \end{pmatrix}$ (n) $\begin{pmatrix} 16\frac{1}{3} & 10\frac{8}{9} \\ -5\frac{4}{9} & -3\frac{17}{27} \end{pmatrix}$

(o) Not possible   (p) $\begin{pmatrix} 16 \\ 4 \end{pmatrix}$

19.3 $\begin{pmatrix} 0 & 0 \\ 0 & 0 \end{pmatrix}$

19.4 (a) $\begin{pmatrix} -1 & 1 & 0 & 0 & 0 & 0 & 0 \\ 1 & 0 & -1 & 0 & 0 & 1 & 0 \\ 0 & 0 & 0 & 0 & 0 & -1 & 1 \\ 0 & -1 & 1 & 1 & -1 & 0 & -1 \\ 0 & 0 & 0 & -1 & 1 & 0 & 0 \end{pmatrix}$

$\begin{pmatrix} -1 & 1 & 0 & 0 & 0 & -1 & 0 & 0 \\ 0 & -1 & 1 & 0 & 0 & 0 & -1 & 0 \\ 0 & 0 & -1 & -1 & 0 & 0 & 0 & -1 \\ 1 & 0 & 0 & 1 & -1 & 0 & 0 & 0 \\ 0 & 0 & 0 & 0 & 1 & 1 & 1 & 1 \end{pmatrix}$

19.5 (a) $(0, 0)$, $(-0.5, 0.866)$, $(-1.366, 0.366)$, $(-0.866, -0.5)$

(b) $(-4, 1)$, $(-3, 1)$, $(-3, 2)$, $(-4, 2)$

(c) $(0, 0)$, $(1, 0)$, $(1, -1)$, $(0, -1)$

(d) $(0, 0)$, $(1, 0)$, $(1, 5)$, $(0, 5)$

(e) $(-2, 3)$, $(-2.5, 3.866)$, $(-3.366, 3.366)$, $(-2.866, 2.5)$

(f) $(-2.098, -2.366)$, $(-2.598, -1.5)$, $(-3.464, -2)$, $(-2.964, -2.866)$

(g) $(0, 0)$, $(-0.5, 4.33)$, $(-1.366, 1.83)$, $(-0.866, -2.5)$

(h) $(0, 0)$, $(0, 1)$, $(1, 1)$, $(1, 0)$

(i) $(0, 0)$, $(3.25, 1.299)$, $(4.549, 3.049)$, $(1.299, 1.75)$

(j) (a) Rotation through $-120°$ about the origin

(b) Translation by $(4, -1)$

(c) Reflection in the x-axis

(e) Translation by $(2, -3)$ followed by rotation through $-120°$ about the origin

(g) Scaling by $\frac{1}{3}$ in the y direction followed by rotation through $-120°$ about the origin

(i) Scaling along a line at an angle of $30°$ to the x-axis by a factor of $\frac{1}{4}$

19.6 (a) Determined, $x=0$, $y=3$ (b) Determined, $x=5$, $y=-2$

(c) Inconsistent   (d) Indeterminate

(e) Determined, $x=1$, $y=-10$ (f) Inconsistent

(g) Inconsistent   (h) Determined, $x=6$, $y=4$

19.7 (a) Determined, $x=-2$, $y=-3$, $z=1$

(b) Determined $x=\frac{400}{21}$, $y=\frac{430}{7}$, $z=\frac{-265}{51}$

(c) Indeterminate

(d) Inconsistent

19.8 (a) $\begin{pmatrix} 1 & 2 \\ 0 & 1 \end{pmatrix}$ (b) $\begin{pmatrix} \frac{2}{5} & -\frac{1}{5} \\ \frac{1}{15} & \frac{2}{15} \end{pmatrix}$ (c) No inverse

(d) $\begin{pmatrix} \frac{3}{4} & \frac{1}{4} & -\frac{1}{4} \\ -\frac{1}{4} & \frac{1}{4} & -\frac{1}{4} \\ \frac{5}{4} & \frac{1}{4} & \frac{1}{4} \end{pmatrix}$ (e) $\begin{pmatrix} \frac{1}{3} & -\frac{1}{3} & \frac{1}{3} \\ \frac{4}{15} & -\frac{1}{15} & -\frac{1}{30} \\ -\frac{1}{3} & \frac{1}{3} & \frac{1}{6} \end{pmatrix}$ (f) No inverse

19.9 (a) -9 (b) 16   (c) 0   (d) 50

19.10 (a) $-6\mathbf{i}-8\mathbf{j}+5\mathbf{k}$ (b) $\frac{3}{2}\mathbf{i}-\frac{3}{2}\mathbf{j}$

19.11 (a) (i) 10   (ii) 0

19.12 $\begin{pmatrix} \sigma_x \\ \sigma_y \\ \sigma_z \end{pmatrix} = \frac{E}{1-3v^2-2v^3} \begin{pmatrix} 1-v^2 & v+v^2 & v+v^2 \\ v+v^2 & 1-v^2 & v+v^2 \\ v+v^2 & v+v^2 & 1-v^2 \end{pmatrix}$

19.13 (a) $6, \begin{pmatrix} 1 \\ 0 \end{pmatrix}, 2, \begin{pmatrix} 1 \\ 4 \end{pmatrix}$ (b) $2, \begin{pmatrix} 5 \\ 1 \end{pmatrix}, -4, \begin{pmatrix} 1 \\ -1 \end{pmatrix}$

(c) $1, \begin{pmatrix} 1 \\ -1 \end{pmatrix}, 4, \begin{pmatrix} 2 \\ 1 \end{pmatrix}$

19.14 $V_R = 4.905I + 1.147$

19.15 (a) Length $= 9.882 + 1.905 \times$ Load   (b) (i) 14.6 cm   (ii) 19.4 cm

19.16 $P = 11.327 - 0.0556T$

19.17 (a) $y = \frac{1}{70}(-82 + 123x + 215x^2)$

(b) $y = 1.29 - 2.223\,33x + 2.033\,33x^2$

CHAPTER 20

20.1 (a) Linear, 1   (b) Linear, 2   (c) Linear, 2

(d) Not linear, 1   (e) Not linear, 2

20.2 (a) $1.6\cos(3t) + 1.2\sin(3t) + t^2 - \frac{1}{2}t + \frac{1}{2}$

(b) $0.8\cos(3(t-10)) + 0.6\sin(3(t-10))$

20.3 (a) $\dfrac{t^2 - 2\sin(t)}{\cos(t)}$

20.4 (a) $y = e^{3t}$ (b) $y = 2e^{t/4} - 4$ (c) $y = \frac{12}{169}e^{-5t} - \frac{12}{169}\cos(12t) + \frac{5}{169}\sin(12t)$
(d) $y = 4e^{-\frac{2}{3}t} - e^{-t}$ (e) $y = 3t^2 + 3$ (f) $y = \frac{-18}{169}e^{-12t} + \frac{18}{169}\cos(5t) + \frac{92}{169}\sin(5t)$

20.5 (a) $x = -0.064e^{-\frac{5}{2}t} - 0.56e^{-t} + 0.624 - 0.56t + 0.2t^2$
(b) $x = (-0.2t + 0.08)e^{-2t} - 0.08\cos(6t) + 0.06\sin(6t)$
(c) $x = e^{-2t}(-0.5\cos(t) - 0.5\sin(t)) + 0.5e^{-t}$

20.6 (a) $q = \frac{-5}{13} \times 10^{-3}\cos(1000t) + \frac{12}{13} \times 10^{-3}\sin(1000t)$
$V_c = \frac{-50}{13}\cos(1000t) + \frac{120}{13}\sin(1000t)$
$V_R = \frac{600}{13}\sin(1000t) + \frac{1440}{13}\cos(1000t)$
$V_L = \frac{300}{13}\cos(1000t) - \frac{720}{13}\sin(1000t)$
(b) $q = -10^{-9} + 10^{-6}t$, $V_c = t - 10^{-3}$, $V_R = 10^{-3}$, $V_L = 0$

20.7 (a) $q = \dfrac{cv_i}{Rcj\omega + 1}e^{j\omega t}$ (i) $V_c = \dfrac{v_i}{Rcj\omega + 1}e^{j\omega t}$ (ii) $V_R = \dfrac{Rcv_j\omega}{Rcj\omega + 1}e^{j\omega t}$

(c) $0.99995, 0.995, 0.707, 0.0995, 0.0099995, 0.001, f_{hc} = \dfrac{10^3}{2\pi}$

20.8 (a) $x_1 = \frac{-1}{10}e^{-t} - \frac{1}{90}e^{-11t} - \frac{1}{9}e^{-2t}$
(b) $x_1 = \frac{8}{15}e^{\frac{-5}{2}t} - \frac{5}{24}e^{-4t} - \frac{13}{40} + \frac{1}{2}t$
20.9 (a) $y_n = 4(\frac{1}{2})^n - 3$ (b) $y_n = \frac{49}{25}(\frac{-1}{4})^n + \frac{1}{25} + \frac{1}{5}n$
(c) $y_n = (-22 - \frac{49}{3}n)(0.6)^n + 25$
(d) $y_n = \frac{767}{390}(0.2)^n - \frac{209}{130}(-0.8)^n + \frac{25}{39}(0.5)^n$
(e) $y_n = \frac{3}{5}(-\frac{1}{2})^n + \frac{2}{5}\cos(\frac{\pi}{2}n) - \frac{1}{5}\sin(\frac{\pi}{2}n)$

CHAPTER 21

21.1 (a) $Re(s) > 4$ (b) $Re(s) > -2$ (c) $Re(s) > 0$ (d) $Re(s) > 0$

21.2 (a) $\dfrac{15}{s^2 + 9}$ (b) $\dfrac{s}{s^2 + \frac{1}{4}}$ (c) $\dfrac{8}{s^5}$ (d) $\dfrac{1}{2(s + 5)}$

21.3 (a) e^{4t} (b) $3e^{-t}$ (c) $4u(t)$ (d) $\dfrac{2}{\sqrt{3}}\sin(\sqrt{3}t)$ (e) $\frac{4}{9}\cos(\frac{2}{3}t)$

(f) $\dfrac{t^2}{2}$ (g) $\dfrac{5t^3}{6}$

21.4 (a) $\dfrac{1}{(s + 3)^2}$ (b) $\dfrac{4}{(s + 2)^2 + 1}$ (c) $\dfrac{8s}{(s^2 - 16)^2}$ (d) $\dfrac{e^{-s}}{s}$ (e) $\dfrac{e^{-\pi s}}{s^2 + 1}$

21.5 (a) $\frac{1}{3}e^{2t}\sin(3t)$ (b) $e^{-4t}\cos(t)$ (c) $e^{-(t-2)}u(t-2)$ (d) $e^t\cos(3t)$

21.6 (a) $u(t) - e^{-t}$ (b) $\frac{1}{2}u(t) - \frac{1}{2}\cos(\sqrt{2}t)$ (c) $\frac{-1}{3}\cosh(2t) + \frac{2}{3}\sinh(2t) + \frac{1}{3}e^{-t}$
(d) $\frac{-2}{5}\cos(t) - \frac{6}{5}\sin(t) + \frac{2}{5}e^{3t}$
21.7 (a) $3e^{-5t}$ (b) $-1 + t + e^{-t}$ (c) $\frac{19}{8}\sin(2t) + \frac{9}{4}t$ (d) $4e^{2t} + 4te^{2t} - 7e^t$

(e) $\cos(t) - 3\sin(t) + t$ (f) $2 - 2e^{-t/2}\cos\left(\dfrac{\sqrt{7}}{2}t\right) - \dfrac{2}{\sqrt{7}}e^{-t/2}\sin\left(\dfrac{\sqrt{7}}{2}t\right)$

21.8 $V_0 c e^{-t/RC}$
21.9 $\frac{1}{2}e^{-t} - e^{-2t} + \frac{1}{2}e^{-3t}$

21.10 (a) $\frac{3}{4}u(t) - \frac{3}{4}e^{-4t}$ (c) (i) $-\frac{4}{3} - \dfrac{s}{s+2} - \dfrac{2s}{3(s+3)}$ (ii) $-2e^{-2t} - e^{-3t}$

21.11 $0.08\delta(t) + t$
21.12 (a) $\frac{1}{689}(-17 - 20j)e^{j5t}$ (b) $\frac{1}{689}(-17\cos(5t) + 20\sin(5t))$, $\frac{1}{689}(-20\cos(5t) - 17\sin(5t))$
21.13 (a) All z (b) $|z| > 1$ (c) $|z| > 3$

21.14 (a) $\dfrac{2z}{z-1} + \dfrac{z}{2(z-1)^2}$ (b) $\dfrac{z^2 - z\cos(3) + 2z\sin(3)}{z^2 - 2z\cos(3) + 1}$

(c) $\dfrac{4z}{z-0.2} - \dfrac{6z}{z-2}$ (d) $\dfrac{2z}{z-e^{j4}}$

21.15 (a) $2u_n + 2\delta_n$ (b) $\frac{1}{2}n + (-3)^n$ (c) $(\frac{3}{2})^n + 2(\frac{-3}{2})^n$
(d) $\cos(n) + \sin(n)$

21.16 (a) $\dfrac{z^3}{z-3} - (z^2 + 3z + 9)$ (b) $\dfrac{z}{2(z-\frac{1}{2})^2}$ (c) $\dfrac{ze^{j}}{(z-e^{j})^2}$

(d) $\dfrac{z^3 + 4z^2 + z}{(z-1)^4}$

21.17 (a) $n4^{n-1}$ (b) $2^n e^{j4n}$ (c) $n(0.4)^{n-1}$ (d) $2^{n+2} - 2$

21.18 (a) $u_n/2 + (-1)^n/2$ (b) $-u_n + (-\sqrt{2})^n/(2+\sqrt{2}) + (\sqrt{2})^n/(2-\sqrt{2})$
(c) $-(0.1)^n/15 + (0.4)^n/15$ (d) $2u_n/9 + n/3 - 2(-2)^n/9$
(e) $(j)^n/2 + (-j)^n/2$

21.19 (a) $3(-5)^{n+1}$ (b) $\frac{1}{4}u_n + \frac{1}{2}n - \frac{1}{4}(-1)^n$ (c) $\frac{9}{5}u_n + \frac{16}{5}(-4)^n$
(d) $17u_n - 32(2)^n + 8n(2)^n$ (e) $\frac{1}{2}u_n + \frac{1}{2}n - \frac{1}{2}\cos(\pi n/2)$
(f) $\frac{1}{3}u_n + \frac{1}{14}(1/2)^n - \frac{11}{105}(-1/5)^n$

21.20 $3\frac{3}{4}(0.3)^n - (0.1)^n + \frac{5}{4}(-0.5)^n$

21.21 (a) $5u_n - 4(0.8)^n$ (c) (i) $\dfrac{z-1}{24(z-0.2)} + \dfrac{5}{24}$

(ii) $\frac{5}{6}(0.5)^n + \frac{2}{3}(0.2)^n$

21.22 $\delta_n + 0.5n$

21.23 (a) $\dfrac{e^{j5(n+1)}}{10e^{j5} - 3}$ (b) $\dfrac{\cos(5(n+1))(10\cos(5) - 3) + \sin(5(n+1))\sin(5)}{(10\cos(5) - 3)^2 + \sin^2(5)}, \dfrac{\sin(5(n+1))(10\cos(5) - 3) - \cos(5(n+1))\sin(5)}{(10\cos(5) - 3)^2 + \sin^2(5)}$

CHAPTER 22

22.1 $\frac{1}{4} + \displaystyle\sum_{n=1}^{\infty} \left(\dfrac{\sin(n\pi/2)\cos(n\pi t/2)}{n\pi} + \dfrac{(1 - \cos(n\pi/2))\sin(n\pi t/2)}{n\pi} \right)$

$\frac{1}{4} + \displaystyle\sum_{\substack{n=-\infty \\ n\neq 0}}^{\infty} \left(\dfrac{\sin(n\pi/2)}{2n\pi} + j\dfrac{(\cos(n\pi/2) - 1)}{2n\pi} \right) e^{jn\pi t/2}$

22.2 $\frac{1}{6} + \displaystyle\sum_{n=1}^{\infty} \left(\dfrac{2}{n^2\pi^2}(-1)^n \cos(n\pi t) + \left(\dfrac{-(-1)^n}{n\pi} + \dfrac{2((-1)^n - 1)}{n^3\pi^3} \right) \sin(n\pi t) \right)$

$\frac{1}{6} + \displaystyle\sum_{\substack{n=-\infty \\ n\neq 0}}^{\infty} \left(\dfrac{(-1)^n}{n^2\pi^2} + j\left(\dfrac{(-1)^n}{2n\pi} - \dfrac{(-1)^n}{n^3\pi^3} + \dfrac{1}{n^3\pi^3} \right) \right) e^{jn\pi t}$

22.3 $\frac{1}{2} + \frac{1}{2}\cos(2t)$
$\frac{1}{2} + \frac{1}{4}e^{jnt}$

22.4 $\frac{1}{4} + \displaystyle\sum_{n=1}^{\infty} \left(\left(\dfrac{(-1)^n - 1}{n^2\pi^2} \right) \cos(n\pi t) - \dfrac{(-1)^n}{n\pi}\sin(n\pi t) \right)$

$\frac{1}{4} + \displaystyle\sum_{\substack{n=-\infty \\ n\neq 0}}^{\infty} \left(\dfrac{1}{2n^2\pi^2}(-1)^n - 1) + \dfrac{j}{2n\pi}(-1)^n \right) e^{jn\pi t}$

22.5 $\frac{1}{2} + \displaystyle\sum_{n=1}^{\infty} \left(\dfrac{2}{n^2\pi^2}((-1)^n - 1)\cos(n\pi t) + \dfrac{4}{n^3\pi^3}((-1)^n - 1)\sin(n\pi t) \right)$

$\frac{1}{2} + \displaystyle\sum_{\substack{n=-\infty \\ n\neq 0}}^{\infty} \left(\dfrac{(-1)^n - 1}{n^2\pi^2} + j\left(\dfrac{2(1 - (-1)^n)}{n^3\pi^3} \right) \right) e^{jn\pi t}$

22.6 $0 + \frac{1}{2}\cos(2\pi t) + \displaystyle\sum_{\substack{n=1 \\ n\neq 2}}^{\infty} \dfrac{n(1 - (-1)^n)}{\pi(n^2 - 4)}\sin(n\pi t)$

22.7 $0 + \frac{1}{4}(e^{j2\pi t} + e^{-j2\pi t}) + \displaystyle\sum_{\substack{n=-\infty \\ n\neq 0,2,-2}}^{\infty} \dfrac{nj((-1)^n - 1)}{2\pi(n^2 - 4)} e^{jn\pi t}$

$0 + \displaystyle\sum_{n=1}^{\infty} 2\left(\dfrac{(-1)^n - 1}{n^2\pi^2} \right)\cos(n\pi t)$

$0 + \displaystyle\sum_{\substack{n=-\infty \\ n\neq 0}}^{\infty} \left(\dfrac{(-1)^n - 1}{n^2\pi^2} \right) e^{jn\pi t}$

22.8 (a) Odd   (b) Even   (c) Odd

22.9 (a) 2π, $c_1 = 1$, $c_3 = 1$, $c_n = 0$ for all other n, $\phi_1 = \frac{\pi}{2}$, $\phi_3 = -\frac{\pi}{2}$, $\sigma_n = 0$ for all other n

(b) π, $c_1 = \dfrac{1}{\sqrt{2}}$, $c_n = 0$ for all other n, $\phi_1 = \frac{\pi}{4}$, $\phi_n = 0$ for all other n

(c) 2π, $c_n = \dfrac{2}{n^2}$, $n \neq 0$, $c_0 = 0$, $\phi_n = 0$ for all n

22.10 (22.1) $c_n = \begin{cases} 0 & \text{when } n = 4m \text{ for some } m \\[2mm] \dfrac{\sqrt{2}}{n\pi} & \text{when } n = 4m+1 \text{ or } n = 4m+3 \text{ for some } m \\[2mm] \dfrac{2}{n\pi} & \text{when } n = 4m+2 \text{ for some } m \end{cases}$

$\phi_n = \begin{cases} 0 & \text{when } n = 4m \text{ for some } m \\ -\pi/4 & \text{when } n = 4m+1 \text{ for some } m \\ -\pi/2 & \text{when } n = 4m+2 \text{ for some } m \\ -3\pi/4 & \text{when } n = 4m+3 \text{ for some } m \end{cases}$

(22.3) $c_1 = \frac{1}{2}$, $c_n = 0$ for all other n, $\phi_n = 0$ for all n,

(22.7) $c_n = \begin{cases} 0 & \text{when } n \text{ is even} \\[2mm] \dfrac{4}{n^2 \pi^2} & \text{when } n \text{ is odd} \end{cases}$

$\phi_n = \begin{cases} 0 & \text{when } n \text{ is even} \\ \pi & \text{when } n \text{ is odd} \end{cases}$

22.12 $\frac{1}{4} + \frac{1}{8}(\cos(2t) + \sin(2t))$

CHAPTER 23

23.1 (a), (e) and (g) are isomorphic. (d) and (i) are isomorphic.

23.3 K_n is planar for $n \leq 4$.

$K_{m,n}$ is planar for $m \leq 2$ or $n \leq 2$.

23.4 (a)

Incidence matrix

$$\begin{pmatrix} 1 & 0 & 0 & 1 & 0 & 1 \\ 1 & 1 & 1 & 0 & 0 & 0 \\ 0 & 1 & 0 & 0 & 0 & 0 \\ 0 & 0 & 1 & 1 & 1 & 0 \\ 0 & 0 & 0 & 0 & 1 & 1 \end{pmatrix}$$

Adjacency matrix

$$\begin{pmatrix} 0 & 1 & 0 & 1 & 1 \\ 1 & 0 & 1 & 1 & 0 \\ 0 & 1 & 0 & 0 & 0 \\ 1 & 1 & 0 & 0 & 1 \\ 1 & 0 & 0 & 1 & 0 \end{pmatrix}$$

$$\begin{pmatrix} 1 & 1 & 1 & 0 & 0 & 0 \\ 1 & 0 & 0 & 1 & 0 & 1 \\ 0 & 1 & 0 & 0 & 1 & 1 \\ 0 & 0 & 1 & 1 & 1 & 0 \end{pmatrix}$$

$$\begin{pmatrix} 0 & 1 & 1 & 1 \\ 1 & 0 & 1 & 1 \\ 1 & 1 & 0 & 1 \\ 1 & 1 & 1 & 0 \end{pmatrix}$$

$$\begin{pmatrix} 1 & 1 & 0 & 0 & 0 & 0 \\ 0 & 0 & 1 & 1 & 0 & 0 \\ 0 & 0 & 0 & 0 & 1 & 1 \\ 1 & 0 & 1 & 0 & 1 & 0 \\ 0 & 1 & 0 & 1 & 0 & 1 \end{pmatrix}$$

$$\begin{pmatrix} 0 & 0 & 0 & 1 & 1 \\ 0 & 0 & 0 & 1 & 1 \\ 0 & 0 & 0 & 1 & 1 \\ 1 & 1 & 1 & 0 & 0 \\ 1 & 1 & 1 & 0 & 0 \end{pmatrix}$$

(c) $(3 \quad 3 \quad 1 \quad 3 \quad 2)$, $(3 \quad 3 \quad 3 \quad 3)$, $(2 \quad 2 \quad 2 \quad 3 \quad 3)$

These represent the degrees of the vertices.

23.5 A minimum spanning tree has total weight 21. There is more than one.

23.6 The shortest path is sbdt. It has length 15.

23.7 Maximum flow is 18.

23.8 Maximum flow is 6.

CHAPTER 24

24.1 99.5625, 8.9496

24.2 $\frac{1}{3}$

24.3 $\frac{2}{3}$

24.4 $\frac{2}{3}$

24.5 $\frac{4}{13}$

24.6 $\frac{1}{12}$

24.7 $\frac{1}{6}$

24.8 $\frac{3}{51}$

24.9 $\frac{1}{100\,00}$

24.10 0.9362

24.11 (a) $\frac{3}{5}$ (b) $\frac{3}{5}$ (c) $\frac{3}{10}$

24.12 0.091

24.13 (a) 0.066 81   (b) 0.022 75   (c) 0.001 35   (d) 0.308 54

24.14 (a) 0.308 54   (b) 0.022 75   (c) 0.158 66   (d) 0.226 63

24.15 (a) 0.990 18   (b) 0.747 51   (c) 0.630 56   (d) 0.630 56
(e) 0.261 12   (f) 0.737 69

24.16 17.8%

24.17 (a) 0.699   (b) 0.139   (c) 0.192   (d) 0.273

24.18 8.858, the councillor is correct

24.19 (a) 0.512   (b) 0.2048   (c) 0.879

BIBLIOGRAPHY

Kreysig, E. (1988) *Advanced Engineering Mathematics*, John Wiley, Chichester.

Poularikas, D. and S. Seely (1988) *Elements of Signals and Systems*, PWS-Kent, Boston.

Senior, T. B. A. (1986) *Mathematical Methods in Electrical Engineering*, Cambridge University Press, Cambridge.

Skvarcius, R. and W. B. Robinson (1986) *Discrete Mathematics with Computer Science Applications*, Benjamin Cummings, New York.

Speigel, M. (1965) *Schaum Outline of Theory and Practice of Laplace Transforms*, McGraw-Hill, Maidenhead.

Wilson, F. A. (1981) *Practical Electronics Calculations and Formulae*, Bernard Babani, Shepherds Bush.

Wilson, F. A. (1988) *Further Practical Electronics Calculations and Formulae*, Bernard Babani, Shepherds Bush.